GRIBOV MEMORIAL VOLUME

Quarks, Hadrons, and Strong Interactions

GRIBOV
MEMORIAL VOLUME

Quarks, Hadrons, and Strong Interactions

Proceedings of the Memorial Workshop Devoted
to the 75th Birthday of V N Gribov

Budapest, Hungary   22 – 24 May 2005

Editors

Yu L Dokshitzer
LPTHE Universités Paris-VI-VII, Paris, France

P Lévai & J Nyíri
MTA KFKI RMKI, Budapest, Hungary

World Scientific

NEW JERSEY • LONDON • SINGAPORE • BEIJING • SHANGHAI • HONG KONG • TAIPEI • CHENNAI

Published by

World Scientific Publishing Co. Pte. Ltd.
5 Toh Tuck Link, Singapore 596224
USA office: 27 Warren Street, Suite 401-402, Hackensack, NJ 07601
UK office: 57 Shelton Street, Covent Garden, London WC2H 9HE

British Library Cataloguing-in-Publication Data
A catalogue record for this book is available from the British Library.

GRIBOV MEMORIAL VOLUME
Quarks, Hadrons, and Strong Interactions
Proceedings of the Memorial Workshop Devoted to the 75th Birthday of V N Gribov

ISBN 981-256-756-9

Printed in Singapore by B & JO Enterprise

PREFACE

Young students often stop in front of a white door with a copper plate announcing that Volodya Gribov was working in that room in his last twenty years. Most of the time they are puzzled. At the university courses they have already heard about the great Russian physicist V.N. Gribov, father of the relativistic Gribov–Regge theory of high energy hadron scattering, author of the Glauber-Gribov model of nucleon–nucleon superposition in heavy ion collisions, scientific brother of L.D. Landau and I.Ya. Pomeranchuk. Students use to place him at St. Petersburg or at least in Russia. However, this white door is in Budapest, at the Theoretical Physics Department of the RMKI Institute. Realising that they just missed a personal meeting and a discussion with him by a few years, make these students more curious; they often ask about him and his work. Fortunately, these days one can offer four books (with a fifth book close to be ready), which cover and summarize Gribov's discoveries in high energy physics and non-Abelian gauge theories as well as his ideas about light quark confinement. However, reading these books one can feel that they contain only the beginning of a story and not the end.

All of us here in Budapest, and around the world, would like to see the continuation of his work, the evolution of his ideas, the discussions about the deepest problems of high energy physics. This wish called into existence the Gribov Memorial Workshop series held in Bonn in 1998, in Paris in 2000 and now, on Gribov's 75th birthday, in Budapest. The participants wished to keep alive the Gribov heritage, and the Gribov style of scientific work, which they widely used in their own research. This purpose created a very warm and fruitful Workshop in Budapest, for what we would like to thank all participants. We all carry our own picture of Volodya, including his achievements and personality; the Workshop offered the possibility to reveal this. Our common belief is that Volodya's influence on high energy physics will only grow and his ideas will appear in many contexts.

We collected the talks of the participants and wanted-to-be participants to give another opportunity to the students and young fellows to understand better the community of which Volodya was an active member. All of us would be glad to see him again, but since this is impossible, then at least to feel his style and personality through the talks of this workshop and the participation of his colleagues.

We thank all participants for having accepted our invitation to Budapest and we hope to meet again at the next Gribov Memorial Workshop.

Yu. Dokshitzer, P. Lévai, J. Nyiri

Memorial Workshop Devoted to the 75th Birthday of V.N. Gribov
Budapest, Hungary, 22–24 May 2005

CONTENTS

GENERAL FIELD THEORY, GRAVITY AND MACRO-WORLD

CONCLUDING REMARKS

REMEMBERING VOLODYA

REMEMBERING VOLODYA*

A. FRENKEL
RMKI Research Institute for Particle and Nuclear Physics,
P.O. Box 49, Budapest, 1525 HUNGARY
E-mail: frenkel@rmki.kfki.hu

Dear Colleagues,

Around 1960 - I do not remember exactly when - I spent a few weeks at the Laboratory of Theoretical Physics of the Joint Institute for Nuclear Research in Dubna. During my visit a conference on particle physics was held there; not an international conference, the talks were in Russian. One afternoon I asked Sasha Filippov whether tomorrow he would go to the morning session. "Of course I shall go, Gribov will be there!" he answered. I did not know then who Gribov was, and in the program of the conference there was not a single Gribov among the speakers. So I did not understand what Sasha meant, but the program seemed interesting and I decided to go. I arrived ten minutes before the beginning. Usually there were free seats, but this time people were sitting even on the stairs of the lecture room. I got a place there, too. Five or ten minutes after the beginning of the first, 45 minutes talk, given, if I remember correctly, by Vitya Ogievetski, somebody from one of the last rows shouted: "May I ask a question?" "Yes, of course" said the speaker. The voice asked something, and got an answer. "But this is not what I meant" said the voice, "may I come to the blackboard for a few minutes?" "OK, Volodya, OK for a few minutes" answered the speaker. Then I saw a slim, black-haired young man running down the stairs, beginning to argue with the speaker, and shortly after running back to his seat. Similar episodes repeated themselves several times during the morning session, with other members of the audience joining the discussion. We run out of time by about an hour, but nobody complained about that. This is how I got to know who was Vladimir Naumovich Gribov.

*Edited version of the talk given at the Gribov-75 Workshop.

Of course, at a conference such discussions cannot be too long. However, at home, in the Institute for Nuclear Physics in Leningrad, Volodya was leading the theoretical physics researches by means of endless, but fruitful discussions at the blackboard on the various problems people were working on. When arguing about physics Volodya could be very harsh, first of all to himself. He was not interested in being right, he was interested in the right answer. On the other hand, in discussions about music or literature he was not harsh at all. He had his own, thoughtful and interesting views, but he admitted that in such fields several distinct opinions may coexist. As far as everyday-life is concerned, Volodya was a very kind, open-minded man.

At the same time, he was one of the outstanding physicists of the last century, a close colleague of L.D. Landau and of I.Ya. Pomeranchuk. After his too early death in August 1997, a few of his former students and his widow undertook the difficult and time-consuming task of the edition of Gribov's scientific heritage. Four thick volumes appeared already, the fifth, which is the last one, will come out shortly. These books contain, among other topics, Gribov's field theoretical Regge calculus, his basic theorems on dispersion relations and on non-Abelian field theory, as well as two important contributions to physics, about which I would like to say a few words.

Shortly after Bruno Pontecorvo's arrival to Dubna in the fifties, he and Volodya became close friends. About ten years later Bruno, who was a world-famous experimental physicist, had the idea that neutrinos may oscillate, but he could not formulate the theory, and asked Volodya's help. The seminal paper of Gribov and Pontecorvo on neutrino oscillation, with the proof that the oscillation appears only if the neutrinos have different masses, i.e. only if at least one of them has non-zero mass, has been published in 1969. As you certainly know, the existence of the electron-muon neutrino oscillation was firmly established experimentally a few years ago. I am convinced that if the Nobel prize could be given posthumus, Bruno an Volodya would get it soon.

From the early eighties Volodya spent about half of his time in our Institute in Budapest, the home town of his wife, Julia Nyiri. During this whole period he was working on the problem of the quark confinement, which he considered to be the most important, and also the most difficult problem he had ever tried to solve.

As I mentioned already, Volodya liked discussions at the blackboard. Also, he was ready to talk about his emerging ideas, and to show his unfinished calculations to anyone interested in the subject. I was interested, and step by step I got a feeling of Volodya's work-style. First, he designed a loose physical picture, the components of which came from different branches of physics, in this case from the theory of supercharged nuclei, from a Dirac see with unconventionally filled levels, and from some aspects of superconductivity. Then he began to work out the details, sometimes modifying the overall picture, more often throwing out many months of work if the detail didn't fit. These ups and downs went on during about two decades, and only in the beginning of '97 Volodya felt that the theory is OK. He told then Julia that soon they will have a good time, calculating the mass spectrum of the hadrons. Unfortunately he had no time to carry out those calculations. The question whether the Gribov theory of confinement is the right one, in which case Volodya deserves a second Nobel prize, remains open.

I would have been pleased if this meeting could be held in Russia, organized by the members of a flourishing there Gribov School. But since Volodya's former students are now dispersed in the world, it is a great honor for our Institute to host the Gribov-75 Memorial Workshop in the town where he spent the last two decades of his life, a town he liked; and I feel honored to be here with you.

Thank you for your attention.

MEMORIES OF VOLODYA

DANIELE AMATI
SISSA, 34014 Trieste, Italy
E-mail: amati@sissa.it

I first met Volodya at the 1962 CERN conference. It was a large international conference, and the first one in the west attended by a real scientific delegation of Soviet physicists. Among them the young Gribov, not yet internationally known but already announced to me, by Pomeranchuk, as an extraordinarily clever physicist.

It was the time of the triumphant S matrix approach with the Regge pole description of high energy processes that the new synchrotrons unveiled. Volodya presented his approach to angular momentum analysis of scattering amplitudes as derived from their analytic properties showing the need to introduce a characteristic (named the Froissart-Gribov representation) that gave rise to what has then been called the signature of Regge trajectories. Subsequently I visited him a few times in Leningrad and I remember giving seminars during which he came up to the blackboard to comment a point and engaging a discussion in which it was no longer clear if the seminar was mine or his and which lasted hours, as is usual in Russia. I sensed in those occasions the strong influence he had in the very approach to physics on the lively group that grew around him that constituted what has been called the Leningrad School. He gave his imprinting in which physics was an endeavour without closed compartments, neither in research fields such as hadrons, weak or electromagnetic processes as customary at the time, nor in methodologies such as S matrix approaches or field theoretical ones in their perturbative or non-perturbative aspects.

Volodya had not only an original approach to physical insights, but also an original way to put the problem and to formalize it. And he would not make big efforts to adapt his attitude to the request of his interlocutor nor to translate into his language. But he would show his generous readiness to guide him, again and again, along his path or along alternative ways compatible with his vision. Without losing his temper, specially if he could detect a real interest to follow him. For what I could understand, despite his quite reserved character, the terrible and absurd loss of his beloved – and admired – son Lonya implied a real disruption of the basic structures he had built in his life. Not of his characteristics as his love for physics, his need of warm human feelings and bindings, his sense of responsibility towards others, etc. But he had to rebuild all the world around him and thus his place and way of functioning, his real and emotional anchorage points. I have been perhaps nearer to him in this more introverted second stage than in the more vivid first one and thus I may be biased against a normal change that aging implies to everybody.

The last long period we spent together is during a visit he and Julia made to SISSA for a few months in the nineties. I then tried hard to understand his non confining QCD theory and I remember his patient private lectures and long discussions on related subjects. I may say that I understood his view even if I didn't succeed in adopting his attitude, i.e. to follow his path or to have a clear translation between his view and mine. But I remember with a deep warmth and with great pleasure his smile and his cigarette when he succeeded in making me grasp an extra concept or illuminating with an example what I had the feeling to have understood. They were living in a small house on a green hill near the railroad and we had to interrupt our discussions, or our chatting, when a train was passing so close that it could have seemed it was between us. A great physicist and a real mensch.

MEMORY SKETCH TO V.N. GRIBOV'S PORTRAIT

ALEXANDER BELAVIN

Landau Institute for Theoretical Physics,
Chernogolovka, 142432, Russia
E-mail: belavin@itp.ac.ru

1. It is very difficult for me to write about Vladimir Naumovich. First, because I understand that I am not able to put into words the bright image of V.N. living in my heart. And second, there are things I just may not remember. But there are moments in my memory that I would like to share with the reader.

2. It was during a lecture of I.Ya. Pomeranchuk at MIFI in 1964 or 1965, when I heard Gribov's name for the first time. Pomeranchuk told us that the next lecture would be skipped, because he was expecting the visit of V.N. Gribov from Leningrad, and they would work together. It was clear that I.Ya. was very happy about this.

3. I wrote my diploma work at ITEP, my advisor was M.V. Terentyev. There was a marvelous atmosphere at the Theory Department of the institute: many discussions, a big seminar every Monday at 15:30, and another small "coffee"-seminar on Thursdays ... One day rumour spread in ITEP that Gribov was coming. I remember the excited expectation of the event and the seminar itself. The lecturer's extraordinary fascination. What was its secret? The mystery of his personality. But also, I think, the process of thinking that took place in front of the audience's eyes, being a pleasure for V.N. himself, charmed us.

4. Around 1968 V.N. has invented the Reggeon diagram calculus. I recall how highly this work was esteemed by Sasha Polyakov. Probably, it was just this paper that initiated (or supported) Polyakov's idea about the bootstrap approach in quantum field theory which he developed not much later.

5. The paper of Gribov and A. Migdal has played an important role in the appearing idea of A. Polyakov and A. Migdal to derive the Kadanov - Potashinsky - Pokrovsky scaling law for critical phenomena from quantum field theory.

6. For me personally, the works of Polyakov and A. Migdal together with the papers of Gribov and Lipatov on the double logarithms have played a crucial role. As a PhD student at ITEP (from 1967 to 1970, my advisor was I.Yu. Kobzarev), I was working on a phenomenological approach to the theory of weak interactions. The above papers persuaded me to switch to more fundamental field theoretical investigations.

7. I remember the LNPI schools in the mid-seventies. During the days there were lectures and seminars; among others, a seminar by L.D. Faddeev on the quantization of solitons. Faddeev had a hard time, facing the temperament of V.N. who did not believe that solitons may survive after quantization.

In the evenings everybody, including V.N., gathered in somebody's room, drinking tea, (small amounts of) vodka, listening and singing songs of Galich and Okudzhava, with Yura Dokshitzer and Mitya Diakonov playing the guitar. And still later, after midnight, sometimes night seminars ("lipatniks") began.

8. I recall that once, being in Leningrad, I attended a seminar of Gribov in the Physics-Technical Institute. Suddenly Ya.B. Zeldovich entered the room. He interrupted the speaker and began to talk about the Hawking effect. And he told us with repentance that Gribov predicted the black hole radiation long before Hawking's paper appeared.

9. Every time when I was in Leningrad I was living at the place of Arkady Aronov, an outstanding condensed matter physicist, a very kind and nice man. The Aronovs' home was on the Engels street, quite close to the Gribovs' house. And every time we made a gift to ourselves: we called Gribov and spent an evening with him.

10. The next piece of memory is an international symposium in the sunny Tbilisi, probably in 1976 or 1977. A.D. Sakharov, not yet in exile, but already in disgrace, participated at the symposium. People, more exactly the soviet colleagues, tried to stay at a distance from him. So the open and friendly conversations between V.N. and A.D. really caught the eye. Gribov was not a dissident, although his relation to the Soviet regime was obvious and that to the human rights movement was, naturally, positive.

I recall that at the same symposium we were sitting in the bar of the hotel where the participants were living. We were talking about various things, and the conversation was partly definitely anti-soviet. So I suggested to be careful - there could be microphones. V.N.'s reaction was that there is no doubt that there are microphones, but it could be guaranteed that they do not work.

11. The instantons in the gauge field theory were discovered in 1975. They were found as the minimum of action in the Euclidean formulation. V.N. did not like the Euclidean description, regarding it as unphysical. He soon he came to the conclusion (and this was not much later done by 't Hooft) that the instantons describe a process of under-barrier tunneling from one classical vacuum to the other.

12. It was in 1977 when Gribov discovered that in non-Abelian gauge theories the local gauge condition does not fix the gauge unambiguously. This phenomenon is now known as "Gribov copies", and is, at the same time, an important mathematical discovery. How very typical for V.N. that he made it while he was thinking about the physical nature of the confinement. The "copies" were important for him not as a mathematical theorem, but as a possible explanation for this phenomenon.

13. In 1980 Gribov moved from Leningrad to Moscow, where he became a member of our Landau Institute for Theoretical Physics. I recall the piety with which our "pillars of science" approached Gribov. Still, their scientific interests were rather far from the main interests of Gribov, although he was ready to discuss any physics, be it condensed matter physics or cosmology. For example, I remember a meeting of the "Scientific Council", where a widely known physicist working on cosmology and gravitation was giving a talk. V.N. jumped to the blackboard (which was on a stage in the hall where the Friday seminars of the institute were held), took the chalk from the speaker, and started to explain why and where he, the speaker, was not right. Then he turned to the audience and said, "It would be nice to invite a specialist in quantum gravity here".

Unfortunately, no close contacts were formed between V.N. and the members of the Quantum field theory group of the institute. Gribov did not care much about the beautiful features of the integrable field theory, he considered them as specific properties of two-dimensional theories. On the other hand, the efforts of V.N. to solve the problem of quark confinement in the framework of quantum chromodynamics did not attract the quantum field theorists who at that time moved away from gauge theories and became interested in strings, integrable models and conformal field theory.

14. In the beginning of the 1980s we started to organize schools on quantum field theory in Odessa (separating them from the schools of the institute). This went on for a few years; V.N. and Julia came every time. V.N. participated intensely in the discussions. Those were happy times ...

15. From the mid-1970s to his last day V.N. was working on the problem of quark confinement. This problem was in the centre of attention in the early 1970s, and the greatest minds have tried to find a solution. Tiring of the lack of success, gradually almost everybody abandoned the problem, switching to other subjects, first of all to string theories. So Gribov, who was convinced that it was absolutely necessary to solve the problem of quark confinement, continued his studies almost alone. This was a hard and heroic task, and I hope that this work of V.N. will find congenial followers to continue it.

THE HIGH ENERGY GRIBOV: SOME RECOLLECTIONS

LARRY MCLERRAN
Physics Department, Brookhaven National Laboratory,
Upton, NY 11973 USA
RIKEN BNL Research Center, Brookhaven National Laboratory,
Upton, NY 11973 USA

I describe through a few anecdotes, Gribov's work on the high energy limit of strong interactions.

1. Introduction

When I was a graduate student in the early 70's, there was much interest in the physics of strong interactions. Among the topics of central interest were: How are cross sections computed in strongly interacting theories? How are particles produced? How does this depend on the energy and species of colliding particles? These were the issues which drove many of my generation into theoretical particle physics.

The directions research took in trying to answer these questions were unexpected. For example, the dual resonance Veneziano model was proposed to explain basic features of high energy interactions [1]. One of the basic results of the Veneziano model was a beautiful and symmetric formula for cross sections of strongly interacting particles. It arose from a duality of the scattering amplitude under interchange of resonances in the s and t channels. The spectrum of these resonances, and the formula for the scattering amplitude could be derived by postulating a set of operators with a Virasoro algebra [1]. These operators had a simple interpretation in terms of the dynamics of a relativistic string. The origins of string theory arose from this attempt to explain measured properties of strong interactions [2].

Modern day string theory has largely evolved away from strong interactions, and is now thought of as an attempt to understand as yet unmeasured properties of gravity at extremely short distance scales. It has become a mathematical structure which survives more by its own arcane beauty than by its relationship with the world as we know it.

On the other hand, in the early 70's QCD had not arrived as a full fledged theory with the potential to describe strong interactions. It arose largely from combinations of attempts to understand masses of baryons and mesons, features of scattering of electrons from nucleons, and the mathematical beauty of Yang-Mills theory [3]. The combination of Bjorken's beautiful and seminal interpretation of electron-nucleon scattering in terms of underlying point-like degrees of freedom [4], quarks and gluons, and Gell-Mann's deep insight into hadron spectroscopy [5], together with 't Hooft's and Veltman's development of techniques which made Yang-Mills theories computable [6], led in the mid-seventies to a revolution in our understanding of the nature of strong interactions. They arose from an underlying Yang-Mills theory of quarks and gluons. Early in this development came Gribov, Bjorken, and Feynman's development of pictures of the space-time evolution of high energy scattering, and their interpretation of it in terms of point-like degrees of freedom [4,7,8]. Another significant contribution was Dokshitzer, Gribov, Lipatov, Altarelli and Parisi's determination of the evolution of the quark structure function in deep inelastic scattering as the resolution scale of the virtual photon exchanged between target and electron varied [9,10,11].

Ultimately, the computation of the dependence of the coupling constant of Yang-Mills theory upon distance scale by Gross, Wilczek, and Politzer, fundamentally changed the way in which we could understand strong interactions, and allowed us to compute a class of phenomena where the interaction strength of QCD is small [12,13]. As a result of being able to compute some class of phenomena in QCD, a hierarchy of classes of problems began to evolve. The "interesting class" was defined to be that which you could reliably compute. This class represented very special processes which occur only at very short distance scales where the coupling constant of Yang-Mills theory is very weak. The "uninteresting" class included the overwhelming majority of phenomena of strong interaction physics: Cross sections, particle production, the quark and gluon distributions inside of nucleons, and the spectrum of baryons and mesons, represent a few examples.

As time has proceeded, QCD has come to be described as a "solved theory", or as "uninteresting" or labeled with other put-downs, implicit or direct. This is because it contains the sector of "interesting" physics that now has been solved; we understand the fundamental interactions in the QCD Hamiltonian. Many people earlier interested in QCD moved on to supersymmetric theories, possibly testable at the LHC. After supersymmetry, there was supergravity and superstrings, and with each theoretical advance we moved further away from observable phenomena and ideas which can be tested against experiment.

Meanwhile, the "uninteresting" problems remained largely not yet understood.

It is the purpose of this short paper to qualitatively explain Gribov's contributions to this "uninteresting" area of theoretical physics. Gribov was involved in deep, seminal and important developments in this understanding. Many of Gribov's contributions arise from deep physical insights, but his insights developed not simply from some mystical vision but also from doing many hard computations. He no doubt travelled many wrong paths, turned back, found another wrong path – yet ultimately found a correct path and proceeded towards deeper understanding. One difference between a creative person and a fool is that while both make many mistakes in reasoning, a creative person understands he or she has made a mistake. Sometimes these "mistakes" become great new ideas.

In this paper, I hope to convince the reader that there is something of fundamental importance to be learned from the class of "uninteresting" problems of particle and nuclear physics, and to introduce some more modern ways of posing, and perhaps understanding, the problems that Gribov was the among the first to conceptualize.

This paper is not about Gribov's contributions to the understanding of "interesting" phenomena" in QCD. That work is perhaps better known, and was essentially the understanding of how quark distribution functions evolve as the resolution scale of the virtual photon changed in deep inelastic scattering. It has had a profound influence on our understanding of QCD. The work which I will describe is related to this and provides some of the intellectual basis for it. It has had a deep and lasting influence on the thinking of me and my colleagues.

I also want to tell a few anecdotes about Gribov. These anecdotes I think can convey some of his vitality and life force. Gribov was a rare man. Those of us fortunate enough to have met him knew his intensity and his strong sense of right and wrong. For Gribov, there was good physics, and nonsense – very little else between. He was a man of intense energy, emotion, and curiosity. He was filled with a passion for physics that seemed to dominate the way he related to others, and directed the way in which he saw the world around him. I never saw a Gribov relaxed or satisfied with himself.

I learned that with my Russian colleagues their deepest compliments were expressed simply, rather than with the use of extravagant adjectives.

One does not use the phrase "a real man" lightly. I greatly admired Volodya, and was not surprised to hear him called a Real Physicist, perhaps the highest compliment you can give to a colleague.

2. The Pomeron and the Reggeon

The Pomeron is the collective excitation of QCD responsible for controlling the high energy limit of strong interactions. It is a single entity, which makes it universal and of fundamental interest. In modern language, it is thought of as a composite excitation made of gluons [14]. In the early days of strong interactions, it was abstracted from theoretical analysis of partial wave sums fro scattering amplitudes. It was discovered that if one replaced the sum over angular momentum by a dispersion integral, then one could analytically continue into the complex angular momentum plane. Poles in the complex plane corresponded to Reggeons, and gave a characteristic power law dependence of the cross section on energy. There was a contribution which gave the leading high power law dependence in energy, and this was the Pomeron, named after the great Russian theorist Pomeranchuk. One can prove that the Pomeron, if it is indeed a simple pole, will lead to a constant asymptotic dependence of the cross section upon energy. Because of this, and the Froissart bound that cross sections can asymptotically rise at most as $log^2(E)$, the Pomeron is the leading Regge pole. This led people to conjecture that cross sections asymptotically become constant.

In the early 70's, the first results from the Serpukhov accelerator at Protvino near Moscow came out for the *pp* cross section. It surprised nearly everyone, showing that cross sections were rising, not asymptotically going to a constant. This led to a conjecture that there was a bare Pomeron, which was a simple Regge pole which gave a positive power law dependence on the energy. This bare Pomeron was only a lowest first order approximation to the theory, and was presumably corrected in higher orders in the strength of Pomeron interactions, and these corrections would, hopefully, makes for a fully interacting theory which had cross sections growing with at most a $log^2(E)$. Thus was the birth of Gribov's Reggeon calculus [15].

I first heard of Gribov as a graduate student at the University of Washington in Seattle. My advisor, Marshall Baker, took a year's sabbatical at ITEP in Moscow where he began collaborating with Karen Ter-Martirosian. When Marshall returned from Moscow he insisted with excitement that I learn everything possible about the Reggeon calculus. The first thing he wanted me to do was to produce a proof of the Abramovsky-Gribov-

Kancheli cutting rules [16]. These were a set of relations between the various imaginary parts of multi-Pomeron exchange diagrams. They follow from rather general conditions on the analyticity properties of scattering amplitudes (as I learned after a year of intense work).

Along the way to understanding these cutting rules, I studied Gribov's space-time picture of high energy interactions, a picture developed by Gribov in the late 60's and early 70's. As far as I can tell it was developed at about the same time as the parton model which was developed by Feynman and Bjorken, and independently of it. I basically learned about this area twice in my life, once from Gribov's papers and once later from listening to Bj. At the time I did not draw a strong connection between the two arguments since Gribov's arguments were dressed in the language of Feynman diagrams involving the exchange of ladders of mesons, and Bjorken's arguments were in terms of classical concepts laced with just the right amount of quantum mechanics.

These related approaches had a very strong effect on me. They led me to my interest in high energy nucleus-nucleus collisions, and to my interest in using these collisions as a tool to make and study new forms of matter. The ideas are at the very core of the problem I have spent the last ten years working on, the nature of matter which controls the high energy limit of QCD, the Color Glass Condensate. The glassy nature of this matter follows directly from time-dilation arguments originated in the work of Gribov, Bjorken and Feynman.

Gribov's Reggeon calculus was a field theoretical attempt to try to sum up multi-Pomeron interactions. It turns out that the basic interaction strength between three Pomerons, the triple Regge vertex, is imaginary. This lead to an amusing field theory, which is basically non-relativistic many-body theory with a complex Hamiltonian. Much work was done in trying to solve such a theory, but the ideas were being developed at the same time that the J/Ψ was discovered. This discovery, when combined with scaling seen in deep inelastic scattering as described by Bjorken, with the discovery that one can compute in non-abelian gauge theories by 't Hooft and Veltman, and with the discovery of asymptotic freedom by Gross, Politzer and Wilczek, led the entire community away from this problem to the fundamental issue of whether or not QCD is the correct theory of nature.

QCD studies drifted away from trying to understand high energy particle interactions. Interesting phenomena became defined to be either those phenomena which take place at high energy AND short distances, or the non-perturbative problems of confinement and mass generation. However, among the Russian community – particularly in the St. Petersburg and ITEP institutes where Gribov's work has been very influential – interest in high energy interactions thrived. In recent years, it has acquired some new dimension, since it appears that the high energy limit where there are very many Pomerons being simultaneously exchanged is also the limit where the gluon density becomes large. This is the Color Glass Condensate which I alluded to before, and now wish to sketch showing how it is built using concepts found in Gribov's work [17]:

- Color: Gluons are colored.
- Glass: According to arguments offered by Bjorken, Gribov and Feynman, during the time that fast moving particles sweep by a target, their wavefunction does not evolve. This is due to time dilation of the fast moving particles. Glasses are systems which one expects to change on natural time scales as does a liquid, but the time evolution is slowed down, and over very long times compared to natural time scales, the glass does not evolve.
- Condensate: Gribov argued that at very high energies, the growing total cross section could be understood by multiple Pomeron exchange. We now understand Pomerons as composite quasi-particles made of gluons. When the density of Pomerons becomes large so does the density of gluons. The density of gluons becomes so high that they condense, and are highly coherent.

The Color Glass Condensate is, I believe, the solution to the problem posed by the Reggeon calculus. The universality of this matter follows from the universality of Pomeron interactions.

3. Meeting Gribov

The first time I met a physicist named Gribov, I met Leonya, not Volodya. This was during my first visit to Russia in 1984. It was a black time then, and many physicists were boycotting going to Russia because of its mistreatment of Sakharov, and its bad human rights record. I had an opportunity to get to know several Russian physicists in Finland a year before, and they encouraged me to visit. I talked things over with my good friend Keijo Kajantie, who had managed to maintain good relations with Russian colleagues, and decided to go. It was a good decision. I began many lifelong friendships as a result of the visit, and discovered that with remarkably few exceptions the Russian theoretical community showed little liking for the policies of their government. I remember Leonya as a tall, dark haired young man of great intensity. He had just finished an epic work with Genya Levin and Misha Ryskin on deep inelastic scattering. This work has proven to be something close to the Bible of deep inelastic scattering for the past 20 years [18].

I think Leonya would have become a major figure in our field had he lived. Unfortunately, while on a hiking trip with Mitya and Misha Diakonov, he fell through a crevasse in the Pamirs.

I suspect the pain of this loss caused a certain alienation of Volodya from the St. Petersburg school which he had built. Contributing to that alienation, perhaps, was the growing tension between various groups of people. People who work together all their lives can develop somewhat strained personal relations. This happens in all groups. but it is very painful to watch in a group which is your creation.

About four years later, I met Volodya Gribov at Fermilab, where I was on the permanent staff. This was the time of Perestroika. Many scientists who had earlier been forbidden to travel – a class that included just about all the active physicists in the Soviet Union – were suddenly able to travel like ordinary human beings. At Fermilab, there was a perpetual delegation of visiting experimentalists. They had their local command structure complete with a nachalnik who was in charge. I came to know the nachalniks quite well, and they all seemed to be decent people. In fact one of my nachalnik friends used the first available opportunity to remain in the US. (The only demonstration of his power that I ever saw him exercise occurred after I drove him down from Independence Pass near Aspen where we had been fishing. I learned that he became so frightened of my driving that he secretly forbade the members of his delegation to ride with me in the mountains.)

However benign the supervision of the local nachalnik might in fact be, Bjorken and I decided it would be more pleasant for Volodya to live off site. Bj formally put on record that there was no housing available on site at the time of Volodya's visit, and arranged that he would be put up in my house.

I went to the airport to pick up our visitor. Bj was going to meet me there, but had not yet appeared when the plane arrived and passengers began to disembark. I had never met Gribov, but was confident I'd be able to identify him. I can usually spot physicists. They have their own iconoclastic way of dressing, often have a dazed look, and seldom is their hair well combed. However, I saw no one that seemed likely to be Gribov.

I searched the gate area for someone who looked lost, but without success.

I was getting worried – could he have missed his plane? Just then Bj walked up with Volodya. Apparently the latter had walked briskly out of the gate expecting to meet us at the luggage carrousel, and Bj had run into him en route. Volodya was wearing a suit and tie, and was impeccably groomed. How was I to recognize such a man as a physicist?

Volodya was a heavy smoker. The first morning in our house when he came downstairs for breakfast, he lit up a cigarette. We began talking and he lit another, having not finished the first. By the time breakfast was started he had three cigarettes going at once. He drank very strong coffee, and a lot of it. I should add that Volodya had no need of the stimulants: he was naturally hyper. In conversation, his baud rate was incredible. He believed in understanding everything under discussion in complete and utter detail. When he was pursuing an idea, he could not be deterred.

At this time, he was interested in seeing if confinement could be explained not by linear forces but by screening of charge [19]. My thesis had been on vacuum polarization to all orders in the strength of an external field, so we had something in common to talk about. Yet somehow we never connected on what should have been a fruitful topic. Perhaps it was because I was at that time interested in electroweak baryon number violation, a process mediated by topological excitations. I tried to explain this interest to Volodya, but his only reaction was that topologically excitations, instantons and their ilk, were all absolute rubbish. I liked Volodya and I think he liked me. It still puzzles me that we never successfully found common intellectual ground.

To give an idea of another perspective on the impression Gribov made during his trip to Fermilab, let me quote from what my wife recorded about that visit:

"When Gribov was visiting Fermilab, we managed to arrange to have him stay in our house, allowing him greater freedom. I was entranced by his energy, and the passionate joy he had in the many things that interested him. What I remember most vividly was an evening when we all went with Gribov and Bjorken to Chicago so he could hear a blues singer he knew would be singing at a jazz club there. (I'm accustomed to having Russian friends be more familiar with American and English literature than most American readers, but Gribov was also impressively knowledgeable about other aspects of our culture!) The singer was indeed wonderful, although the band was amplified to a level that threatened permanent hearing damage. Still, Gribov's pleasure in the music was contagious, and we stayed on.

We finally left, and found that outside the club the music coming through the walls was at last at comfortable hearing level. Gribov strained happily to hear every last note, almost dancing with enthusiasm as we made our way back to the car."

Years later, I had moved to the University of Minnesota. I was able successfully to hire a number of good Russian theoretical physicists there. This was late in the Gorbachev years when it was apparent that Russian science was going to suffer badly in the new economic environment. We were lucky at Minnesota because we acted quickly and got really good people. In particle physics, there is Arkady Vainshtein, Misha Voloshin and Misha Shifman. We hired a number of excellent condensed matter people as well.

We had resources for visitors, and we invited Volodya and Julia Nyiri. Julia, a Hungarian, was Volodya's second wife. At this time, Volodya's affiliation was in Budapest. Life with Julia seemed to have effected a subtle change in him that was pleasing to observe. The energy level was still as before, yet combined now with occasional overtones of something almost like relaxation.

During these visits I was developing my interest in the Color Glass Condensate, and I very much hoped to explain the ideas to Volodya. I remember Arkady Vainshtein tried very hard to help us to discuss physics successfully this time – he virtually shut us together in my office for long periods. Yet once again, in spite of our mutual wish to do so, we could not connect. This time some of our difficulty may have been due to a technical disagreement. Gribov could not accept the ideas that a classical coherent field could have anything to do with a quantum process like high energy scattering. But part of the difficulty probably stemmed from our difference in style. Volodya was very rigorous in his standards; he wanted well-developed mathematics. At that time, many of the basic ideas of what I was working on were in place, but mathematical structure of the sort he expected was still the the future – it took us several more years to develop it. That I could be comfortable working within such an imprecise framework, excited by ideas still nebulous by Gribov's standards, may well have struck him as uncomfortably close to bizarre. I regret we could not find a way to reach across that divide and connect – that I never managed to share successfully the work I was doing with a physicist I liked so well, a physicist whose own work had contributed to concepts on which I was then enthusiastically building.

We never had another chance. Gribov died not too long after the second of his visits to Minnesota.

4. Summary

Obviously I could not know Gribov as well as his lifelong colleagues did. As I have confessed, our attempts to discuss physics together were frustratingly unsuccessful. I nevertheless felt his influence throughout my career, and valued him highly both as a physicist and a human being. I hope this account makes that admiration clear, and adds to the record of the depth of his accomplishments and influence.

Acknowledgments

I thank Julia Nyiri and Peter Levai for their work in putting together the celebration of Gribov's physics and collecting the anecdotal accounts of his person. I thank Alice McLerran for adding her reminiscences, and help in editing this manuscript.

This manuscript has been authorized under Contract No. DE-AC02-98H10886 with the U. S. Department of Energy.

References

1. G. Veneziano, *Phys. Rept.* **9**, 199 (1974); *Nuovo. Cim.* **A57**, 190 (1968).
2. P. Goddard, J. Goldstone, C. Rebbi and C. Thorn, *Nuc. Phys.* **B56**, 109 (1973).
3. C.N. Yang and R.L. Mills, *Phys. Rev.* **96**, 191 (1954).
4. J.D. Bjorken, *Phys. Rev.* **179** , 1547 (1969); J.D. Bjorken and E. Paschos, *Phys. Rev.* **185**, 1975 (1969).
5. M. Gell-Mann, *Phys. Lett.* **8**, 214 (1964).
6. G. 't Hooft and M.J.G. Veltman, *Nuc. Phys* **B44**, 189 (1972).
7. V.N. Gribov, *Sov. Phys. JETP* **29**, 483 (1969); *Zh. Eksp. Teor. Fiz.* **56**, 892 (1969).
8. R.P. Feynman, *Phys. Rev. Lett.* **23**, 1415 (1969).
9. V.N. Gribov and L.N. Lipatov, *Phys. Lett.* **B37**, 78 (1971); *Sov. J. Nucl. Phys.* **15**, 675 (1972); *Yad. Fiz.* **15**, 1218 (1972); *Sov. J. Nucl. Phys.* **15**, 438 (1972); *Yad. Fiz.* **15**, 781 (1972).
10. Yu.L. Dokshitzer, *Sov. Phys. JETP* **46**, 641 (1977); *Zh. Eksp. Teor. Fiz.* **73**, 1216 (1977).
11. G. Altarelli and G. Parisi, *Nuc. Phys.* **B126**, 298 (1977).
12. D.J. Gross and F. Wilczek, *Phys. Rev. Lett.* **30**, 1343 (1973).
13. H.D. Politzer, *Phys. Rev. Lett.* **30**, 1346 (1973).
14. E.A. Kuraev, L.N. Lipatov and V.S. Fadin, *Sov. Phys. JETP* **45**, 199 (1977); *Zh. Eksp. Teor. Fiz.* **72** 377 (1977); I.I. Balitsky and L.N. Lipatov, *Sov. J. Nucl. Phys.* **28**, 822 (1978); *Yad. Fiz.* **28**, 1597 (1978).
15. V.N. Gribov, *Sov. Phys. JETP*, **26**, 414 (1968); *Zh. Eksp. Teor. Fiz*, **53**, 654 (1967).

16. V.A. Abramovsky, V.N. Gribov and O.V. Kancheli, *Yad. Fiz.* **18**, 595 (1973); *Sov. J. Nuc. Phys.* **18**, 308 (1974).
17. L. Mclerran and R. Venugopalan, *Phys. Rev.* **D49**, 2233 (1994); 3352 (1994); E. Iancu, A. Leonidov and L. McLerran, *Nucl. Phys* **A692**, 583 (2001); E. Iancu, E. Ferreiro, A. Leonidov and L. McLerran, *Nucl. Phys.* **A703**, 489 (2002).
18. L.V. Gribov, E.M. Levin and M.G. Ryskin, *Phys. Rept.* **100**, 1 (1983).
19. V.N. Gribov, *Eur. Phys. J.* **C10**, 91 (1999).

QCD AND HADRONS AT HIGH ENERGIES

DYNAMICS OF *pp* COLLISIONS AT SMALL IMPACT PARAMETERS, UNIVERSALITY OF CROSS SECTIONS OF HADRON-HADRON COLLISIONS AND CRITICAL PHENOMENA AT SUPERHIGH ENERGIES

L. FRANKFURT
Department of Physics and Astronomy
Tel Aviv University, Ramat Aviv
Tel Aviv, Israel
E-mail: frankfur@lev.tau.ac.il

M. STRIKMAN
104 Davey Lab,
Penn State University,
University Park, PA 16802, U.S.A.
E-mail: strikman@phys.psu.edu

Studies of the small x dynamics at HERA energies allow to quantify the range of momentum transfers and energies for which the strength of the interaction approaches the maximum strength allowed by unitarity in the gluon channel (black disk limit - BDL). Implications for the proton-proton interactions at collider energies (Tevatron, LHC) include a dominance of BDL up to transverse momenta of hadrons (jets) in the final state $\sim$ few GeV for collision at small impact parameters, and explanation of the proximity of the impact factor of *pp* interaction to one at small impact parameters, leading to universality of cross sections at superhigh energies. We discuss briefly the theoretical challenges in the applications of pQCD to small x phenomena, the similarity between small x physics and critical phenomena, and new QCD phenomena related to this puzzle.

1. Transverse distribution of partons in a nucleon

To ensure effective searches for new particle production at LHC one needs to account properly for the presence of the hard QCD interactions which rapidly increases with energy. The practical way to meet this challenge is to develop a realistic model for the QCD interactions at the LHC energies which would incorporate information about the nucleon structure and small x dynamics revealed in the analyses of the QCD phenomena at HERA. One key element is the information about transverse structure of the parton distributions in nucleon as given by generalized parton distributions. This information was extracted from the analysis of exclusive vector meson production which satisfies the QCD factorization theorem for exclusive processes [1,2]. The second effect is rapidly growing with energy in the kinematical region, where the leading twist approximation for the interaction of partons up to rather large virtualities (approach to the black disk limit - BDL) breaks down. This effect is probably manifested in the well known problem for the current Monte Carlo event generators which have to introduce a minimum transverse momentum cutoff for p_t of the generated jets which is a strong function of pp energy. Another effect which seems to be relevant is the presence of the transverse correlations between medium x partons [a].

It is well known that Gribov diffusion in the impact parameter space leads to an increase of the size of the fast hadron with the increase of energy. The use of hard exclusive processes allows to address the question of the transverse distribution of partons for finite x. Our analysis of the J/ψ photoproduction at fixed target energies demonstrates that the transverse size of $x \geq 0.1$ gluons is substantially smaller than the average proton transverse size [4]. A natural explanation is that the average size of the proton gets a substantial contribution ($\approx 20\%$) from the pion cloud, while its contribution to the gluon destribution at $x \geq 0.1$ is strongly suppressed [4,5]. At smaller x the gluon transverse distribution increases with the decrease of x, however the rate of the increase is still significantly smaller than for the soft partons. This leads to a two scale picture of the nucleon - hard parton concentrating at significantly smaller distances from the center than the soft ones.

2. How close is the interaction of small dipoles with nucleons to the black disk limit at HERA?

The increase of the gluon pdf at small x leads to the breakdown of the leading twist approximation at sufficiently small x. A simple way to see it is to consider elastic scattering of a small colorless dipole off a nucleon. One can calculate this process in pQCD within the DGLAP approximation. An alternative method is to use information from the analyses of data on DIS scattering within the approximation where structure functions are described in the proton rest frame as the scattering of colorless dipole whose

[a] For a detailed discussion of the issues covered in this talk and extensive references see review [3].

interaction is constrained to match perturbative QCD calculations justified for dipoles of a sufficiently small size, d. Such an analysis gave a possibility to determine $\sigma_{tot}(d,x)$ for the interaction of $q\bar{q}$, $2g$ dipoles with nucleons, and yielded information on the t-dependence from hard exclusive vector meson production to estimate the t-dependence of the scattering amplitude, $A^{dp}(s,t)$. Next we introduce the profile function $\Gamma^{dp}(s,b)$ which is related to $A^{dp}(s,t)$

$$\Gamma^{dp}(s,b) = \frac{1}{2is}\frac{1}{(2\pi)^2}\int d^2\vec{q}\exp(i\vec{q}\vec{b})A^{dp}(s,t), \tag{1}$$

where $t = -\vec{q}^{\,2}$, $\mathcal{I}\Updownarrow A^{dp}(t=0) = \sigma_{tot}$. The results of this estimate are presented in Fig.1 both for the case of $q\bar{q}$ and gg dipoles. In the latter case we assumed that as in pQCD the ratio of $gg - N$ and $q\bar{q} - N$ cross sections is given by the ratio of the Casimir operators for the corresponding representations of the $SU(3)$ gauge group - 9/4.

In the case of interaction induced by the inelastic scattering, S-channel unitarity leads to the limit $\Gamma \leq 1$, with $\Gamma = 1$ corresponding to the BDL of elastic scattering in which it is equal to 1/2 of the total cross section. One can see from the figure that $q\bar{q} - N$ interaction is rather far from BDL for the dipole size $d \sim 0.3$ fm corresponding to the virtuality $Q \sim 2$ GeV. On the other hand, in the case of gluon dipoles the interaction does approach BDL for $Q \sim 2$ GeV for a rather large range of b.

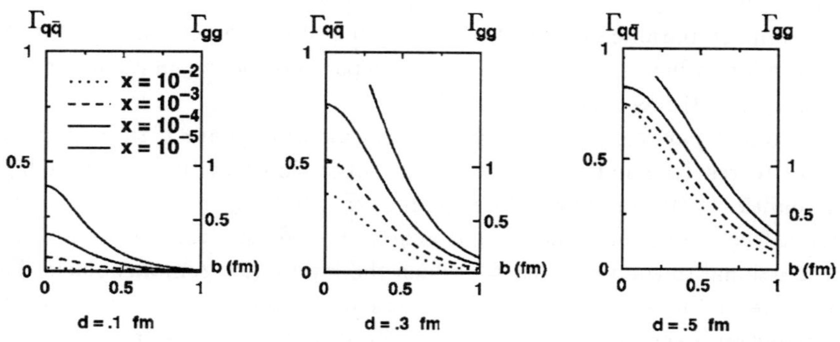

Figure 1. The profile function of dipole–nucleon scattering, Γ^{dp}, as a function of the impact parameter, b, for various values of the dipole size, d, and x.

If the interaction of colorless gluon dipoles with nucleons is close to the maximal strength at $x \sim 10^{-4}$, $Q \sim 2$ GeV the probability of the hard diffraction in the interaction with a nucleon for a probe which couples to gluons should be close to 1/2. Indeed, we find that the gluon diffractive pdfs extracted from the analysis of the diffractive data by H1 correspond to the probability of the gluon induced diffraction in the discussed kinematics of the order 0.3-0.4. Also, the QCD DGLAP evolution should break down. Indeed, though the NLO fits allow the description of the DIS, the price is a decrease of the power $\lambda(Q^2)$ in the fit $xg(x,Q) \propto x^{-\lambda}$ to the values smaller than for the soft regime for $Q \sim 1.5$ GeV^2. Such a behavior leads to a breakdown of soft matching to the nonperturbative regime and hence provides another hint that the perturbative description breaks for this kinematics. Since the gluon densities in pQCD strongly increase with the decrease of x, the BDL regime should extend to higher and higher virtualities with the decrease of x. It is worth noting here that the breakdown of the pQCD description occurs in the x-range where $\ln x$ effects are still not large – due to energy conservation one needs to take into account mostly terms $\propto \ln x$ which are taken care of in the NLO approximation.

3. Inelastic pp scattering at small impact parameters

In the case of "hard probe"-nucleon scattering both interactions at small and large impact parameters contribute. Interactions at small impact parameters should deviate from the leading twist regime at lower energies (larger x) than the peripheral ones. In the case of high energy *pp* scattering the large difference between the transverse spread of soft partons and hard partons with $x \geq 0.01$ leads to a strong difference between the distribution over impact parameters for the generic inelastic interactions and for the interactions where a hard collisions of two partons (or two pairs of partons) with $x_i \geq 0.01$ has occurred [6], see Fig.2.

Incident partons in the projectile nucleon which have high enough energies to resolve small x partons in the target interact (up to rather high virtualities) with the target gluon fields in the BDL regime. As a result they lose a fraction of their energy and acquire significant transverse momenta proportional to the virtuality for which the interaction remains black. The estimate of the average transverse momenta involves taking into account distribution over impact parameters, determining what is the minimal x which is resolved in the second proton at a given virtuality Q^2 by a parton with a given x_1:

$$x = 4p_{\perp}^2/x_1 s,\; Q^2 = 4p_{\perp}^2. \tag{2}$$

Next, to estimate at what virtualities the interaction is black, we can treat the leading parton as one of the constituents of a small dipole scattering

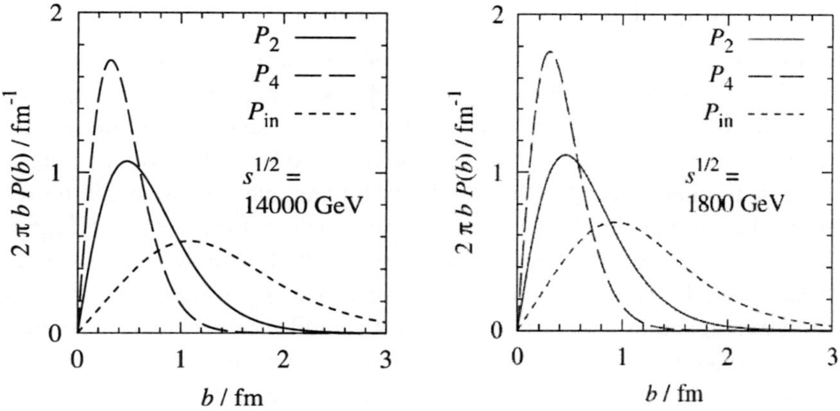

Figure 2. Comparison of the impact parameter distributions for the minimal bias events - dashed curves, events with dijet trigger (solid curve) and a trigger on production of four jets in two binary collisions.

from the target and use our results for the interaction of small dipoles with nucleons to estimate when Γ becomes comparable to one. We find that at LHC energies at central impact parameters average transverse momenta for both leading quarks and gluons should exceed a few GeV. The same conclusion holds for the dijet (four jet trigger) since average b for such triggers are small.

As a result, associated hadron production in the LHC events with a dijet trigger should strongly differ from the production in minimal bias events: (i) The leading particle spectrum will be strongly suppressed compared to minimal bias events since each parton fragments independently. This effect should be especially pronounced for leading nucleons: for $z > 0.1$ the differential multiplicity of pions should exceed that of nucleons (this is on the top of the strong suppression of the pion spectrum). (ii) A large fraction of the dijet tagged events will have no particles with $z \geq 0.05$. This suppression will occur simultaneously in both fragmentation regions, corresponding to the emergence of long–range rapidity correlations between the fragmentation regions, leading to a large energy release at rapidities below $y \sim 6$. (iii) Average transverse momenta of the leading particles will exceed one GeV/c. Since new particles are supposed to be produced in collisions of $x \geq 0.01$ partons, the discussed effects will strongly increase the semi-hard activity in such events as compared to naive expectations.

4. Universality of the total cross sections

The discussed pattern of the interaction of the leading quarks and gluons allows to estimate impact factor (Eq. 1) for elastic pp scattering using unitarity relation which leads to a probability of the inelastic interaction at a given b equal to $1 - |1 - \Gamma(b)|^2$. When a single leading quark gets a transverse momentum, $p_\perp$, the probability for the nucleon to remain intact is approximately given by the square of the nucleon form factor, $F_N^2(p_\perp^2)$, which is ≤ 0.1 for $p_\perp > 1\,\mathrm{GeV}/c$, *i.e.*, very small. One may thus conclude that the probability of the survival averaged over $p_\perp$ should be less than 1/2 (on average, half of the time the quark should receive a transverse momentum larger than the average transverse momentum, $p_\perp > 1\,\mathrm{GeV}/c$). Because there are six leading quarks (plus a number of leading gluons), the probability for the two protons to stay intact, should go as a high power of the survival probability for the case of single parton removal, and thus be very small. Hence $\Gamma(b) \approx 1$, which agrees with the Tevatron data and expectations for LHC. Consequently, the interaction at small b is not reduced to single + double + ... Pomeron exchanges – it is likely that a transition to a new hard regime takes place. Within the same space-time picture we find that the range of the impact parameters where interaction is practically black should grow as $\propto \ln s$, leading to Froissart type behavior of the total hadron–hadron(nucleus) cross sections, $\sigma_{\gamma^* T} \propto \ln^3 s$, and the t-slope $\propto \ln^2 s$. At extremely high energies the memory of the nature of colliding particles is lost as a consequence of the blackening of the interaction. As a result, the ratio of total cross sections for any colliding particles should become equal to one. Universality reflects the universal dependence on the impact parameters of the Pomeron exchange which dominates soft peripheral interactions and the universal dependence on the impact parameters of the two pion tail of a hadron wave function.

5. Hard processes at small x

A challenging problem is to calculate amplitudes of hard processes at small x directly in QCD. The most popular approach is the DGLAP approximation. This approximation can be formally justified within the leading twist approximation in QCD by the smallness of running coupling constant $\alpha_s(Q^2)$ and proved validity of the renormalization group for the moments of the structure functions. The origin of fundamental problems for the DGLAP approximation is the appearance of a new large parameter $\ln(x_o/x)$ which radically changes the ranking of the leading terms at sufficiently small

x and therefore destroys applicability of DGLAP approximation.

An alternative approximation, which sums leading $\alpha_s \ln(x_o/x)$ terms – LO BFKL approximation – produces amplitudes too fastly increasing with energy. The account of the energy-momentum conservation strongly reduces the amplitude and its increase with energy as compared to the LO BFKL approximation. The second effect is the lack of scale separation, i.e. an uncontrolable contribution of soft, nonperturbative QCD. The third effect is a neglect by the decrease of invariant charge with virtuality.

Calculated NLO corrections to BFKL approximation are large and negative [9] since it restores to some extent energy-momentum conservation. Account of both the running coupling constant and the energy-momentum conservation makes the x dependence of structure functions close to that obtained before within the DGLAP approximation [8]. It appears that LO+NLO DGLAP approximations to pQCD are valid down to $x \approx 10^{-7 \div -8}$ where an LO+NLO BFKL type behaviour of amplitudes of the processes seems to appear [12]. Thus, it seems now that pQCD calculation is under control if HT effects are neglected. Therefore the BDL occurs at significantly larger x those for which the DGLAP approximation loses its credibility. The puzzle occurs mostly because of a significant gluon distribution in nonperturbative support and fast increase of the amplitudes with energy [3] and discussion in the previous sections.

6. Aligned jet model and disappearance of the color transparency phenomenon

It has been understood long ago [10,11] that at moderately small x there is conspirancy between two competing contributions to the structure functions of a hadron target which are hidden in the infinite momentum frame description in the initial condition for the evolution with Q^2. In the target the rest frame virtual photon transforms into a $q\bar{q}$ pair. The contribution of quarks with small transverse to photon momentum direction momenta k_t corresponds to the phase volume $1/Q^2$ but strong interaction. On the contrary, contribution of $q\bar{q}$ pairs with large transverse momentum $\approx Q/2$ corresponds to almost the whole permitted phase volume but cross section of interaction with target which is suppressed by $1/k_t^2 \propto 1/Q^2$ as a result of color screening – this is one of color transparency phenomena. At sufficiently small x cross section of hard interaction becomes large, so the transparency of hadron matter disappears and hard interaction becomes dominant because of larger phase volume. This new regime where

xG, $F_2 \propto Q^2 \ln^3(x_o/x)$ [13] corresponds to complete disappearance of color transparency and related Bjorken scaling.

A recent evaluation of the triple reggeon vertex within LO BFKL approximation found that this vertex is $\propto 1/t$ [14]. This is another example of the important role of soft QCD = the aligned jet model: LO BFKL approximation does not separate hard and soft scales. Thus it includes an uncontrollable piece of soft QCD physics which can not be evaluated legitimately by the methods of pQCD. Account of the color confinement will remove spurious singularity at small t and spontaneously broken chiral symmetry will screen terms $1/m_q^2$ where m_q is the mass of a light quark. In the theoretical description of hard processes where a separation of scales is performed, the interaction of three hard ladders in a triple reggeon kinematics has properties which are evidently different from that obtained within the LO BFKL approximation. In particular, gluons(quarks) will be strongly off mass shell as a result of the account of Sudakov form factors. Moreover, at sufficiently small x where cross sections of interaction of spatially small dipoles becomes sufficiently large (near the unitarity limit) soft QCD contribution becomes a correction.

7. Fundamental challenges of pQCD

It is explained above that approximate descriptions of hard phenomena in pQCD predicts a fast increase of the amplitude with energy. To visualize this property we parametrize $xG \propto x^{-\mu(Q^2)}$ where $\mu(Q^2 = 10) \approx 0.2$. At sufficiently small x μ weakly increases with the increase of Q^2 [8]. At the same time near the BDL, higher twist effects blow up: calculation of shadowing "corrections" due to rescatterings produces series gives: $xG = \sum_1^\infty (c_n/Q^2(n-1))x^{-n\mu}$. Here c_n depends on $\ln(Q^2/Q_o^2)$ and fastly increasing with n – this is well known property of rescattering diagrams cf. Ref. [15] resulting from the t-channel iterations of pQCD ladders – c_2 is calculable in QCD [16]. Existence of the term with $n = 2$ has been established experimentally at HERA as cross section of diffractive vector meson production in DIS. Different approximations differ in the value of c_n. In particular, it is easy to show that coefficients $c_n = \frac{f(n, N_c\alpha_s)}{N_c^{n-1}}$ are decreasing with the number of colors N_c. Thus at sufficiently small x leading N_c approximation is violated.

To summarize, the applications of the pQCD methods lead to series of terms increasing with the increase of n and with the decrease of x. At sufficiently small x, series over powers of coupling constant are beyond the radius of convergency or of applicability of asymptotic series because effective parameter of decomposition becomes large. So pQCD series become divergent and therefore incomplete. To calculate the limiting behaviour of hard processes directly in QCD it is necessary to evaluate terms related to divergency of such series, i.e. to find and evaluate terms which can not be decomposed into series over powers of coupling constant α_s. So manifold of functions to work with should include functions like $\exp(1/\alpha_s)$. This new nonperturbative phenomenon is unrelated to the well known nonperturbative phenomena such as confinement and spontaneously broken chiral symmetry.

Following the V. Gribov *Reggeon Field Theory* [17] one can easily build effective theory in the complex angular momentum plane or in the plane of energy and impact parameter. Within such an approach singularities are located at $j-1=n\mu$ i.e. in the region forbidden by causality and nonlinear relations between Green functions. Thus pQCD predicts existence of a tachyon. In this respect the challenging problem of the perturbative pQCD small x regime resembles emergence of the tachyons in the superconductivity and in the string theory in 26 dimensions. pQCD vacuum is a false vacuum and it is necessary to find physical vacuum and to build pQCD expansion around the physical vacuum. Thus expected new phenomenon is the transition from a false to a physical vacuum. This transition has been accounted for in [18] for single scale hard processes. New QCD phenomena color inflation, massles "phonons" and color network were found.

To summarize: it looks now that small x phenomena in QCD are a promising method to produce new forms of the QCD matter.

This work is supported by DOE and BSF grants.

References

1. S.J. Brodsky, L. Frankfurt, J.F. Gunion, A.H. Mueller and M. Strikman, *Phys. Rev.* **D50**, 3134 (1994).
2. J.C. Collins, L. Frankfurt and M. Strikman, *Phys. Lett.* **B56**, 2982 (1997).
3. L. Frankfurt, M. Strikman and C. Weiss, *Ann. Rev. Nucl. Part. Sci.* **55**, 403 (2005).
4. L. Frankfurt and M. Strikman, *Phys. Rev.* **D66**, 031502 (2002).
5. M. Strikman and C. Weiss, *Phys. Rev.* **D69**, 054012 (2004).
6. L. Frankfurt, M. Strikman and C. Weiss, *Phys. Rev.* **D69**, 114010 (2004).
7. L. Frankfurt, M. Strikman and M. Zhalov, *Phys. Lett.* **B616**, 59 (2005).
8. M. Ciafaloni and G. Camici *Phys. Lett.* **B430**, 349 (1998); M. Ciafaloni and D. Colferai *Phys. Lett.* **B452**, 372 (1999).
9. V.S. Fadin, L.N. Lipatov, *Phys. Lett.* **B429**, 127 (1998).

10. J.D. Bjorken, SLAC-PUB-1017. *Presented at Conf. on Particle Physics*, Irvine, Calif., Dec 3-4, 1971.
11. L.L. Frankfurt and M.I. Strikman, *Phys. Rep.* **160**, 235 (1988).
12. G.P. Salam, arXiv:hep-ph/0510090.
13. L. Frankfurt, V. Guzey, M. McDermott and M. Strikman, *Phys. Rev. Lett.* **87**, 192301 (2001).
14. J. Bartels, M.G. Ryskin and G.P. Vacca, *Eur. Phys. J.* **C27**, 101 (2003).
15. V.N. Gribov and A.A. Migdal, *Sov. Phys. JETP* **78**, 784 (1969); *Sov. J. of Nucl. Phys.* **8**, 583 (1969).
16. H. Abramowicz, L. Frankfurt, M. Strikman, *Surveys High Energ. Phys.* **11**, 51 (1997), also in SLAC Summer Inst. 1994:0539-574.
17. V.N. Gribov, *Theory of complex angular momenta: Gribov lectures on theoretical physics*, Cambridge Univ. Press, 2003.
18. B. Blok and L. Frankfurt, arXiv:hep-ph/0508218, arXiv:hep-ph/0510345.

HIDDEN QCD SCALES AND DIQUARK CORRELATIONS

ARKADY VAINSHTEIN

William I. Fine Theoretical Physics Institute,
University of Minnesota, Minneapolis, MN 55455

Numerically large QCD scales were discussed first in connection with a number of phenomena mostly related to vacuum quantum numbers and $0^{\pm}$ glueball and quark channels. We present arguments regarding the possible role of the larger scale in diquarks important for low-energy hadron phenomenology. Good diquarks, i.e. the 0^+ states of two quarks, are argued to have a two-component structure with one of the components peaking at distances several times shorter than a typical hadron size (a short-range core).

The 75th birthday of Volodya Gribov is a good occasion to think once more about novel ideas he introduced into high energy physics. I mean not just his famous achievements which are well recognized but a set of ideas he developed in his efforts to solve the problem of confinement in QCD. One of the aspects of his picture of the world refers to the existence of a point-like structure in the pion. I would like to relate this to a possibility of hierarchical scales in QCD, a subject of my talk. On the other hand, having an experience of numerous discussions with Volodya, I am virtually sure that he would not agree with me. It is a great sadness that I cannot try.

1. Introduction

The presentation is based on the work [1] with Misha Shifman where we study an impact of hierarchical scales in QCD on diquark features and related phenomena.

Let us start with the observation that the constituent quark model enjoy a remarkable success in qualitative and semiquantitative descriptions of a huge body of data on traditional hadron spectroscopy, static parameters and other regularities At the same time, since the advent of QCD it has been known that some important hadronic phenomena cannot be easily understood in this model. Probably, the most clear-cut example of this type is the masslessness of pion (in the chiral limit), which can only occur due to a "superstrong" attraction in the 0^- quark-antiquark channel. In the 1980's, a similar strong attraction was argued to exist in the 0^+ quark-quark channel. This conjecture was supported by occasional observations of a special role played by diquarks in low-energy hadronic phenomenology.

It was argued [2,3] that an instanton-induced interaction is sufficient to form a bound antitriplet scalar diquark. Then, there were suggestions [4] (see also the earlier analysis [5] and the recent discussion in [3]) that diquark correlations are important in the enhancement of $\Delta I = 1/2$ amplitudes of strangeness-changing decays.

Recently, the issue was revitalized by experimental evidence for existence of exotic pentaquark baryon, the state which was predicted to be narrow in Ref. [6] based on the chiral-quark-soliton model. Diquarks were suggested as an alternative explanation for pentaquark in a number of papers [7,8,9,10] which also discussed other phenomenological evidences. Although the existence of pentaquark is now strongly doubtful, implications of diquark correlations could be instrumental in excited and exotic hadron spectroscopy and in other aspects of hadronic physics. "Good" diquarks in color-flavor locked antisymmetric combination, due to a strong attraction, are probably bound to the extent that the "mass" of a good diquark roughly coincides with that of a constituent quark.

It is natural to suggest that the characteristic size of the good diquark as well as that of the constituent quark is considerably smaller than the nucleon size. In the case of the constituent quark a relevant momentum scale in the $\bar{q}q$ channel is of order of a few GeV. We will argue that a similar scale appears in the diquark qq channel. A basic tool here is the observation that if $SU(3)_{\rm color}$ is replaced by $SU(2)_{\rm color}$, diquarks become well-defined gauge-invariant objects, related by symmetry to conventional Goldstone bosons (pions).

The existence of hierarchical scales in QCD was discussed long ago in [11]. The largest of these scales shows up in $0^\pm$ glueball channels. QCD with massless flavors (the chiral limit) classically contains no scale and is conformally invariant. Quantum effects break this invariance, and the physical scale $\Lambda_{\rm QCD}$ appears due to so called dimensional transmutation,

$$\Lambda_{\rm QCD} = M_{\rm UV} \cdot {\rm e}^{-\frac{8\pi^2}{b g_{\rm UV}^2}} . \tag{1}$$

It is in one-loop approximation for the running coupling g, $M_{\rm UV}$ is the ultraviolet cut-off, $b = \frac{11}{3}N_c - \frac{2}{3}N_f$ and $g_{\rm UV} = g(M_{\rm UV})$.

Phenomenologically $\Lambda_{\rm QCD}$ is of the order of a few hundred MeV. In terms of hadronic masses the one of the meson can be taken as a characteristic hadronic scale,

$$m_\rho = \text{const} \times \Lambda_{\text{QCD}} = 770 \text{ MeV}\,. \qquad (2)$$

Although parametrically Λ_{QCD} is the only scale in QCD indeed there are some evidences for an existence of *numerically* much larger hadronic scales.

- $\Delta I = 1/2$ enhancement in weak nonleptonic decays. The $\Delta I = 1/2$ amplitudes are about 10 times larger than the $\Delta I = 3/2$ ones. Related is an enhancement in the ratio ϵ'/ϵ of the parameters of CP violation in K decays.
- A number of phenomena in heavy flavor decays, including a value of the Λ_b lifetime which is 10-15 % shorter than expected, large rates for $B \to \eta' X$ decays, importance of penguins in $b \to s$ transitions and problem with quark-hadron duality in $b \to u$ transitions.
- Large mass (958 MeV) of η'. Theoretically $m^2_{\eta'}/m^2_N \propto 1/N_c^3$. In reality this ratio is close to 1.
- Enhancement in comparison to perturbative QCD estimates in many exclusive processes with large momentum transfer.
- Large duality interval in 0^-, 0^+ glueball and quark channels.
- Large s-wave $\pi\pi$ interaction without prominent resonances.
- A small value of α', a slope of the pomeron, the vacuum Regge trajectory.
- Jet production in e^+e^- annihilation – the total energy should be around 10 GeV to see clear-cut jets.
- Striking (and unexpected) experimental evidence from heavy ion collisions at RHIC. Instead of weakly coupled quark-gluon plasma experimentalists see a hydrodynamical flow.

2. Glueballs and large momentum scale

Breaking of conformal invariance in QCD by quantum effect can be expressed as an anomaly in the trace of energy-momentum tensor $\theta_{\mu\nu}$,

$$\theta^\mu_\mu = \frac{\beta(\alpha_s)}{4\alpha_s}\, G^a_{\mu\nu} G^{a\,\mu\nu}\,, \qquad \beta(\alpha_s) = \mu\,\frac{\mathrm{d}\alpha_s(\mu)}{\mathrm{d}\mu} = -b\,\frac{\alpha_s}{2\pi} + \dots\,. \qquad (3)$$

The dilation feature of θ^μ_μ leads to the low-energy theorem [11],

$$\int \mathrm{d}x\, \left\langle 0 | T\left\{ \frac{\alpha_s}{\pi}\, G^2_{\mu\nu}(x), \frac{\alpha_s}{\pi}\, G^2_{\gamma\delta}(0) \right\} | 0 \right\rangle = \frac{32}{b} \left\langle 0 \left| \frac{\alpha_s}{\pi}\, G^2_{\mu\nu}(0) \right| 0 \right\rangle\,. \qquad (4)$$

Numerically, using the phenomenological value for gluon condensate [12]

$$\left\langle 0 \left| \frac{\alpha_s}{\pi} G^2_{\mu\nu}(0) \right| 0 \right\rangle = (330\ \mathrm{MeV})^4\,, \tag{5}$$

we find a large momentum scale λ_G at which perturbative and nonperturbative effects are comparable,

$$\lambda_G^2 = 20\ \mathrm{GeV}^2\,. \tag{6}$$

Compare this with the scale in the 1^- isovector quark channel

$$\lambda_\rho^2 = m_\rho^2 = 0.6\ \mathrm{GeV}^2\,. \tag{7}$$

The order of magnitude difference is quite significant.

3. Two-component structure of pion

We know about "abnormally strong" interaction in the channel where the light pion exists. One can present a strong argument for a short-range core in the pion. Compare two matrix elements,

$$\begin{aligned} \langle 0|\bar{d}\gamma_\mu\gamma_5 u|\pi^+\rangle &= i f_\pi\, q_\mu\,, \\ \langle 0|\bar{d}\gamma_5 u|\pi^+\rangle &= i f_\pi\, \frac{m_\pi^2}{m_u+m_d}\,. \end{aligned} \tag{8}$$

The second matrix element is enhanced by the ratio

$$\frac{m_\pi^2}{m_u+m_d} \approx 1.8\ \mathrm{GeV}\,. \tag{9}$$

It has a smooth chiral limit $m_q \to 0$ but rather large numerically. Within QCD sum rules [12] it shows as a considerable difference in duality intervals,

$$\begin{aligned} \bar{d}\gamma_\mu\gamma_5 u\ &: \qquad 0.6\ \mathrm{GeV}^2, \\ \bar{d}\gamma_5 u\ &: \qquad 1.9\ \mathrm{GeV}^2. \end{aligned} \tag{10}$$

How can one interpret this fact? In the quantum-mechanical approach the matrix elements (8) can be viewed as a value of wave function at zero separation, $\psi(0)$. We have two of such wave functions: in the leading twist, $t=2$ for $\bar{d}\gamma_\mu\gamma_5 u$ channel and in the nonleading twist, $t=3$, for $\bar{d}\gamma_5 u$,

$$\psi = \alpha_2\psi_{t=2} + \alpha_3\psi_{t=3}\,. \tag{11}$$

While the coefficient α_3 is small the large value of $\psi_{t=3}(0)$ for the $\bar{d}\gamma_5 u$ component implies the existence of a smaller size.

We try to illustrate this in Fig. 1. The factor $(1/3)^2$ for a core size can be interpreted in the framework of the instanton liquid model [13]. The model

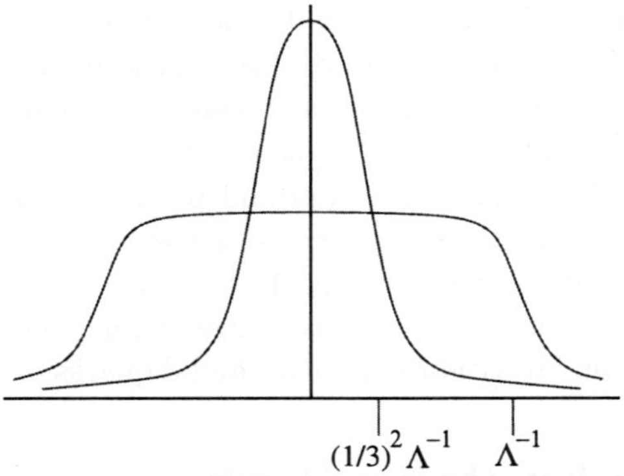

Figure 1. Double-component structure with a core.

operates with two parameters, the average instanton size $\rho = 0.48\,\Lambda_{\rm QCD}^{-1}$ which is a factor of ~ 3 smaller than the average instanton separation $R = 1.35\,\Lambda_{\rm QCD}^{-1}$. Instantons are Euclidean objects and to relate their parameters with our Minkowski world, note that $R^{-1} \sim \Lambda_{\rm QCD}$ is a typical hadronic scale while ρ^{-1} is the geometrical mean between $\Lambda_{\rm QCD}$ and the higher glueball scale $\Lambda_{\rm gl}$. Hence, we conclude that in the glueball world $\Lambda_{\rm gl} \sim 3^2\Lambda_{\rm QCD}$, which explains $(1/3)^2$ in Fig. 1.

4. Diquarks' progenitors

What about diquarks' sizes? They can be related to the pion ones when the color SU(3) is replaced by SU(2). The flavor group becomes SU(4),

$$\chi^i = \big(u_L,\, d_L\,, \bar d_R\,, -\bar u_R\big) \tag{12}$$

and the the pattern of the spontaneous breaking is [14,15].

$$\mathrm{SU}(4) \to \mathrm{SO}(5)\,. \tag{13}$$

Indeed, a nonvanishing vacuum average of the six bilinears $\chi^i\chi^j = -\chi^j\chi^i$ can be aligned as

$$\langle \bar u u + \bar d d\rangle = \langle \bar u_R u_L + \bar u_L u_R + \bar d_R d_L + \bar d_L d_R\rangle \neq 0\,. \tag{14}$$

Acting on $\bar u u + \bar d d$ by SU(4) generators we determine five Goldstone bosons corresponding to operators

$$\bar u_R u_L - \bar d_R d_L - (R \leftrightarrow L) = -\bar u\gamma_5 u + \bar d\gamma_5 d\,,$$

$$\bar u_R d_L - (R \leftrightarrow L) = -\bar u\gamma_5 d\,, \qquad \bar d_R u_L - (R \leftrightarrow L) = -\bar d\gamma_5 u\,, \tag{15a}$$

$$u_L d_L + (R \leftrightarrow L) = -uC\gamma_5 d\,, \qquad \bar d_L \bar u_L + (R \leftrightarrow L) = -\bar d C\gamma_5 \bar u\,. \tag{15b}$$

In addition to the triplet of massless 0^- pions (first two lines) we get two "baryon" Goldstone bosons (the third line) which are 0^+ diquark states.

Thus, we have more than the conventional triplet of pions; there emerge two extra diquark states related to pions by exact symmetry. Since in $\mathrm{SU(2)_{color}}$ theory baryonic Goldstones and pions are related by an exact symmetry, their spatial structure is the same. In particular, the two-component structure depicted in Fig. 1 is shared by both. As we pass to $\mathrm{SU(3)_{color}}$, the exact symmetry no longer holds, but an approximate similarity in the spatial structure of pions and diquarks is expected to hold.

5. Approaching from the large-N_c side

A powerful tool in qualitative and semiquantitative analysis of QCD is 't Hooft's $1/N_c$ expansion [16]. A complementary "orientifold" large-N expansion was suggested recently [17]. This expansion is similar in spirit — but not technically — to expansion which goes under the name of *topological expansion* and was suggested long ago [18]. (Topological expansion assumes that the number of flavors N_f scales as N in the large-N limit, so that the ratio N_f/N_c is kept fixed. A quark-diquark representation of nucleon is natural in topological expansion.)

The orientifold large-N_c extrapolation treats quarks as Dirac fields in the two-index antisymmetric representation of the $\mathrm{SU}(N_c)$ color group. At $N_c = 3$ this representation is equivalent to fundamental; hence, the starting point for both extrapolations — orientifold and 't Hooft's — is bona fide $N_c = 3$ QCD, and both are equally suitable for $1/N$ expansion.

In most instances orientifold and 't Hooft large-N expansions lead to overlapping predictions. There are notable exceptions, however, of which the most important are:

(i) quark-mass dependence of vacuum energy density, and
(ii) the Okubo–Zweig–Iizuka (OZI) rule.

In the former case, according to 't Hooft, quark-mass dependence should be suppressed by a factor $1/N_c$ as all quark loops scale as $1/N_c$ at large N_c. On the other hand, in the orientifold theories quark loops are *not* suppressed, and one should expect quark-mass dependence of a "natural" order of magnitude (see below). The very same property, $1/N_c$ for each "additional" quark loop in the 't Hooft limit, is supposed to explain inhibition of transitions between quarks of distinct flavors (the OZI rule). In the orientifold large-N_c limit quark loops do not carry $1/N_c$, and, correspondingly, transitions between quarks of distinct flavors are not suppressed by $1/N_c$.

At large N_c 't Hooft and orientifold extrapolations present distinct theories, and, therefore, it is not surprising that some predictions do not coincide. There is one point, $N_c = 3$, where these theories do coincide, and, hence, must lead to the same results. Logically there are two options that would lead to a reconciliation at $N_c = 3$: a numerical suppression of appropriate amplitudes in orientifold theory or a numerical enhancement in 't Hooft's theory. We will argue that it is the latter option that is realized in certain important cases.

On the phenomenological side, both points (i) and (ii) above, rather than being successful, present a serious challenge to the 't Hooft approach. Indeed, two fundamental relations

$$\frac{d}{dm_q}\,\mathcal{E}_{\rm vac} = \langle \bar{q}\, q\rangle \tag{16}$$

and

$$\mathcal{E}_{\rm vac}\Big|_{m_q=0} = \frac{\beta(\alpha_s)}{16\alpha_s}\,\left\langle G^a_{\mu\nu}\, G^{\mu\nu\, a}\right\rangle \tag{17}$$

being combined and evaluated numerically [11] imply that in QCD with light quarks the vacuum energy density $\mathcal{E}_{\rm vac}$ changes by a factor of ~ 2 as the strange quark mass increases from zero to its actual value ~ 150 MeV. Moreover, while the extra quark-loop suppression inherent in the 't Hooft N_c counting seemingly does explain the OZI rule in the 1^- channel, where the ϕ meson is an almost pure $\bar{s}s$ state, it badly fails in $0^\pm$ channels where flavor mixing is fully operative: say, the composition of η' meson only slightly deviates from $\bar{u}u + \bar{d}d + \bar{s}s$.

It was suggested [11] that the 't Hooft-defying enhancement in these two cases is due to "direct" instantons. Instantons are theoretical objects existing in Euclidean space-time. It is natural to ask what actual physics lies behind this phenomenon. What is the relevant Minkowski-space picture? As was discussed in Sect. 2, a new numerically large scale in hadronic physics shows in phenomena where vacuum quantum numbers (e.g. the $0^\pm$ channels) were involved, a hidden energy scale, larger than the conventional ~ 300 MeV by a factor of ~ 10 or so.

The space-time structure of good diquarks is (in Weyl notation): $\chi^{[if}_\alpha\, \chi^{jg]\,\alpha}$ where square brackets denote antisymmetrization with respect to both color indices i, j and flavor indices f, g; and α is the spinor index. Good diquarks are scalars for which the flavor mixing is maximal. They appear, right from the start, as two-index antisymmetric objects in color representation. This naturally matches them with orientifold theory.

If N_c is treated as a free parameter, then diquarks propagating in loops with virtual momenta in the interval ($\sim R_h^{-1}$, $\sim R_{dq}^{-1}$) have no $1/N_c$ suppression. This simple observation suggests that they play a crucial role in strong quark-mass dependence of vacuum energy as well as in the failure of the OZI rule in $0^{\pm}$ channels, acting as Minkowski-space "representatives" of the Euclidean instantons much in the same way as sphalerons and monopoles are Minkowski signatures of instantons at weak coupling. Note that, being composite in flavor, diquark loops (with diquarks treated as pointlike objects) also negate the notion of planarity.

At $N_c = 3$ the failure of the OZI rule in $0^{\pm}$ channels should look like a numerical enhancement with regards to the 't Hooft counting. For instance, the VEV $\langle \bar{s}s \rangle$ is contributed by loops of (su) and (sd) diquarks. This provides a numerical enhancement factor ~ 4. Moreover, if this picture is correct, then in 0^+ channels at low energies flavor mixing has no suppression whatsoever (the same is true for 0^-). An immediate consequence is that the correlation functions

$$\langle \bar{q}(x)q(x)\, ,\, \bar{q}(y)q(y)\rangle\, , \qquad \langle \bar{q}(x)\gamma^5 q(x)\, ,\, \bar{q}(y)\gamma^5 q(y)\rangle \tag{18}$$

in addition to two-quark mesons in the intermediate state, must also exhibit four-quark exotic mesons.

Why then does flavor transition inhibition work so well in 1^- channels (φ meson decays)? The answer may be purely kinematical. In the $J = 1$ case the diquark pair must be in the P wave which increases effective energies to the extent that diquarks no longer can be viewed as local; they are essentially destroyed and do not contribute in loops. The destruction of diquarks resurrects planarity.

6. Diquarks and weak nonleptonic decays of heavy baryons

As was mentioned in Sect. 1, the short-distance diquark component is presumably responsible for (a part of) the enhancement of nonleptonic decays of strange hyperons[a]. It is natural to ask whether a similar enhancement occurs for baryons containing heavy c and b quarks.

---

[a] In addition to diquark correlations, the enhancement is also due to the quark-antiquark correlations, i.e. the penguin mechanism which, simultaneously, explains the enhancement in K meson decays [5].

Let us start from b-containing baryons. The phenomenological issue here is that diquarks could play a role in reconciling experimental trends with theoretical expectations in the problem of lifetimes of b-containing hadrons, in particular, Λ_b. This issue has a long and dramatic history. Asymptotically (at $m_b \to \infty$) lifetimes of all b-containing hadrons must be equal. At finite but large m_b deviations of the ratio $\tau(\Lambda_b)/\tau(B_d)$ from unity can be calculated as an expansion in powers of $1/m_b$. A (formally) leading power correction m_b^{-2} is due to the dimension-5 chromomagnetic operator parametrized by μ_G^2. It contributes at the level ~ -0.02, see e.g. [19] and is rather unimportant. Moreover, this correction is flavor blind. Flavor-dependent and more important numerically are dimension-6 four-quark operators whose contribution is suppressed by m_b^{-3}. The original estimates [20] of four-quark operators in heavy baryon lifetimes were at the level ~ -0.02.

At the same time, experimental measurements conducted in the 1990's seemingly indicated that the ratio $\tau(\Lambda_b)/\tau(B_d)$ could be as small as ~ 0.8, i.e. preasymptotic corrections as large as ~ -0.2. This apparent discrepancy ignited a renewed theoretical effort and extensive debate in the literature. It is worth emphasizing that there are two distinct mechanisms responsible for m_b^{-3} corrections in $\tau(\Lambda_b)/\tau(B_d)$, namely, *Pauli interference* and *weak scattering*. Pauli interference works in the "unfavorable" direction, increasing the ratio $\tau(\Lambda_b)/\tau(B_d)$, while weak scattering tends to decrease it. Stretching estimates of the matrix elements of relevant four-quark operators in the "favorable" direction one typically gets ~ 0.03 and ~ -0.07 for Pauli interference and weak scattering, respectively, so that there is significant compensation [21].

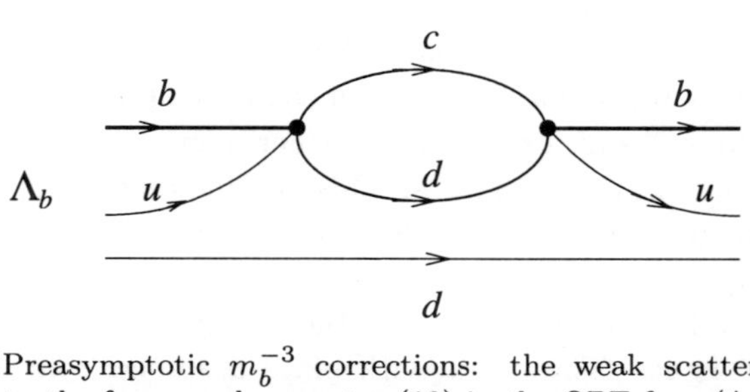

Figure 2. Preasymptotic m_b^{-3} corrections: the weak scattering mechanism giving rise to the four-quark operator (19) in the OPE for $\tau(\Lambda_b)$.

For our purposes it is important to note that the four-quark operators emerging in Pauli interference and weak scattering have spatial structures which do not coincide for these two mechanisms. In weak scattering the only relevant operator occurring in OPE (Fig. 2) has a structure "good diquark density times good diquark density," of the type mentioned for O_1 in Ref. [5], namely,[b]

$$O_- = 2\,(j_k)^\dagger\,(j_k)\,, \qquad j_k = \varepsilon_{kji}\, b^j C\, \frac{1-\gamma_5}{2}\, u^i\,, \tag{19}$$

where i, j, k are color indices. The assumption that a strong (positive) diquark correlation persists in the 0^+ system of one heavy and one light quark, would result in a further enhancement of the weak scattering contribution in Λ_b, effectively destroying the cancellation between Pauli interference and weak scattering. In this case theoretical prediction $\tau(\Lambda_b)/\tau(B_d) \sim 0.9$ would become natural.

Note that recent experimental measurements of the ratio $\tau(\Lambda_b)/\tau(B_d)$ tend to shift the central value of the ratio from ~0.8 up to ~0.9. Note also that the argumentation above is applicable to Ξ_b baryons as well.

One can give another argument in favor of a certain enhancement of the Λ_b matrix element of the operator (19), compared to the naive estimates of the 1980's. This argument goes along the lines discussed in Sect. 4. There, reducing the color group from SU(3) to SU(2), we were able to relate 0^+ light diquarks to pions. If we follow the same strategy here, we will relate the (bu) good diquark to B mesons. Extensive analyses in the 1990's firmly established the fact that f_B is quite large, $f_B \sim 200$ MeV, a factor of ~ 1.5 larger than was previously believed. In $SU(2)_{\rm color}$ the factor of 2 enhancement of f_B^2 immediately translates into a similar enhancement of $\langle \Lambda_b | O_- | \Lambda_b \rangle$.

Now let us briefly comment on the situation with c-containing baryons. This case is marginal: on the one hand the charmed quark can be viewed as heavy, and heavy quark-mass expansion could be applied. As seen from explicit calculations [20], this expansion is rather poorly convergent and can be used only for qualitative estimates. This fact indicates that the c quark is not heavy enough. The hidden scale quoted in Eq. (9) is actually of the same order as m_c.

In this marginal situation one can try the opposite description, treating the c quark as "light." The cd diquark plays a similar role in c-baryon nonleptonic decays as the ud diquark in hyperon decays. What is different is that the penguin mechanism plays no significant role in total widths of

[b] Strictly speaking j_k represents a combination of the 0^+ and 0^- diquarks.

c-containing hadrons. That is why one can qualitatively expect a relative enhancement of all c-baryon decays over meson ones. This expectation is indeed confirmed by experimental data.

7. Conclusions

We discussed various approaches allowing one to attempt to quantify the role of diquarks in hadronic physics. The most solid consideration, albeit somewhat remote from bona fide QCD, is that based on $SU(2)_{color}$. Reducing the gauge group from SU(3) to SU(2) allows one to relate diquarks and pions through a global symmetry which exists only for $SU(2)_{color}$. Diquarks become well-defined gauge-invariant objects, which share with pions a two-component structure with a relatively short-range core. Then one can speculate, qualitatively or, with luck, semiquantitatively on what remains of this symmetry upon lifting $SU(2)_{color}$ to $SU(3)_{color}$. It is worth noting that all instanton-based calculations carry a strong imprint of the above symmetry since basic instantons are, in essence, $SU(2)_{color}$ objects.

In the case of pions a hidden large scale is known to exist based on various arguments. Our task was to investigate consequences of the existence of this scale in diquarks. We argue that short-range diquark correlations play an important role in QCD vacuum structure: they help resolve the apparent contradiction with large-N expectations for vacuum structure. We also discuss consequences of diquarks in nonleptonic baryon decays. In all cases — hyperons, c- and b-containing baryons — a short-range diquark core seems to be instrumental in understanding existing phenomenology.

The issue of diquarks was raised and discussed mostly in application to hadron spectroscopy in Refs. [7,9,10]. We extended the discussion by including for consideration a hidden large scale implying a short-range core in diquarks. We tested the idea in a range of applications other than hadronic spectroscopy. The idea survived these tests. However, our considerations are essentially qualitative; more quantitative analyses are most welcome.

Acknowledgments

I would like to express to Julia Nyiri, Yuri Dokshitzer and Peter Levai my appreciation of their remarkable organization of the Workshop.

This work was supported in part by DOE grant DE-FG02-94ER408.

References

1. M. Shifman and A. Vainshtein, *Phys. Rev.* **D71**, 074010 (2005).
2. T. Schäfer, E.V. Shuryak and J.J.M. Verbaarschot, *Nucl. Phys.* **B412**, 143 (1994).
3. M. Cristoforetti, P. Faccioli, E.V. Shuryak and M. Traini, *Phys. Rev.* **D70**, 054016 (2004).
4. B. Stech, *Phys. Rev.* **D36**, 975 (1987);
   H.G. Dosch, M. Jamin and B. Stech, *Z. Phys.* **C42**, 167 (1989);
   B. Stech and Q.P. Xu, *Z. Phys.* **C49**, 491 (1991);
   M. Neubert and B. Stech, *Phys. Rev.* **D44**, 775 (1991).
5. M.A. Shifman, A.I. Vainshtein and V.I. Zakharov, *Sov. Phys. JETP* **45**, 670 (1977).
6. D. Diakonov, V. Petrov and M.V. Polyakov, *Z. Phys.* **A359**, 305 (1997).
7. M. Karliner and H.J. Lipkin, hep-ph/0307243; hep-ph/0307343; *Phys. Lett.* **B575**, 249 (2003); **B612**, 197 (2005).
8. R.L. Jaffe and F. Wilczek, *Phys. Rev. Lett.* **91**, 232003 (2003); *Phys. Rev.* **D69**, 114017 (2004); *Phys. World* **17**, 25 (2004).
9. E.V. Shuryak and I. Zahed, *Phys. Lett.* **B589**, 21 (2004).
10. R.L. Jaffe, *Exotica*, *Phys. Rep.* **409**, 1 (2005).
11. V.A. Novikov, M.A. Shifman, A.I. Vainshtein and V.I. Zakharov, *Nucl. Phys.* **B191**, 301 (1981).
12. M.A. Shifman, A.I. Vainshtein and V.I. Zakharov, *Nucl. Phys.* **B147**, 385 (1979); **B147**, 448 (1979).
13. E.V. Shuryak, *Nucl. Phys.* **B203**, 93, 116, 140 (1982); **B302**, 559, 574, 599, 621 (1988);
    D. Diakonov and V.Y. Petrov, *Nucl. Phys.* **B272**, 457 (1986); **B245**, 259 (1984);
14. S. Dimopoulos, *Nucl. Phys.* **B168**, 69 (1980);
    M.E. Peskin, *Nucl. Phys.* **B175**, 197 (1980).
15. I. Kogan, M. Shifman and M. Vysotsky, *Sov. J. Nucl. Phys.* **42**, 318 (1985) [*Yad. Fiz.* **42**, 504 (1985)].
16. G. 't Hooft, *Nucl. Phys.* **B72**, 461 (1974).
17. A. Armoni, M. Shifman and G. Veneziano, *Phys. Rev. Lett.* **91**, 191601 (2003); *Phys. Lett.* **B579**, 384 (2004).
18. G. Veneziano, *Nucl. Phys.* **B117**, 519 (1976).
19. I. Bigi, B. Blok, M. Shifman, N. Uraltsev and A. Vainshtein, in: *B decays, revised*, Ed. S. Stone, 2nd edition, World Scientific, Singapore, 1994, p. 132;
    G. Bellini, I.I.Y. Bigi and P.J. Dornan, *Phys. Rep.* **289**, 1 (1997);
    I. Bigi, M. Shifman and N. Uraltsev, *Ann. Rev. Nucl. Part. Sci.* **47**, 591 (1997).
20. M.A. Shifman and M.B. Voloshin, *Sov. J. Nucl. Phys.* **41**, 120 (1985) [*Yad. Fiz.* **41**, 187 (1985)]; *Sov. Phys. JETP* **64**, 698 (1986) [*Zh. Eksp. Teor. Fiz.* **91**, 1180 (1986)].
21. N.G. Uraltsev, *Phys. Lett.* **B376**, 303 (1996);
    D. Pirjol and N. Uraltsev, *Phys. Rev.* **D59**, 034012 (1999).

AGK RULES IN PERTURBATIVE QCD

JOCHEN BARTELS

II.Institut für Theoretische Physik, Universität Hamburg
Luruper Chaussee 149, 22761 Hamburg, Germany

I summarize the present status of the AGK cutting rules in perturbative QCD. Particular attention is given to the application of the AGK analysis to diffraction and multiple scattering in DIS at HERA and to pp collisions at the LHC. I also discuss the bootstrap conditions which appear in pQCD.

1. Introduction

Multiple parton interactions play an important role both in electron proton scattering at HERA and in high energy proton proton collisions at the LHC. At HERA, the linear QCD evolution equations provides, for not too small Q^2, a good description of the F_2 data (and of the total $\gamma^* p$ cross section, $\sigma_{tot}^{\gamma^* p}$). This description corresponds to the emission of partons from a single chain (Fig.1a). However, at low Q^2 where the transition to nonperturbative strong interaction physics starts, this simple picture has to supplemented with corrections. First, there exists a class of models [1] which successfully describe this transition region; these models are based upon the idea of parton saturation: they assume the existence of multiple parton chains (Fig.1b) which interact with each other, and they naturally explain the observed scaling behavior, $F_2(Q^2, x) \approx F_2(Q^2/Q_s^2(x))$ with $Q_s^2(x) = Q_0^2(1/x)^\lambda$. Next, in the photoproduction region, $Q^2 \approx 0$, direct evidence for the presence of multiple interactions also comes from the analysis of final states [2]. A further strong hint at the presence of multi-chain configurations comes from the observation of a large fraction of diffractive final states in deep inelastic scattering at HERA. In the final states analysis of the linear QCD evolution equations, it is expected that the produced partons are not likely to come with large rapidity intervals between them. In the momentum-ordered single chain picture (Fig.1a), therefore, diffractive final states should be part of the initial conditions (inside the lower blob in Fig.1a), i.e. they should lie below the scale Q_0^2 which seperates the parton

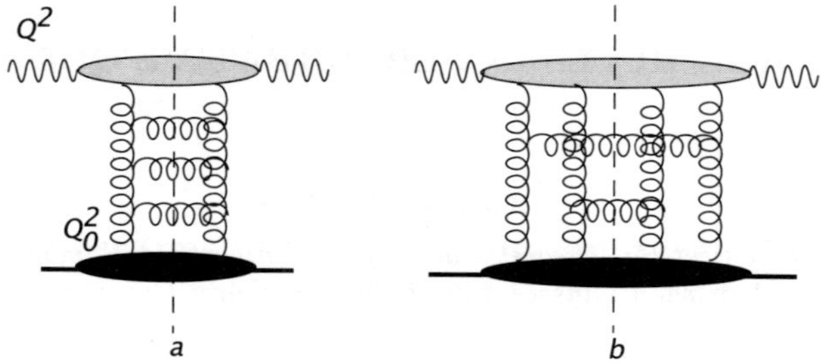

Figure 1. Contributions to the total cross section $\sigma_{tot}^{\gamma^* p}$: (a) the single chain representing the linear QCD evolution equations; (b) gluon production from two different gluon chains.

description from the nonperturbative strong interactions. This assignment of diffractive final states, however, cannot be complete. First, data have shown that the Pomeron which generates the rapidity gap in DIS diffraction is harder than in hadron - hadron scattering; furthermore, there are specific diffractive final states with momentum scales larger than Q_0^2, e.g. vector mesons built from heavy quarks and diffractive dijets: the presence of such final states naturally requires corrections to the single chain picture. These 'diffractive' corrections, viewed from a t-channel point of view, belong to the same class of corrections as shown Fig.1b: they are characterized by four gluon states in the t-channel.

In proton-proton collisions corrections due to multiple interactions should be important in those kinematic regions where parton densities for small momentum fractions values and for not too large momentum scales are being probed, e.g. jet production near the forward direction. Another place could be the production of multijet final states: multiple jets may come from different parton chains, and these contributions may very well affect the background to new physics beyond the standard model. Moreover, the modelling of multijet configurations will be necessary for understanding the underlying event structure in pp collidions [3].

From the point of view of collinear factorization, multiple interactions with momentum ordered parton chains are higher-twist effects, i.e they are suppressed by powers of the hard momentum scale. At small x, however, this suppression is compensated by powers of the large logarithms, $\ln 1/x$: multiple interactions, therefore, are mainly part of small-x physics. In this kinematic region the Abramovsky-Gribov-Kanchelli (AGK) [4] rules can be applied to the analysis of multi-gluon chains, and it is the aim of this article to present a brief overview about the current status of AGK rules in pQCD.

As we will discuss below, in the analysis of multiple parton chains the couplings of n gluons to the proton play an essential role. Regge factorization suggests that these coupling should be universal, i.e. the couplings in $\gamma^* p$ collisions at HERA are the same as those in pp scattering at the LHC. Therefore, a thorough analysis of the role of multiple interactions in deep inelastic electron-proton scattering at HERA should be useful for a solid understanding of the structure of events at the LHC.

2. Basics of the AGK cutting rules

The original AGK paper [4], which was written before the advent QCD, addresses the question of how, in the total hadron-hadron cross section, the presence of multi-Pomeron exchanges leads to observable effects in the final states: Based upon a few model-independent assumptions on the couplings

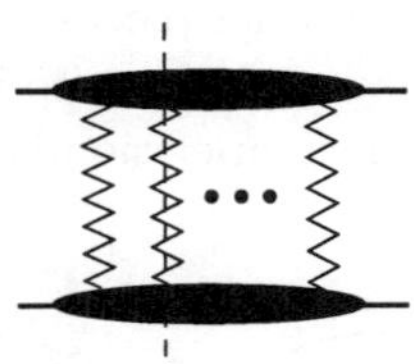

Figure 2. s-cut through a multi-Pomeron exchange: the zig-zag lines stand for nonperturbative Pomerons.

of multi-Pomeron exchanges to the proton, the authors derived simple 'cutting rules': different contributions to the imaginary part belong to different cuts across the multi-Pomeron diagrams, and each cut has its own, quite distinct, final state characteristics. As a result, the authors found counting rules for final states with different particle multiplicities, and they proved cancellations among rescattering corrections to single-particle and double-particle inclusive cross sections.

In the QCD description of hard (or semihard) final states a close analogy appears between (color singlet) gluon ladders and the nonperturbative Pomeron: multiple parton chains (for example, the two chains in Fig.1b) can be viewed as cuts through two perturbative BFKL Pomerons. In the same way as in the original AGK paper, the question arises how different cuts through a QCD multi-ladder diagram can be related to each other. In the following we briefly describe how AGK cutting rules can be derived in pQCD [5,6]. In the subsequent section we will present a few new results which come out from pQCD, going beyond the original AGK rules.

One of the few assumptions made in the original AGK paper states that the coupling of the Pomerons to the external particle are (i) symmetric under the exchange of the Pomerons (Bose symmetry), and (ii) that they remain unchanged if some of the Pomerons are beeing cut. These properties also hold in pQCD, but they have to be reformulated: (i') the coupling of (reggeized) gluons to external particles is symmetric under the exchange of reggeized gluons, and (ii') it remains unchanged if we introduce cutting lines between the gluons. In QCD, however, the color degree of freedom also allows for another possibility: inside the n-gluon state (with total color zero), a subsystem of two gluons can form an antisymmetric color octet state: in this case the two gluons form a bound state of a reggeized gluon (bootstrap property). For the case of $\gamma^*\gamma^*$ scattering, explicit calculations [7] have shown that the coupling of n gluons to virtual photons can be written as a sum of several pieces: the fully symmetric ('irreducible') one which satisfies (i') and (ii'), and other pieces which, by using the bootstrap property, can be reduced to symmetric couplings of a smaller number of gluons ('cut reggeons'). This decomposition is illustrated in Fig.3.

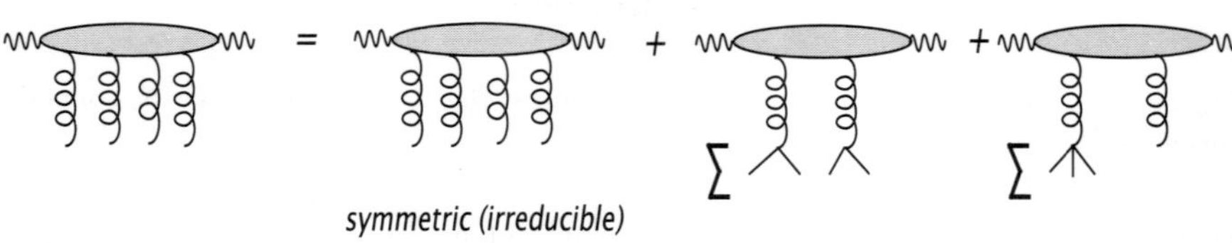

Figure 3. Decomposition of the coupling of four gluons to a virtual photon. In the last two terms on the rhs it is understood that we have to sum over different pairings of gluons below.

Since the bootstrap property is related to the regeization of the gluon and, therefore, is expected to be valid to all orders perturbation theory, also these properties of the couplings of multi-gluon states to external particles should be of general validity. In this short review we will mainly concentrate on the symmetric couplings.

As an illustrative example, we consider the coupling of four gluons to a proton. The simplest model of a symmetric coupling is a sum of three pieces, each of which contains only the simplest color structure.

In order to be able to generalize and to sum over an arbitrary number of gluon chains, it is necessary to be more specific. Inserting the eikonal

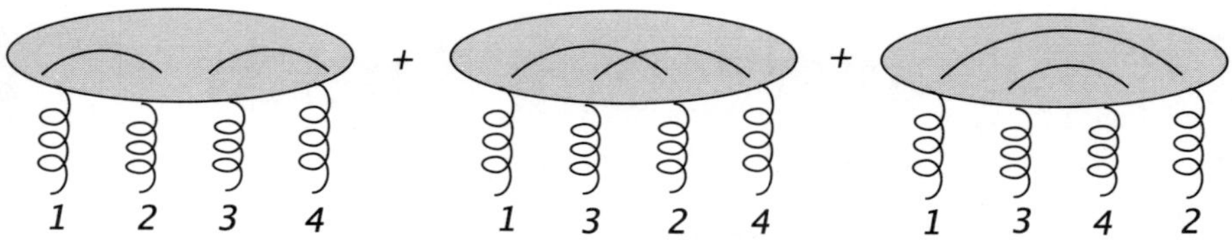

Figure 4. The symmetric coupling of four gluons to an external particle. The lines inside the blob denote the color connection, e.g. the first term has the color structure $\delta_{a_1a_2}\delta_{a_3a_4}$.

ansatz:

$$N_{2n}^{A}(\mathbf{k_1},\mathbf{a_1};\ldots;\mathbf{k_{2n}},\mathbf{a_{2n}};\omega) = \frac{\mathbf{1}}{\sqrt{(\mathbf{N_c^2}-\mathbf{1})^{\mathbf{n}}}} \cdot \tag{1}$$

$$\cdot\left(\sum_{Pairings} \phi^A(\mathbf{k_1},\mathbf{k_2};\omega_{\mathbf{12}})\delta_{\mathbf{a_1a_2}}\cdot\ldots\cdot\phi^{\mathbf{A}}(\mathbf{k_{2n-1}},\mathbf{k_{2n}};\omega_{\mathbf{2n-1,2n}})\delta_{\mathbf{a_{2n-1}a_{2n}}}\right)$$

into the hadron - hadron scattering amplitude, using the large-N_c approximation, and switching to the impact parameter representation, one obtains, for the contribution of k cut gluon ladders, the well-known formula:

$$ImA_k = 4s\int d^2\mathbf{b}\mathbf{e}^{\mathbf{iqb}}\mathbf{P}(\mathbf{s},\mathbf{b}) \tag{2}$$

where

$$P(s,\mathbf{b}) \;=\; \frac{[\boldsymbol{\Omega}(\mathbf{s},\mathbf{b})]^{\mathbf{k}}}{\mathbf{k}!}\mathbf{e}^{-\boldsymbol{\Omega}(\mathbf{s},\mathbf{b})}, \tag{3}$$

and Ω stands for the (cut) two-gluon ladder.

Another result [6] which follows from the symmetry properties of the n gluon-particle coupling is the cancellation of rescattering effects in single and double inclusive cross sections. In analogy with the AGK results on the rescattering of soft Pomerons, it can be shown that the sum over multi-chain contributions and rescattering corrections cancels (Fig.5), leaving only the single-chain contribution (in agreement with the factorization obtained in the collinear analysis). This statement, however, holds only for rescattering between the two projectile: it is not valid for the multiple exchanges between the tagged jet and the projectile which require a seperate discussion [8].

All these results can be generalized to include also the soft Pomeron: all one needs to assume is that the couplings of soft Pomerons and reggeized gluons are symmetric under interchanges, and they are not altered if cutting lines are introduced.

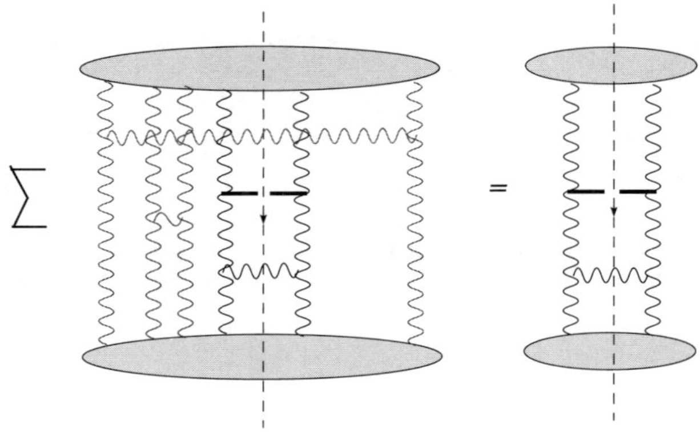

Figure 5. AGK cancellations in the one-jet inclusive cross section.

3. New results

Explicit calculations in QCD lead to futher results on multiple interactions. First, in the four gluon exchange there are other configurations than two-ladder exchanges; for example, the pairing of gluon chains switches from (14)(23) in the upper part (= upper rapidity interval) to (12)(34) in the lower part (= lower rapidity interval). One can show that the ratio $1 : 2 : -4$ holds for each rapidity interval. In [6] this has been generalized to an arbitrary number of exchanged gluon lines.

Next a few comments on reggeization and cut reggeons. Clearly there are more complicated configurations than those which we have discussed so far; an example appears in $\gamma^* p$ scattering (deep inelastic electron proton scattering). In contrast to pp scattering, the coupling of multi-gluon chains to the virtual photon can be computed in pQCD, and the LO results, for the case of $n = 4$ gluons, are illustrated in Fig.6. It turns out that we have two alternative possibilities: in the completely inclusive case (total cross section), it is convenient to chose Fig.6a, i.e. the sum of all contributions can be decomposed into two sets of diagrams. In the first set, at the top of the diagram two gluons couple to the quark-antiquark pair, and the subsequent transition to the four-gluon state goes via the pQCD triple Pomeron vertex. This vertex, as a function of the 4 gluons below, has the symmetry properties described above. As a result, we can apply the cutting rules to the four gluon state, as discussed before. However, there is also the second term in Fig.6a, which consists of a two gluon state only: this is the reggeizing contribution we have mentioned before. As indicated in

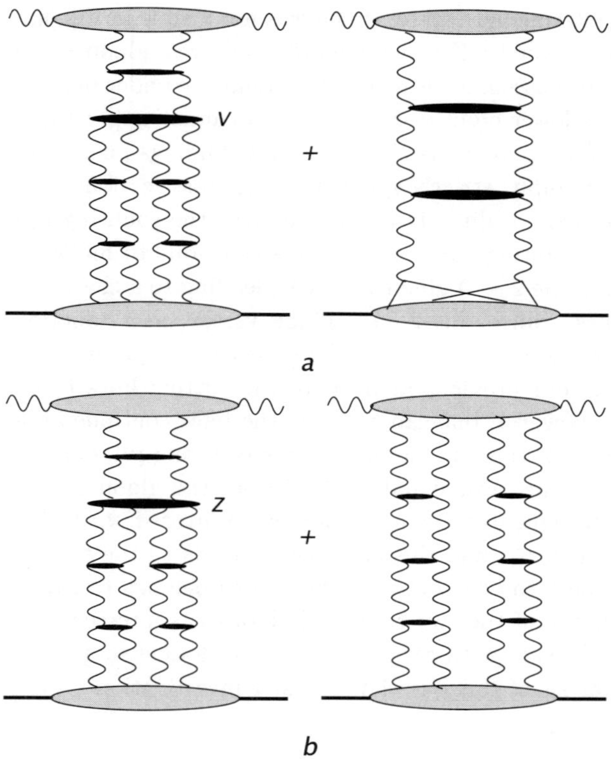

Figure 6. Four-gluon contributions to $\gamma^* p$ proton scattering: two equivalent ways of summing over all contributions. (a) the decomposition of Fig.6 with the pQCD triple Pomeron vertex. (b) an alternative way of summation which explicitly shows the coupling of two Pomerons to the photon vertex and which leads to a new vertex Z.

the figure, the splitting of the reggized gluons at the bottom amounts to a change in the (nonperturbative) coupling. We want to stress that, because of the inclusive nature of this set of diagrams, the triple Pomeron vertex V in Fig.6a, similar to the BFKL kernel, contains both real and virtual contributions. For this reason, the decomposition in Fig.6a is applicable to inclusive cross sections, and it is not convenient for investigating specific final states such as, for example, diffractive final states with a fixed number of quarks and gluons in the final state.

There exists an alternative way of summing all contributions (Fig.6b) which is completely equivalent to Fig.6a but allows to keep track of diffractive $q\bar{q}$, $q\bar{q}g$,... final states: this form is illustrated in Fig.6b. One recognizes the 'elastic intermediate state' which was not visible in Fig.6a, and the new triple Pomeron vertex Z which contains only real gluon production. This vertex Z, as discussed in [9] is no longer symmetric under permutations of the gluons at the lower end; consequently, we cannot apply the AGK cutting rules to the four gluon states below.. These findings for multiple scattering effects in DIS imply, strictly speaking, that cross sections for diffractive $q\bar{q}$ or $q\bar{q}g$ states cannot directly be inserted into the counting rules.

Also pp scattering will contain corrections due to multiple interactions which contain the $2 \to 4$ gluon vertex V, leading to a change of the number of gluon lines. Since this $2 \to 4$ gluon vertex, as a function of the four gluons below the vertex, satisfies the symmetry requirements listed above, we can apply our previous analysis to the cutting lines below the vertex. In addition, however, one can ask how the lines continue above the $2 \to 4$ gluon vertex. For this two gloun state the counting rules are quite different: in contrast to the even-signature Pomeron, the gluon is a odd-signature reggeon. Consequently, the cut gluon is suppressed w.r.t. he uncut gluon by one power in α_s, and this suppression leads to the following hierarchy of cutting lines: the cut between the gluons belongs to leading order, the cut through one of the two reggeized gluons is supressed by one power in α_s, the cut through both reggeized gluons is double suppressed (order α_s^2). A closer analysis of this question is under investigation [8].

4. Bootstrap conditions

Apart from their usefulness in modelling multiparton contributions to total and to semi-inclusive cross sections, the AGK rules provide deep insight into s-channel unitarity properties of scattering amplitudes. In this context, the most fundamental feature of pQCD calculations are the bootstrap properties; they allow, in a multi-Pomeron diagram, to examine cuts through a reggeized gluon. For the LO BFKL ladder, it is well-known that in the antisymmetric color octet t-channel the ladder diagrams 'collapse' into a single reggeized gluon (Fig.7). It has been shwon [10] that this bootstrap condition is fulfilled also in the next-to-leading order; this proves that the reggeization of the gluon, together with the bootstrap equations, are fundamental features of QCD which can be expected to hold in all orders perturbation theory.

The bootstrap property of the BFKL Pomeron can be generalized to multi-Pomeron exchanges. As we have mentioned before, couplings of n reggeized gluons to a virtual photon, D_n, have been calculated up to $n = 6$

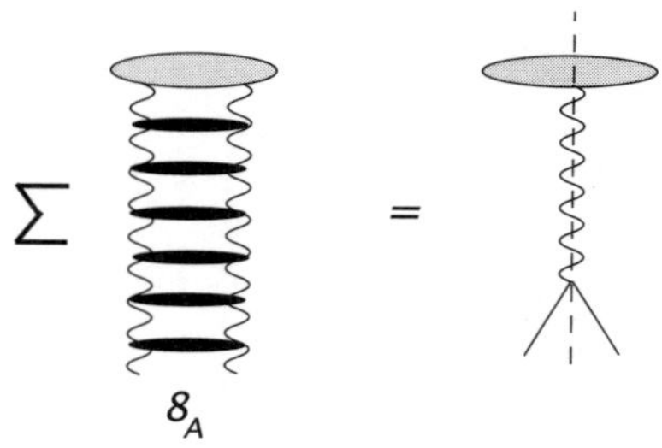

Figure 7. Bootstrap property of the BFKL ladder diagrams.

(in leading order) [11]. From the results of these calculations a simple pattern emerges. Namely, these functions D_n which are derived from multiple energy discontinuities of suitable multiparticle scattering amplitudes, initially, are not symmetrized and contain different color representations (an example is illustrated in Fig.3). In a sequence of steps of 'reduction', the functions D_n are decomposed into terms where subsystems of reggeized gluons are in the antisymmetric 8_A representation and 'collapse' into a single reggeized gluon. The coupling function then coincides with a function $D_{n'}$ with $n' < n$. After the complete removal of such contributions, the remaining 'irreducible' function D_n^{ir} is completely symmetric under the interchange of any two reggeons (where the interchange referes to both color and momenta). We illustrate this decomposition in Fig.8. Note that, on the rhs of Fig.8, the sums extend over all possible decays of the reggeized gluons into 2, 3,... elementary gluons and over different permutations of decay products. For example, on the rhs of the fourth line, either two reggeized gluons decay into pairs of elementary gluons or, alternatively, one reggeized gluon decays into three elementary gluons. Each decay of a reggeized gluon is accompanied by a color tensor which is universal, i.e. it only depends upon the number of decay products and not the 'environment' of the reggeized gluon [12]. On the rhs of Fig.8, the last term in the third line (D_4^{ir}) contains the triple Pomeron vertex, the last term in the sixth line (D_6^{ir}) contains the Pomeron $\to$ 2 Odderon vertex.

With this set of generalized bootstrap conditions we are in the position to analyse a diagram with two cut reggeized gluons (Fig.9). Invoking the reggeon unitarity equations for the t-channel four gluon state, we use, in Fig.8, the first terms on the third line. We identify this Fig.9 as the contribution to the absorptive part of the BFKL ladder with two cut reggeized gluons; a simple power counting confirms that, for the exchange of two

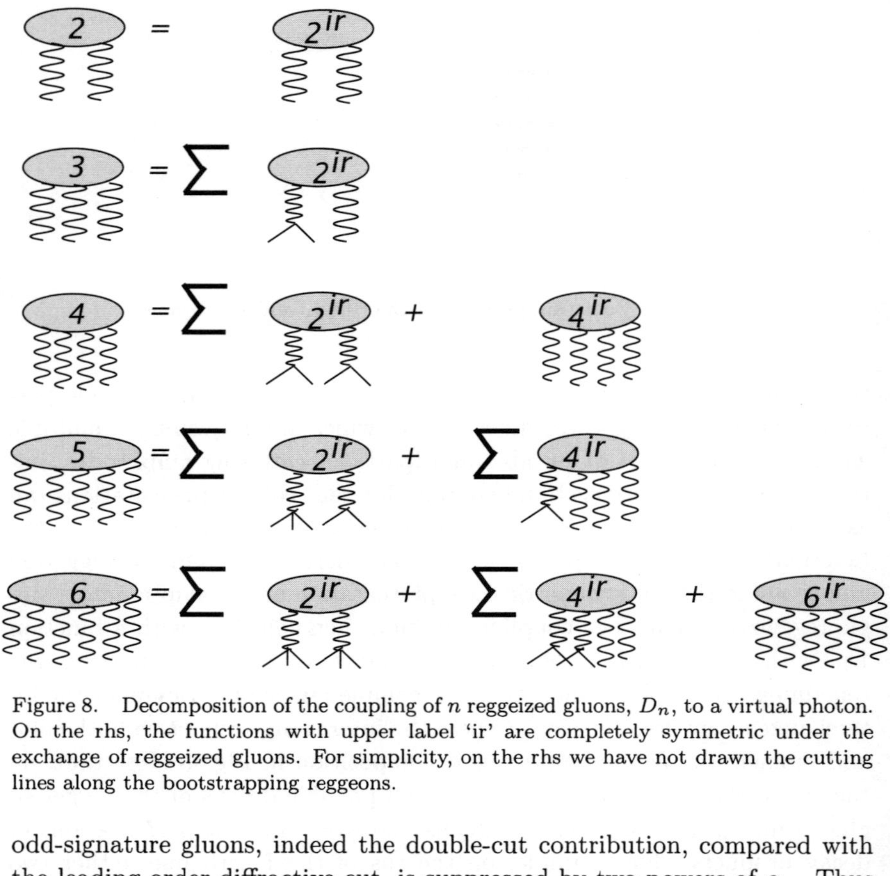

Figure 8. Decomposition of the coupling of n reggeized gluons, D_n, to a virtual photon. On the rhs, the functions with upper label 'ir' are completely symmetric under the exchange of reggeized gluons. For simplicity, on the rhs we have not drawn the cutting lines along the bootstrapping reggeons.

odd-signature gluons, indeed the double-cut contribution, compared with the leading order diffractive cut, is suppressed by two powers of α_s. Thus it is of the same order as the two Pomeron exchange.

5. Conclusions

In this paper we have given an overview over the present status of AGK cutting rules. In the first part we have addressed phenomenological aspects. Corrections due to multiple interactions seem to be important in DIS at small x and low Q^2; they are expected to play a significant role also in multijet production in pp scattering. The study of the AGK rules in pQCD then provides help in understanding the systematics of multiple gluon chains. Results described in this review represent the beginning of a systematic analysis. We have listed a few questions which require further work.

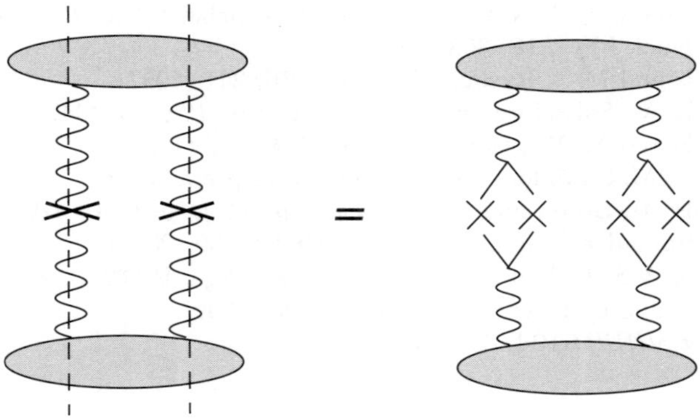

Figure 9. The double cut-reggeon contribution of the BFKL amplitude. The crosses ('mass-shell conditions') indicate that we are considering discontinuities: on the lhs, a discontinuity across the two-reggeon cut; on the rhs the four particle cut.

As an immediate application, we believe that a quantitative analysis of multiple scattering at HERA will provide a useful input to the modelling of final states at the LHC.

An question of practical importance which we have not adressed at all is the hadronization of partonic final states. All statements on ratios of 'particle densities in the final states' refer to the parton (gluon) level. The hadronization of events which, for example, belong to a double-cut ladder configuration may be quite different from the one obtained by applying just the normal single-chain hadronization to each chain seperately. The answer to this question [a] goes beyond the AGK analysis discussed in this paper.

In the second part we have discussed aspects of the bootstrap equations which appear in the couplings of multi-reggeon states to external particles. They provide insight into the s-channel unitarity content of reggeon diagrams, and there is evidence that these conditions are valid also beyond the leading order.

References

1. K. Golec-Biernat and M. Wusthoff, *Phys. Rev.* **D59**, 014017 (1999); *Phys. Rev.* **D60**, 114023 (1999);
   J. Bartels, K. Golec-Biernat and H. Kowalski, *Phys. Rev.* **D66** 014001 (2002).
2. C. Gwenlan, *Acta Phys. Polon.* **B35**, 377 (2004).
3. T. Sjostrand and P.Z. Skands, *Eur. Phys. J.* **C39**, 129 (2005) and references therein.

[a] I thank G.Gustafson for a very useful discussion on this point.

4. V.A. Abramovsky, V.N. Gribov and O.V. Kanchelli, *Yad. Fiz.* **18**, 595 (1973) [Sov. J. Nucl. Phys. **18**, 308 (1974)].
5. J. Bartels and M.G. Ryskin, *Z. Phys.* **C76**, 241 (1997).
6. J. Bartels, M. Salvadore and G.P. Vacca, *Eur. Phys. J.* **C42**, 53 (2005).
7. J. Bartels and M. Wüsthoff, *Z. Phys.* **C66**, 157 (1995).
8. J. Bartels, M. Salvadore and G.P. Vacca, in preparation.
9. J. Bartels, M. Braun and G.P. Vacca, *Eur. Phys. J.* **C40**, 419 (2005).
10. V.S. Fadin and A. Papa, *Nucl. Phys.* **B640**, 309 (2002);
J. Bartels, V.S. Fadion and R. Fiore, *Nucl. Phys.* **B672**, 329 (2003).
11. J. Bartels and C. Ewerz, *JHEP* **9909**, 026 (1999).
12. C. Ewerz, *JHEP* **0104**, 031 (2001).

COMPLEX ANGULAR MOMENTA AND THE PROBLEM OF EXOTIC STATES

YA.I. AZIMOV

Petersburg Nuclear Physics Institute, Gatchina, 188300, Russia

After some brief personal recollections about V.N. Gribov, I demonstrate that his results and ideas on complex angular momenta may be applied in unfamiliar directions. As an example, it is shown, that any strong interaction amplitude, satisfying dispersion relation (in momentum transfer), has infinite number of energy-plane poles, both for exotic and non-exotic quantum numbers. This result ensures the necessary condition for existence of exotic hadrons. However, without more detailed knowledge of dynamics one cannot secure sufficient conditions for the exotics existence.

1. Introduction

I was lucky to work for 20 years in the close contact with V.N. Gribov, in the Ioffe Physico-Technical Institute and in the Leningrad (now Petersburg) Nuclear Physics Institute. It seems that I was the youngest of his colleagues who called him by his Russian nickname, Volodya. When he changed his home institution, to the Landau Theoretical Physics Institute, our contacts still continued, though became irregular. Later, after his death in 1997, I looked through his list of publications and, quite unexpectedly, was disappointed. For me, both the list, and publications themselves, could hardly reflect his bright personality.

V.N. Gribov was a rare person, with whom one could discuss practically any problem of physics (and not only physics). Discussions with him were always very intensive and passionate. Sometimes he was not right after all, but even then he suggested such non-trivial arguments for his position, which were difficult to reject. Similar discussions always emerged also at our theoretical seminars, even if the presented work had been earlier discussed with Gribov personally. I think that was why many physicists disliked (feared?) to talk at our seminars. However, being very difficult, discussions with Gribov were always very interesting and very useful. They gave to the author a better understanding of own work. Moreover, they provided a beautiful training how to present results of the work.

References to Gribov's works are rather frequent today in the literature. But they mostly refer to QCD problems only - Gribov copies, Gribov horizon, and, of course, DGLAP equations for the parton evolution (here G means just Gribov). Very intensive Gribov's activity on Regge theory seems to be nearly forgotten now. Meanwhile, his numerous papers in this direction may still generate ideas and arguments on various problems, even if the problems look totally unrelated to the Reggeology. In the present note I demonstrate this for the problem of exotic states, *i.e.*, hadrons with such quantum numbers that cannot be composed of 3 quarks (for baryons) or one quark-antiquark pair (for mesons). The corresponding result has been published earlier [1]. Its derivation and meaning will be discussed here in more detail.

2. Dispersion relations and energy-plane poles

As is well-known, amplitude of any 2-particle-into-2-particle process depends on two kinematical variables. They may be taken to be $s = W^2$ and $z = \cos\theta$, with W and θ being the c.m. energy and scattering angle, respectively. It is convenient in many cases, instead of z, to use the invariant quantity, squared c.m. momentum transfer. There are two such variables: t, linear in z, and u, linear in $-z$ (of course, the sum $t + u$ is independent of z).

As function of z, the invariant amplitude $A(s, z)$ can be decomposed into partial-wave amplitudes $f_j(s)$, that correspond to the process at a definite value of the total angular momentum j, the same in initial and final states. Physical values of j are (half)integer. Of course, one could define $f_j(s)$ as an analytical function for arbitrary values of j, real or even complex, such that coincides with physical amplitudes at physical values of j. Such procedure is always possible, but is generally ambiguous (*e.g.*, when continuing from positive integer points one could add any term proportional to $1/\Gamma(1-j)$). However, under some conditions, the partial-wave amplitudes $f_j(s)$ admit an unambiguous analytical continuation to non-physical values of j.

In what follows, we use three assumptions: 1) possibility of the unambiguous analytical continuation in the angular momentum j; 2) absence of massless physical hadrons; 3) unitarity condition. Let us consider their meaning.

For relativistic amplitudes, the possibility of unambiguous analytical continuation in j is usually connected with dispersion relations in z (see, *e.g.*, the monograph [2]). Generally, dispersion relations have not been deduced mathematically from axioms of the Quantum Field Theory. They have not been proved also in terms of the more specific Quantum Chromodynamics (QCD), which is believed now to provide the physical basis for strong interactions. However, not all details of QCD have been understood up to now. Even relation between quarks and gluons, on one side, and mesons and/or baryons, on the other, cannot be described today in a model-independent way. That is why various dispersion relations (or their analogs/generalizations) are widely applied to phenomenological treatment of strong interactions. The dispersion relations are used as input in modern Partial-Wave Analyses (PWA; see, *e.g.*, the latest pion-nucleon PWA [3]). They are operative to pick out such strong interaction parameters as meson-baryon coupling constants, and/or the pion-nucleon σ-term. Analytical properties, corresponding to dispersion relations, provide basis for various sum rules. Analogous analytical properties of inelastic amplitudes are assumed when extracting the meson-meson scattering amplitude from data on processes like $\pi \to 2\pi$. Up to now, phenomenological applications of dispersion relations have not revealed any inconsistencies, neither theoretical nor experimental, and we still may consider them to be true.

Dispersion relations lead to the integral Gribov-Froissart formula [4,2] for the continued partial-wave amplitudes, at least at large positive $\mathrm{Re}\, j$. Note, however, that the dispersion relations provide a sufficient, but not necessary condition for the analytical continuation of $f_j(s)$ in j. The continuation survives under much weaker conditions on singularities of the invariant amplitudes in t and/or u, and the continued partial-wave amplitudes still conserve essentially the same Gribov-Froissart form.

Now, consider the 2nd condition. Let us begin, for simplicity, with the amplitude for elastic scattering of two spinless particles, where the total angular momentum j coincides with the orbital momentum l. If there are no massless hadrons, the physical partial-wave amplitudes, being described by the Gribov-Froissart formula, reveal the familiar threshold behavior

$$f_l(s) \sim k^{2l} \quad \text{at} \quad k \to 0\,, \tag{1}$$

where k is the c.m. momentum. The same threshold behavior is true also for the continued amplitudes $f_l(s)$ with any, real or complex, angular momenta, at least till the Gribov-Froissart formula is applicable. This result has very simple physical meaning. Non-massless exchange, in the non-relativistic limit, induces the Yukawa potential of the form $\exp(-\mu r)/r$, with non-vanishing μ and the limited radius of action $R \sim 1/\mu$. As is well-known, every quantum mechanical potential with final radius R generates amplitudes with the threshold behavior $\sim (kR)^{2l}$. In contrast, a massless exchange, in the same limit, would correspond to the Coulomb potential, with infinite radius and more complicated threshold behavior.

Considering unitarity condition, we can again begin with the simple case of purely elastic interaction near the threshold. Then the unitarity relation, continued to any real l, conserves the same form as for physical values of l:

$$f_l(s) - f_l^*(s) = 2ik\, f_l(s)\, f_l^*(s)\,. \tag{2}$$

Evidently, the threshold behavior (1) at real l may be consistent with the unitarity relation (2) only if $l \geq -1/2$. This means that near the threshold (*i.e.*, at $s \to s_{\rm th}$, $k^2 \to 0$) the Gribov-Froissart formula cannot stay unchanged (with its integral convergent) at $\mathrm{Re}\, l \leq -1/2$.

The analysis, made by Gribov and Pomeranchuk [5], shows that at $k^2 \to +0$ the analytically continued partial-wave amplitude $f_l(s)$ should reveal unlimited number of poles in l at $\mathrm{Re}\, l > -1/2$. They are Regge poles (or reggeons), with energy-dependent positions. Further, from more detailed consideration of the small-k^2 region, Gribov and Pomeranchuk demonstrated [5] that the reggeons in the vicinity of $l = -1/2$ have trajectories

$$l_n(s) \approx -\frac{1}{2} + \frac{i\pi n}{\ln(R\sqrt{|k^2|})} + O(\ln^{-2}(R\sqrt{|k^2|}))\,, \tag{3}$$

with R being the effective interaction radius. The number n takes any positive and negative integer values, $n = \pm 1, \pm 2, \ldots$. We see that there are infinitely many reggeons accumulating to $l = -1/2$ at $k^2 \to 0$.

The structure of this accumulation, at real or even complex energies, may be investigated more explicitly in the non-relativistic quantum mechanics with final range potential, *e.g.*, with the Yukawa potential [6,7]. Below the threshold (at $k^2 < 0$), the set of accumulating poles consists of infinitely many pairs of reggeons in the left half-plane $\mathrm{Re}\, l \leq -1/2$. The reggeons of each pair have the same $|n|$, and their trajectories are complex conjugate to each other. At $k^2 \to -0$, they tend to the limiting point -1/2. Such simple correlation becomes destroyed above the threshold (at $k^2 > 0$), and some reggeons appear in the right half-plane $\mathrm{Re}\, l \geq -1/2$.

Up to now we have neglected particle spins. Accounting for them changes the situation, but only quantitatively. For the elastic threshold of two particles with spins σ_1 and σ_2, there appear several accumulation points in the j-plane, the rightmost one is at

$$j = -1/2 + \sigma_1 + \sigma_2\,, \tag{4}$$

instead of $j = -1/2$ [8]. The reason is simple: the accumulation points still correspond to $l = -1/2$, but particle spins provide several possible j-values for any fixed l-value, and *vice versa.* Correspondingly, structure of the accumulations in the j-plane, at thresholds of two spinning particles, are still described by trajectories (3), with the shifted limiting points (4). The case of multi-particle thresholds has never been really studied (though hypothesized by Gribov and Pomeranchuk [5]).

The above consideration was applied to a purely elastic two-particle case, and the discussed reggeons were seen as poles for the corresponding elastic partial-wave amplitudes. But at other (higher) energies the unitarity relation connects the elastic scattering with different, inelastic processes. After all, we arrive at the conclusion, that partial-wave amplitudes of all processes, which can be related by rescattering, should couple to the same reggeons. In particular, they should contain those accumulating Regge poles.

Essential for our present purpose is the infinite number of the accumulating Regge poles [5], which implies that the total number of Regge poles is certainly infinite as well. Their positions depend on energy and are determined by some relation of the form

$$F(s,\, j) = 0\,. \tag{5}$$

Each of its solution corresponds to a pole of a partial-wave amplitude and may be considered in two ways: either as a pole in j with the position depending on energy (on s), or as a pole in energy (in s) with the position depending on the total angular momentum j. This provides one-to-one correspondence between reggeons and energy-plane poles. Therefore, the infinite number of reggeons corresponds to the infinite number of poles in the energy plane. The same conclusion is true for the non-relativistic case of a finite-range potential, in particular, for the Yukawa potential. This allows to investigate more clearly the pole structure of amplitudes.

At first sight, the infinite number of energy-plane poles for any finite-range potential looks impossible, since the energy-plane poles are related to bound states or resonances. As well known, the bound or resonance states may exist, say, in the Yukawa potential, only if it is attractive and sufficiently strong. However, one can ask here an interesting question: whether every complex-energy pole should be considered as possibly related with a resonance? The problem is that, in terms of complex angular momenta, the Regge poles participating in the Gribov-Pomeranchuk accumulations are clearly separated from physical points, which are, for the j-plane, only integer (or half-integer) non-negative points. Thus, the accumulating poles cannot provide physically meaningful (bound or resonance) states, at least near the corresponding threshold. This could be true for non-threshold energies as well.

Moreover, it might be that the total set of Regge poles is split into two (or more?) different subsets: one related, say, with the Gribov-Pomeranchuk accumulations, the other with the bound and/or resonance states. Then there could be infinite number of poles of the former type, while only few (if any) poles of the latter type. This would be so, *e.g.*, if Eq.(5) could be factorized as

$$F_1(s,j)\,F_2(s,j) = 0\,.$$

However, explicit expressions of Ref.[6] show non-factorizable structure of Eq.(5). Moreover, there is no basic difference between various Regge poles [7]: all Regge trajectories appear to be different branches of the same multi-valued analytical function. Formally, this fact is related to non-trivial analytical properties of the Regge trajectories as functions of the energy: their singularities (branch points) come not only from physical thresholds, but also from coincidence of two (or more) reggeons [7], where those reggeons can be interchanged.

When considering the poles as energy-plane poles, one should have in mind that the energy plane generally has many Riemann sheets. An energy-plane pole corresponds to a bound state or resonance only if it is placed at the physical sheet or near the physical region. Poles placed far from the physical region or at far Riemann sheets are "not seen".

The infinite number of reggeons for the Yukawa potential can be "visualized". Evidently, the limit $\mu \to 0$ transforms the Yukawa potential $\exp(-\mu r)/r$, with the finite radius of interaction $\sim 1/\mu$, into the Coulomb potential, with the infinite radius. As traced in Ref.[6], this limit simultaneously transforms the Yukawa potential reggeons, that realize the Gribov-Pomeranchuk accumulation, into Coulomb reggeons, that realize orbital and radial Coulomb excitations. Therefore, the infinite number of energy-plane poles in the Yukawa potential becomes seen in the limit $\mu \to 0$ as the infinite number of the Coulomb levels. The Gribov-Pomeranchuk accumulation of reggeons near the threshold energy transforms in this limit into the well-known accumulation of the Coulomb bound states to the threshold (note that the double limiting transition $\mu \to 0$, $k \to 0$ is not equivalent here to the similar, but reversed limit $k \to 0$, $\mu \to 0$).

Returning from the non-relativistic quantum mechanics to relativistic amplitudes, we emphasize that the above arguments have not assumed any specific quantum numbers in the scattering channel. Therefore, their conclusion on the infinite number of energy-plane poles should be equally applicable (or non-applicable) to both bosonic and fermionic hadron poles, having any flavor quantum numbers (exotic or non-exotic).

Thus, if we study $2 \to 2$ strong interaction amplitudes, we should admit existence of (infinite number of) complex-energy poles with any exotic quantum numbers, both mesonic and baryonic. Alternatively, one could assume that analytical properties of strong interaction amplitudes, having exotic quantum numbers (at least, for one of the physical channels, s-, t-, or u-channel), are essentially different from those of totally non-exotic amplitudes. Such assumption looks very unnatural, and should be rejected. Thus, we obtain an argument for existence of exotic states, absent in previous publications.

Of course, existence of the energy-plane poles is only a necessary condition for the physical existence of exotic hadrons. To guarantee fulfillment of the sufficient condition, *i.e.*, existence of energy pole(s) with particular quantum numbers near the physical region, one should have more detailed knowledge of dynamics.

It is interesting to note that the above consideration uses the reggeons in a non-standard manner. Usually, to apply the complex angular momenta approach for obtaining results in the s-channel (the channel where the invariant s has the meaning of the squared c.m. energy), one begins from the crossed channel, where t and s are, respectively, the squared energy and squared momentum transfer (see Ref. [2]). Analytical continuation of partial-wave amplitudes in this t-channel allows, after returning into the s-channel, to study behavior of the invariant amplitude (and cross section) at high energy s at fixed value of momentum transfer t. Now, to obtain the conclusion about energy-plane poles in the s-channel, we use complex angular momenta in the same s-channel.

To summarize, earlier results on complex angular momenta imply that any 2-hadron interaction amplitude, under standard assumptions, has infinite number of poles in the energy plane. Those poles may have exotic, as well as non-exotic, quantum numbers. However, the poles may be placed at far Riemann sheets of the energy plane. Therefore, more detailed dynamical information is necessary to guarantee real existence of exotic hadrons.

Acknowledgments

In preparing this work I was strongly influenced by numerous past discussions with V.N. Gribov. I also thank R.A. Arndt, I.I. Strakovsky, R.L. Workman, and K. Goeke, my coauthors in paper [1], for more recent stimulating discussions. This text is based on the work partly supported by the U. S. Department of Energy Grant DE–FG02–99ER41110, by the Jefferson Laboratory, by the Southeastern Universities Research Association under DOE Contract DE–AC05–84ER40150, by the Russian-German Collaboration (DFG, RFBR), by the COSY-Juelich-project, and by the Russian State grant RSGSS-1124.2003.2.

References

1. Ya.I. Azimov, R.A. Arndt, I.I. Strakovsky, R.L. Workman, and K. Goeke, *Eur. Phys. J.* **A26**, 79 (2005); hep-ph/0504022.
2. P.D.B. Collins, *An Introduction to Regge Theory and High Energy Physics*, Cambridge Univ. Press, 1977.
3. R.A. Arndt, W.J. Briscoe, I.I. Strakovsky, R.L. Workman, and M.M. Pavan, *Phys. Rev.* **C69**, 035213 (2004); nucl-th/0311089.
4. V.N. Gribov, *ZhETF* **41**, 1962 (1961) [translated in *Sov. Phys. JETP* **14**, 1395 (1962)]; M. Froissart, *Phys. Rev.* **123**, 1053 (1961).
5. V.N. Gribov and I.Ya. Pomeranchuk, *Phys. Rev. Lett.* **9**, 238 (1962).
6. Ya.I. Azimov, A.A. Anselm, and V.M. Shekhter, *ZhETF* **44**, 361 (1963) [translated in *Sov. Phys. JETP* **17**, 246 (1963)].
7. Ya.I. Azimov, A.A. Anselm, and V.M. Shekhter, *ZhETF* **44**, 1078 (1963) [translated in *Sov. Phys. JETP* **17**, 726 (1963)].
8. Ya.I. Azimov, *Phys. Lett.* **3**, 195 (1963).

FROM $\pi \to 2\pi$ TO HANDEDNESS IN $\pi \to 3\pi$

ANATOLI V. EFREMOV*
JINR, Dubna 141980 Russia
E-mail: efremov@theor.jinr.ru

1. Introduction

Being invited to this memorial Conference, I recall one of the first works of BH (Russian abbreviation for Vladimir Naumovich, as people called him). "On the possibility of determining the amplitude for charge exchange pion pion scattering from an analysis of the $\pi^- + P \to N + \pi^+ + \pi^-$ reaction near the threshold" [1] where a correlation between pions was first proposed to extract an information on the scattering length of $\pi\pi$-interaction. Notice, that in those times there was no direct proof for the strong interactions of pions. His idea was first realized experimentally in Dubna by the group of nuclear emulsionists [2] with one friend of mine among them. From him I know, that, when in Dubna, BH liked to visit their room, saying: "Here my fame came from!"

I recall this story in connection with another pion correlation which would allow measurements of the polarization of a resonance and of quark or gluon jets. This correlation was named the HANDEDNESS.

The measurement of polarization at high energies is a complicated problem, since the observation of a spin or of a total angular momentum projection is, as a rule, a non-direct one, and explores features such as angular distribution in secondary scattering or in decay process. For a strong interaction process parity and Lorentz invariance requires that at least three momenta of particles (either spinless or spin-averaged) in a final state were measured.

Some years ago the concept of handedness was introduced[a] as a measure of polarization of parent partons (or decaying hadrons) [6]. It was defined as

*Work partially supported by grant 03-02-16816 of the Russian Foundation for Basic Research.

[a]In fact, an idea similar to the handedness was earlier proposed in works [3]. Its application

an asymmetry of a process probability W with respect to a spatial component of an axial 4-vector $n_\mu \propto \epsilon_{\mu\nu\sigma\rho} k_1^\nu k_2^\sigma k^\rho$, where k^ρ is the 4-momentum of a particle (or a system) in question ($k = k_1 + k_2 + k_3 + \cdots$), with respect to some direction $\mathbf{i}$ ($n_i = \mathbf{n} \cdot \mathbf{i}$)

$$H_i = \frac{W(n_i > 0) - W(n_i < 0)}{W(n_i > 0) + W(n_i < 0)} = \alpha_i P_i \ , \tag{1}$$

which was shown to be proportional to polarization P_i (at least for spin 1/2 and spin 1), provided the analyzing power α_i is not zero. The direction $\mathbf{i}$ could be chosen either as longitudinal (L) or as transversal ($T1$ or $T2$) with respect to the momentum $\mathbf{k}$.

2. Handedness in diffractive production of pion triples

In our earlier publication [7] the attention was drawn to the fact that in diffractive production of pion triples [8]

$$\pi^- + A \to (\pi^-\pi^+\pi^-) + A \ , \tag{2}$$

by a 40 GeV/c π^- beam from a nucleus A, a noticeable asymmetry with respect to the triple production plane (transversal handedness H_{T1}) was observed.

For reaction (2), one can define the normal to the plane of production of a secondary pion triple ($\pi_f^-\pi^+\pi_s^-$) (see Fig. 1)

$$\mathbf{N} = (\mathbf{v}_{3\pi} \times \mathbf{v}_b) \ , \tag{3}$$

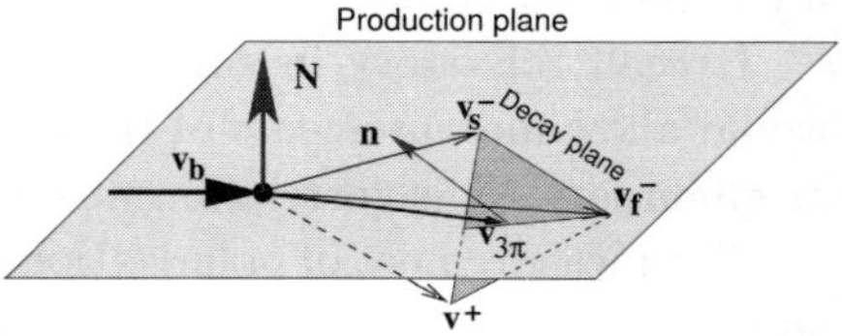

Figure 1. The kinematics of diffractive pion triple production.

where $\mathbf{v}_b = \mathbf{k}_b/\epsilon_b$, $\mathbf{v}_{3\pi} = \mathbf{k}_{3\pi}/\epsilon_{3\pi}$ are velocities of the initial π^- beam and the center of mass of the triple in Lab. r.f. and the indices f and s label fast and slow π^-'s. The normal to the "decay plane" of the triple in its center of mass is defined as

$$\mathbf{n} = (\mathbf{v}_f^- - \mathbf{v}^+) \times (\mathbf{v}_s^- - \mathbf{v}^+) \ , \tag{4}$$

where $\mathbf{v}_{f(s)}^-$ or $\mathbf{v}^+$ are velocities of the fast (slow) π^- or π^+.

---

to certain heavy quark decays was studied in Ref. [4]. Similar technique, which nowadays is known as "interference fragmentation", was also studied in work [5].

The transversal handedness according to (1) is[b]

$$H_{T1} = \frac{W(\mathbf{N}\cdot\mathbf{n} > 0) - W(\mathbf{N}\cdot\mathbf{n} < 0)}{W(\mathbf{N}\cdot\mathbf{n} > 0) + W(\mathbf{N}\cdot\mathbf{n} < 0)} \,. \tag{5}$$

Two other components of the handedness connected with $\mathbf{n}\cdot\mathbf{v}_{3\pi}$ and $\mathbf{n}\cdot(\mathbf{v}_{3\pi}\times\mathbf{N})$ are forbidden by the parity conservation in the strong interaction.

3. Experimental results and discussion

In the work [7] the experimental material of Bologna–Dubna–Milan Collaboration for diffraction production of $40\,GeV/c\ \pi^-$ into three pions was used [8].

The transversal handedness (5) was measured for a wide sample of nuclear targets.The dependence of H_{T1} on the atomic number A is presented in Fig. 2. One can see that the handedness systematically decreases with the increase of A, which resembles a depolarization effect in multiple scattering. An argument in favor of this is the decrease of the effect as, approximately, the inverse radius of the nucleus.

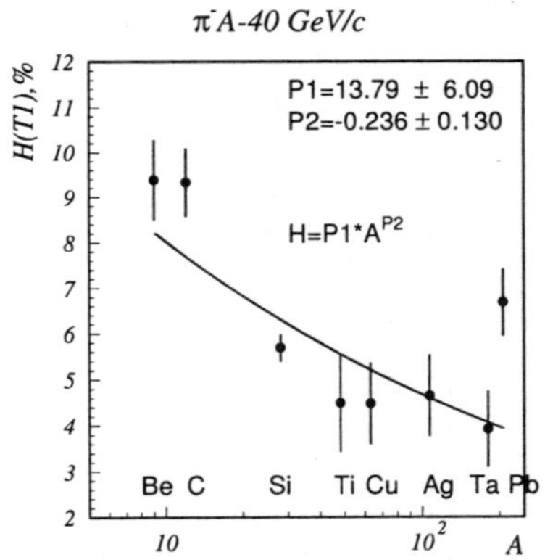

Figure 2. The A-dependence of the transversal handedness H_{T1}.

The value of the asymmetry (5), averaged over all nuclei is

$$H_{T1} = (5.96 \pm 0.21)\% \tag{6}$$

Statistically, this is a highly reliable verification of the existence of correlation of the triple production and the decay planes in the process (2).

The values of two other asymmetries with respect to correlations $\mathbf{n}\cdot\mathbf{v}_{3\pi}$ and $\mathbf{n}\cdot(\mathbf{v}_{3\pi}\times\mathbf{N})$ have been found to be comparable to zero from the same statistical material: $H_L = (0.25 \pm 0.21)\%$ and $H_{T2} = (0.43 \pm 0.21)\%$ respectively. This is by no means surprising, since they are forbidden by the parity conservation in process (2). Also they show the order of magnitude of systematic errors.

[b]It is easy to show that this quantity is in fact Lorentz-invariant.

The Monte-Carlo simulation of reaction (2) show no transversal handedness effect

$$H_{T1}^{\mathrm{MC}} = (0.20 \pm 0.28)\% \tag{7}$$

Thus, the effect (6) cannot be explained by the kinematics or by the influence of the apparatus.

To understand the nature of the effect observed, the dependence of the handedness (5) on the Feynmann variable $X_F = k_f/k_b$ of the leading π^-, on the triple transversal momentum k_T on the invariant mass of the triple $m_{3\pi}$ and its neutral subsystem $m_{\pi^+\pi^-}$ was studied. From Fig. 3a one can see that the handedness (5) increases with X_F, which resembles the behavior of the single spin asymmetry (e.g. the pion asymmetry or the Λ-polarization [9]).

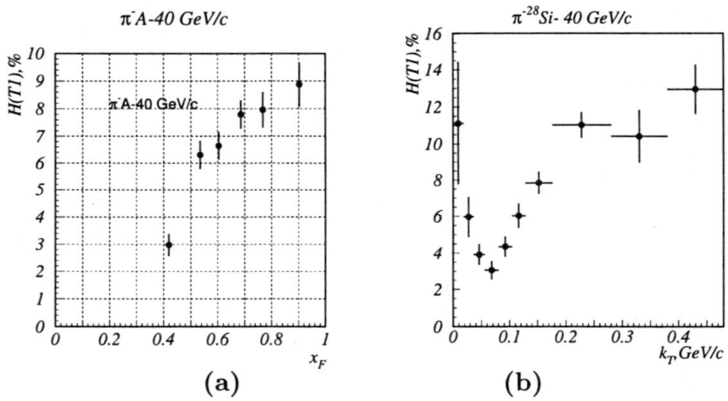

Figure 3. The handedness dependence: **(a)** on X_F of the leading π^- and **(b)** on k_T of the pion triple.

A complicated picture of the k_T-dependence with a sharp dip at $k_T = 0.05\text{–}0.07\,GeV/c$ (Fig. 3b) reflects by all means the fact of interference of the resonant and non-resonant processes in the triple production. With the further increase of k_T the handedness increases; this resembles also the single spin asymmetry behavior.

The dependence of H_{T1} on the triple invariant mass (Fig. 4a) is especially interesting. It clearly indicates two different sources of H_{T1} with comparable contributions: a resonant and a non-resonant one. The resonance contribution is clearly seen at the mass region of $a_2(1320)$ and $\pi_2(1670)$ and

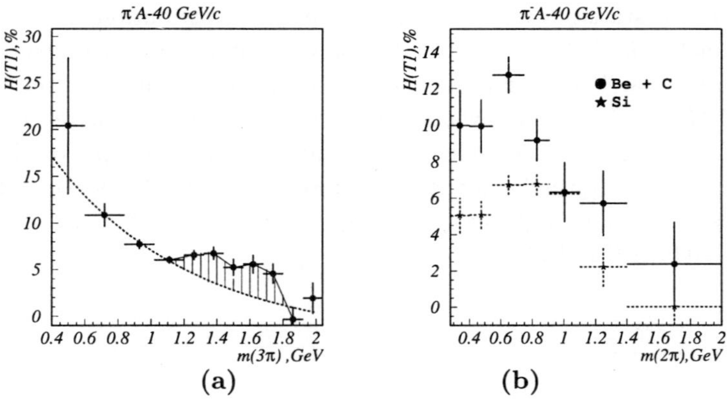

Figure 4. The handedness dependence on invariant mass: **(a)** $m_{3\pi}$ and **(a)** $m_{\pi^+\pi^-_f}$.

by all means is due to a non-zero polarization of the resonances. The non-resonant background could also be polarized, provided that the 3π system is predominantly in a state with a total angular momentum $J \neq 0$, e.g. if a neutral pair $m_{\pi^+\pi^-}$ was predominantly produced from a ρ-decay. Some indication of this can be seen from Fig. 4b. In this context, the growth of H_{T1} in the region of small $m_{3\pi}$, i.e. in the region of small relative momenta of pions, looks quite intriguing.

So, one can conclude that a rather large handedness transversal to the production plane was definitely observed in the diffractive production of $(\pi^-\pi^+\pi^-)$ triples in the π^-–beam dissociation region. The phenomenon has a clear dynamical origin and in some features resembles the single spin asymmetry behavior.

4. Handedness from VES Collaboration data

We continue to investigate this problem in more detail [10] using the high statistics of the VES collaboration [11] for the process (2).

Data are produced by the 37.0 GeV/c π^--beam on a Beryllium target and collected by general-purpose multi-particle magnet spectrometer with full identification of final states by lead-glass calorimeter and multichannel Čerenkov counter. We have used events in the $m_{3\pi}$ interval of 0.6–2.5 GeV and in the $|t'| = |t - t_{max}|$ interval of 0.0–1.0 GeV^2/c^2. A more detailed description of the VES setup can be found in [11]. The total of about $9 \cdot 10^6$ events was used in the analysis.

Such rich statistics allow us to see the $m_{3\pi}$ distribution with a good resolution 50 MeV. Raw $m_{3\pi}$ distribution without acceptance corrections is given in Fig. 5a. You see nothing special except for a shoulder in the region of 1600 MeV. The same events corrected for acceptance were analyzed using Exp. (5) for the handedness H_{T1} transverse to the triple production plane and similar formulae for H_L and H_{T2}. The handedness H_L and H_{T2} becomes compatible with zero (average over $m_{3\pi}$ $\overline{H_L} = 0.09 \pm 0.13\%$ and $\overline{H_{T2}} = -0.02 \pm 0.13\%$).

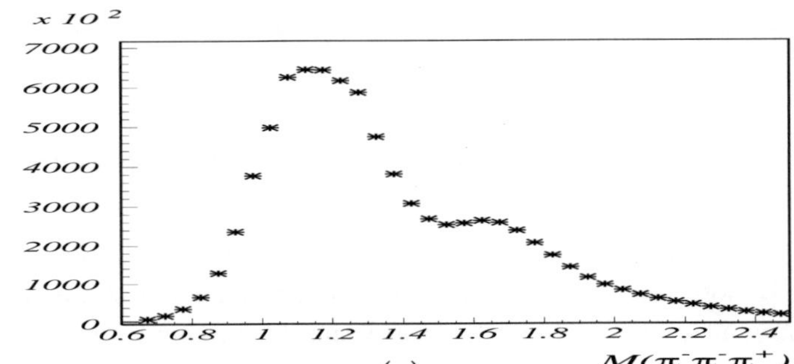

(a)

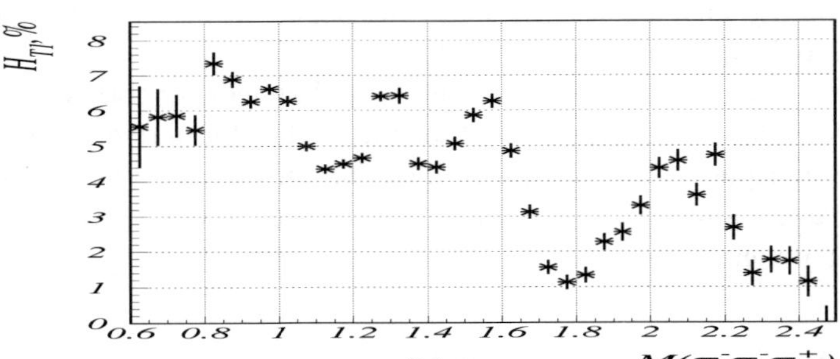

(b)

Figure 5. **(a):** The raw data invariant mass spectrum of the pion triples. **(b):** Spectrum of the transverse handedness H_{T1} after acceptance corrections.

The result of this analysis for H_{T1} is presented in Fig. 5b. We were really shocked when we saw this picture! (BH, I am sure, would be also excited.) In contrast to the $m_{3\pi}$ spectrum (Fig. 5a), where no prominent resonance peak is seen, the transverse handedness spectrum H_{T1} (Fig. 5b) reveals distinct peaks in the region of $a_2(1320)$, $\pi_1(1670)$ and $a_4(2040)$.

Does it really mean that all this resonances are produced transversally polarized? A strange peak is also seen in the region of 800 and 2200 MeV, since no 3π-resonance is known in this region. To what spin-parity does it correspond? Curious enough is the absence of a prominent peak in the region of $a_1(1230)$. All this questions can be probably answered by the future Partial Wave Analysis.

Even much richer statistics for the process 2 can be obtained in the frame of hadronic programm of COMPASS collaboration and it would be very interesting to see what the handedness analysis shows there.

5. Handedness of jets

The longitudinal handedness in jet fragmentation was also studied [12] in the process

$$e^+e^- \to Z^0 \to q\bar{q} \to 2\text{jets} \tag{8}$$

using the DELPHI data since the quarks produced in this process are longitudinally polarized due to parity non-conservation. Regrettably we can not find noticeable effect beyond two standard deviation at the level 1%.

An interesting indication to a break of CP-conjugation was however seen in **handedness correlation** [13] of quark and antiquark jets in the process (8). The sign of this correlation was opposite to those predicted by spin correlation, although the effect do not exceeds two standard deviations also.

It was hypothesized [13,14] that this effect could be connected with a nonzero vacuum chromomagnetic field fluctuation that influenced to the fragmenting quark and antiquark. If it is so, it would be the most direct way to measure the vacuum fluctuation of the gluon field.

References

1. A.A. Anselm and V.N. Gribov, *Sov. Phys. JETP* **37**, 354 (1960) [*Zh. Eksp. Teor. Fiz.* **76**, 501 (1959)].
2. Y.A. Batusov *et al.*, *Sov. Phys. JETP* **39**, 506 (1960); *Yad. Fiz.* **9**, 378 (1969).
3. O. Nachtmann, *Nucl. Phys.* **B127**, 314 (1977); A.V. Efremov, *Sov. J. Nucl. Phys.* **28**, 83 (1978).
4. R.H. Dalitz, G. Goldstein and R. Marshall, *Z. Phys.* **C42**, 441 (1989).
5. J.C. Collins, *Nucl. Phys.* **B396**, 161 (1993); J.C. Collins *et al.*, *Nucl. Phys.* **B420**, 565 (1994); X. Artru and J.C. Collins, *Z. Phys.* **C69**, 277 (1996).
6. A.V. Efremov, L. Mankiewicz and N. Törnqvist, *Phys. Lett.* **B284**, 394 (1992).
7. A.V. Efremov, Y.I. Ivanshin, L.G. Tkachev and R.Y. Zulkarneev, *Phys. Atom. Nucl.* **63**, 445 (2000) [*Yad. Fiz.* **63**, 515 (2000)];

A.V. Efremov, Y.I. Ivanshin, L.G. Tkachev and R.Y. Zulkarneev, *Prepared for the 13th International Symposium on High-Energy Spin Physics (SPIN 98), Protvino, Russia, 8-12 Sep. 1998* In *Protvino 1998, High energy spin physics*, p.579-583.

8. G. Bellini *et al.*, *Nucl. Phys.* **199**, 1 (1982); *Phys. Rev. Lett.* **48**, 1697 (1982); *Let. Nuovo Cim.* **38**, 433 (1983); *Nuovo Cim.* **A79**, 282 (1984).
9. See e.g. talks of K. Heller and S.B. Nurushev, *In Proc. of 12th Int. Symp. on High Energy Spin Physics, Amsterdam, 1996.*
10. A.V. Efremov, Y.I. Ivanshin, L.G. Tkachev, R.Y. Zulkarneev and I.A. Kachaev, JINR-E1-2000-322.
11. G. Beladidze *et al.* (VES Collaboration), *Phys.Lett.* **B268** 137, (1991).
12. A.V. Efremov, I.K. Potashnikova, L.G. Tkatchev and L.S. Vertogradov, report DELPHI 94-11 PHYS 355;
    K. Abe *et al.* (SLD-Collaboration), *Phys. Rev. Lett.* **74**, 1512 (1995);
    A.V. Efremov, I.K. Potashnikova and L.G. Tkatchev, *Proc. Ranc. de Moriond*, Meribel, 1994; See also *Proc. of 11th Int. Symp. on High Energy Spin Physics*, Bloomington 1994, *AIP Conf. Proc.* **343**, 821 (1995) and *Proc. of 27th Int. Conf. on High Energy Physics*, Glasgow 1994, IOP, London, 1995, p.875.
13. A.V. Efremov and L.G. Tkachev, *Acta Phys. Polon.* **B29**, 1385 (1998);
    A.V. Efremov, I.K. Potashnikova and L.G. Tkachev (DELPHI Collaboration), *Chin. J. Phys.* **34**, 949 (1996);
    A.V. Efremov, I.K. Potashnikova and L.G. Tkachev, JINR-E1-95-417; A.V. Efremov and L.G. Tkachev, *Prepared for the 11th International Symposium on High-energy Spin Physics and the 8th International Symposium on Polarization Phenomena in Nuclear Physics (SPIN 94)*, Bloomington, Indiana, 15-22 Sep. 1994, published in *Bloomington 1994, High energy spin physics*, p.821-833;
    A.V. Efremov and L.G. Tkachev, *Prepared for the 12th International Symposium on High-energy Spin Physics (SPIN 96)*, Amsterdam, Netherlands, 10-14 Sep. 1996, published in *Amsterdam 1996, Spin 96*, p.292-296.
14. A.V. Efremov and D. Kharzeev, *Phys. Lett.* **B366**, 311 (1996).

MATCHING NLO QCD WITH PARTON SHOWERS*

B.R. WEBBER
Cavendish Laboratory,
Madingley Road,
Cambridge CB3 0HE, U.K.
E-mail: webber@hep.phy.cam.ac.uk

The MC@NLO program combines the precision of next-to-leading-order (NLO) perturbative QCD with the detailed final-state modelling of the HERWIG Monte Carlo (MC) event generator, without double counting. After a toy model exposition of the underlying method, results on a number of hadron collider processes, including vector boson pair, heavy flavour and associated and unassociated Higgs production, are reviewed.

1. Introduction

The reliable prediction of cross sections and final-state distributions for QCD processes is important not only as a test of QCD but also for the design of collider experiments and new particle searches[a].

All systematic approaches to this problem are based on perturbation theory, usually truncated at next-to-leading order (NLO).

For the description of exclusive hadronic final states, perturbative calculations have to be combined with a model for the conversion of partonic final states into hadrons (hadronization). Existing hadronization models are in remarkably good agreement with a wide range of data, after tuning of model parameters.

However, these models operate on partonic states with high multiplicity and low relative transverse momenta, which are obtained from a parton shower Monte Carlo (MC) approximation to QCD dynamics and not from

---

*Work supported in part by the UK Particle Physics and Astronomy Research Council and by the EU Fourth Framework Programme 'Training and Mobility of Researchers', Network 'Quantum Chromodynamics and the Deep Structure of Elementary Particles', contract FMRX-CT98-0194 (DG 12 - MIHT).

[a]For reviews and more complete references, see Refs. [1,2].

fixed-order calculations. Therefore one needs a combination of the fixed-order and parton-shower approaches in which there is no double counting of terms order-by-order in perturbation theory.

1.1. *Objectives*

Our aim [3,4] is to develop a *practical* method for combining *existing* parton shower MC programs with NLO perturbative calculations (MC@NLO [5]). We require MC@NLO to have the following characteristics:

- The output is a set of events, which are fully exclusive.
- Total rates are accurate to NLO.
- NLO results for all observables are recovered on expansion in α_S.
- Hard emissions are treated as in NLO computations.
- Soft/collinear emissions are treated as in the MC.
- The matching between hard- and soft-emission regions is smooth.
- MC hadronization models are adopted.

For alternative approaches to similar goals, see Refs. [6,7,8,9,10]

2. Toy Model

Consider first a toy model [3] that allows a simple discussion of the key features of NLO, of MC, and of matching between the two.

- Assume a system can radiate massless "photons", energy x, with $0 \le x \le x_s \le 1$, x_s being the energy of the system before radiation.
- After radiation, the energy of the system is $x'_s = x_s - x$.
- The system can undergo further emissions, but photons themselves cannot radiate.

The task of predicting an infrared-safe observable O to NLO amounts to computing the quantity

$$\langle O \rangle = \lim_{\epsilon \to 0} \int_0^1 dx\, x^{-2\epsilon} O(x) \left[\left(\frac{d\sigma}{dx}\right)_{\rm B} + \left(\frac{d\sigma}{dx}\right)_{\rm V} + \left(\frac{d\sigma}{dx}\right)_{\rm R} \right], \quad (1)$$

where the *Born*, *virtual* and *real* contributions are, respectively,

$$\left(\frac{d\sigma}{dx}\right)_{\rm B} = B\delta(x)\,, \quad \left(\frac{d\sigma}{dx}\right)_{\rm V} = a\left(\frac{B}{2\epsilon} + V\right)\delta(x)\,, \quad \left(\frac{d\sigma}{dx}\right)_{\rm R} = a\frac{R(x)}{x}\,, \quad (2)$$

a is the coupling constant, and $\lim_{x\to 0} R(x) = B$.

In the *subtraction method*, the real contribution is written as:

$$\langle O\rangle_{\mathrm{R}} = aBO(0)\int_0^1 dx\,\frac{x^{-2\epsilon}}{x} + a\int_0^1 dx\,\frac{O(x)R(x)-BO(0)}{x^{1+2\epsilon}}\,. \tag{3}$$

The second integral is non-singular, so we can set $\epsilon = 0$ there:

$$\langle O\rangle_{\mathrm{R}} = -a\frac{B}{2\epsilon}O(0) + a\int_0^1 dx\,\frac{O(x)R(x)-BO(0)}{x}\,. \tag{4}$$

Therefore the NLO prediction is:

$$\langle O\rangle_{\mathrm{sub}} = BO(0) + a\left[VO(0) + \int_0^1 dx\,\frac{O(x)R(x)-BO(0)}{x}\right]\,. \tag{5}$$

We rewrite this in a slightly different form more suited to discussion of the matching with the MC:

$$\langle O\rangle_{\mathrm{sub}} = \int_0^1 dx\left[O(x)\frac{aR(x)}{x} + O(0)\left(B + aV - \frac{aB}{x}\right)\right]\,. \tag{6}$$

2.1. *Toy Monte Carlo*

In a treatment based on Monte Carlo methods, the system can undergo an arbitrary number of emissions (branchings), with probability controlled by the *Sudakov form factor*, defined for our toy model as follows:

$$\Delta(x_1, x_2) = \exp\left[-a\int_{x_1}^{x_2} dz\frac{Q(x)}{x}\right]\,, \tag{7}$$

where $Q(x)$ is a monotonic function with the following properties:

$$0 \le Q(x) \le 1, \quad \lim_{x\to 0} Q(x) = 1, \quad \lim_{x\to 1} Q(x) = 0\,. \tag{8}$$

$\Delta(x_1, x_2)$ is the probability that no photon be emitted with energy x such that $x_1 \le x \le x_2$.

2.2. *Modified Subtraction*

We want to interface NLO to MC. A naive first try might be:

$$O(0) \Rightarrow \text{start the MC with 0 real emissions: } \mathbf{F}^{(0)}_{\mathrm{MC}}\,,$$
$$O(x) \Rightarrow \text{start the MC with 1 emission at } x\text{: } \mathbf{F}^{(1)}_{\mathrm{MC}}(x)\,,$$

where $\mathbf{F}^{(i)}_{\mathrm{MC}}$ is the *generating functional* representing final states that are generated starting from a configuration with i real emissions. Then the overall generating functional would be

$$\int_0^1 dx\left[\mathbf{F}^{(0)}_{\mathrm{MC}}\left(B + aV - \frac{aB}{x}\right) + \mathbf{F}^{(1)}_{\mathrm{MC}}(x)\frac{aR(x)}{x}\right]\,. \tag{9}$$

This would be *wrong*: an MC starting with no emissions will generate emission, with NLO distribution

$$\left(\frac{d\sigma}{dx}\right)_{\text{MC}} = aB\frac{Q(x)}{x} \,. \tag{10}$$

In addition, the NLO probability of no emissions will be underestimated, by an amount equal to the integral of this distribution. Therefore we should subtract (10) from the second term in the integrand of (9), and add it to the first:

$$\mathbf{F}_{\text{MC@NLO}} = \int_0^1 dx \left[\mathbf{F}_{\text{MC}}^{(0)}\left(B + aV + \frac{aB[Q(x)-1]}{x}\right) + \mathbf{F}_{\text{MC}}^{(1)}(x)\frac{a[R(x)-BQ(x)]}{x}\right] . \tag{11}$$

This prescription has several good features:

- $\mathbf{F}_{\text{MC}}^{(0)} = \mathbf{F}_{\text{MC}}^{(1)}$ at order a^0, so the added and subtracted terms are equal to order a^1;
- The coefficients of $\mathbf{F}_{\text{MC}}^{(0)}$ and $\mathbf{F}_{\text{MC}}^{(1)}$ are now separately finite;
- The same resummation of large logs is performed in $\mathbf{F}_{\text{MC}}^{(0)}$ and $\mathbf{F}_{\text{MC}}^{(1)}$. Hence $\mathbf{F}_{\text{MC@NLO}}$ gives the same resummation as $\mathbf{F}_{\text{MC}}^{(0)}$, normalised to the correct NLO cross section.

Note, however, that some events may have *negative weight.*

2.3. *Toy Model Observables*

As an example of an "exclusive" observable, we consider the energy y of the hardest photon in each event. The NLO and MC predictions are

$$\left(\frac{d\sigma}{dy}\right)_{\text{NLO}} = a\frac{R(y)}{y} \,, \quad \left(\frac{d\sigma}{dy}\right)_{\text{MC}} = aB\frac{Q(y)}{y}\Delta(y,1) \,. \tag{12}$$

As an "inclusive" observable, consider the fully inclusive distribution of photon energies, z:

$$\left(\frac{d\sigma}{dz}\right)_{\text{NLO}} = a\frac{R(z)}{z} \,, \quad \left(\frac{d\sigma}{dz}\right)_{\text{MC}} = aB\frac{Q(z)}{z} \,. \tag{13}$$

The toy model results below are for

$$a = 0.3 \,, \quad B = 2 \,, \quad V = 1 \,, \quad R(x) = B + x(1 + x/2 + 20x^2) \tag{14}$$

so that the total cross section is

$$\sigma_{\text{tot}} = B + a\left(V + \frac{95}{12}\right) \,. \tag{15}$$

Note that this implies a large K-factor, $\sigma_{\text{tot}}/B = 2.34$. For the MC we assumed a "dead zone" $Q(x) = 0$ for $x > 0.6$ with three different possible forms when $x < 0.6$, as shown in the upper right pane of Fig. 1. In spite of the large discrepancies between the MC and NLO at large x, we see in the lower panes a smooth behaviour of the observable y and z distributions, with only small differences between the three choices of $Q(x)$ functions, corresponding to uncontrolled NNLO contributions.

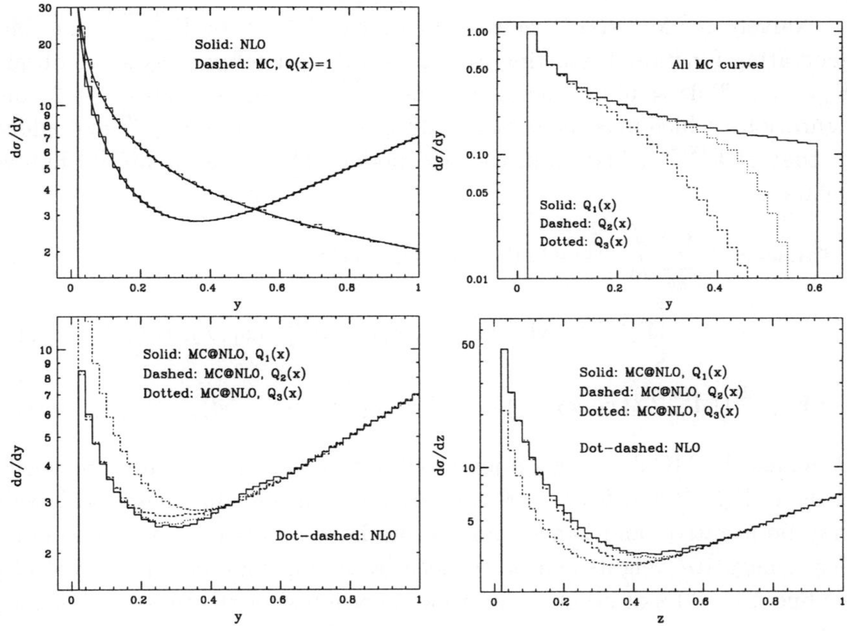

Figure 1. Toy model observables.

3. Modified Subtraction for Real QCD

Consider a hadron collider process which is $2 \to 2$ at LO, e.g. W^+W^- or $Q\overline{Q}$ pair production. A schematic expression for any observable O, evaluated by the subtraction method, is

$$\langle O\rangle_{\text{sub}} = \sum_{ab}\int_0^1 dx_1\,dx_2\,d\phi_3\,f_a(x_1)f_b(x_2)\Big[O^{(2\to 3)}\mathbf{M}_{ab}^{(h)}(x_1,x_2,\phi_3)+$$

$$+\,O^{(2\to 2)}\left(\mathbf{M}_{ab}^{(b,v,c)}(x_1,x_2,\phi_2)-\mathbf{M}_{ab}^{(c.t.)}(x_1,x_2,\phi_3)\right)\Big]\,, \qquad (16)$$

where $\mathbf{M}_{ab}^{(h)}$ is the NLO real-emission contribution, $\mathbf{M}_{ab}^{(b,v,c)}$ are the LO Born, NLO virtual and collinear (finite parts), and $\mathbf{M}_{ab}^{(c.t.)}$ are counter-terms which cancel the divergences of $\mathbf{M}_{ab}^{(h)}$.

Naively, for MC@NLO we would replace $O^{(2\to 2,3)}$ by $\mathbf{F}_{\text{MC}}^{(2\to 2,3)}$ (i.e. MC generating functionals starting from $2\to 2,3$ hard subprocesses), to obtain $\mathbf{F}_{\text{MC@NLO}}$. This would be *wrong* because $\mathbf{F}_{\text{MC}}^{(2\to 2)}$ also generates $2\to 3$ configurations, which must be subtracted from the weight of $\mathbf{F}_{\text{MC}}^{(2\to 3)}$ and added to that of $\mathbf{F}_{\text{MC}}^{(2\to 2)}$, just as in the toy model. Therefore for MC@NLO we define

$$\mathbf{F}_{\text{MC@NLO}} = \sum_{ab}\int_0^1 dx_1\,dx_2\,d\phi_3\,f_a(x_1)f_b(x_2)\,\cdot$$

$$\cdot\Big[\mathbf{F}_{\text{MC}}^{(2\to 3)}\left(\mathbf{M}_{ab}^{(h)}(x_1,x_2,\phi_3)-\mathbf{M}_{ab}^{(\text{MC})}(x_1,x_2,\phi_3)\right)+ \qquad (17)$$

$$+\mathbf{F}_{\text{MC}}^{(2\to 2)}\left(\mathbf{M}_{ab}^{(b,v,c)}(x_1,x_2,\phi_2)-\mathbf{M}_{ab}^{(c.t.)}(x_1,x_2,\phi_3)+\mathbf{M}_{ab}^{(\text{MC})}(x_1,x_2,\phi_3)\right)\Big]\,.$$

Provided the MC does a good job in all soft and collinear limits, the coefficients of $\mathbf{F}_{\text{MC}}^{(2\to 2)}$ and $\mathbf{F}_{\text{MC}}^{(2\to 3)}$ are now *separately finite.* But these coefficients may be negative, and hence some events have *negative weight.* The number of negative weights can be reduced by tuning counterterms. Typically we find $10-20\%$ of events have negative weights for the processes studied below.

4. Vector Boson Pair Production

The first process to be implemented in MC@NLO was the hadroproduction of pairs of electroweak vector bosons [3].

4.1. *WW Observables*

Figure 2 shows the results on WW correlations at LHC, compared to NLO and to HERWIG MC results rescaled by the K-factor for the total WW cross section. These correlations are problematic: soft and hard emissions

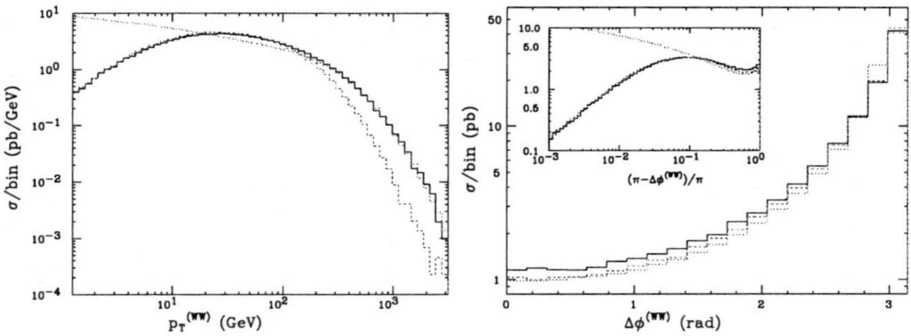

Figure 2. WW observables. Solid: MC@NLO. Dashed: HERWIG$\times\frac{\sigma_{NLO}}{\sigma_{LO}}$. Dotted: NLO.

are both relevant. MC@NLO does well, resumming large logarithms, and yet handling the large-scale physics correctly. In the distribution of the azimuthal difference at small values of $\Delta\phi$, MC@NLO provides the multiple soft emission missing from NLO and hard emission missing from MC, so that the prediction lies above both NLO and MC.

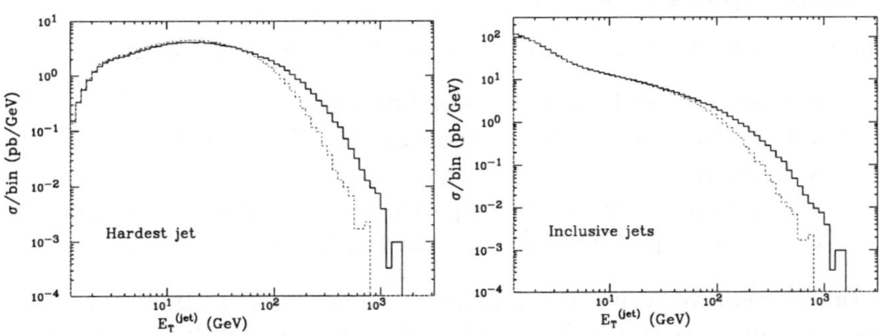

Figure 3. WW jet observables. Solid: MC@NLO. Dashed: HERWIG$\times\frac{\sigma_{NLO}}{\sigma_{LO}}$.

In Fig. 3, jets have been reconstructed with a k_T-algorithm. Note that both distributions shown here are given at NLO by the dashed curve in Fig. 2 (left pane). It is striking that the inclusive jet distribution displays the same behaviour as in the toy model: the MC@NLO and MC agree as $p_t \to 0$, after the latter is rescaled by the overall K-factor.

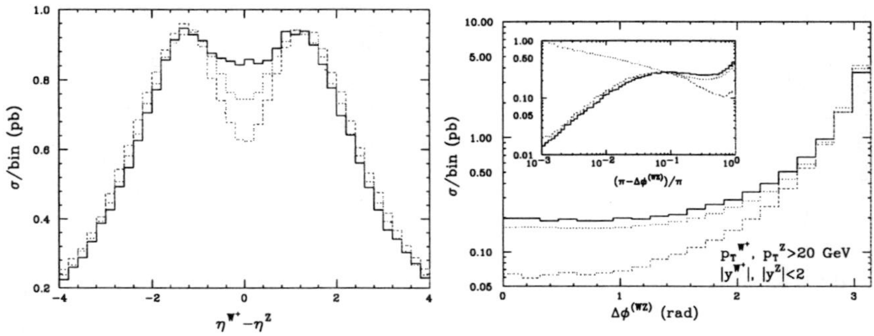

Figure 4. WZ observables. Solid: MC@NLO. Dashed: HERWIG$\times \frac{\sigma_{NLO}}{\sigma_{LO}}$. Dotted: NLO.

4.2. *WZ Observables*

Figure 4 shows some results for WZ production at LHC. It is interesting that the MC@NLO fills further the kinematic dip in the pseudorapidity distribution at $\eta_W - \eta_Z = 0$. The difference between MC@NLO and MC at small $\Delta\phi$ is enhanced here by the transverse momentum cuts.

5. Heavy Quark Production

The modified subtraction formula above can be used for any process:

- Take the standard subtraction formula;
- Calculate analytically *exactly* what the MC does at NLO;
- Insert $\mathbf{M}_{ab}^{(\mathrm{MC})}(x_1, x_2, \phi_3)$ terms;
- Generate $2 \to 2$ and $2 \to 3$ parton configurations and weights;
- Feed into the MC (e.g. using the Les Houches interface [11]).

In this Section we review the application to heavy quark production. The most difficult part is calculating what the MC does. Fuller details can be found in Ref. [4].

5.1. *MC Heavy Quark Production*

The MC starts from a $2 \to 2$ subprocess and *momentum reshuffling* is done after real emission to conserve 4-momentum. The relationship between invariants and shower variables depends on which leg emits. Some contributions are shown in Fig. 5.

A colour SU(N) structure is assigned (for showering and hadronization)

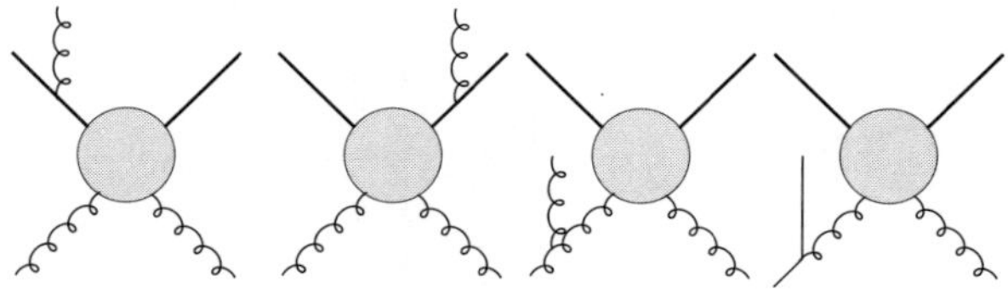

Figure 5. Real-emission contributions to heavy quark hadroproduction.

according to the $N \to \infty$ limit (Fig. 6)

$$\text{Prob}_i = |\sum_j M_j^{(N=3)}|^2 \, |M_i^{(N=\infty)}|^2 / \sum_j |M_j^{(N=\infty)}|^2 \tag{18}$$

This is the best we can do, since different colour structures interfere at higher orders in $1/N$.

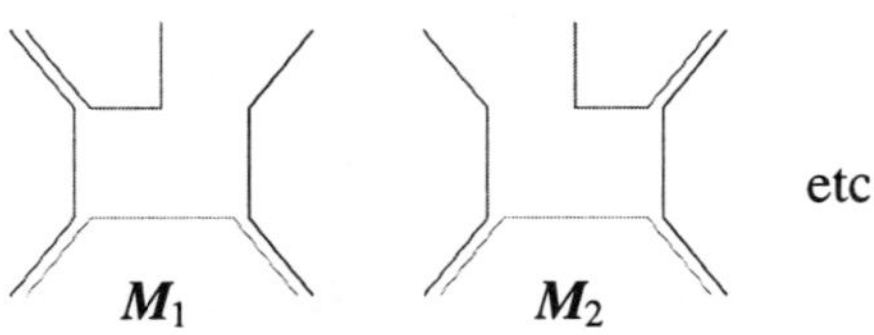

Figure 6. Alternative colour structures in $gg \to Q\overline{Q}g$.

5.2. $t\bar{t}$ *Observables at Colliders*

For the distributions shown in Fig. 7, resummation is not important and so MC@NLO is similar to NLO. Notice, however, that the MC@NLO results are more stable numerically, owing to the improvements in the event weight distribution discussed above.

5.3. *Top Rapidity Asymmetry at Tevatron*

A new feature is the $Q\overline{Q}$ asymmetry at the Tevatron. This is an NLO effect, not present in the MC alone but preserved in MC@NLO (Fig. 8).

5.4. $b\bar{b}$ *Production*

In $b\bar{b}$ production there are additional complications because the b quark mass scale is not so large. In parton shower MC's, three classes of pro-

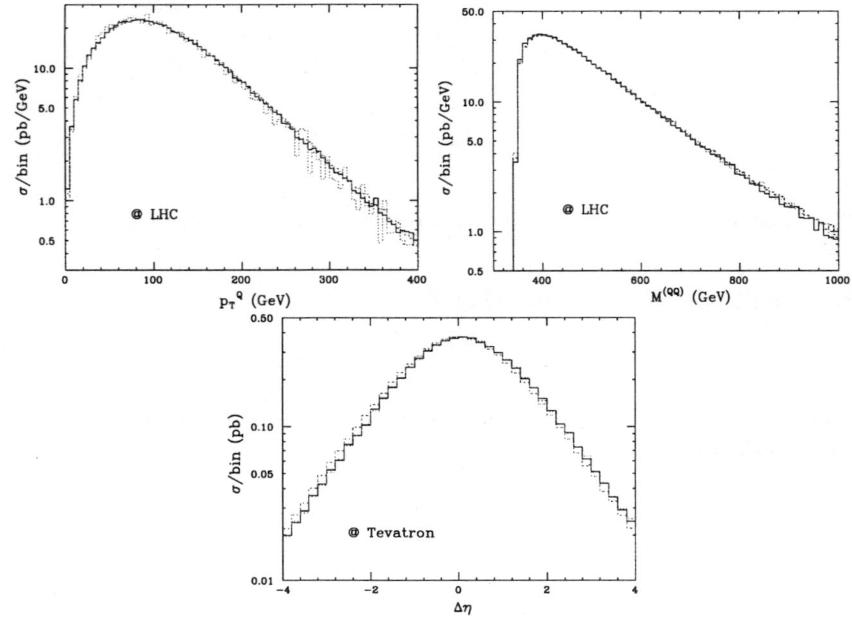

Figure 7. Collider $t\bar{t}$ observables. Solid: MC@NLO. Dashed: HERWIG$\times\frac{\sigma_{NLO}}{\sigma_{LO}}$. Dotted: NLO.

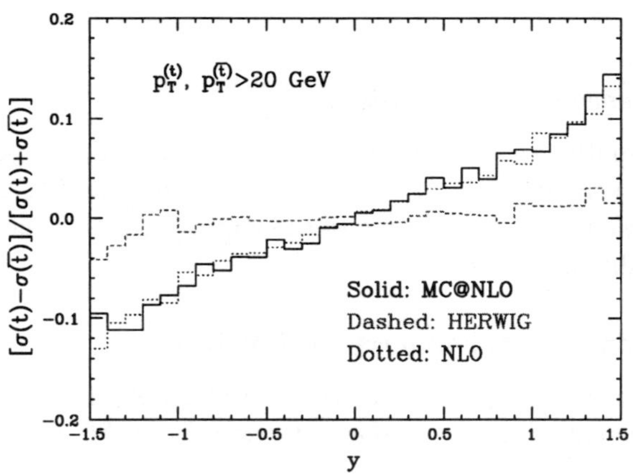

Figure 8. Tevatron $t\bar{t}$ asymmetry.

cesses make important contributions: flavour creation (FCR), gluon splitting (GSP) and flavour excitation (FEX).

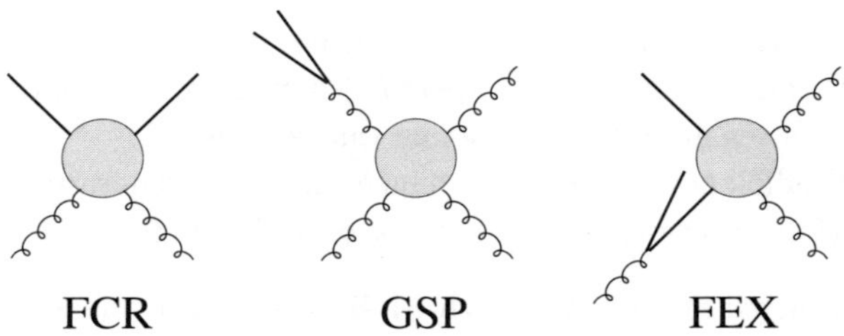

All are needed to get close to the data (Fig. 9, from Ref. [12]).

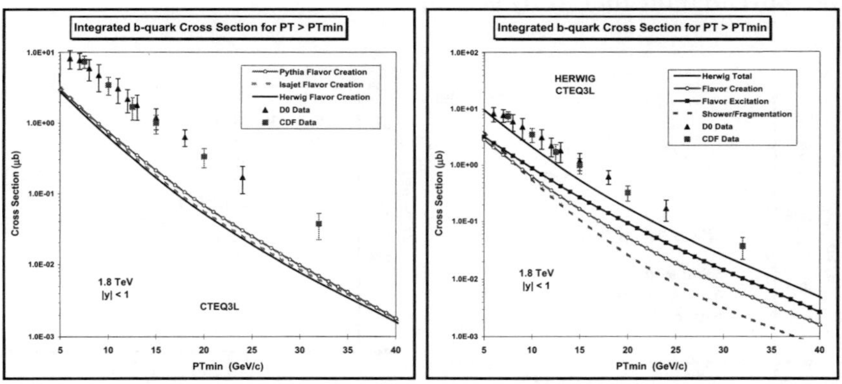

Figure 9. MC contributions to Tevatron $b\bar{b}$ production.

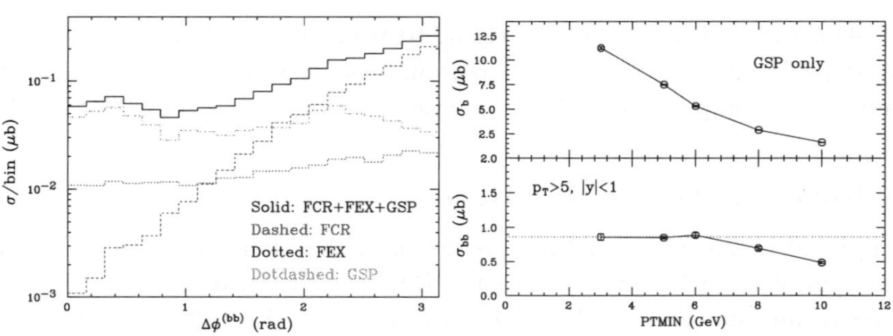

Figure 10. HERWIG MC contributions to Tevatron $b\bar{b}$ production.

The three processes make rather different contributions to the $b\bar{b}$ azimuthal correlation (Fig. 10, left pane). However,

- GSP, FEX and FCR are complementary and all must be generated;
- GSP taken alone is cutoff (PTMIN) dependent (right pane);
- FEX is sensitive to the bottom quark PDF;
- GSP efficiency is very poor, since one must generate all gluon production processes and wait for a $g \to b\bar{b}$ splitting.

All these problems are avoided with MC@NLO, since the FCR, GSP and FEX subprocesses are already present at NLO level.

5.5. $b\bar{b}$ *Correlations at Tevatron*

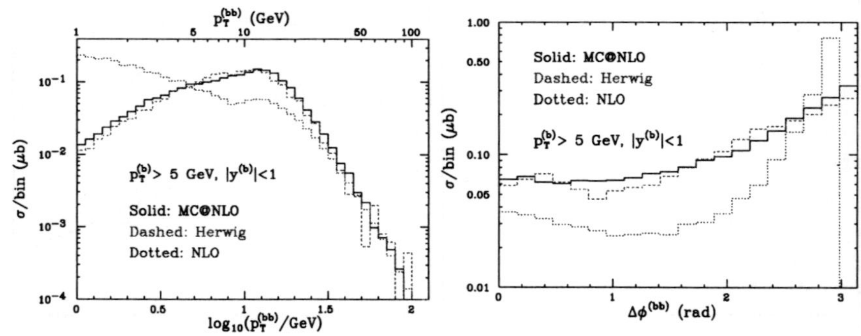

Figure 11. Tevatron $b\bar{b}$ observables. Solid: MC@NLO. Dashed: HERWIG (no K-factor). Dotted: NLO.

As shown in Fig. 11, MC@NLO and HERWIG give similar predictions after suitable cuts. However, for the reasons mentioned above, HERWIG needs much more computation: 14 million events versus 1 million for MC@NLO.

6. NLO + k_T-kick vs MC@NLO

Empirically, smearing the NLO prediction with a gaussian transverse momentum distribution for the $b\bar{b}$ pair (NLO + k_T-kick) with $\langle k_T \rangle = 4$ GeV gives results similar to MC@NLO at the Tevatron (Fig. 12). This does *not* mean that there is an 'intrinsic' $\langle k_T \rangle = 4$ GeV inside the proton: it simply emulates the effect of initial-state parton showers.

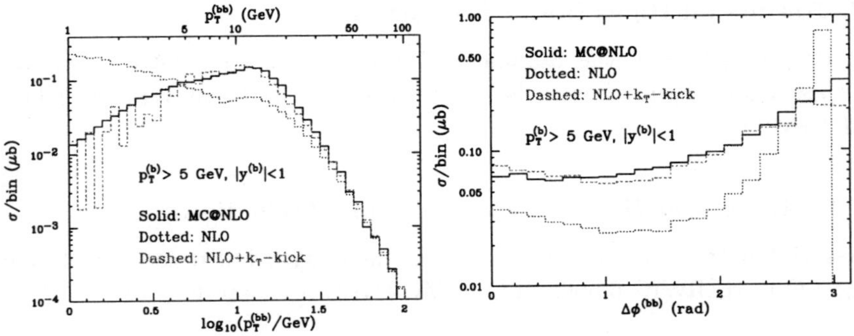

Figure 12. Tevatron $b\bar{b}$ production: NLO + k_T-kick vs MC@NLO.

6.1. *Hadron-level Results on B production*

$B \to J/\psi$ results from Tevatron Run II [13] provide the most accurate current data on B hadron production (Fig. 13). There is good agreement between the data, MC@NLO and resummation matched to fixed order (FONLL [14,15]). Consequently there is now no serious discrepancy between the data and the best available QCD predictions.

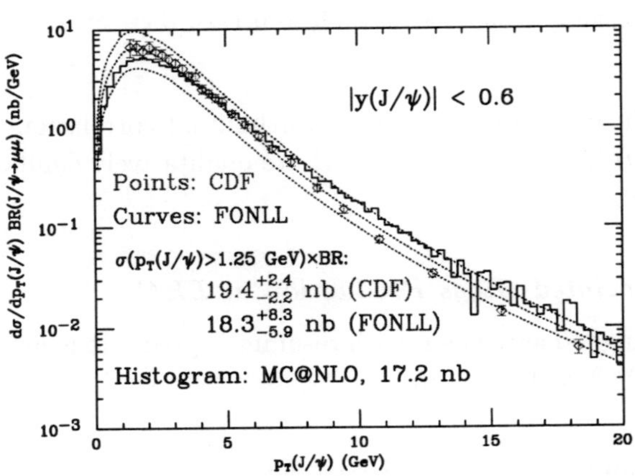

Figure 13. Tevatron Run II $B \to J/\psi$ production.

7. Results on Higgs Boson Production

Hadroproduction of the Standard Model Higgs boson, with or without an associated vector boson, is implemented in the latest version (3.1) of MC@NLO [5,16]

7.1. *Associated Higgs + Vector Boson Production*

Associated Higgs production is implemented with full decay correlations (to LO in HERWIG 6.506, NLO in MC@NLO 3.1 – see Fig. 14).

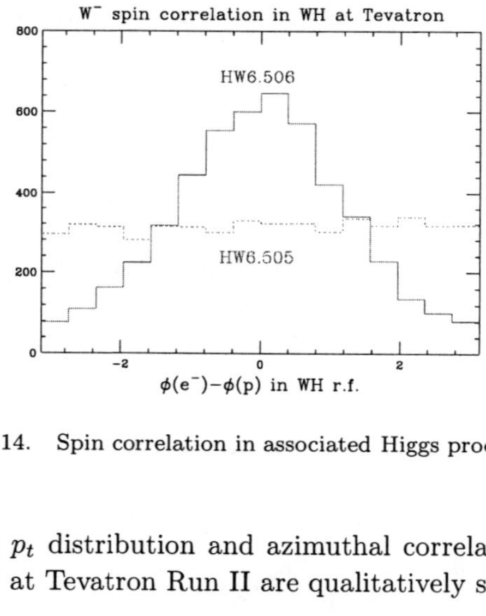

Figure 14. Spin correlation in associated Higgs production.

The predicted p_t distribution and azimuthal correlation of WH pairs in $\bar{p}p \to W^+H^0X$ at Tevatron Run II are qualitatively similar to those for WW (Fig.15).

7.2. *Unassociated Higgs Production at LHC*

Results are in good agreement with resummed predictions matched to fixed order (NNLO+NNLL, Fig. 16).

8. Conclusions and Future Prospects

A practical method for combining NLO calculations with HERWIG parton showers now exists, in the form of the MC@NLO program. This permits a full hadron-level simulation of final states, while retaining NLO accuracy for observables computable at that order. The method has been implemented

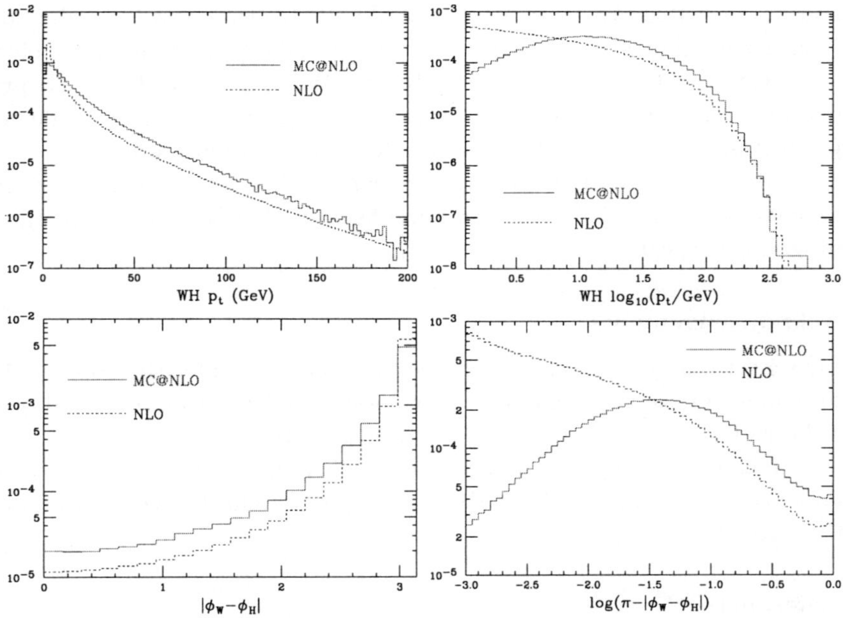

Figure 15. Tevatron WH observables.

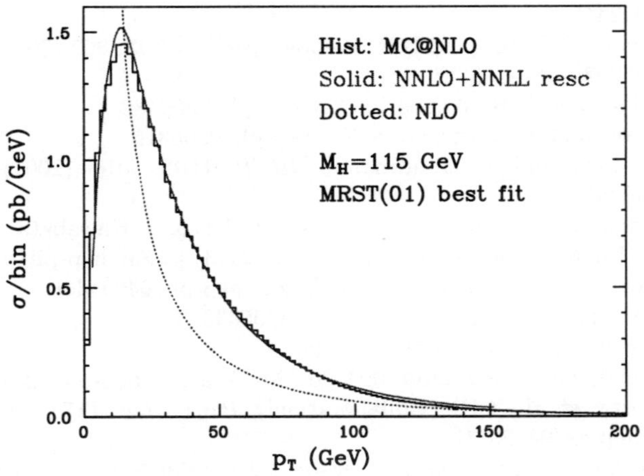

Figure 16. Transverse momentum of unassociated Higgs bosons at LHC.

and works well for W, Z, H, WW, WZ, ZZ, WH, ZH, $t\bar{t}$ and $b\bar{b}$ production. Negative weights, at the level of $\sim 10\%$ ($t\bar{t}$) to 20% ($b\bar{b}$), are not a problem.

Decay correlations are implemented for W, Z, WW, WH and ZH production, but not yet for other processes. Implementation of single top production is in progress [17]. Dijet production needs more work.

If one is prepared to modify the parton showers, avoiding negative weights looks possible [9]. A general interface to NLO (subtraction method) programs is feasible.

Acknowledgements

I am most grateful to Yuri Dokshitzer and Peter Lévai for the opportunity to commemorate the profound contributions and inspiring personality of Volodya Gribov in this beautiful and historic city.

I acknowledge my collaborators on the MC@NLO project, led by Stefano Frixione, and thank the CERN Theory Group for frequent hospitality throughout the project.

References

1. S. Catani *et al.*, arXiv:hep-ph/0005025.
2. M.A. Dobbs *et al.*, arXiv:hep-ph/0403045.
3. S. Frixione and B.R. Webber, *JHEP* **0206**, 029 (2002) [arXiv:hep-ph/0204244].
4. S. Frixione, P. Nason and B.R. Webber, *JHEP* **0308**, 007 (2003) [arXiv:hep-ph/0305252].
5. S. Frixione and B.R. Webber, arXiv:hep-ph/0506182.
6. C. Friberg and T. Sjostrand, arXiv:hep-ph/9906316.
7. J.C. Collins and F. Hautmann, *JHEP* **0103**, 016 (2001) [arXiv:hep-ph/0009286].
8. Y. Kurihara, J. Fujimoto, T. Ishikawa, K. Kato, S. Kawabata, T. Munehisa and H. Tanaka, *Nucl. Phys.* **B654**, 301 (2003) [arXiv:hep-ph/0212216].
9. P. Nason, *JHEP* **0411**, 040 (2004) [arXiv:hep-ph/0409146].
10. Z. Nagy and D.E. Soper, arXiv:hep-ph/0503053.
11. E. Boos *et al.*, arXiv:hep-ph/0109068.
12. R.D. Field, *Phys. Rev.* **D65**, 094006 (2002) [arXiv:hep-ph/0201112].
13. D. Acosta *et al.* [CDF Collaboration], *Phys. Rev.* **D71**, 032001 (2005) [arXiv:hep-ex/0412071].
14. M. Cacciari, M. Greco and P. Nason, *JHEP* **9805**, 007 (1998) [arXiv:hep-ph/9803400].
15. M. Cacciari *et al.*, *JHEP* **0407**, 033 (2004) [arXiv:hep-ph/0312132].
16. V. Del Duca, S. Frixione, C. Oleari and B.R. Webber, paper in preparation.
17. S. Frixione, E. Laenen, P. Motylinski and B.R. Webber, paper in preparation.

SOFT GLUON EMISSION AT LARGE ANGLES

GIUSEPPE MARCHESINI
Dipartimento di Fisica, Università di Milano-Bicocca and INFN
Sezione di Milano, Italy

I review recent results on soft gluon radiation at large angles.

1. Gribov's legacy on soft radiation

Although the major interest of Vladimir Gribov [1] in QCD was on non-perturbative aspects (notably colour confinement), his work on soft photons or soft pions emission in the pre-QCD era [2] is the basis for understanding general features of perturbative QCD dynamics. In principle soft emission is "trivial", it reduces essentially to classical physics. That is, it depends only on external charges which are treated as classical currents (asymptotic, large-time states), thus it does not reveal the internal short distance structure of emitting systems.

In spite of this "trivial" aspects, universality of soft emission allows one to compute multi-soft gluon emission distributions and thus to obtain important information on the QCD properties in general. The key point is that in computing multi-soft gluon amplitudes one deals with multi-soft scales. The softest gluon does not "see" any internal structure. However the next-to-soft gluon can "see" the softest one. In this way one can recursively reconstruct the detailed deep structure of the emitting system up to the primary hard emitting system. Therefore, while the result is "trivial" in QED since only the primary emitting charges are involved in soft photon emission, in QCD instead, due to direct gluon interaction, also the successive emitted gluons are involved and then successively resolved.

In this talk I discuss various results obtained in the course of years by exploiting soft gluon emission properties obtained in this way.

2. Multi-soft gluon amplitudes

I start by recalling how one obtains [3] (see also [4]) the amplitude for the emission of n soft gluons $q_1 \cdots q_n$, off a colour singlet dipole of primary hard quark antiquark p_b, p_b as emitted for instance in e^+e^- annihilation at high c.m. energy Q (generalization to other dipoles is straightforward). The general colour structure is given as sum of Chan-Paton factors

$$\mathcal{M}_n(p_a, p_b; q_1 \cdots q_n) = \sum_{\text{perm.}} \{\lambda^{b_{i_1}} \cdots \lambda^{b_{i_n}}\}_{\beta\bar{\beta}} \, M_n(p_a q_{i_1} \cdots q_{i_n} p_b) \,, \quad (1)$$

with $\beta, \bar{\beta}$ and b_i the quark, antiquark and gluon colour indices (in $SU(N)$ in general). In M_n (*colour ordered amplitudes*) the soft gluon momenta q_i are ordered as in the Chan-Paton factor.

In the soft limit M_n can be obtained by recurrence relations. Take q_j as the softest among the soft gluons. One uses eikonal emission (including three-gluon vertex) and colour conservation and obtains

$$M_n(\cdots q_\ell \, q_j \, q_{\ell'} \cdots) = g_s \, M_{n-1}(\cdots q_\ell q_{\ell'} \cdots) \cdot \left(\frac{q_\ell^{\mu_j}}{(q_\ell q_j)} - \frac{q_{\ell'}^{\mu_j}}{(q_{\ell'} q_j)} \right) . \quad (2)$$

Here μ_j is the Lorentz index of softest gluon and $q_\ell, q_{\ell'}$ are the momenta "near in colour" to q_j in the permutation considered (they are soft gluons or primary quark or antiquark). At this stage no information is needed for the deeper internal structure of M_{n-1}.

The factorization structure (2) is completely general and can be iterated. At each stage $m < n$ one factorizes the dipole contribution of the softest gluon in M_m and recovers the entire colour amplitudes M_{m-1}. The multi-soft gluon amplitude M_n is finally obtained and expressed in terms of successive (energy ordered) dipole factors for all soft gluons and the Born amplitude M_0 for the primary emitting system.

A direct consequence of the fact that at each stage one recovers the *entire* colour amplitudes is that one does not need to specify the gauge frame since, at each stage, one is dealing with gauge invariant amplitudes and dipole emission factors.

By construction M_n is the leading soft part of the amplitude, proportional to the inverse of all soft gluon energies $M_n \sim (\omega_1 \cdots \omega_n)^{-1}$. It is valid in any angular configuration even away from collinear configurations. Therefore the factorization form (2) is adequate for our discussion on physics of large angle soft radiation.

Next-to-leading corrections in the soft limit are known [5], see also [6]. They are expressed in terms of a factorized 2-soft gluon emission current.

These corrections are needed in order to reconstruct the argument of the running coupling α_s entering the distribution as the soft gluon transverse momentum in the center of mass of the emitting dipole.

For the square of the colour ordered amplitude M_n the recurrence (2) gives [3]

$$|M_n(p_a q_1 \cdots q_n p_b)|^2 = \left(\frac{\alpha_s}{2\pi}\right)^n W_{ab}(q_1 \cdots q_n) \cdot |M_0|^2,$$
$$W_{ab}(q_1 \cdots q_n) = \frac{(p_a p_b)}{(p_a q_1) \cdots (q_n p_b)} \,. \tag{3}$$

The distribution $W_{ab}(1 \cdots n)$ contains collinear singularities when any two partons which are near in colour become parallel. However, this expression is valid in any angular configuration, even at large angles (hedgehog configurations). There are three remarkable properties for this colour distribution.

The first is that in any energy-ordered configuration for the soft gluons one obtains the same expression for the distribution (3). This in spite of the fact that the expression of the colour ordered amplitude M_n, obtained from the recurrence relation (2), depends on the soft gluon energy-ordering.

The second property is the factorization structure

$$W_{ab}(1 \cdots n) = W_{ab}(\ell) \cdot W_{a\ell}(1 \cdots \ell{-}1)\, W_{\ell b}(\ell{+}1 \cdots n) \,, \tag{4}$$

with ℓ an arbitrary soft gluon. This branching structure allows the use of parton language. The colour-ordered distribution factorizes into the emission of ℓ off the primary ab-dipole with the remaining soft gluons emitted by two multi-dipole with "primary" partons $a\ell$ and ℓb.

The third property refers to the multi-soft-gluon distribution obtained in the same way [3] in the pure Yang-Mills case (primary dipole partons $p, \bar{p}$ are hard gluons and the Chan-Paton factor in (1) is a trace). One finds that the square colour amplitude obtained in the soft limit coincides with the square of the exact MHV amplitude obtained by Parke-Taylor [7].

The full distribution $|\mathcal{M}_n|^2$ is obtained by squaring (1). To leading order in N it is given by the sum of permutations of colour ordered distributions (3) (colour indices organized in the planar way). The non-planar contributions (subleading in N) are given by the product of two colour amplitudes for different permutations. They are easily computed from (2) for any fixed n, however there is not a close expression valid for arbitrary values of n. As a general feature, non-planar corrections depend in general on the energy ordering and are less collinear singular than the planar contribution (3), one of the collinear singularities cancels [3]. For these reasons, in studies in which collinear singularities are relevant, it is justified to neglect non-planar contributions. For studies of soft radiation at large angle in which collinear enhancements are not relevant neglecting non-planar contribution could be less justified.

In the following I schematically report results on physics of large angle soft emission. First, using the multi-soft gluon distributions in the planar approximation, I deduce the generating functional which provides a "complete" description of soft radiation emission. It can be formulated as a Markov process and then I describe the corresponding Monte Carlo simulation procedure. I describe the use of the generating functional to study the inclusive distribution of energy deposited in a region away from jets. This distribution takes contributions, to leading order, from successive soft gluon branching thus involves the full structure of the multi-soft gluon distribution. Finally I discuss distributions for which the planar approximation is not necessary. In particular I discuss the fifth form factor in hadron collision and the non-relativistic heavy quark multiplicity.

3. Generating functional (soft and planar limit)

Complete information on the emission of soft gluons $\{q_1 \dots q_n\}$ off the dipole $p_a p_b$ is contained in the generating functional

$$\Sigma_{ab}[Q,u] = \sum_n \frac{1}{n!} \int \frac{d\sigma_{ab}^{(n)}}{\sigma_{ab\,T}} \prod_{i=1}^{n} u(q_i)\,, \tag{5}$$

with Q the hard scale, σ_{ab} the dipole distributions and $u(q)$ the source which serves to specify the radiation observable considered. In the planar approximation the multi-soft gluon distribution $d\sigma_{ab}^{(n)}(q_1 \cdots q_n)$ is proportional to $W_{ab}(q_1 \cdots q_n)$. Taking the derivative with respect to Q (with energy ordered phase space) and using the branching factorization property (4) one obtains the evolution equation [8]

$$Q\partial_Q \Sigma_{ab}[Q,u] = \int \frac{d\Omega_q}{4\pi}\, \bar{\alpha}_s w_{ab}(q) \big\{ u(q)\, \Sigma_{aq}[Q,u] \cdot \Sigma_{qb}[Q,u] - \Sigma_{ab}[Q,u] \big\}, \tag{6}$$

with

$$\bar{\alpha}_s = \frac{N\,\alpha_s}{\pi}\,, \qquad w_{ab}(q) = \omega_q^2 \cdot W_{ab}(q) = \frac{\xi_{ab}}{\xi_{aq}\xi_{qb}}\,, \qquad \xi_{ij} = 1 - \cos\theta_{ij}\,. \tag{7}$$

The initial condition is set at a scale Q_0 corresponding to no emission, $\Sigma_{ab}[Q_0,u] = 1$. Virtual corrections (the last term in (6)) are included to satisfy unitarity $\Sigma[Q, u=1] = 1$. This realizes the real-virtual cancellation of the collinear singularity of $w_{ab}(q)$ for $\xi_{aq}\to 0$ or $\xi_{qb}\to 0$. This evolution equation reflects the factorization branching structure (4) of the multi-dipole distribution.

The generating functional (5) fully describes soft emission even at large angles and thus it is valid beyond the collinear singularity approximation. However is difficult to generalize the evolution equation (6) beyond soft approximation including non-soft emission and recoil.

In the next Sections I will report applications exploiting (6).

4. Large angle soft emission and Monte Carlo

The evolution equation (6) for Σ_{ab} can be solved numerically by Monte Carlo simulation. In the simulation one generates events with any number of soft gluons which can be used to compute any final state observables. First one needs to resum virtual corrections. To split the real and virtual terms in (6) one introduces a cutoff Q_0 in the dipole transverse momentum. The resummation of virtual corrections gives rise to the Sudakov form factor

$$\ln S_{ab}(Q,Q_0) = -\int_{Q_0}^{Q}\frac{d\omega_q}{\omega_q}\frac{d\Omega_q}{4\pi}\,\bar{\alpha}_s(q_{ab\,t})\,w_{ab}(q)\,\cdot\theta(q_{ab\,t}-Q_0)\,, \tag{8}$$

where $q_{ab\,t}^2 = 2\omega_q^2/w_{ab}(q)$. The evolution equation (6) can be written as

$$\Sigma_{ab}[Q,u] = S_{ab}(Q,Q_0) + \int_{Q_0}^{Q} dP_{ab}(q)\,u(q)\,\Sigma_{aq}[\omega_q,u]\cdot\Sigma_{qb}[\omega_q,u])\,, \tag{9}$$

with $dP_{ab}(q)$ the splitting probability which can be casted in the following factorized form

$$dP_{ab}(q) = d\left(\frac{S_{ab}(Q,Q_0)}{S_{ab}(\omega_q,Q_0)}\right)\cdot dR_{ab}(\Omega_q)\,, \qquad \omega_q < Q\,, \tag{10}$$

with

$$\frac{dR_{ab}(\Omega_q)}{d\Omega_q} = \frac{\bar{\alpha}_s(q_{ab\,t})\,w_{ab}(q)}{N_{ab}(\omega_q)}\cdot\theta(q_{ab\,t}-Q_0)\,, \qquad \int dR_{ab}(\Omega)=1\,. \tag{11}$$

From (9) one has that the Sudakov form factor $S_{ab}(Q,Q_0)$ can be interpreted as probability for no-emission of a soft gluon with dipole transverse momentum larger than Q_0.

Using this branching structure and the factorized form of the splitting probability (10) one can construct a Monte Carlo event generator (see also [9]). To generate an event (with resolution Q_0) one proceeds as follows. Starting from the ab-dipole with virtuality Q, one tries to emit a soft gluon q using (10). To this end one first sorts a random number $0<r<1$:

- for $r < S(Q, Q_0)$ the dipole does not branch (with Q_0 cutoff);
- for $r > S(Q, Q_0)$ a soft gluon is emitted with energy ω_q obtained by solving

$$S_{ab}(\omega_q, Q_0) \cdot r = S_{ab}(Q, Q_0)\,. \tag{12}$$

Then, using (11), one finds the direction Ω_q. If the emission takes place, one has to deal with the two dipoles aq and qb which are further emitting soft gluons with energy smaller than ω_q. The procedure is repeated for each dipole till no dipole is emitting (withing the cutoff Q_0). Such an event generator can be used to study radiation quantities in the leading soft limit. Due to strong ordering, no recoil is considered. To make a realistic simulation valid beyond soft limit one needs to include parton recoil, full collinear structure and quark branching. Finally hadronization had to be accounted for.

A way to account for non-soft emission consists in taking the collinear limit of (3) and one obtains [3] the evolution equation in angles (instead of in energy). Using angular evolution one can easily introduce recoil, the full splitting functions (including quarks and the full N dependence) and initial state emission. In the angular evolution large angle soft emission are accounted for only in an average sense (coherence through angular ordering). The angular evolution is the basis for the Monte Carlo simulations [10] presently used. There is a continuous upgrading of these Monte Carlo simulations [11]. It would be important to match the angular evolution with the one which fully accounts for large angle soft emission as given in (6).

5. Large angle soft emission and jet characteristics

Measuring final state characteristics in hard interactions supplements the overall hardness scale Q with the second scale $Q_0 \ll Q$ that quantifies small deviation of the final state system from the Born kinematics. The ratio of these two scales being a large parameter calls for analysis and resummation of double (DL) and single logarithmic (SL) radiative corrections in all orders. Logarithmically enhanced (both DL and SL) contributions of *collinear* origin take contributions from bremsstrahlung radiation off primary hard partons and are resummed into exponential Sudakov form factors. SL effects due to soft gluons radiated at large angles, when present, pose more problems. In the following I discuss two cases in which these SL contributions are actually present.

The first case refers to the so called *non-global observables* that acquire contributions from a restricted phase space window. Examples are particle energy flow $E = Q_0$ in a given inter-jet direction, Sterman-Weinberg distribution (energy in a cone), photon isolation, rapidity cuts in DIS, hadron-hadron inter jet string/drag effects, profile of a separate jet (current hemisphere).

The second case consists in measuring final state characteristics in hard hadron–hadron interactions.

5.1. *Non-global logarithms*

Non-global observables acquire SL contributions from ensembles of n-energy ordered gluons radiated at arbitrary (large) angles which are resummed [12], at the planar level, by the soft gluon evolution equation (6). The simplest case is the study in e^+e^- of the radiation emitted in the region "out" away from the thrust axis as shown here

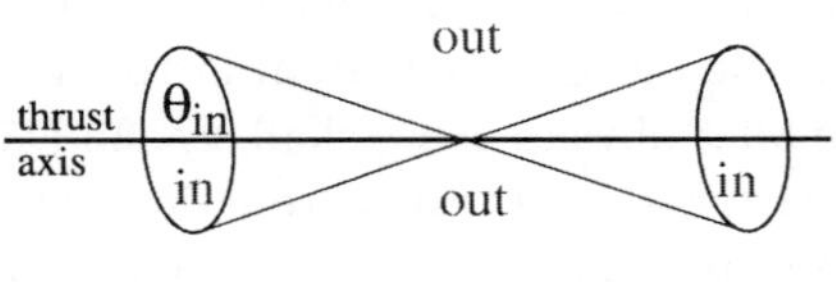

Most of emission is contributing to the two jets around the thrust axis. Therefore, to study the distribution in energy E_{out} away from jet

$$\Sigma_{\text{out}}(Q, E_{\text{out}}) = \sum_n \int \frac{d\sigma_n}{\sigma_T}\, \Theta\Big(E_{\text{out}} - \sum_{\text{out}} q_{ti}\Big) \tag{13}$$

one needs a careful description of soft gluons emitted at large angle, that is the transverse momentum evolution equation (6). Generalizing to a the emission off a general ab-dipole and specializing the soft gluon source $u(q)$ to this specific observable, one obtains [8]

$$\partial_\tau \Sigma_{ab}(\tau) = -(\partial_\tau R_{ab})\cdot\Sigma_{ab}(\tau) + \int_{\text{in}} \frac{d\Omega_q}{4\pi}\, w_{ab}(q)\,[\Sigma_{aq}(\tau)\cdot\Sigma_{qb}(\tau) - \Sigma_{ab}(\tau)]\,, \tag{14}$$

where, at SL accuracy, the distribution $\Sigma_{ab}(\tau)$ is a function of the two scales Q and E_{out} via the logarithmic integral of the running coupling

$$\tau = \int_{Q_0}^{Q} \frac{dq_t}{q_t} \bar{\alpha}_s(q_t)\,, \qquad Q_0 = E_{\text{out}}\,, \tag{15}$$

and $R_{ab}(\tau)$ the ab-radiator evaluated in the region away from the jets (region "out")

$$R_{ab}(\tau) = \int_{E_{\text{out}}}^{Q} \frac{dq_t}{q_t} \bar{\alpha}_s(q_t) \int_{\text{out}} \frac{d\Omega_q}{4\pi}\, w_{ab}(q) \simeq 2\tau \ln \frac{2}{\theta_{\text{in}}}\,. \tag{16}$$

In (14) the radiator term corresponds to bremsstrahlung emission while the integral term corresponds to the soft gluon branching which takes place inside the jet region (region "in") and gives rise to "non-global logs". The solution of (14) has the form

$$\Sigma_{ab}(\tau) = e^{-R_{ab}(\tau)} \cdot C_{ab}(\tau). \tag{17}$$

The first is the Sudakov form factor. The second factor C_{ab} comes from the branching inside the jet region[a]. Collinear singularities in $w_{ab}(q)$ partially cancel between real and virtual contributions. At large τ virtual corrections are overwhelming real contributions and one finds the Gaussian behaviour

$$C_{ab}(\tau) \sim e^{-\frac{c}{2}\tau^2}, \qquad c = 4.883... \tag{18}$$

The asymptotic beheviour (18) is independent on the geometry of the recorded region away from the jet. The value of the constant c is related to universal features of the evolution equation in the asymptotic regime and enters also in the small x physics as we shall see. In (17) the Sudakov form factor is negligible with respect to the correlation function at large τ.

Berger, Kúcs and Sterman [13] have formulated a programme of how to avoid non-global logarithms in a measurement of transverse energy flow away from jets. They suggested to squeeze the jets by taking the thrust T close to one and thus suppress multi-gluon branching inside the jet region. They introduced the *jet/shape correlation* $\Sigma_{fs}(Q, V, E_{\rm out})$ in the two variables, the energy-flow $E_{\rm out}$ and the global jet-shape V, for instance $V = 1 - T$. They showed that for $V \sim E_{\rm out}/Q \ll 1$ the flow/shape correlation reduces to the global jet-shape distribution $\Sigma(Q, V)$. This result has been extended [14] for general (small) values V and $E_{\rm out}/Q$. At SL accuracy one finds the factorized result

$$\Sigma_{fs}(Q, V, E_{\rm out}) \simeq \Sigma(Q, V) \cdot \Sigma_{\rm out}(VQ, E_{\rm out}), \tag{19}$$

so that the associated measure is equivalent to the two independent ones, the global one $\Sigma(Q, V)$ at the scale at Q and the non-global one at VQ.

Analysis of non-global observables can be generalized (in planar approximation) to hard hadron-hadron collisions. In planar approximation, the hard square matrix element could be expressed as a superposition of dipoles and each ab-dipole contributes with the distribution $\Sigma_{ab}(\tau)$ in (17). In hadron collision, contributions beyond planar approximations have been

[a]The integral term in (14), and then the correlation C_{ab}, is absent for global observables which involve the full phase space.

obtained only for global jet-shape observables as we shall discuss in the following.

5.2. *Hadron interaction and the fifth form factor*

We discuss now global jet characteristics in hard hadron–hadron interaction with the underlying parton scattering process $p_1, p_2 \to p_3, p_4$ at a scale Q. Examples are out-of-event-plane energy production, near-to-backward particle correlations, inter-jet energy flows, etc. They are characterized by a scale $Q_0 \ll Q$.

For global observables all enhanced contributions at DL and SL level comes from bremsstrahlung radiation off the four primary partons. This implies that the problem reduces to the analysis of *virtual corrections* due to multiple gluons with $q_t > Q_0$ attached to primary hard partons only. They can be treated iteratively and fully exponentiated.

In addition to the standard collinear contributions leading to Sudakov for factors associated to the four primary partons p_i, one has SL enhancements coming from large angle soft emission. This is due to the fact that soft gluon emission changes the *colour state* of the hard parton system which in turn affects successive radiation of a softer gluon. Resummation of these soft logarithms generates a fifth form factor in addition to the Sudakov form factors.

The programme of resumming soft SL effects due to large angle gluon emission in hadron–hadron collisions was pioneered by Botts and Sterman [15] and discussed in general also in [16]. The treatment for finite N of large angle gluon radiation in global final state characteristics is based on the observation that the square of the eikonal current for emission of a soft gluon q off the ensemble of four hard partons p_i can be represented as

$$
\begin{aligned}
-j^2(q) = T_1^2\, W_{34}^{(1)}(q) + T_2^2\, W_{34}^{(2)}(q) + T_3^2\, W_{12}^{(3)}(q) + T_4^2\, W_{12}^{(4)}(q) \\
+\; T_t^2 \cdot A_t(q) \;\; + T_u^2 \cdot A_u(q) \,.
\end{aligned}
\tag{20}
$$

Here T_i^2 is the $SU(N)$ "colour charge" of parton p_i and the two operators $T_t^2 = (T_1 - T_3)^2$ and $T_u^2 = (T_1 - T_4)^2$ are the charges exchanged in the t- and u-channels. In the first line one has the combination $W_{ab}^{(c)} = w_{ac} + w_{bc} - w_{ab}$ with the singular integral

$$
\int \frac{d\Omega}{4\pi}\, W_{ab}^{(c)} \;=\; \ln \frac{Q^2}{2m^2}\,, \qquad Q^2 = \frac{tu}{s} = s \sin^2 \Theta_s\,, \tag{21}
$$

with $s = (p_1 + p_2)^2$, $t = (p_1 - p_3)^2$ and $u = (p_1 - p_4)^2$. The distribution $W_{ab}^{(c)}$ is collinear singular *only* when q is parallel to p_c Here m^2 is a collinear cutoff which is replaced by the proper observable dependent scale Q_0 when real and virtual contributions are taken together. Its exponentiation leads to the Sudakov form factors $F_c(Q_0, Q)$ with colour charge of parton c and common hard scale Q in (21).

The last two terms in (20) are collinear regular

$$A_t = w_{12} + w_{34} - w_{13} - w_{24}\,, \qquad \int \frac{d\Omega}{4\pi} A_t(q) \;=\; 2 \ln \frac{s}{-t}\,, \tag{22}$$

and similarly for A_u. Non-Abelian Coulomb corrections due to virtual gluons exchanges between the two incoming or two outgoing partons can be simply incorporated by replacing $-t, -u$ with t, u in (22). Contrary to the first four DL contributions, this additional contribution originates from coherent gluon radiation at angles *larger than the cms scattering angle* Θ_s. It gives rise to the *fifth form factor* which is a single logarithmic function of τ given in (15) with $E_{\rm out}$ replaced by Q_0.

In summary, the two scale distributions ais given by

$$\Sigma(Q_0, Q) \;=\; \Sigma^{\rm coll}(Q_0, Q) \cdot \mathcal{S}_X(\tau), \tag{23}$$

Here $\Sigma^{\rm coll}$ embodies the first four terms in (20) as well as collinear logarithms from parton distribution functions. $\mathcal{S}_X(\tau)$ is the *soft factor* coming from the fifth form factor.

In the following I illustrate the solution in some relevant special cases for gluon–gluon scattering in $SU(N)$ that was first treated in [17]. The answer for the soft factor $\mathcal{S}_X$ is expressed in terms of the following suppression factors

$$\chi_t(\tau) \;=\; \exp\left\{-2N\tau \cdot \ln \frac{s}{-t}\right\}, \quad \chi_u(\tau) \;=\; \exp\left\{-2N\tau \cdot \ln \frac{s}{-u}\right\}. \tag{24}$$

Scattering at 90°. In this kinematical configuration ($t = u = -s/2$) one has

$$\mathcal{S}_X(\tau) = \frac{\chi^2}{3}\left[\frac{4\,\chi^2}{N^2-1} + \chi + \frac{N-3}{N-1}\,\chi^{\frac{2}{N}} + \frac{N+3}{N+1}\,\chi^{-\frac{2}{N}}\right], \quad \chi(\tau) \;=\; e^{-2N\tau \ln 2}\,. \tag{25}$$

$N \to \infty$ limit. The soft factor becomes

$$\mathcal{S}_X(\tau) \;=\; \chi_t\,\chi_u\,\frac{(m_t + m_u)^2 + (m_s - m_u)^2\,\chi_t + (m_s + m_t)^2\,\chi_u}{(m_t + m_u)^2 + (m_s - m_u)^2 + (m_s + m_t)^2}\,. \tag{26}$$

Here m_s, m_t, m_u are pieces of the Born gg scattering matrix element each containing the gluon exchange diagram in the corresponding channel (together with the piece of the four-gluon vertex with the same colour structure). One has

$$(m_t + m_u)^2 \;=\; 1 - \frac{st}{u^2} - \frac{us}{t^2} + \frac{s^2}{tu},$$

with the other two obtained by crossing symmetry.

Regge limit In the case of close to forward scattering, $|t| \ll s \simeq |u|$. Since for $t \to 0$ scattering in the Born approximation is dominated by t-channel one gluon exchange, we are left with

$$S(\tau) \;=\; \chi_t(\tau) \;=\; \left(\frac{s}{t}\right)^{-2N\tau}, \tag{27}$$

which exponent coincides with the (twice) Regge trajectory of the gluon.

The eigenvalues of the anomalous dimension matrix possesses a weird symmetry which interchanges of internal (colour group) and external (scattering angle) degrees of freedom:

$$\frac{\ln s^2/tu}{\ln t/u} \quad\Longleftrightarrow\quad N\,. \tag{28}$$

In particular, this symmetry relates 90-degree scattering, $t = u$, with the large-N limit of the theory. Giving the complexity of the expressions involved, such a symmetry being accidental looks highly improbable. Its origin remains mysterious and may points at existence of an enveloping theoretical context that correlates internal and external variables (string theory?).

6. Is there a BFKL Pomeron in jet emission?

It is surprising that large angle soft emission in jet physics has some relation with high-energy scattering and BFKL dynamics. This in spite of the difference in the relevant multi-soft gluon kinematical configurations. In the following I recall two examples in which the two dynamics are (formally) related.

Kovchegov equation. The evolution equation (14) for the jet observable $\Sigma_{ab}(\tau)$ has some resemblance with the Kovchegov equation. To make the resemblance more striking, take a ab-dipole with small angle $\theta_{ab} \ll 1$ and introduce the two-dimensional "angular" variable $\vec{\theta}_{ab}$ and $\Sigma_{ab}(\tau) = \Sigma(\tau, \vec{\theta}_{ab})$. The evolution equation (6) becomes

$$\partial_\tau \Sigma(\tau, \vec{\theta}) = \int \frac{d^2\theta'}{2\pi} \frac{\theta'^2}{\theta'^2 (\vec{\theta} - \vec{\theta}')^2} \Big[\Sigma(\tau, \vec{\theta}')\Sigma(\tau, \vec{\theta}' - \vec{\theta}) - \Sigma(\tau, \vec{\theta})\Big]. \tag{29}$$

Here also the angular integration has been replaced by the two-dimensional integration over $\vec{\theta}'$. One recognizes that (29) formally coincides with the Kovchegov equation for elastic S-matrix at high energy, $S(\tau, \vec{x})$ where $\vec{x}$ is the impact parameter and $\tau = N\alpha_s/\pi Y$ with Y the rapidity and $\bar{\alpha}_s$ the fixed coupling. For large τ one has (see the asymptotic behaviour of the correlation in (18))

$$\Sigma(\tau, \vec{\theta}) \sim e^{-\frac{c}{2}\tau^2}, \quad \text{for } \theta > \theta_{\rm crit}(\tau) \sim e^{-\frac{c}{2}\tau}, \quad c = 4.883... \tag{30}$$

so that virtual corrections are overwhelming above $\theta_{\rm crit}$ (buffer region for the jet distribution [12]). The asymptotic behaviour (30) corresponds, in the case of the Kovchegov equation, to the saturation for the S-matrix with $1/\theta_{\rm crit}(\tau) \to Q_s(\tau)$, the saturation scale. The parameter c is determined [18] by the characteristic function of the BFKL equation which is the linear limit of (29), as I shall discuss next.

BFKL equation. An other case of a formal connection between high energy and jet physics is encountered [19] considering the multiplicity of non-relativistic heavy $Q\bar{Q}$ pair of total mass $\mathcal{M}$ emitted by a ab-dipole. This observable $N(\rho_{ab}, \tau)$, with $\rho_{ab} = (1 - \cos\theta_{ab})/2$ and τ given in (15) for Q_0 replaced by $\mathcal{M}$, satisfies the linearized version of (6) which, after azimuthal integration, is given by

$$\partial_\tau N(\rho,\tau) = \int_0^1 d\eta \frac{(\eta^{-1}N(\eta\rho,\tau) - N(\rho,\tau))}{1-\eta} + \int_\rho^1 d\eta \frac{(N(\eta^{-1}\rho,\tau) - N(\rho,\tau))}{1-\eta}. \tag{31}$$

The only formal difference between (31) and the BFKL equation for the T-matrix at high energy is the presence of the lower limit $\eta > \rho$ in the second integral which ensures that the argument of $N(\rho/\eta, \tau)$ remains within the physical region $\rho/\eta < 1$. We have[b]

$$N(\rho,\tau) \sim e^{\tau \cdot 4\ln 2} \cdot \frac{e^{-\ln^2\rho/2D\tau}}{\tau\sqrt{2\pi D\tau}\ln\rho}. \tag{32}$$

Here $4\ln 2$ and $D = 28\zeta(3)$ are related to the BFKL characteristic function. The difference of this asymptotic behaviour with the one for the T-matrix

[b] The well known result for the multiplicity in the collinear limit [3,4] is obtained by taking the collinear piece in (31) giving (for fixed α_s)

$$\partial_\tau N_{\rm coll}(\rho,\tau) = \int_0^\rho \frac{d\rho'}{\rho'} N_{\rm coll}(\rho',\tau), \qquad N_{\rm coll}(Q) = N_0 \exp\sqrt{2\bar{\alpha}_s}\,\ln Q/Q_0,$$

with $Q^2 = E^2\rho$ and Q_0 a collinear cutoff.

in BFKL equation is given by $1/\tau$ in the prefactor which is the result of the fact that ρ is an angular variable with constraint $\rho < 1$. For details on the jet physics solution and explanation on the origin of the additional $1/\tau$ suppression pre-asymptotic factor in $N(\rho, \tau)$, see [19].

The basis for the relation between the equations in jet-physics ((29), (31)) and the Kovchegov and BFKL ones in high-energy scattering is of course the (same) multi-soft gluon-distribution. However the dominant contributions in the jet and high energy case come from very different kinematical configurations as we shall discuss now.

Jet-physics case. Here all angles of emitted gluons are of same order. This is due to the fact that the observables considered in this Section do not contain collinear singularities. Moreover, in the (leading) infrared limit soft gluon energies are ordered so that also the emitted transverse momenta are ordered. The ordered variable q_t enters the argument of the running coupling so that τ is never too large.

High-energy scattering case. Here all intermediate soft gluon transverse momenta are of same order (no singularities for vanishing transverse momentum differences). On the other hand, energy ordering implies in this case that intermediate gluon angles are ordered. Contrary to the previous case, the running coupling depends on transverse momenta which are all of same order. Therefore, in first approximation, one can take α_s fixed and so that τ becomes large at high energy.

7. Concluding remarks

I recalled recent results in the studies of soft gluon emission at large angles (hedgehog configurations). Here I present some of the possible developments of these studies.

Previous studies on non-perturbative corrections (power corrections) to perturbative results has been analysed for quantities involving emission in the collinear regions. It would be important to study power corrections for observables involving hedgehog soft gluon configurations (without collinear enhancements) in which takes place neutralization of colour between jets takes place.

Present Monte Carlo simulations involves contributions of large angle soft emission only in an approximate sense (angular ordering). It would be important to include exact contributions from hedgehog configuration. This requires to merge the Monte Carlo simulation here described with the one presently used.

Multi-soft gluon distributions $|\mathcal{M}_n|^2$ are known for arbitrary n only in the planar approximation. It would be important to have a close expression of the distribution for any n and arbitrary N or at least the subleading $1/N$ corrections. They are specially important for a correct description of hedgehog configurations.

Similarities and differences in the dynamics of high energy scattering and jet-physics (with non-global logs) should be further exploited in order to have new insights in both fields.

References

1. V.N. Gribov, *Gauge theories and quark confinement*, Phasis, Moskow 2002.
2. V.N. Gribov, Yad. Fiz. **5** (1967) 399 [Sov. J. Nucl. Phys. **5** (1967) 280].
3. A. Bassetto, M. Ciafaloni, and G. Marchesini, Phys. Rep. **100** (1983) 201; F. Fiorani, G. Marchesini, and L. Reina, Nucl. Phys. **B 309** (1988) 439.
4. Yu.L. Dokshitzer, V.A. Khoze, A.H. Mueller, and S.I. Troian, *Basics of perturbative QCD*, Gif-sur-Yvette, France: Ed. Frontieres (1991).
5. S. Catani, Phys. Lett. **B 427** (1998) 161
6. Y.L. Dokshitzer, G. Marchesini, and G. Oriani, Nucl. Phys. **B 387** (1992) 675.
7. S. J. Parke and T. R. Taylor, Phys. Rev. Lett. **56** (1986) 2459.
8. A. Banfi, G. Marchesini, and G. Smye, J. High Energy Phys. **08** (2002) 006.
9. L. Lonnblad, Comp. Phys. Commun. **71** (1992) 15.
10. G. Marchesini and B.R. Webber, Nucl. Phys. **B 238** (1984) 1; T. Sjostrand, Comp. Phys. Commun. **82** (1994) 74.
11. S.Frixione, P.Nason, and B.R. Webber, J. High Energy Phys. **08** (2003) 007.
12. M. Dasgupta and G.P. Salam, Phys. Lett. **B 512** (2001) 323 and J. High Energy Phys. **03** (2002) 017.
13. C.F. Berger, T. Kucs, and G. Sterman, Phys. Rev. **D 68** (2003) 014012.
14. Yu.L. Dokshitzer and G. Marchesini, J. High Energy Phys. **03** (2003) 040.
15. J. Botts and G. Sterman, Nucl. Phys. **B 325** (1989) 62.
16. R. Bonciani, S. Catani, M.L. Mangano, and P. Nason, Phys. Lett. **B 575** (2003) 268;
    A. Banfi, G.P. Salam, and G. Zanderighi, Phys. Lett. **B 584** (2004) 298.
17. N. Kidonaqis, G. Oderda and G. Sterman, Nucl. Phys. **B 531** (1998) 365.
18. A.H. Mueller and D.N. Triantafyllopoulos, Nucl. Phys. **B 640** (2002) 331.
19. G. Marchesini and A.H. Mueller, Phys. Lett. **B 575** (2003) 37; see also G. Marchesini and E. Onofri, J. High Energy Phys. **07** (2004) 031.

WHEN PARTONS AREN'T PARTICLES: THE COLLINS MECHANISM FOR SINGLE SPIN ASYMMETRIES

ELLIOT LEADER

Imperial College
Prince Consort Road
London SW7 2BW, UK.
E-mail: e.leader@imperial.ac.uk

The large single-spin asymmetries found in hard semi-inclusive reactions involving a transversely polarized nucleon remained unexplained for several decades, until the advent of the Collins mechanism, which provides a non-perturbative explanation. However, the mechanism vanishes in the naive parton model, where a quark is regarded as an essentially "free" particle. We give an intuitive analysis which highlights the difference between the naive picture and the realistic one and show how the Collins mechanism arises when the quark is described as an off-shell particle by a field in interaction.

1. Introduction

If I were presenting this as a talk I would have begun by saying that the majority of particle physicists still think about partons as in the pre-QCD picture, as little point-like almost free particles. At this juncture, if Volodya Gribov had been in the audience, he would have shaken his head vigorously and interrupted with a scathing remark, something like "Not me! I've certainly never had such a naive viewpoint", and, of course, that would have been true, because Volodya had such a remarkable ability to avoid superficial explanations and to get to the really important and fundamental aspects of any physical problem.

The Collins mechanism [1] provides a possible explanation of the very large single-spin asymmetries – sometimes as big as 40% – found in semi-inclusive reactions like $pp^{\uparrow} \to \pi X$ or $ep^{\uparrow} \to e\pi X$, where $p^{\uparrow}$ is a transversely polarized proton. The existence of these asymmetries has been known for decades and one of the great puzzles has been the fact that it is quite impossible to produce such large asymmetries via perturbative QCD mechanisms. The Collins mechanism is a "soft" non-perturbative effect describing the fragmentation of a transversely polarized quark into e.g. $\pi + X$. It makes the angular distribution of the pion with respect to the direction of motion of the quark depend upon the direction of the quark's transverse polarization. But this feature is itself puzzling, since it is well known [2,3] that in the decay of a transversely polarized particle or resonance $R^\uparrow \to \pi + X$ the angular distribution $W(\theta, \phi)$ of the pion cannot depend on the direction of the transverse polarization of R , in any parity conserving theory. Being aware of this, and, I must confess, being one of those who tended to think of partons as "particles", I was for a long time convinced that the Collins mechanism cannot exist. Indeed, as we shall see explicitly, the Collins mechanism does vanish in the naive parton model where the quark is visualized as a "free" particle, but the Collins effect can exist when care is exercised in treating the quark as an off-shell object described by an interacting field.

In this paper we give an intuitive analysis of the Collins mechanism based upon rotational invariance, which makes clear the difference between particle or resonance decay, which *cannot* have a Collins effect, and quark fragmentation, which can.

In Section 2 we revisit the standard QCD picture for hard semi-inclusive reactions, stressing the difference between treating the active quark as "free" and as an off-shell interacting particle. Section 3 derives the angular distribution, contrasting the cases of "free" and interacting quarks. The results are discussed in Section 4.

2. QCD Analysis of $ep^\uparrow \to e\pi X$

For concreteness we shall present our analysis in the context of semi-inclusive deep inelastic lepton-proton scattering (see the Figure below) in which an unpolarized lepton interacts with a transversely polarized proton and the final state pion is detected at angles (θ, ϕ) in the "γ"p CM frame, with the proton moving along the OZ axis.

Regarding the Figure as a Feynman diagram, the amplitude for the hadronic part of the process is given by

$$H^\mu = F_\beta (\gamma^\mu)_{\beta\alpha} A_\alpha \tag{1}$$

where A_α is the amplitude for the proton, momentum P, to emit a quark with four-momentum p and Dirac index α, and F_β is the amplitude for a quark of four-momentum k and Dirac index β to fragment into a pion and some other particle X. As is usual in the parton model, both A_α and F_β are defined to include their adjacent quark propagators.

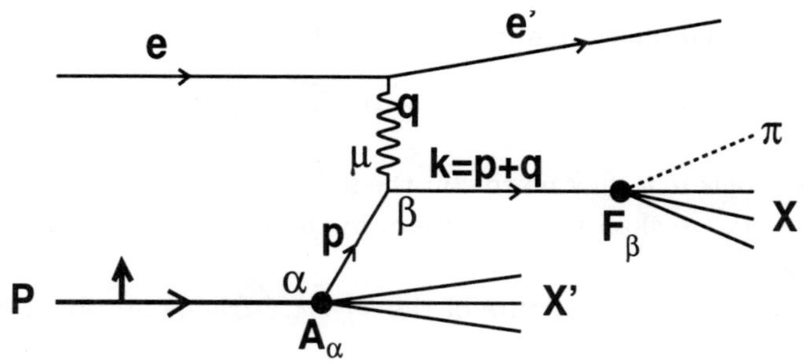

Figure 1. Feynman diagram for semi-inclusive scattering.

In QCD these amplitudes are given by:

$$A_\alpha = \int d^4x e^{ip.x} < X' \mid \Psi_\alpha(x) \mid P > \tag{2}$$

$$F_\beta = \int d^4z e^{-ik.z} < \pi, X \mid \bar{\Psi}_\beta(z) \mid 0 > \tag{3}$$

where k_μ is a time-like four-vector, $k^0 > 0$, and the hadronic tensor $W^{\mu\nu}$ is then

$$W^{\mu\nu} = \sum_{X,X'} H^\mu H^{\nu *} \tag{4}$$

Note that the πX final state is, strictly speaking, an *out* state, but this is irrelevant in our analysis.

In these and all further expressions we are only interested in the general structure of the amplitudes, so we will ignore numerical and other inessential factors. Also, since we deal with unpolarized leptons, we are only concerned with the part of $W^{\mu\nu}$ symmetric under $\mu \leftrightarrow \nu$.

Substituting Eq. (1) into Eq. (4) we obtain

$$W^{\mu\nu} = \rho^{\mu\nu}_{\alpha\beta} \sum_X F_\alpha \bar{F}_\beta \tag{5}$$

where, as usual,

$$\bar{F}_\beta = F^*_\tau \gamma^0_{\tau\beta} \tag{6}$$

and

$$\rho^{\mu\nu}_{\alpha\beta} = (\gamma^\mu \Phi \gamma^\nu)_{\alpha\beta} \tag{7}$$

with Φ the usual quark-quark correlation function

$$\Phi_{\alpha\beta} = \int d^4x e^{-ip.x} < P \mid \bar{\Psi}_\beta(x)\Psi_\alpha(0) \mid P > \tag{8}$$

Since the labels μ, ν play no role in the following, aside from the symmetry under $\mu \leftrightarrow \nu$, we shall suppress them in the following, simply writing the LHS of Eq. (7) as $\rho_{\alpha\beta}$.

Consider now the structure of the $q \to \pi X$ fragmentation amplitude

$$\begin{aligned} F_\beta &\equiv F_\beta(p_\pi; X, \lambda_X; k) \\ &= \int d^4z e^{-ik.z} < p_\pi; X, \lambda_X \mid \bar{\Psi}_\beta(z) \mid 0 > \end{aligned} \tag{9}$$

where λ_X is the helicity of X.

At this point it is important to stress the difference between the case of a free field and an interacting one.

If $\bar{\Psi}(z)$ were a free field $\bar{\Psi}^0(z)$ we could make the usual Fourier expansion into creation and annihilation operators

$$\begin{aligned} \bar{\Psi}^0_\beta(z) \sim \sum_\lambda \int \frac{d^3p}{E_p} \{ &\bar{u}_\beta(\boldsymbol{p}, \lambda) a_0^\dagger(\boldsymbol{p}, \lambda) e^{ip.z} \\ &+ \bar{v}_\beta(\boldsymbol{p}, \lambda) b_0(\boldsymbol{p}, \lambda) e^{-ip.z} \} \end{aligned} \tag{10}$$

where $E_p = \sqrt{\boldsymbol{p}^2 + m^2}$, and the subscript 0 on the operators signifies free field operators.

When $\bar{\Psi}(z)$ is a field in interaction we can still make a 3-dimensional Fourier expansion at some instant of time, say $t = 0$. Then one finds [4], at arbitrary times t,

$$\begin{aligned} \bar{\Psi}_\beta(z) \sim \sum_\lambda \int \frac{d^3p}{E_p} \{ &\bar{u}_\beta(\boldsymbol{p}, \lambda) a^\dagger(\boldsymbol{p}, \lambda; t) e^{-i\boldsymbol{p}.\boldsymbol{z}} \\ &+ \bar{v}_\beta(\boldsymbol{p}, \lambda) b(\boldsymbol{p}, \lambda; t) e^{i\boldsymbol{p}.\boldsymbol{z}} \} \end{aligned} \tag{11}$$

where now the time dependence is non-trivial, being controlled by the Hamiltonian H:

$$\begin{aligned} a^\dagger(\boldsymbol{p}, \lambda; t) &= e^{iHt} a^\dagger(\boldsymbol{p}, \lambda) e^{-iHt} \\ b(\boldsymbol{p}, \lambda; t) &= e^{iHt} b(\boldsymbol{p}, \lambda) e^{-iHt} \end{aligned} \tag{12}$$

This is to be contrasted with the free field case where

$$\begin{aligned} a_0^\dagger(\boldsymbol{p}, \lambda; t) &= e^{iE_p t} a_0^\dagger(\boldsymbol{p}, \lambda) \\ b_0(\boldsymbol{p}, \lambda; t) &= e^{-iE_p t} b_0(\boldsymbol{p}, \lambda) \end{aligned} \tag{13}$$

Of course the interacting $a^\dagger(\boldsymbol{p},\lambda;t)$ and $b(\boldsymbol{p},\lambda;t)$ are not simply one-particle creation and annihilation operators. However, the spinors occurring in Eqs. (10) and (11) are the same as in the free case i.e. are mass m spinors.

Bearing in mind Eq. (11) we have for the field operator appearing in Eqs. (3) and (9)

$$\begin{aligned}\bar{\Psi}_\beta(k) \equiv \int d^4 z e^{-ik.z}\bar{\Psi}_\beta(z) = \frac{1}{E_k}\sum_\lambda \{&\bar{u}_\beta(\boldsymbol{k},\lambda)A^\dagger(k,\lambda) \\ &+ \bar{v}_\beta(-\boldsymbol{k},\lambda)B(k,\lambda)\}\end{aligned} \tag{14}$$

where

$$\begin{aligned} A^\dagger(k,\lambda) &\equiv \int dt e^{-ik_0 t} a^\dagger(\boldsymbol{k},\lambda;t) \\ B(k,\lambda) &\equiv \int dt e^{-ik_0 t} b(-\boldsymbol{k},\lambda;t)\end{aligned} \tag{15}$$

As stressed earlier, these are not single particle creation and annihilation operators. However, and this will be crucial in the later argument, acting on the vacuum state $| \, 0 >$ they produce states which can be shown to be eigenstates of momentum and energy with definite helicity and definite parity. Thus we shall define

$$| \, k^\mu, \lambda)_+ \equiv A^\dagger(k,\lambda) \, | \, 0 > \tag{16}$$

and

$$| \, k^\mu, \lambda)_- \equiv B(k,\lambda) \, | \, 0 > \tag{17}$$

and the key point will be the fact that the states in Eqs. (16) and (17) have *opposite* parity, as indicated by the labels +/-. We use the notation $| \quad)$ to emphasize the fact that these states are not necessarily one particle states. Hence we may write Eq. (9) in the form

$$\begin{aligned} F_\beta(p_\pi; X, \lambda_X; k) = \frac{1}{E_k}\sum_\lambda \{&\bar{u}_\beta(\boldsymbol{k},\lambda) < p_\pi; X, \lambda_X \mid k^\mu, \lambda)_+ \\ &\bar{v}_\beta(-\boldsymbol{k},\lambda) < p_\pi; X, \lambda_X \mid k^\mu, \lambda)_-\}\end{aligned} \tag{18}$$

This is to be contrasted with the expression one obtains for the decay of a *particle* of momentum $\boldsymbol{k}$ (or for the case of a free quark field) namely

$$F^0_\beta(p_\pi; X, \lambda_X; k) = \frac{1}{E_k}\sum_\lambda \bar{u}_\beta(\boldsymbol{k},\lambda) < p_\pi; X, \lambda_X \mid k, \lambda > \tag{19}$$

The analogue of the second term on the RHS of Eq.(18) is here absent, because, via Eq. (10), it involves

$$\int dte^{-ik_0t}e^{-iE_pt} \propto \delta(k_0 + E_p) = 0 \tag{20}$$

since $k_0 > 0$.

3. The Hadron Angular Distribution: Free Particle *vs* Interacting Quark

It will suffice to consider the decay

$$\text{“}q\text{”} \longrightarrow \pi + X$$

in the CM system of the π and X i.e. where $\boldsymbol{p}_\pi + \boldsymbol{p}_X = 0$, since the angular distribution in any other frame can be derived using the properties of the states $\mid k, \lambda)_{+/-}$.

3.1. *Free Particle Decay*

Let us look first at the decay of a spin1/2 particle (equivalently, the case of a “free” quark) where Eq. (19) holds. From Eq. (5) and (19)

$$W^{\mu\nu} \propto \rho^{uu}_{\lambda'\lambda} \sum_X < p_\pi; X, \lambda_X \mid k, \lambda >< p_\pi; X, \lambda_X \mid k, \lambda' >^* \tag{21}$$

where

$$\rho^{uu}_{\lambda'\lambda} = \bar{u}_\alpha(k, \lambda')\rho_{\alpha\beta}u_\beta(k, \lambda) \tag{22}$$

Rotational and parity invariance allow us to write, for the matrix element in Eq. (21), in the frame where $\boldsymbol{p}_\pi = -\boldsymbol{p}_X$

$$< p_\pi; X, \lambda_X \mid k, \lambda >= M(\lambda_X)e^{i\lambda\phi}d^{1/2}_{\lambda\lambda_X}(\theta) \tag{23}$$

where (θ, ϕ) are the polar angles of $\boldsymbol{p}_\pi$ and the reduced matrix element $M(\lambda_X)$ is independent of λ [3], and

$$M(-\lambda_X) = \eta M(\lambda_X) \qquad (\eta = \pm 1)\ . \tag{24}$$

Then Eq. (21) becomes

$$W^{\mu\nu} \propto \rho^{uu}_{\lambda'\lambda}e^{i\phi(\lambda'-\lambda)} \sum_{\lambda_X} \mid M(\lambda_X) \mid^2 d^{1/2}_{\lambda'\lambda_X}(\theta)d^{1/2}_{\lambda\lambda_X}(\theta)\ . \tag{25}$$

For a spin-1/2 particle moving along OZ the dependence on the transverse polarization (P_x, P_y) is contained in the off diagonal helicity matrix elements

$$\rho^{uu}_{+-} = \rho^{uu*}_{-+} = \frac{1}{2}(P_x - iP_y) \tag{26}$$

The contribution from ρ^{uu}_{+-} to Eq. (25) is

$$\rho^{uu}_{+-} e^{i\phi} \{| M(1/2) |^2 \, d^{1/2}_{1/2,1/2}(\theta) d^{1/2}_{-1/2,1/2}(\theta) + | M(-1/2) |^2 \, d^{1/2}_{1/2,-1/2}(\theta) d^{1/2}_{-1/2,-1/2}(\theta)\} \tag{27}$$

$$= \rho^{uu}_{+-} e^{i\phi} \, | M(1/2) |^2 \;\; \{d^{1/2}_{1/2,1/2}(\theta) \, d^{1/2}_{-1/2,1/2}(\theta) + d^{1/2}_{-1/2,-1/2}(\theta) d^{1/2}_{1/2,-1/2}(\theta)\} \tag{28}$$

where we have used Eq. (24).

Now

$$d^J_{-1/2,-1/2}(\theta) = d^J_{1/2,1/2}(\theta), \qquad d^J_{1/2,-1/2}(\theta) = -d^J_{-1/2.1/2}(\theta) \tag{29}$$

so that the expression in parenthesis in Eq. (28) vanishes. Similarly the term arising from ρ^{uu}_{-+} vanishes.

Hence, in the decay of a particle or a "free" quark "q" $\to \pi + X$, the angular distribution of the pion is independent of the transverse polarization of the decaying particle, i.e. the Collins mechanism vanishes.

The crucial point leading to the above conclusion was our ability to factor out the reduced matrix element terms $| M(1/2) |^2 = | M(-1/2) |^2$ in going from Eq. (27) to Eq. (28). It is this step that fails in the case of an interacting quark field, as we shall now see.

3.2. *Fragmentation of an Interacting Quark*

In this case the the analogoue of Eq. (21) is, on putting Eq. (18) into Eq. (5),

$$W^{\mu\nu} \propto \rho^{uu}_{\lambda'\lambda} \sum_{\lambda_X} \mathcal{M}_+ \mathcal{M}'^*_+ + \rho^{vv}_{\lambda'\lambda} \sum_{\lambda_X} \mathcal{M}_- \mathcal{M}'^*_- + \rho^{uv}_{\lambda'\lambda} \sum_{\lambda_X} \mathcal{M}_- \mathcal{M}'^*_+ + \rho^{vu}_{\lambda'\lambda} \mathcal{M}_+ \mathcal{M}'^*_- \tag{30}$$

where, for brevity we have used

$$\mathcal{M}_{+/-} \equiv < p_\pi; X, \lambda_X \mid k, \lambda)_{+/-}$$
$$\mathcal{M}'_{+/-} \equiv < p_\pi; X, \lambda_X \mid k, \lambda')_{+/-} \tag{31}$$

and where, analogously to Eq. (22),

$$\begin{aligned}\rho^{vv}_{\lambda'\lambda} &= \bar{v}_\alpha(k,\lambda')\rho_{\alpha\beta}v_\beta(k,\lambda)\\ \rho^{uv}_{\lambda'\lambda} &= \bar{u}_\alpha(k,\lambda')\rho_{\alpha\beta}v_\beta(k,\lambda)\\ \rho^{vu}_{\lambda'\lambda} &= \bar{v}_\alpha(k,\lambda')\rho_{\alpha\beta}u_\beta(k,\lambda)\end{aligned} \tag{32}$$

The analogues of Eqs. (23) and (24) are now

$$\mathcal{M}_{+/-} =< p_\pi; X, \lambda_X \mid k, \lambda)_{+/-} = M_{+/-}(\lambda_X)e^{i\lambda\phi}d^{1/2}_{\lambda,\lambda_X}(\theta) \tag{33}$$

with

$$M_+(-\lambda_X) = \eta M_+(\lambda_X) \qquad M_-(-\lambda_X) = -\eta M_-(\lambda_X) \tag{34}$$

Consider now the dependence of the angular distribution upon the transverse polarization of the quark i.e. upon the terms in Eq. (30) involving ρ_{+-} and ρ_{-+}. For the terms containing $\rho^{uu}_{+-}, \rho^{uu}_{-+}, \rho^{vv}_{+-}$ and ρ^{vv}_{-+} the argument is identical to the free particle case and the Collins mechanism vanishes. But, for example, for the ρ^{uv}_{+-} term one has

$$\begin{aligned}&\rho^{uv}_{+-}e^{i\phi}\sum_{\lambda_X} M^*_+(\lambda_X)M_-(\lambda_X)d^{1/2}_{-1/2,\lambda_X}(\theta)d^{1/2}_{1/2,\lambda_X}(\theta) =\\ &\rho^{uv}_{+-}e^{i\phi}\{M^*_+(1/2)M_-(1/2)d^{1/2}_{-1/2,1/2}(\theta)d^{1/2}_{1/2,1/2}(\theta) +\\ &M^*_+(-1/2)M_-(-1/2)d^{1/2}_{-1/2,-1/2}(\theta)d^{1/2}_{1/2,-1/2}(\theta)\}\end{aligned} \tag{35}$$

$$\begin{aligned}= \rho^{uv}_{+-}e^{i\phi}M^*_+(1/2)M_-(1/2)\{&d^{1/2}_{-1/2,1/2}(\theta)d^{1/2}_{1/2,1/2}(\theta)\\ &-d^{1/2}_{-1/2,-1/2}(\theta)d^{1/2}_{1/2,-1/2}(\theta)\}\end{aligned} \tag{36}$$

where we have used Eq. (34).

Note the crucial difference: the minus sign in the parenthesis in Eq. (36) compared with (28). Consequently the angular factor in Eq. (36) does *not* vanish and the Collins mechanism survives. The other contributions, from $\rho^{uv}_{-+}, \rho^{vu}_{+-}$ and ρ^{vu}_{-+} also do not vanish, and there is no cancellation between them.

Hence for the interacting, off-shell quark the Collins mechanism does not vanish, and the fundamental reason is that the operator in Eq. (14) creates a superposition of states with opposite parity, whereas in the case of a free field it creates a state with just one definite parity.

4. Discussion

We have, for simplicity, discussed the case of the semi-inclusive production of pions, but, in fact, the results hold equally well for any hadron h. Thus, in the fragmentation of a transversely polarized off-shell quark "q" $\rightarrow h+X$ where the quark is described by an interacting QCD field, the angular distribution of the h *can* depend on the direction of the quark transverse polarization i.e. the Collins mechanism is non-zero.

On the contrary, for a decaying particle, or when the quark is regarded as a "free" on-shell particle, the Collins mechanism vanishes.

The key difference between the two cases is that the operator involved in the fragmentation amplitude

$$\int d^4ze^{-ik.z} < h; X, \lambda_X \mid \bar{\Psi}_\beta(z) \mid 0 > \qquad (k_0 > 0) \tag{37}$$

i.e. the operator

$$\bar{\Psi}_\beta(k) = \int d^4ze^{-ik.z}\bar{\Psi}_\beta(z) \tag{38}$$

when acting on the vacuum creates a superposition of states with opposite parity when it is an interacting field,

$$\bar{\Psi}_\beta(k) \mid 0 >= \frac{1}{E_k}\sum_\lambda\{\bar{u}_\beta(\boldsymbol{k},\lambda) \mid k,\lambda)_+ + \bar{v}_\beta(-\boldsymbol{k},\lambda) \mid k,\lambda)_-\} \tag{39}$$

as can be seen from Eqs. (14) to (17), whereas it creates a state with definite parity when it is a free field. The existence of a definite parity for the decaying particle is crucial for the cancellation that causes the Collins mechanism to vanish.

While thinking about this problem I was suddenly reminded of a puzzle that had baffled me as a graduate student. I had been given the exercise of calculating the Feynman diagram for $\pi^-p \rightarrow \pi^-p$ elastic scattering, in the Born approximation i.e. via the s-channel diagram

$$\pi^-p \rightarrow \text{``}n\text{''} \rightarrow \pi^-p$$

using old fashioned pseudoscalar $\pi - N$ field theory. I then had to project out expressions for the s- and p-wave scattering amplitudes.

Now the parity of a $\pi - N$ state with orbital angular momentum l is $(-1)^{(l+1)}$, so that the s-state has parity (-1) and the p-state parity $(+1)$. Since the intermediate state neutron has parity $(+1)$ I was convinced that I should get zero for the s-wave amplitude, in contradiction to what was emerging from my calculation.

It took me sometime to realize that an off-shell intermediate neutron does not have a definite parity! And it was then very pretty to notice that the ratio of s-wave to p-wave amplitudes was proportional to $(s - m^2)$ i.e. that the s-wave amplitude vanished as the momentum of the intermediate neutron approached the mass shell.

Of course, as I suggested in the Introduction, all this would have been perfectly obvious to Volodya Gribov!

References

1. J.C. Collins, *Nucl. Phys.* **B396**, 161 (1993).
2. See for example C. Bourrely, E. Leader and J. Soffer, *Phys. Rep.* **59**, 95 (1980).
3. E. Leader, *Spin in Particle Physics*, Cambridge University Press, UK, 2001, Section 8.2.
4. J.D. Bjorken and S.D. Drell, *Relativistic Quantum Fields*, McGraw-Hill, New York, 1965, Section 15.4.

DIFFRACTIVE PROCESSES AT THE LHC

M.G. RYSKIN, A.D. MARTIN AND V.A. KHOZE

Petersburg Nuclear Physics Inst., Gatchina, St. Petersburg, 188300, Russia
Inst. for Particle Physics Phenomenology, University of Durham, Durham, UK
and
University of Minnesota, Minneapolis, MN 55455, USA

We consider diffractive processes which can be measured at the LHC. Analysis of diffractive events will give unique information about the high energy asymptotics of the hadron scattering. In semihard diffraction one may study the partonic structure of the Pomeron. Central Exclusive Diffractive production provides a possibility to investigate the new particles (Higgs bosons, SUSY particles,...) in an exceptionally clean environment.

1. Introduction

It was shown about 40 years ago that the behaviour of the diffractive cross sections, together with the crucial role played by unitarity constraints, determine the high energy asymptotics of hadron-hadron interactions. The LHC collider is the first accelerator which will have enough energy to produce the events with a few ($n = 2 - 4$) large rapidity gaps. This will be the first time we could measure diffractive and multi-Pomeron processes sufficiently close to their asymptotic regime.

Our discussion is based on the ideas (and publications) of V.N. Gribov. Remarkably, Gribov played the pivotal role in investigating almost all aspects of this field. He introduced the Reggeon diagram technique [1]. He discussed in detail the possible high energy behaviours of hadron interaction amplitudes [2,3]. He found the relations between the diffractive and inelastic cross sections [4] — the AGK cutting rules. He discovered the constraints on the diffractive amplitudes coming from the unitarity conditions, and the specific asymptotic behaviour of the elastic cross section [5].

Our brief review is divided into three parts. First, we discuss pure *soft* interactions: the behaviour of the total cross section and the slope of differential elastic cross section, the t-dependence of the diffractive dissociation and the survival probability of one (or a few) large rapidity gaps. Next, we consider the *semihard* diffractive processes, where high E_T jets or a heavy boson (e.g. J/ψ, W, Z, ...) may be produced. Finally, we describe the advantages of studying New Physics (in particular Higgs bosons) in Central Exclusive Diffractive (CED) events.

2. Soft diffractive events

2.1. *Unitarity constraints for elastic amplitude*

It is known that the present data on high energy elastic hadron-hadron scattering are well described by a simple parametrization proposed by Donnachie and Landshoff (DL) [6]. The nucleon-nucleon amplitude is written as

$$A(s,t) \;\simeq\; i\sigma_0 \; F_1^2(t)\; (s/s_0)^{\alpha_{I\!P}(t)}, \tag{1}$$

where s is the square of the incoming c.m. energy ($s_0 = 1$ GeV^2), $F_1(t)$ is the electromagnetic proton form factor (t is the momentum transfer squared), and the Pomeron trajectory $\alpha_{I\!P}(t) = 1.08 + t \cdot 0.25$ GeV^{-2}. The corresponding cross section is

$$\sigma = (1/s)\mathrm{Im}A(s,0) \propto \sigma_0 (s/s_0)^{0.08}. \tag{2}$$

It is convenient to transform (1) to impact parameter, b_t, space

$$A(s,b_t) \;=\; \frac{-i}{2s(2\pi)^2} \int A(s,t=-q_t^2)\; e^{i\vec{q}_t\cdot\vec{b}_t}\; d^2q_t \;. \tag{3}$$

If the power growth ($\sigma \propto s^{0.08}$) continues, then the amplitude, which at the Tevatron reaches $A(s_{\rm Tev}, b_t = 0) \simeq 0.96$, will exceed the black-disk limit at the LHC, $A(s_{\rm LHC}, b_t = 0) > 1$.

In the DL approach the high energy amplitude is described by single Pomeron ($I\!P$) exchange only. To account for the unitarity constraints, we must include multi-Pomeron diagrams. In particular, the KMR model [7], which accounts for multi-Pomeron contributions, and successfully describes the same data in the ISR – Tevatron energy range, predicts $A(b_t = 0) < 1$ (but close to 1) at the LHC energy. However in such models [7,8], due to the multi-Pomeron effects, the interaction radius R^2 (that is the elastic slope $B(t = 0)$) grows faster than that in DL model. At the LHC energy, $B_{\rm KMR} \sim 22$ GeV^{-2}, while $B_{\rm DL} = 19$ GeV^{-2} (see [9] for more details). Thus it will be important to measure the elastic slope (i.e. interaction radius) at the LHC, since the unitarity-induced corrections caused by the multi-Pomeron cuts, first reveal themselves in the value of the slope B.

2.2. *Possible asymptotics of the high energy amplitude*

Detailed analyses performed at the end of 60's [2,3] had showed that there could be a few different regimes with different energy dependences of the cross section $\sigma_{\rm tot}$ as $s \to \infty$.

(a) $\sigma \to$ const. – the so-called 'weak coupling' regime [2].
In this case, as $s \to \infty$, the major contribution comes from $I\!P$-pole exchange, while the $I\!P$-cut contribution dies out; the elastic slope $B \propto \ln s$.

(b) $\sigma \sim (\ln s)^\epsilon$, $B \sim (\ln s)^\eta$ with $0 < \epsilon \le \eta < 2$ – the 'critical Pomeron' [3].

(c) $\epsilon = \eta = 2$ – the 'supercritical Pomeron', which leads to the Froissart regime.
In terms of the bare $I\!P$-pole, the critical and/or supercritical Pomeron amplitudes are $I\!P$-cuts in which the single $I\!P$-contribution is completely screened by multi-Pomeron rescatterings.

(d) Finally, it may be possible for the cross section to first grow as $\sigma_{\rm tot} \propto s^\epsilon$, but then at larger energies to decrease like $s^{-\epsilon}$, due to the increasing role of Pomeron-Pomeron self-interactions (see for example [10]).
Surprisingly, this last possibility is still allowed by the present data! The Tevatron energy is not sufficient to reject it, since the transition to the $\sigma \sim s^{-\epsilon}$ behaviour is expected to occur only after the possible size of the rapidity gap becomes much larger than $1/\epsilon \sim 12$ (for $\epsilon = 0.08$).

2.3. *Diffractive dissociation and the triple-* $I\!P$ *vertex*

The diffractive dissociation into a high mass (M_X) state is described by the triple-Pomeron diagram. If we assume that the triple-Pomeron vertex $G_{3I\!P}$ is just a small constant, then we face a problem. In this 'naive' approximation the cross section of diffractive dissociation $\sigma^{\rm SD}$ grows faster than $\sigma_{\rm tot}$. This applies to cases (a), (b) and (c) above. Indeed, even for $\alpha_{I\!P} = 1$, after the integration over the mass M_X ($\int^s dM_X^2/M_X^2 \sim \ln s$), the cross section $\sigma^{\rm SD} \propto (\ln s)\sigma_{\rm tot}^2 > \sigma_{\rm tot}$. An analogous, and more severe, problem occurs in processes with so-called multi-Reggeon kinematics. The cross section of the events with a few large rapidity gaps grows faster than the total inelastic cross section [11,12].

In the 'weak coupling' regime, (a), the resolution of the problem is the vanishing of the triple-Pomeron coupling (that is of the vertex $G_{3I\!P}$) at small transverse momenta, $t \to 0$. The same vanishing[a] is predicted [5] for any diffraction dissociation vertex, say $V(N \to N^*) \to 0$ as $t \to 0$. Observation of this 'vanishing' will be a strong argument in favour of the 'weak coupling' asymptotic regime (a), where at extremely high energies all the cross section are predicted to become equal to each other – $\sigma(aa) = \sigma(ab) = \sigma(bb)$ – independent of the type of each incoming hadron [5].

However, in reality, the situation is not so simple. Besides single $I\!P$-exchange, there is the $I\!P$-cut contribution, which does not vanish as $t \to 0$. At relatively low energies we practically do not see dissociation due to the single $I\!P$-pole; the $I\!P$-cut contributions dominate.

An exception is the 'weak coupling' regime, (a), where the $I\!P$-cut terms die out with increasing energy. At the LHC we would get a chance to observe a dissociation amplitude arising almost entirely from single-Pomeron-exchange. The cut corrections are much smaller than those occurring in the lower (ISR – S$p\bar{p}$S – Tevatron) energy range. Thus in case (a) we expect that the cross section of diffractive dissociation in forward direction (i.e. $d\sigma^{\rm SD}(s,t)/dt$ as $|t| \to 0$) to decrease with energy. Also, the diffractive minimum caused by the destructive interference between the pole and cut contributions will take place at a rather small $|t| = t_0$. Moreover the value of t_0 will decrease with energy faster than the position of the diffractive dip in elastic scattering.

However, the t-behaviour of $d\sigma^{\rm SD}/dt$ has, not as yet, been studied with enough precision. On the other hand, it is crucial to know this t-behaviour in order to construct a realistic model for the high-energy asymptotics of the strong interaction amplitude.

There is another solution to the problem (that $\sigma^{\rm SD} > \sigma_{\rm tot}$), which should be realized for the more realistic (b,c) scenarios. That is to assume that, due to the screening corrections arising from the multi-loop Pomeron graphs, the strength of the 'effective' triple-Pomeron vertex $G_{3I\!P,{\rm eff}}$ decreases with energy, or with the size of the gap (Δy). Using modern terminology, we would say that the gap survival factor $S^2 \to 0$ when $s \to \infty$ (or $\Delta y \to \infty$).

Thus, it is important to study the dependence of S^2 on the initial energy $\sqrt{s}$, on the gap size (Δy), and on the number of the gaps. The LHC is the

[a]In terms of the Good-Walker model [13] (or Additive Quark Model) the vanishing is provided by the orthogonality of different diffractive eigenstates – if the Pomeron couples to a single parton then, at $t = 0$, one-Pomeron-exchange does not change the distributions of partons in incoming wave function.

first collider with the sufficient energy to produce 2,3 (or may be even 4) large rapidity gaps[b]. Note that the cross section for 3-gap formation at the LHC is not too small. In Ref. [15] is was evaluated to be $\sigma \sim 1 - 3\mu b$.

Finally, recall that diffractive dissociation comes mainly from the periphery of the interaction disk, due to stronger absorptive corrections in the center of the disk. This has the important consequence that the mean transverse momenta of the secondaries produced in a diffractive dissociation process should be *smaller* than the transverse momenta of secondaries coming from an ordinary inelastic collision at the corresponding energy $s_{\rm inel} = M_X^2$ [16].

3. Semihard diffractive dissociation

The formation of a system of mass M_X by diffractive dissociation of the proton may be considered as an inelastic proton-Pomeron interaction.

3.1. *Partonic structure of the Pomeron and the LHC*

The production of high E_T jets, or heavy bosons (W, Z) or heavy quarks (J/Ψ, $b\bar{b}$,...) within the diffractive system M_X can be described in the usual way as the convolution of the incoming parton distributions (in the proton and in the Pomeron) with the cross section of the 'hard' subprocess. The parton distributions in a proton are well known. Therefore, by selecting events with a high E_T or heavy quarks (or bosons) in the diffractive dissociation (M_X), we have the possibility to study the internal partonic structure of the Pomeron. The expected cross sections are typically of the order of 1 - 10 nb [17].

Recall that the global parton analyses of the HERA deep inelastic scattering data indicate that at low $Q^2 = 1 - 2$ GeV^2 we have Pomeron-like sea quarks (behaving as $xq \sim x^{-0.2}$, but valence-like gluons (typically of the form $xg \sim \sqrt{x}$); see Refs. [18,19]). Does this mean that the Pomeron is built up of quarks and not gluons, or the Pomeron at low scales does not couple to gluons but only to quarks? LHC has to answer this question!

[b]Contrary to the model of Ref. [14], for the case of supercritical Pomeron (c), the survival factor S^2 decreases as the number of gaps n grows. Note that the KMR model [7] already predicts a lower S^2 for 'central exclusive' production (with 2 gaps) than that for single dissociation (with one gap).

3.2. *The 'direct' hard Pomeron interaction*

Note, however, that contrary to the Ingelman-Schlein ansatz [20], the Pomeron is not a hadron-like object of more or less fixed size. In perturbative QCD the Pomeron singularity is not an isolated pole, but a branch cut which may be regarded as a continuum series of poles in the complex angular momentum plane [21]. That is, the Pomeron wave function consists of a continuous number of components. Each component i has its own size, $1/\mu_i$. A 'direct' hard interaction of a small-size component of the Pomeron with the parton coming from the beam proton will violate conventional collinear factorization. At first sight such a contribution would appear to be suppressed by the form-factor-like dependence of the effective Pomeron flux $f_{I\!P}(x_{I\!P},\mu^2) \sim 1/\mu^2$. Here $x_{I\!P}$ is the longitudinal momentum fraction of the beam particle transfered through the Pomeron, and μ is the scale corresponding to the specific component of the Pomeron.

On the other hand, at small $x_{I\!P}$ this power-like suppression coming from the form factor is compensated by a large gluon density $g(x_{I\!P},\mu^2)$ which grows as $(\mu^2)^\gamma$, with an anomalous dimension which behaves as $\gamma \to 1/2$ as $x_{I\!P} \to 0$ [21]. Thus, at very low $x_{I\!P}$ the integral over the Pomeron size $1/\mu_i$ takes the logarithmic form

$$f_{I\!P} \propto \frac{1}{x_{I\!P}} \int \left[\frac{\alpha_s}{\mu} x_{I\!P}\, g(x_{I\!P},\mu^2)\right]^2 \frac{d\mu^2}{\mu^2} \sim \frac{1}{x_{I\!P}} \int \frac{d\mu^2}{\mu^2}. \tag{4}$$

Actually at sub-asymptotic energies the anomalous dimension γ is a bit less than 1/2 and the integral (4) is convergent at large μ^2, but numerically we cannot neglect this 'direct' Pomeron-parton hard interaction at the LHC (or even at HERA) energies (see [22] for more details, including a discussion of other sub-asymptotic violations of collinear factorization).

Note that high E_T dijets (or heavy quarks, or bosons) produced in 'direct' hard Pomeron interactions carry away the whole momentum of the Pomeron. For example we have $\gamma I\!P \to jj$ fusion in deep inelastic scattering, or $gI\!P \to jj$ fusion in pp-collisions which corresponds to the process $pp \to Xjj + p$, where the + sign denotes a large rapidity gap. In these examples, jj indicates a pair of high E_T jets. In fact, the 'direct' hard interaction of the small-size components of the Pomeron is the origin of the so-called 'extra hard' component, $\propto \delta(x-1)$, in the parton distributions of the Pomeron, which was proposed in Ref. [23] to describe the events in which high E_T dijets carry away Pomeron momentum fractions x close to 1. The identification of such a component is important for the extraction of diffractive parton densities from diffractive deep inelastic data [22]. Clearly, at the LHC, it will be informative to measure dijets (jj) in the Pomeron fragmentation region.

4. CED probes of New Physics at the LHC

Central Exclusive Diffractive (CED) reactions offer the opportunity to study the New Physics (such as Higgs bosons, SUSY particles,...) in an exceptionally clean environment. These new objects produced in CED events are expected to be rather heavy. Thanks to this large scale, the process can be described within the framework of perturbative QCD. It was shown in [15] that the CED cross section may be calculated as the convolution of the effective (diffractive) gluon luminosity $\mathcal{L}(gg^{I\!PI\!P})$, and the square of the matrix element of the corresponding hard subprocess.

4.1. *An example – Higgs production*

As an example, we consider the production of the SM Higgs boson by the CED process

$$pp \to p + H + p \tag{5}$$

at the LHC, where, again, the + signs denote large rapidity gaps. Let us take the mass range, $M \lesssim 140$ GeV, where the dominant decay mode is $H \to b\bar{b}$. Demanding such an exclusive process (5) leads to a small cross section [24]. At the LHC, we predict

$$\sigma_{\rm excl}(H) \ \sim \ 10^{-4} \ \sigma_{\rm incl}^{\rm tot}(H)\ . \tag{6}$$

In spite of this, the exclusive reaction (5) has the following advantages:

(a) The mass of the Higgs boson (and in some case the width) can be measured with high accuracy (with mass resolution $\sigma(M) \sim 1$ GeV) by measuring the missing mass to the forward outgoing *tagged* protons.

(b) The leading order $b\bar{b}$ QCD background is suppressed by the P-even $J_z = 0$ selection rule [25,26], where the z axis is along the direction of the proton beam. Therefore one can observe the Higgs boson via the main decay mode $H \to b\bar{b}$. Moreover, a measurement of the mass of the decay products must match the 'missing mass' measurement. It should be possible to achieve a signal-to-background ratio of the order of 1. For an integrated LHC luminosity of $\mathcal{L} = 300$ fb^{-1} we predict about 100 observable Higgs events, *after* acceptance cuts [27].

(c) The quantum numbers of the central object (in particular, the C- and P-parities) can be analyzed by studying the azimuthal angle distribution of the tagged protons [28]. Due to the selection rules, the production of 0^{++} states are strongly favoured.
(d) There is a very clean environment for the exclusive process – the soft background is strongly suppressed.
(e) Extending the study to SUSY Higgs bosons, there are regions of SUSY parameter space were the signal is enhanced by a factor of 10 or more, while the background remains unaltered. Indeed, there are regions where the conventional Higgs signals are suppressed and the CED signal is enhanced, and even such that both the h and H 0^{++} bosons may be detected [29].

4.2. *A 'gluon factory'*

In some sense, the CED processes may be considered as a filter which suppresses the production of the light quark dijets, of unnatural parity objects, and of some meson states made of quarks. For example, in the SUSY Higgs sector, the production of the pseudoscalar A boson is suppressed in comparison to the scalars, h and H. Another example, is the suppression of the production of non-relativistic 2^+ quarkonia.

On the other hand, CED reactions are a good place to search for 'glueballs' or for studying gluon dijets. Indeed the CED production of high E_T dijets, via the $I\!PI\!P \to jj$ hard subprocess, may be used as an excellent 'gluon factory' [26]. This is a good way to study the properties of gluon jets – multiplicity, jet shape,... – without an admixture of quark jets.

5. Conclusion

The study of diffractive processes at the LHC can be very rich and fruitful[c]. We have a chance to answer a number of important and outstanding questions. What is the high energy asymptotic behaviour of the strong interaction amplitude? Does Nature select the weak Pomeron-Pomeron coupling regime, or do we have a 'critical' or a 'supercritical' Pomeron? In the last case, the total cross section reveals a Froissart-like behaviour, while diffractive dissociation, and events with a few large rapidity gaps, are suppressed by small gap survival factors S^2.

[c]Indeed, already the TOTEM collaboration [30] is geared to study various aspects of soft and semihard physics at the LHC. Also novel aspects of diffractive studies are included in the physics case for forward proton tagging at 420 m at the LHC [31].

What are the parton distributions generated by the Pomeron? Just as we can obtain universal parton distributions from global analyses of data for deep inelastic and related hard scattering processes, so we can obtain universal diffractive parton distributions from the analysis of diffractive data. However, in the latter case the analysis is more subtle and we have to take into account violations of collinear factorization. Here we have seen that studies at the LHC can give important information.

One interesting possibility to consider, concerns the LHCb experiment, where the detector covers the rapidity region of $2 < \eta < 5$. The LHCb detector will have a high track reconstruction efficiency and good π/K separation [32], which may be very useful for glueball searches. It is going to operate at a luminosity 2×10^{32} cm^{-2}s^{-1}, for which there will be usually a single collision per bunch crossing, and hence practically no 'pile-up' problems. Thus installing a forward detector at LHCb would offer the possibility of observing asymmetric events, with one very large rapidity gap, and so probe the region of very small $x_{I\!P} \sim 10^{-5}$ or even less.

Finally, we emphasize again that Central Exclusive Diffractive production provides a unique opportunity to search for New Physics in a very clean experimental environment. Recall that for such an experiment we need detectors to tag the outgoing forward protons, as well as using the main central detector to observe secondaries produced in the central region.

Acknowledgments

We thank Aliosha Kaidalov for numerous valuable discussions concerning diffractive processes and Risto Orava for discussions of experimental issues of forward physics at the LHC. ADM thanks the Leverhulme Trust for an Emeritus Fellowship. This work was supported by the Royal Society, the UK Particle Physics and Astronomy Research Council, by grants INTAS 00-00366, RFBR 04-02-16073, and by the Federal Program of the Russian Ministry of Industry, Science and Technology SS-1124.2003.2.

References

1. V.N. Gribov, *Sov. Phys. JETP* **26**, 414 (1968).
2. V.N. Gribov and A.A. Migdal, *Sov. J. Nucl. Phys.* **8**, 583 (1969).
3. V.N. Gribov and A.A. Migdal, *Sov. Phys. JETP* **28**, 784 (1969).
4. V.A. Abramovsky, V.N. Gribov and O.V. Kancheli, *Sov. J. Nucl. Phys.* **18**, 308 (1974).
5. V.N. Gribov, *Sov. J. Nucl. Phys.* **17**, 313 (1973).

6. A. Donnachie and P.V. Landshoff, *Phys. Lett.* **B296**, 227 (1992).
7. V.A. Khoze, A.D. Martin and M.G. Ryskin, *Eur. Phys. J.* **C18**, 167 (2000).
8. A.B. Kaidalov, L.A. Ponomarev and K.A. Ter-Martirosyan, *Sov. J. Nucl. Phys.* **44**, 468 (1986);
   E. Gotsman, E.M. Levin and U. Maor, *Phys. Rev.* **D60**, 09451 (1999), and references therein.
9. V.A. Khoze, A.D. Martin and M.G. Ryskin, *Nucl. Phys.* **B** *(Proc. Suppl.)* **99**, 213 (2001).
10. K.G. Boreskov, arXiv:hep-ph/0112325.
11. K.A. Ter-Martirosyan, *Sov. Phys. JETP* **17**, 233 (1963); *Nucl. Phys.* **68**, 591 (1964);
    I.A. Verdiev, O.V. Kancheli, S.G. Matinyan, A.M. Popova and K.A. Ter-Martirosyan, *Sov. Phys. JETP* **19**, 148 (1964).
12. J. Finkelstein and K. Kajantie, *Phys. Lett.* **B26**, 305 (1968); *Nuovo Cim.* **56A**, 658 (1968).
13. M.L. Good and W.D. Walker, *Phys. Rev.* **120**, 1857 (1960).
14. K. Goulianos, *Phys. Lett.* **B358**, 379 (1995).
15. V.A. Khoze, A.D. Martin and M.G. Ryskin, *Eur. Phys. J.* **C23**, 311 (2002).
16. M.G. Ryskin, *Sov. J. Nucl. Phys.* **50**, 288 (1989).
17. A.B. Kaidalov, V.A. Khoze, A.D. Martin and M.G. Ryskin, *Eur. Phys. J.* **C21**, 521 (2001).
18. A.D. Martin, R.G. Roberts, W.J. Stirling and R.S. Thorne, *Phys. Lett.* **B604**, 61 (2004).
19. J. Pumplin, D.R. Stump, J. Huston, H.L. Lai, P. Nadolsky and W.-K. Tung, *JHEP* **0207**, 012 (2002).
20. G. Ingelman and P.E. Schlein, *Phys. Lett.* **B152**, 256 (1985).
21. L.N. Lipatov, *Sov. Phys. JETP* **63**, 904 (1986); *Phys. Rep.* **286**, 131 (1997).
22. A.D. Martin, M.G. Ryskin and G. Watt, *Eur. Phys. J.* **C44**, 69 (2005).
23. A. Brandt *et al.* (UA8 Collaboration), *Phys. Lett.* **B297**, 417 (1992).
24. V.A. Khoze, A.D. Martin and M.G. Ryskin, Phys. Lett. **B401** (1997) 30.
25. V.A. Khoze, A.D. Martin and M.G. Ryskin, *In Proc. of 8th Int. Workshop on Deep Inelastic Scattering and QCD (DIS2000)*, Liverpool, eds. J. Gracey and T. Greenshaw (World Scientific, 2001), p.592. ArXiv:hep-ph/0006005.
26. V.A. Khoze, A.D. Martin and M.G. Ryskin, *Eur. Phys. J.* **C19**, 477 (2001); Erratum: *ibid.* **C20**, 599 (2001).
27. A. De Roeck, V.A. Khoze, A.D. Martin, R. Orava and M.G. Ryskin, *Eur. Phys. J.* **C25**, 391 (2002).
28. A.B. Kaidalov, V.A. Khoze, A.D. Martin and M.G. Ryskin, *Eur. Phys. J.* **C31**, 387 (2003).
29. A.B. Kaidalov, V.A. Khoze, A.D. Martin and M.G.Ryskin, *Eur. Phys. J.* **C33**, 261 (2004).
30. TOTEM Collaboration: TOTEM-TDR-001, CERN-LHCC-2004-002.
31. FP-420: M. Albrow *et al.*, CERN-LHCC-2005-025, LHCC-I-015.
32. LHCb Collaboration: CERN-LHCC/98-4; CERN-LHCC-2003-030.

WHY THE REAL PART OF THE PROTON-PROTON SCATTERING AMPLITUDE SHOULD BE MEASURED AT THE LHC

C. BOURRELY AND J. SOFFER

Centre de Physique Théorique, UMR 6207, CNRS-Luminy, Case 907, F-13288 Marseille Cedex 9, France

N.N. KHURI

Physics Department, Rockefeller University, New York, NY 10021-6399, USA

ANDRÉ MARTIN

Theory Division, CERN, 1211 Geneva 23, Switzerland

T.T. WU

Gordon Mc Kay Laboratory, Harvard University, Cambridge, MA 02138, USA

We stress the importance of testing dispersion relations at the LHC. A violation might be a sign for new physics.

We all know that, up to now, scattering amplitudes of PHYSICAL, strongly interacting particles (i.e. Baryons and Mesons) appear to satisfy dispersion relations, while since in the case of quarks which have no free asymptotic states, this statement would be meaningless. The most general versions of local quantum field theory lead to proving dispersion relations [1] and, more generally, analyticity properties in two variables in a rather large domain, if one makes use of the positivity properties of the absorptive part of the scattering amplitudes [2]. Furthermore one can prove the bound

$$\sigma_{total} < \text{Const} \cdot \log^2(s) \;, \tag{1}$$

where s is the square of the center of mass energy.

The question is whether this still makes sense now that we know that protons and mesons are made of quarks. From a conversation of one of us (A.M.) with Volodya Gribov in 1996, it seems that he was on the side of people like Klaus Hepp who thought that all this should be abandoned. I apologize to Volodya, but we have the opposite opinion, following, for instance, Wolfhart Zimmermann who, long ago proved the very curious but undeniable fact that a local field operator could be attached to a composite particle [3]. Zimmermann thinks that a construction following the same lines as his could be made for protons constituted of quarks. This is what we shall assume in what follows. Anyway, it appears empirically that, in the past, dispersion relations have appeared to be always satisfied, with energies reached by fixed target machines (pion-nucleon scattering for instance) or colliders (the ISR collider, the SP$\bar{\text{P}}$S collider and the Tevatron). The question is now if this will continue at higher energies accessible only by the LHC. First, it could be that Volodya is right, after all. Second, many theoreticians prefer String Theory to Field Theory and believe that besides the 4-dimensional Minkowski space, there are extra compact dimensions. These extra dimensions could introduce a non locality in the theory of the size of these extra dimensions, which, according to, for instance, Ignatios Antoniadis [4], could correspond to a scale which is not very far above the one accessible by the LHC. This is why we think that it is essential to check the validity of dispersion relations at the LHC. Testing dispersion relations can be done in two ways:

— use an explicit model which reproduces very well all existing data, and satisfies, by construction, dispersion relations, such as the model of Bourrely-Soffer-Wu [5] which is inspired by the work of Cheng and Wu [6].

— use fits of existing data, the most recent one being done by the UA4.2 collaboration [7].

The superiority of the first approach is that there is no flexibility in the predictions, while in the second it is essential to take a smooth fit, depending on few parameters, because, otherwise, the predictive power is lost.

At the LHC, the Bourrely-Soffer-Wu model predicts a real part

$$\rho = 0.122 \quad \text{with} \quad \sigma_{total} = 103.6 \text{ mb}. \tag{2}$$

The UA4.2 fit predicts

$$\rho = 0.13 \pm 0.018 \quad \text{with} \quad \sigma_{total} = 109 \pm 8 \text{ mb}. \tag{3}$$

We see that these are compatible.

If experiment gives numbers compatible with those above it will mean that the scale of violation is very much above the LHC energy or that the corresponding minimal size is much smaller. It will also mean that the predictions of the model and of the fit are valid, for what concerns cross-sections at much higher energies and this allows to have an idea of what cross sections look like at energies which might never be accessible except by cosmic ray experiments.

We would like to turn now to the case where there is a violation of dispersion relations. Because of the fact that no violation is seen at low energies, the violation must be progressive. It will turn out, in our proposal, that the violation is controlled by a single parameter. Our proposal is to assume that the initial analyticity domain obtained by local field theory (without the extension due to positivity, which needs polynomial boundedness) is still valid but that polynomial boundedness is violated in unphysical — in particular complex — regions of the analyticity domain [8]. It is not so easy to implement this violation, for instance, if one assumes a growth like $\exp(\sqrt{s})$ in complex directions one falls back, in the end on a polynomial bound, back on ordinary dispersion relations and back on the standard bound on the total cross section. The first case of non trivial violation of dispersion relations is when the scattering amplitude is allowed to behave like $\exp(s/s_0)$ in unphysical and/or complex directions. If we assume that this bound also holds inside the "Lehmann Ellipse" at fixed energy, we can prove, using unitarity, that the physical forward and negative transfer amplitudes are bounded by s^4. This means that we can write a dispersion integral with 4 subtractions. However THIS IS NOT the scattering amplitude which differs from the dispersion integral by an entire function of order 1. Another way is to multiply the scattering amplitude by a convergence factor which guarantees that the modified amplitude has no exponential growth in complex directions. Such a factor is

$$\exp\left(\frac{-1}{s_0}\sqrt{(4m^2-s)}\sqrt{(4m^2-u)}\right). \tag{4}$$

Because of the fact that the real part of the scattering amplitude looks small in existing data, models, and fits, the effects of this exponential growth are very visible even if the scale of the exponential, s_o, is much higher than the square of the LHC energy.

For instance at LHC (14 TeV c.m. energy) we expect naively:

$$\rho = 0.12 \text{ to } 0.13\,. \tag{5}$$

With a scale $s_0 = 50$ TeV^2 we would get

$$\rho = 0.21\,. \tag{6}$$

This means that we do not even need a very accurate measurement to see the effect. This is why we are delighted to learn that ρ could be measured by the ATLAS detector at CERN, as a by-product of a luminosity measurement using Coulomb scattering. We do not see why this method, to put in evidence extra compact dimensions would be worse than any other one.

Acknowledgments

We would like to thank Maurice Haguenauer for very stimulating discussions. UMR 6207 is Unité Mixte de Recherche du CNRS and of Universités Aix-Marseille I and Aix-Marseille II and of Université du Sud Toulon-Var, laboratoire affilié à la FRUMAN.

References

1. H. Epstein, V. Glaser, and A. Martin, *Comm. Math. Phys.* **13**, 257 (1969).
2. A. Martin, *Nuovo Cimento* **42**, 930 (1966); *ibid.* **44**, 1219 (1966).
3. W. Zimmermann, *Nuovo Cimento* **10**, 597 (1958).
4. I. Antoniadis, *Eur. Phys. J.* **C33**, 914 (2004).
5. C. Bourrely, J. Soffer, and T.T. Wu, *Eur. Phys. J.* **C28**, 97 (2003); *Nucl. Phys.* **B247**, 15 (1984).
6. H. Cheng and T.T. Wu, *Phys. Rev. Lett.* **24**, 1456 (1970).
7. C. Augier *et al.*, *Phys. Lett.* **B316**, 448 (1993).
8. N.N. Khuri, *Proceedings of Les Rencontre de Physique de la Vallée d'Aoste: Results and Perspectives in Particle Physics* (M. Greco, ed.), pp 701-708, Editions Frontières, Gif-sur-Yvette, France, 1994.

DIFFRACTIVE PROCESSES AS A TOOL FOR SEARCHING FOR NEW PHYSICS

V.A. KHOZE[1,2], A.B. KAIDALOV[1,3], A.D. MARTIN[1], M.G. RYSKIN[1,2] AND W.J. STIRLING[1]

[1] *Institute for Particle Physics Phenomenology, University of Durham, DH1 3LE, UK*

[2] *Petersburg Nuclear Physics Institute, Gatchina, St. Petersburg, 188300, Russia*

[3] *Institute of Theoretical and Experimental Physics, Moscow, 117259, Russia*

We show that the addition of detectors to tag the outgoing forward protons, at the LHC, will significantly enlarge the potential of studying New Physics. A topical example is Higgs production by the exclusive double-diffractive process, $pp \to p+H+p$. We discuss the production of Higgs bosons in both the SM and MSSM. We show how the predicted rates may be checked at the Tevatron by observing the exclusive double-diffractive production of dijets, or χ_c or χ_b mesons, or $\gamma\gamma$ pairs.

1. Introduction

The use of forward proton detectors as a means to study Standard Model (SM) and New Physics at the LHC has only been fully appreciated within the last few years; see, for example [1,2,3,4,5] and references therein. By detecting protons that have lost less than about 2% of their longitudinal momentum, a rich QCD, electroweak, Higgs and BSM programme becomes accessible, with a potential to study phenomena which are unique to the LHC, and difficult even at a future linear collider [6].

In particular, the so-called central exclusive production (CEP) processes may provide a very friendly environment to search for, and identify the nature of, new particles at the LHC; in particular, Higgs bosons. There is also a potentially rich, more exotic, physics menu including (light) gluino and squark production, gluinonia, radions, and indeed any object which has 0^{++} (or 2^{++}) quantum numbers and couples strongly to gluons [1]. By central exclusive, we mean the process $pp \to p+X+p$, where the + signs denote the absence of hadronic activity (that is, the presence of a rapidity gap) between the outgoing protons and the decay products of the central system X.

It is a pleasure to recall that the whole strategy of predicting diffractive phenomena, and, in particular, of CEP processes, is based on the ideas developed by V.N. Gribov. We list only some of these: Regge poles in particle physics, the vacuum pole (Pomeron) and its shrinkage, Glauber-Gribov theory of multiple scattering, Gribov's reggeon calculus, Gribov's factorization, the Abramovsky-Gribov-Kancheli cutting rules, Gribov's theorem for bremsstrahlung at high energies, the Gribov-Lipatov (DGLAP) evolution equations, the Frolov-Gorshkov-Gribov-Lipatov approach to Regge processes in gauge theories, and much, much more.

There are three main reasons why CEP is especially attractive for searches for new heavy objects. First, if the outgoing protons remain intact and scatter through small angles then, to a very good approximation, the primary active di-gluon system obeys a $J_z = 0$, C-even, P-even, selection rule [7,8]. Here J_z is the projection of the total angular momentum along the proton beam axis. This selection rule readily permits a clean determination of the quantum numbers of the observed new (for example, Higgs-like) resonance, which will be dominantly produced in a scalar state. Secondly, because the process is exclusive, the energy loss of the outgoing protons is directly related to the mass of the central system, allowing a potentially excellent mass resolution, irrespective of the decay mode of the produced particle[a]. Thirdly, a signal-to-background ratio of order 1 (or even better) is achievable This ratio becomes significantly larger for the lightest Higgs boson in certain regions of the MSSM parameter space [10].

Moreover, in some MSSM Higgs scenarios CEP provides an opportunity for a lineshape analysis [10,11]. Another attractive feature is the ability to directly probe the CP-structure of the Higgs sector by measuring the azimuthal asymmetry of the outgoing tagged protons [12]. A different strategy, to explore the manifestation of explicit CP-violation in the Higgs sector, was recently studied by Ellis et al. [11].

It is worth mentioning that, by tagging both of the outgoing protons, the LHC is effectively turned into a gluon-gluon collider. This will open up a rich, 'high-rate' QCD physics menu (especially concerning diffractive phenomena), which will allow the study of the skewed, unintegrated gluon densities, as well as the details of rapidity gap survival; see, for example [1,13]. Note that CEP provides a source of practically pure gluon jets; that is we

[a]Recent studies suggest that the missing mass resolution σ will be of order 1% for a 140 GeV central system, assuming both the outgoing protons are detected at 420 m from the interaction point [2,6].

effectively have a 'gluon factory' [8]. This can be an ideal laboratory in which to study the properties of gluon jets, especially in comparison with the quark jets, and will even allow a search for glueballs. The forward-proton-tagging approach also offers a unique programme of high-energy photon-interaction physics at the LHC; see, for example [3,14].

The 'benchmark' CEP process, for these new physics searches, is Higgs production. In the mass range around 115-130 GeV, its detection at the LHC will not be an easy task. There is no obvious perfect detection process, but rather a range of possibilities, none of which is compelling on its own. *Either* large signals are accompanied by a huge background, *or* the processes have comparable signal and background rates for which the number of Higgs events is rather small. The predicted cross section for the CEP production of a SM Higgs, with mass 120 GeV, at the LHC is 3 fb, falling to 1 fb for a mass of 200 GeV; see [15].

From an experimental perspective, the WW decay channel is the simplest way to observe the SM Higgs in the tagged-proton approach [16,17]. The $b\bar{b}$ decay channel is more challenging from a trigger perspective, although, in this case, the 'useful' event rate is more favourable for masses below about 130 GeV. Moreover, the latter decay mode becomes extremely important in the so-called intense coupling regime [18] of the MSSM, where CEP is likely to be the discovery channel [10]. In this case, we expect about 10^3 exclusively produced double-tagged Higgs bosons for 30 fb^{-1} of delivered luminosity. About 100 would survive the experimental cuts [9], with a signal-to-background ratio of the order of 10.

In the case of the exclusive process, $pp \to p+H+p$, a major experimental task is to provide a set-up in which the bulk of the proton-tagged Higgs signal is deposited in a smallest possible missing-mass window; $\Delta M_{\text{missing}}$ of about 3 GeV should be achievable. Note that for the $b\bar{b}$ channel the CEP process allows the mass of the Higgs to be measured in two independent ways. First, the tagged protons give $M_H = M_{\text{missing}}$ and second, via the $H \to b\bar{b}$ decay, we have $M_H = M_{b\bar{b}}$, although now the resolution is much poorer, with $\Delta M_{b\bar{b}} \simeq 10$ GeV or more. The existence of matching peaks, centered about $M_{\text{missing}} = M_{b\bar{b}}$, is a unique feature of the exclusive diffractive Higgs signal. Besides its obvious value in identifying the Higgs, the mass equality also plays a key role in reducing background contributions. Another crucial advantage of the exclusive process $pp \to p + H + p$, with $H \to b\bar{b}$, is that the leading order $gg \to b\bar{b}$ background subprocess is suppressed by the $J_z = 0$, P-even selection rule [8,9].

2. Calculation of the exclusive Higgs signal

The basic mechanism for the exclusive process, $pp \to p + H + p$, is shown in Fig. 1. Since the dominant contribution comes from the region $\Lambda^2_{\rm QCD} \ll Q_t^2 \ll M_H^2$, the amplitude may be calculated using perturbative QCD techniques [8,15]

$$\mathcal{M}_H \simeq N \int \frac{dQ_t^2 \, V_H}{Q_t^6} \, f_g(x_1, x_1', Q_t^2, \mu^2) f_g(x_2, x_2', Q_t^2, \mu^2), \qquad (1)$$

where the overall normalization constant N can be written in terms of the $H \to gg$ decay width [1,15], and where the $gg \to H$ vertex factors for CP $= \pm 1$ Higgs production are, after azimuthal-averaging,

$$V_{H(0^+)} \simeq Q_t^2, \quad \text{and} \quad V_{A(0^-)} \simeq (\vec{p}_{1t} \times \vec{p}_{2t}) \cdot \vec{n}_0, \qquad (2)$$

Expressions (1,2) hold for small p_{it}, where the $\vec{p}_{it}$ are the transverse momenta of the outgoing protons, and $\vec{n}_0$ is a unit vector in the beam direction. The f_g's are the skewed unintegrated gluon densities at the hard scale μ, taken to be $M_H/2$. Since $(x' \sim Q_t/\sqrt{s}) \ll (x \sim M_H/\sqrt{s}) \ll 1$, it is possible to express $f_g(x, x', Q_t^2, \mu^2)$, to single log accuracy, in terms of the conventional integrated density $g(x)$. The f_g's embody a Sudakov suppression factor T, which ensures that the gluon does not radiate in the evolution from Q_t up to the hard scale $M_H/2$, and so preserves the rapidity gaps. The apparent infrared divergence of (1) is nullified for $H(0^+)$ production by these Sudakov factors[b]. However, the amplitude for $A(0^-)$ production is much more sensitive to the infrared contribution. Indeed, let us consider the case of small p_{it} of the outgoing protons. Then we see, from (2), that the dQ_t^2/Q_t^4 integration for $H(0^+)$ is replaced by $p_{1t}p_{2t}dQ_t^2/Q_t^6$ for $A(0^-)$, and now the Sudakov suppression is not enough to prevent a significant contribution from the low Q_t^2 domain.

The radiation associated with the $gg \to H$ hard subprocess is not the only way to populate and to destroy the rapidity gaps. There is also the possibility of soft rescattering in which particles from the underlying event populate the gaps. The probability, $S^2 = 0.03$ at the LHC, that the gaps survive the soft rescattering was calculated using a two-channel eikonal

[b]Note also that the Sudakov factor inside the loop integration induces an additional strong decrease (roughly as M^{-3} [10]) of the cross section as the mass M of the centrally produced hard system increases. Therefore, the price to pay for neglecting this suppression effect would be to considerably overestimate the CEP cross section at large masses.

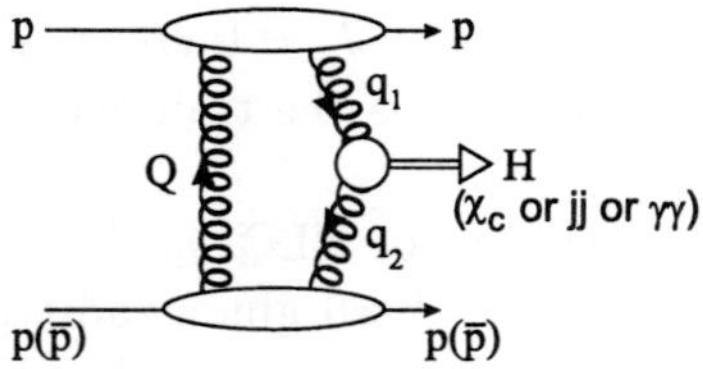

Figure 1. Schematic diagram for central exclusive production, $pp \to p + X + p$. The presence of Sudakov form factors ensures the infrared stability of the Q_t integral over the gluon loop. It is also necessary to compute the probability, S^2, that the rapidity gaps survive soft rescattering.

model, which incorporates high mass diffraction [19]. Including this factor, and the NLO K factor, the cross section is predicted to be [1,15]

$$\sigma(pp \to p + H + p) \simeq 3\,\text{fb} \tag{3}$$

for the production of a SM Higgs boson of mass 120 GeV at the LHC. We evaluated that there may be a factor of 2.5 uncertainty (up or down) in this prediction [10].

If we include a factor 0.6 for the efficiency associated with proton tagging, 0.67 for the $H \to b\bar{b}$ branching fraction, 0.6 for b and $\bar{b}$ tagging, 0.5 for the $b, \bar{b}$ jet polar angle cut, $60° < \theta < 120°$, (necessary to reduce the $b\bar{b}$ QCD background) [9], then, for a luminosity of $\mathcal{L} = 30\,\text{fb}^{-1}$, the original $3 \times 30 = 90$ events are reduced to an observable signal of 11 events.

3. Background to the exclusive $H \to b\bar{b}$ signal

The advantage of the $p + (H \to b\bar{b}) + p$ signal is that there exists a $J_z = 0$ selection rule, which requires the leading order $gg^{PP} \to b\bar{b}$ background subprocess to vanish in the limit of massless quarks and forward outgoing protons. (The PP superscript is to note that each gluon comes from colour-singlet gg t-channel exchange.) However, in practice, LO background contributions remain. The prolific $gg^{PP} \to gg$ subprocess may mimic $b\bar{b}$ production since we may misidentify the outgoing gluons as b and $\bar{b}$ jets. Assuming the expected 1% probability of misidentification, and applying $60° < \theta < 120°$ jet cut, gives a background-to-signal ratio $B/S \sim 0.18$. (Here, for reference, we assume that the mass window over which we collect the signal, $\Delta M \sim 3\sigma = 3$ GeV).

Secondly, there is an admixture of $|J_z| = 2$ production, arising from non-forward going protons which gives $B/S \sim 0.24$. Thirdly, for a massive quark there is a contribution to the $J_z = 0$ cross section of order m_b^2/E_T^2, leading to $B/S \sim 0.18$, where E_T is the transverse energy of the b and $\bar{b}$ jets[c].

Next, we have the possibility of NLO $gg^{PP} \to b\bar{b}g$ background contributions, which for large angle, hard gluon radiation does not obey the selection rules. Of course, the extra gluon may be observed experimentally and these background events eliminated. However, there are exceptions. The extra gluon may go unobserved in the direction of a forward proton. This background may be effectively eliminated by requiring the equality $M_{\text{missing}} = M_{b\bar{b}}$. Moreover, soft gluon emissions from the initial gg^{PP} state factorize and, due to the overriding $J_z = 0$ selection rule, these contributions to the QCD $b\bar{b}$ production are also suppressed. The remaining danger is large angle hard gluon emission which is collinear with either the b or the $\bar{b}$ jet, and, therefore, unobservable. If the cone angle needed to separate the g jet from the b (or $\bar{b}$) jet is $\Delta R \sim 0.5$, then the expected background from unresolved three-jet events leads to $B/S \simeq 0.18$. The NNLO $b\bar{b}gg$ background contributions are found to be negligible (after requiring $M_{\text{missing}} \simeq M_{b\bar{b}}$), as are soft Pomeron-Pomeron fusion contributions to the background (and to the signal) [9]. Also note that radiation off the screening gluon, in Fig. 1, is numerically small [20].

4. The signal-to-background ratio for $H \to b\bar{b}$ mode

So, in total, for the exclusive production of a 120 GeV (SM) Higgs boson at the LHC with the integrated luminosity $\mathcal{L} = 30$ fb^{-1}, after cuts and acceptances we can expect about 10 events, with a signal-to-background ratio S/B of the order of 1. In the case of a Gaussian missing mass distribution of width σ, about 87% of the signal is contained in a bin $\Delta M_{\text{missing}} = 3\sigma$, that is $M_{\text{missing}} = M_H \pm 1.5\sigma$.

We could consider Higgs production in other diffractive channels, such as diffractive production accompanied by proton dissociation ($pp \to M_1 + H + M_2$), or central inelastic production ($pp \to p + (M \to HX) + p$) [1]. However, we do not gain much as compared to the usual totally inclusive production – there is no precise missing mass measurement, no selection rule to suppress the $b\bar{b}$ background and more serious pile-up problems. The somewhat smaller density of soft secondary hadrons in the Higgs rapidity

[c]There are reasons to hope that, due to higher-order QCD effects, this particular background contribution will be a few times smaller.

region does not compensate for the much smaller statistics (cross sections) in diffractive processes.

5. Exclusive SUSY $H \to b\bar{b}$ signals

To be specific, we discuss the three neutral Higgs bosons of the MSSM model: h, H with CP=1 and A with CP=–1 [21]. There are regions of MSSM parameter space, for example, the intense coupling regime [18], where the conventional signals ($\gamma\gamma, WW, ZZ$ decays) are suppressed, but where the exclusive subprocess $gg \to H \to b\bar{b}$ is strongly enhanced [10]. This is evident from Fig. 2, which shows the cross section for CEP production of h, H bosons as function of their mass for $\tan\beta = 30$. Here, and in what follows, we use version 3.0 of the HDECAY code [22].

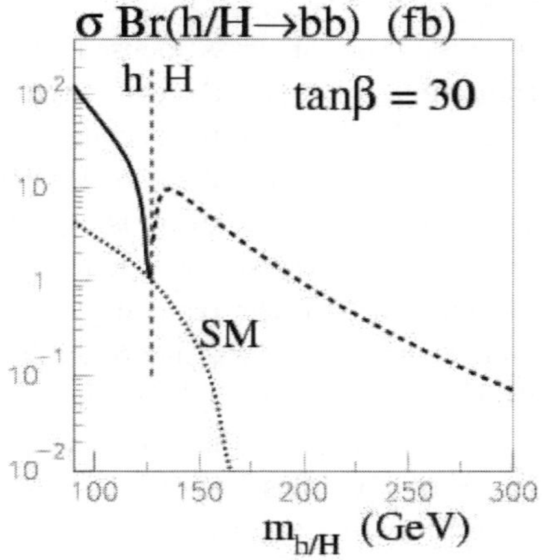

Figure 2. The cross section times the $b\bar{b}$ branching ratio for central exclusive production of h and H MSSM Higgs bosons at the LHC for $\tan\beta = 30$. Also shown (by the dotted curve) is the cross section times the branching ratio for SM Higgs production.

Taking, for example, for $M_A = 130$ GeV and $\tan\beta = 50$, we have $M_h = 124.4$ GeV with $S/B = 71/9$ events, $M_H = 135.5$ GeV with $S/B = 124/6$ events and $M_A = 130$ GeV with $S/B = 0.17$, so both h and H should be clearly visible. (Again, for reference, we assume that $\Delta M_{\text{missing}} = 3$ GeV can be achieved.) Let us emphasize that the intense coupling regime of the MSSM [18] is especially forward proton friendly, and in this particular case the tagged-proton approach may well be *the* discovery channel.

The decoupling regime ($M_A \gtrsim 2M_Z$ and $\tan\beta \gtrsim 5$) is another example where the exclusive signal is of great value. In this case h is indistinguishable from a SM Higgs, and so the discovery of H is crucial to establish the underlying dynamics.

If the exclusive cross sections for scalar and pseudoscalar Higgs production were comparable, it would be possible to separate them readily by a missing mass scan, and by the study of azimuthal correlations between the outgoing protons. Unfortunately, pseudoscalar exclusive production is strongly suppressed by the P-even selection. Maybe the best chance to identify the $A(0^-)$ boson is through the double-diffractive dissociation process, $pp \to X + A + Y$, where both protons dissociate [10].

6. Detecting the Higgs in the *WW* channel

The analysis in the previous sections was focused primarily on light SM and MSSM Higgs production, with the Higgs decaying to 2 b−jets. The potentially copious b−jet (QCD) background is controlled by a combination of the spin-parity selection rules and the mass resolution from the forward proton detectors. The missing-mass resolution is especially critical in controlling the background, since poor resolution would allow more background events into the mass window around the resonance.

Whilst the $b\bar{b}$ channel is very attractive theoretically, allowing direct access to the dominant decay mode of the light Higgs boson, there are some basic problems which render it challenging from an experimental perspective, see [17] for details. First, it relies heavily on the quality of the mass resolution from the proton taggers to suppress the background. Secondly, triggering on the relatively low-mass dijet signature of the $H \to b\bar{b}$ events is a challenge for the Level 1 triggers of both ATLAS and CMS. And, thirdly, this measurement requires double b−tagging, with a corresponding price to pay for the tagging efficiency.

In [16,17], attention was turned to the WW^* decay mode of the light Higgs, and for a Higgs mass above the WW threshold, to the WW decay mode. This channel does not suffer from any of the above problems: suppression of the dominant backgrounds does not rely so strongly on the mass resolution of the detectors, and, certainly, in the semi-leptonic decay channel of the WW system, the Level 1 triggering is not a problem. The advantages of forward proton tagging are, however, still explicit. Even for the double leptonic decay channel (i.e. with two leptons and two final state neutrinos), the mass resolution will be very good, and, of course, observation of the Higgs in the double-tagged channel immediately establishes its quantum numbers. It is worth mentioning that the mass resolution should improve with increasing Higgs mass. Moreover, the semileptonic 'trigger cocktail' may allow a combination of signals, not only from $H \to WW$ decays, but also from the $\tau\tau$, ZZ and even the semileptonic $b-$decay channels.

In Fig. 3 we show the cross section for the process $pp \to p + H + p \to p + WW + p$ as a function of the Higgs mass M_H at the LHC. The increasing branching ratio to $WW^{(*)}$ as M_H increases (see for example [21]) compensates for the falling central exclusive production cross section. For comparison, we also show the cross section times branching ratio for $pp \to p + H + p \to p + b\bar{b} + p$. For reference purposes, the cross sections in Fig. 3 are normalized in such a way that $\sigma_H = 3$ fb for $M_H = 120$ GeV. In Fig. 3 we show also the results for $\tan\beta = 2, 3, 4$. Evidently the expected CEP yield is also promising in the low $\tan\beta$ region.

Experimentally, events with two W bosons in the final state fall into 3 broad categories — fully-hadronic, semi-leptonic and fully-leptonic — depending on the decay modes of the W's. Events in which at least one of the Ws decays in either the electron or muon channel are by far the simplest, and Ref. [17] focuses mainly on these semi- and fully-leptonic modes. As mentioned above, one of the attractive features of the WW channel is the absence of a relatively large irreducible background, in contrast to the large central exclusive $b\bar{b}$ QCD background in the case of $H \to b\bar{b}$, suppression of which relies strongly on the experimental missing-mass resolution and di-jet identification. The primary exclusive backgrounds for the WW channel can be divided into two broad categories:

(1) central production of a WW^* pair $pp \to p + (WW^*) + p$ from either the (a) $\gamma\gamma \to WW^*$ or (b) $gg^{PP} \to WW^*$ subprocess,
(2) the W-strahlung process $pp \to p + Wjj + p$ originating from the $gg^{PP} \to Wq\bar{q}$ subprocess, where the W^* is 'faked' by the two quarks.

As shown in [17], over a wide region of Higgs masses the photon-photon initiated backgrounds are strongly suppressed if we require that the final leptons and jets are central and impose cuts on the transverse momenta of the protons in the taggers. Moreover, our estimates show that the QCD quark-box-diagram contribution from the $gg^{PP} \to WW^*$ subprocess is very small.

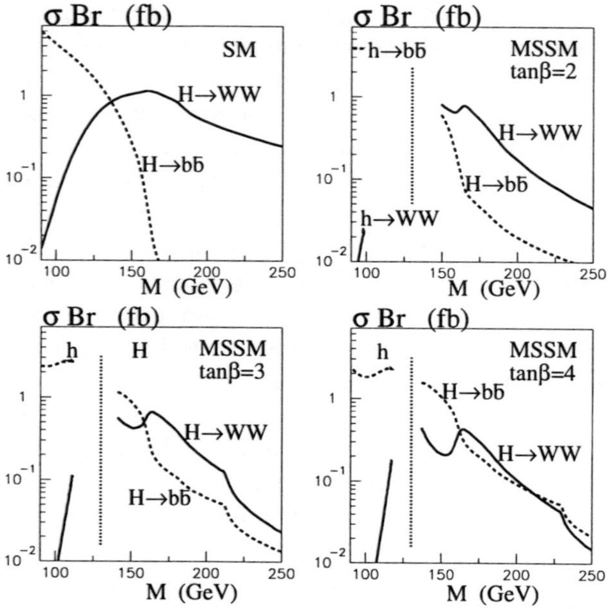

Figure 3. The cross section times branching ratio for the central exclusive production of the MSSM Higgs boson (for three values of $\tan\beta = 2, 3, 4$), as a function of the Higgs mass, in the WW and $b\bar{b}$ decay channels. The cross section for the CEP production of a SM Higgs boson is also shown.

The most important background, therefore, comes from the second category, i.e. from the W-strahlung process. Here we have to take into account the $J_z = 0$ projection of this amplitude, which requires a calculation of the individual helicity amplitudes. This was done in [16] using the spinor technique of Ref. [23]. The analysis in [16,17] shows that this background can be manageable with carefully choosen experimental cuts. For $M_H = 140$ GeV we expect 19 exclusive $H \to WW$ events for an LHC luminosity of 30 fb^{-1}. Note that the largest loss of events in the WW case is caused by the Level 1 trigger efficiency, and we expect significant improvements here.

7. Related processes: checks of the predicted exclusive Higgs yield

As discussed above, the exclusive Higgs signal is particularly clean, and the signal-to-background ratio is favourable. However, the expected number of events in the SM case is low. Therefore it is important to check the predictions for exclusive Higgs production by studying processes mediated by the same mechanism, but with rates which are sufficiently high, so that they may be observed at the Tevatron (as well as at the LHC). The most obvious examples are those in which the Higgs is replaced by either a dijet system, a χ_c or χ_b meson, or by a $\gamma\gamma$ pair, see Fig. 1.

First, we discuss the exclusive production of a pair of high E_T jets, $p\bar{p} \to p + jj + \bar{p}$ [1,15]. This would provide an effective gg^{PP} 'luminosity monitor' just in the kinematical region of the Higgs production. The corresponding cross section was evaluated to be about 10^4 times larger than that for the SM Higgs boson. Thus, in principle, this process appears to be an ideal 'standard candle'. The expected cross section is rather large, and we can study its behaviour as a function of the mass of the dijet system. This process is being studied by the CDF collaboration. Unfortunately, in the present CDF environment, the separation of exclusive events is not unambiguous. At first sight, we might expect that the exclusive dijets form a narrow peak, sitting well above the background, in the distribution of the ratio

$$R_{jj} = M_{\text{dijet}}/M_{\text{PP}} \tag{4}$$

at $R_{jj} = 1$, where M_{PP} is the invariant energy of the incoming Pomeron-Pomeron system. In reality the peak is smeared out due to hadronization and the jet-searching algorithm. Moreover, since M_{dijet} is obtained from measuring just the two-jet part of the exclusive signal, there will be a 'radiative tail' extending to lower values of R_{jj}.

For jets with $E_T = 10$ GeV and a jet cone $R < 0.7$, more than 1 GeV will be lost outside the cone, leading to (i) a decrease of the measured jet energy of about 1-2 GeV, and, (ii) a rather wide peak ($\Delta R_{jj} \sim \pm 0.1$ or more) in the R_{jj} distribution. The estimates based on Ref. [1] (see also [25]) give an exclusive cross section for dijet production with $E_T > 10, 25, 35, 50$ GeV, with values which are rather close to the recent CDF limits [24]. The comparison is shown in Fig. 4. In particular, for $E_T > 50$ GeV, we predict an exclusive cross section of about 1 pb [1], which agrees well with the current CDF upper limit obtained from events with $R_{jj} > 0.8$. As discussed above, one should not expect a clearly 'visible' peak in the CDF data for

R_{jj} close to 1. It is worth mentioning that the CDF measurements have already started to reach values of the invariant mass of the Pomeron-Pomeron system in the SM Higgs mass range.

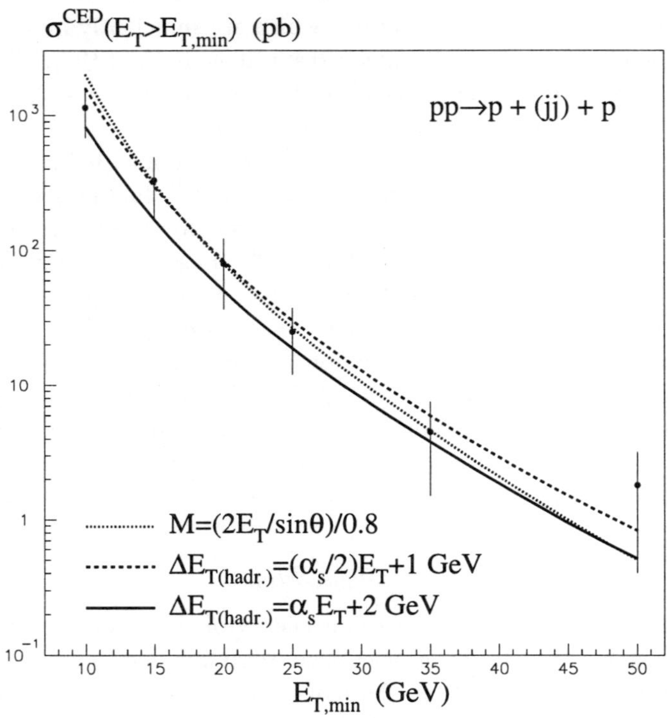

Figure 4. The cross section limit for 'exclusive' dijet production at the Tevatron as a function $E_{T,\min}$ as measured by CDF [26]. These preliminary CDF data correspond to the cross section integrated over the domain $R_{jj} = M_{\rm dijet}/M_{\rm PP} > 0.8$ and $E_T > E_{T,\min}$. The curves are the pure exclusive cross section calculated [1] using the CDF event selection. Different hadronization corrections were applied. The solid curve is obtained assuming that, after the hadronization, the measured jet transverse energy E_T is less than the parton (gluon) transverse energy by $\Delta E_T = E_{T,\rm gluon} - E_T = \alpha_s(E_T)E_T + 2$ GeV; while for the dashed curve it is assumed that ΔE_T is halved, i.e. $\Delta E_T = (\alpha_s(E_T)/2)E_T + 1$ GeV. The dotted curve is calculated assuming $E_{T,\rm gluon} = E_T$, but with the mass of the whole central system (which determines the incoming gluon-gluon luminosity) enlarged according to the R_{jj} ratio – $M_{\rm PP} = (2E_T/\sin\theta)/0.8$, where θ is the jet polar angle in the dijet rest frame.

An alternative 'standard candle' process is exclusive double-diffractive $\gamma\gamma$ production with high E_T photons, that is $p\bar{p} \to p + \gamma\gamma + \bar{p}$ [1,25]. Here there are no problems with hadronization or with the identification of the jets. On the other hand, the exclusive cross section is rather small. The predictions of the cross section for exclusive $\gamma\gamma$ production are shown in Fig. 5.

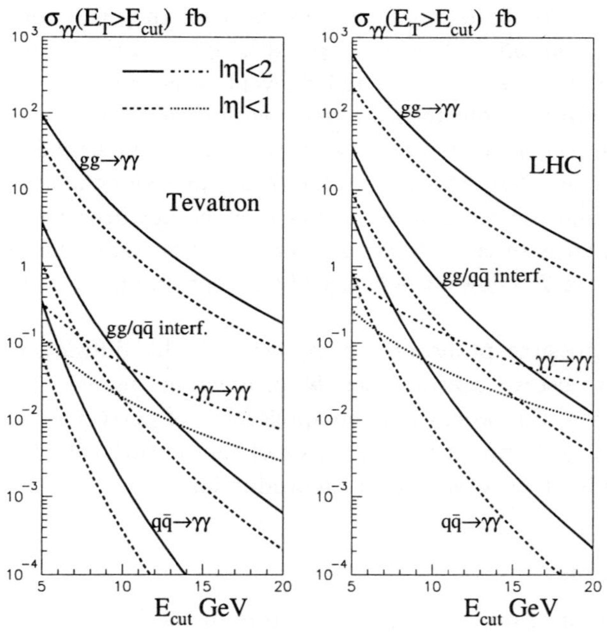

Figure 5. The contributions to the cross section for exclusive $\gamma\gamma$ production from gg and $q\bar{q}$ exchange at the Tevatron and the LHC. Also shown is the contribution from the QED subprocess $\gamma\gamma \to \gamma\gamma$. For each component we show the cross section restricting the emitted photons to have $E_T > E_{\rm cut}$ and to lie in the centre-of-mass rapidity interval $|\eta_\gamma| < 1$ (or $|\eta_\gamma| < 2$). The figure is taken from Ref. [25].

The CDF collaboration has reported [24] a preliminary result for exclusive χ_c production. Although it is consistent with perturbative QCD expectations [26], the mass of the χ_c-boson, which drives the scale of the process, is too low to justify just the use of perturbative QCD[d]. Therefore, it is intriguing that the qualitative features of the observed p_t and azimuthal angular distributions appear to be in good agreement with the perturbatively based expectations [28]. However, in Ref. [26], it was found that both a Regge formalism and perturbative QCD predict essentially the same qualitative behaviour for the central double-diffractive production of 'heavy' $\chi_c(0^{++})$ and $\chi_b(0^{++})$ mesons[e]. Due to the low scale, $M_\chi/2$, there is a relatively small contribution coming from the process, in which the incoming protons dissociate. Therefore simply selecting events with a rapidity gap on either side of the χ, almost ensures that they will come from the exclusive reaction, $p\bar{p} \to p\ +\ \chi\ +\ \bar{p}$. Although exclusive χ production is expected to dominate, the predicted [26] event rates are large enough to select double-diffractive dissociation events with large transverse energy flows in the proton fragmentation regions. Such events are particularly interesting. First, in this case, the large value of E_T provides the scale to justify the validity, and the reasonable accuracy, of the perturbative QCD calculation of the cross section. Next, by measuring the azimuthal distribution between the two E_T flows, the parity of the centrally produced system can be determined.

Another possible probe of the exclusive double-diffractive formalism would be to observe central open $b\bar{b}$ production; namely $b, \bar{b}$ jets with $p_t \gtrsim m_b$. Again, this would put the application of perturbative QCD on a sounder footing. It would allow a check of the perturbative formalism, as well as a study of the dynamics of $b\bar{b}$ production.

8. Conclusion

The installation of proton-tagging detectors in the forward region around ATLAS and/or CMS would add unique capabilities to the existing LHC experimental programme. The current calculations of the rates of CEP processes show that there is a real chance that new heavy particle production could be observed in this channel. For the Standard Model Higgs, this would amount to a direct determination of its quantum numbers, with

[d]Even lower scales correspond to the fixed target central double diffractive meson resonance production observed by the WA102 collaboration at CERN [27].

[e]Note that the results for $\chi_c(0^{++})$, given in [26], should be decreased by a factor of 1.5 due to the new value of the total $\chi_c(0^{++})$ width in PDG-2004 [29]. Thus, the predicted cross section for $\chi_c \to J/\psi + \gamma \to \mu\mu\gamma$ is now about 300 pb; with the CDF experimental cuts, it becomes about 50 pb.

an integrated luminosity of order 30 fb^{-1}. For certain MSSM scenarios, the tagged-proton channel may even be the discovery channel. At higher luminosities, proton tagging may provide direct evidence of CP-violation within the Higgs sector. There is also a rich QCD, electroweak, and more exotic, physics menu. This includes searches for extra dimensions, gluino and squark production, gluinonia, and, indeed, any object which has 0^{++} or 2^{++} quantum numbers and couples strongly to gluons [1].

Here we focused on the unique advantages of CEP Higgs production. The missing mass, $M_{\rm missing}$, measured by the forward proton detectors can then be matched with the mass $M_{b\bar{b}}$ from the main decay mode, $H \to b\bar{b}$. Moreover the QCD $b\bar{b}$ background is suppressed by a $J_z = 0$ selection rule. The events are clean, but the predicted yield is low: about 10 events, after cuts and acceptance, for an integrated luminosity of $\mathcal{L} = 30$ fb^{-1}. The signal-to-background ratio is about 1, depending crucially on the accuracy with which $M_{\rm missing}$ can be measured.

From the experimental perspective, the simplest channel to observe a Higgs Boson of mass between 140 GeV and 200 GeV is the WW decay mode. According to studies in [17], there will be a detectable signal, and the backgrounds should be controllable.

We have emphasized the importance of checking the perturbative QCD predictions by observing analogous CEP processes, with larger cross sections, at the Tevatron.

Acknowledgments

This work was supported by the UK PPARC, by a Royal Society FSU grant, by grants INTAS 00-00366, RFBR 04-02-16073 and 01-02-17383 and by the Federal Program of the Russian Ministry of Industry, Science and Technology SS-1124.2003.2. ADM thanks the Leverhulme Trust for an Emeritus Fellowship.

References

1. V.A. Khoze, A.D. Martin and M.G. Ryskin, *Eur. Phys. J.* **C23**, 311 (2002).
2. B.E. Cox, `arXiv:hep-ph/0409144`.
3. K. Piotrzkowski, *Phys. Rev.* **D63** (2001) 071502.
4. V.A. Petrov, R.A. Ryutin A.E. Sobol and J.-P. Guillaud, *JHEP* **0506**, 007 (2005) and references therein.
5. M. Bonnekamp *et al.*, `arXiv:hep-ph/0506275` and references therein.
6. M. Albrow *et al.* (FP-420), CERN-LHCC-2005-025, LHCC-I-015.
7. V.A. Khoze, A.D. Martin and M.G. Ryskin, `arXiv:hep-ph/0006005`, in *Proc.*

of 8th Int. Workshop on Deep Inelastic Scattering and QCD (DIS2000), Liverpool, eds. J. Gracey and T. Greenshaw (World Scientific, 2001), p.592.
8. V.A. Khoze, A.D. Martin and M.G. Ryskin, *Eur. Phys. J.* **C19** (2001) 477, erratum **C20**, 599 (2001).
9. A. De Roeck, V.A. Khoze, A.D. Martin, R. Orava and M.G. Ryskin, *Eur. Phys. J.* **C25**, 391 (2002).
10. A.B. Kaidalov, V.A. Khoze, A.D. Martin and M.G. Ryskin, *Eur. Phys. J.* **C33**, 261 (2004).
11. J.R. Ellis, J.S. Lee and A. Pilaftsis, *Phys. Rev.* **D71**, 075007 (2005).
12. V. A. Khoze, A. D. Martin and M. G. Ryskin, *Eur. Phys. J.* **C34**, 327 (2004).
13. M.G. Ryskin, A.D. Martin and V.A. Khoze, *these proceedings,* `arXiv:hep-ph/0506272`.
14. V.A. Khoze, A.D. Martin and M.G. Ryskin, *Eur. Phys. J.* **C24**, 459 (2002).
15. V.A. Khoze, A.D. Martin and M.G. Ryskin, *Eur. Phys. J.* **C14**, 525 (2000).
16. V.A. Khoze, M.G. Ryskin and W.J. Stirling, *Eur. Phys. J.* **C44**, 227 (2005).
17. B.E. Cox *et al.*, `arXiv:hep-ph/0505240`.
18. E. Boos, A. Djouadi and A. Nikitenko, *Phys.Lett.* **B578**, 384 (2004).
19. V.A. Khoze, A.D. Martin and M.G. Ryskin, *Eur. Phys. J.* **C18**, 167 (2000).
20. V.A. Khoze, A.D. Martin and M.G. Ryskin, *Eur. Phys. J.* **C26**, 229 (2002).
21. M. Carena and H. Haber, *Prog. Part. Nucl. Phys.* **50**, 63 (2003);
G. Degrassi, S. Heinemeyer, W. Hollik, P. Slavich and G. Weiglein, *Eur. Phys. J.* **C28**, 133 (2003);
A. Djouadi, `arXiv:hep-ph/0503172`.
22. A. Djouadi, J. Kalinowski and M. Spira, *Comput. Phys. Com.* **108**, 56 (1998), `arXiv:hep-ph/9704448`.
23. R. Kleiss and W.J. Stirling, *Nucl. Phys.* **B262**, 235 (1985).
24. M. Gallinaro (for the CDF Collaboration), `arXiv:hep-ph/0311192`; `arXiv:hep-ph/0505159`.
25. V.A. Khoze, A.D. Martin, M.G. Ryskin and W.J. Stirling, *Eur. Phys. J.* **C38**, 475 (2005).
26. V.A. Khoze, A.D. Martin, M.G. Ryskin and W.J. Stirling, *Eur. Phys. J.* **C35**, 211 (2004).
27. D. Barberis *et al.* (WA102 Collaboration), *Phys. Lett.* **B467**, 165 (1999); *ibid.* **B474**, 423 (2000); *ibid.* **B484**, 198 (2000); *ibid.* **B488**, 225 (2000); *ibid.* **B453**, 305, 316 (1999).
28. V.A. Khoze, A.D. Martin and M.G. Ryskin, *Eur. Phys. J.* **C24**, 581 (2002).
29. S. Eidelman *et al.* (Particle Data Group), *Phys. Lett.* **B592**, 1 (2004).

NON-PERTURBATIVE YANG–MILLS FROM SUPERSYMMETRY AND STRINGS, OR, IN THE JUNGLES OF STRONG COUPLING

M. SHIFMAN

William I. Fine Theoretical Physics Institute, University of Minnesota, Minneapolis, MN 55455, USA

I summarize some recent developments in the issue of planar equivalence between supersymmetric Yang-Mills theory and its orbifold/orientifold daughters. This talk is based on works carried out in collaboration with Adi Armoni, Sasha Gorsky and Gabriele Veneziano.

1. Introduction

Unlike some theories whose relevance to nature is sill a big question mark, quantum chromodynamics and other similar strongly coupled gauge theories will stay with us forever. QCD is a very rich (and quite old) theory supposed to describe the widest range of strong interaction phenomena: from nuclear physics to Regge behavior at large E, from color confinement to quark–gluon matter at high temperatures/densities (neutron stars), the vastest horizons of hadronic physics: chiral dynamics, glueballs, exotics, light and heavy quarkonia and mixtures of thereof, exclusive and inclusive phenomena, interplay between strong forces and weak interactions, ... That's why I do not expect theoretical developments to culminate in full analytic solution of QCD. And yet, in spite of its age, advances in QCD continue. The most recent advances are due to proliferation of supersymmetry and string-inspired methods. I will summarize some recent results which, to my mind, are most promising, and pose some stimulating questions.

2. Planar Equivalence

The main stumbling block in theoretical understanding of strongly coupled gauge theories is the absence of obvious expansion parameters. A hidden parameter which might serve the purpose, N (the number of colors) was suggested by 't Hooft long ago [1]. It governs expansion in topologies. The leading order at $N \to \infty$ corresponds to planar topology. Recently it was realized [2,3] that the very same parameter can be used to parametrize deviations of certain non-supersymmetric theories, close relatives of QCD, from supersymmetric Yang–Mills (SYM) theory. These relatives — they go under the name orbifold and orientifold gauge field theories — are obtained from supersymmetric gluodynamics by means of orbifolding and orientifolding, procedures well known in string theory. For our purposes we do not need to delve in string-theoretic aspects since all results we need can be readily formulated in field-theoretic language. They are shown in Fig. 1. SYM theory is assumed to be SU($2N$) or SU(N) gauge theory. The first case is pertinent to Z_2 orbifolding, the second to orientifolding. Then the Z_2 orbifold daughter [a] has the gauge group SU(N)×SU(N), and the fermion sector consisting of one bifundamental Dirac spinor. The gauge coupling of the orbifold daughter is adjusted as follows

$$g_D^2 = 2\, g_P^2\,, \tag{1}$$

where the subscripts D and P mark the daughter and parent theories. For historic reasons the first SU(N) is often referred to as "electric" (and marked by e), while the second as "magnetic" (and marked by m).

The orientifold gauge theory is even closer to QCD. Its gauge group is SU(N), the same as in the parent SU(N) SYM theory. The gauge couplings of the parent SYM and its orientifold daughter are identical too. The fermion sector consists of one Dirac fermion either in two-index symmetric (orienti-S) or two-index antisymmetric (orienti-A) representation of SU$(N)_{\rm color}$. In fact, at $N = 3$ orienti-A is identical to one-flavor QCD.

Both daughter theories, in the limit $N \to \infty$, were shown [3] to be perturbatively equivalent to their parent, supersymmetric gluodynamics. In other words, all planar Feynman graphs of the daughter theories that can be mapped onto the parent theory are in one-to-one correspondence with the parent planar graphs.

This remarkable observation motivated [2] a non-perturbative orbifold (NPO) conjecture, according to which the above planar equivalence holds beyond perturbation theory, non-perturbatively, in a common sector, i.e.

[a] Z_n orbifold daughters with $n > 2$ will not be discussed since these theories are chiral and, hence, cannot be considered as close relatives of QCD. Two other lines of research that are marginally related to my main topic are left aside, namely (i) orbifold pairs with both theories, parent and daughter, supersymmetric; and (ii) orbifolding with one or more compact spatial dimensions. In both cases there are special circumstances whose consideration will lead me far astray.

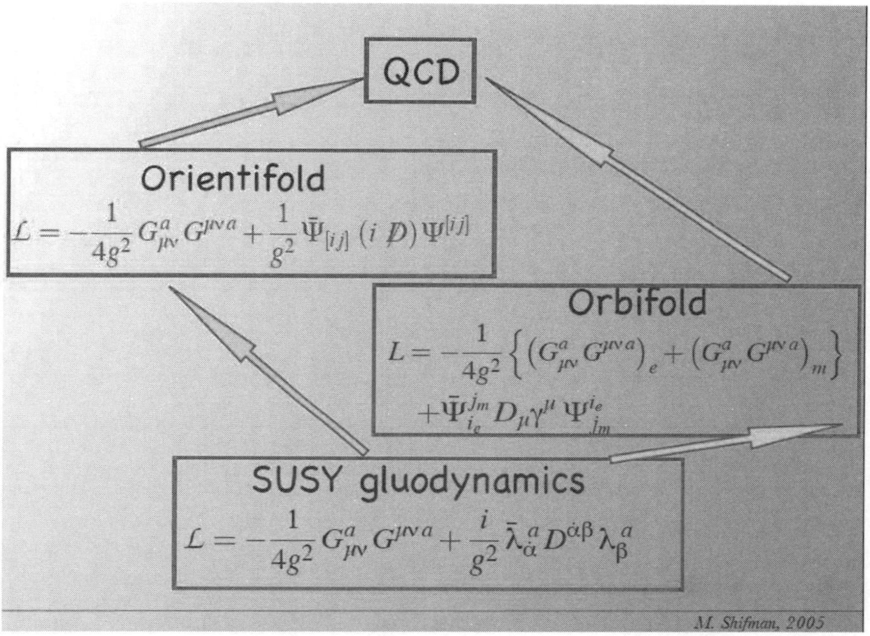

Figure 1. Orbifold and orientifold (orienti-A) daughters are obtained from SYM theory by orientifolding and orbifolding, respectively.

the sector of both theories, orbifold and SYM, which admits mapping in both directions.

As we will see shortly, radical distinctions in the vacuum structure of orbifold and SYM theories make NPO planar equivalence unlikely. At the same time, planar equivalence between orientifold daughter and its supersymmetric parent was solidly demonstrated, see [4,5] and the review paper [6], with quite a few far-reaching consequences that ensued almost immediately. Corresponding results were reported a year ago at various conferences, and I will not discuss them now (except for a few marginal remarks), referring the interested reader to [6]. Instead, I will dwell on the Z_2 orbifold daughter, a theory whose dynamics is rich and interesting irrespective of its (highly probable) non-perturbative non-equivalence to SYM theory.

Concluding this section I would like to display the 't Hooft large-N diagrammar for supersymmetric gluodynamics and its daughters (Fig. 2) which makes the perturbative proof of planar equivalence almost self-evident.

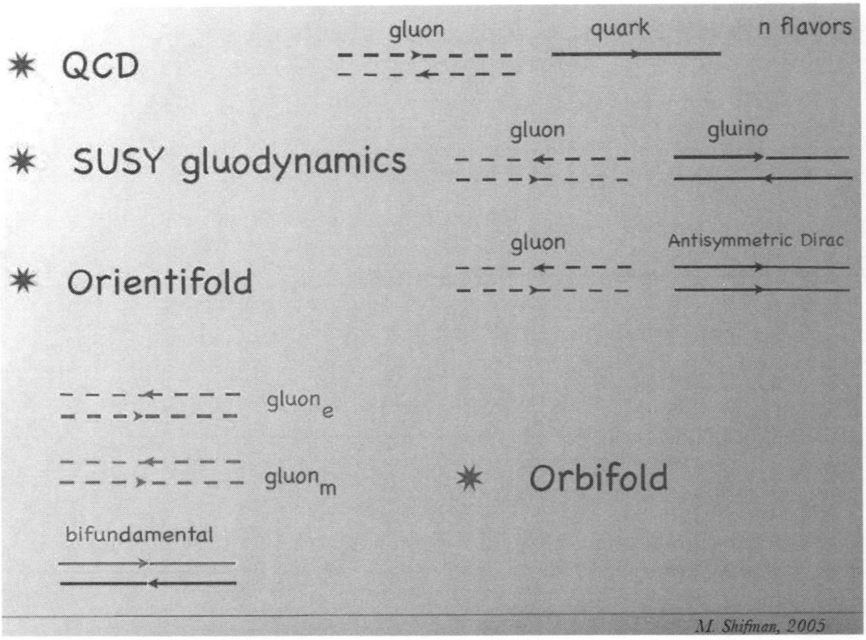

Figure 2. Feynman rules in QCD, SYM theory and orbifold and orientifold (orienti-A) daughters in the 't Hooft notation.

3. Non-perturbative planar equivalence: what does it mean?

As we will see shortly, SYM theory and its orbifold daughter have distinct vacuum structures even at $N = \infty$. The number of underlying (short-distance) degrees of freedom is also different. Under the circumstances one should carefully define what is expected to be equivalent. To calibrate the answer to this question it is instructive to consider an example where the answer is known, namely, let us compare SYM theory with *itself* in two cases:

$$N_c = 2N, \quad g_P^2 \equiv g^2 \tag{2}$$

$$N_c = N, \quad g_D^2 \equiv 2g^2 \,. \tag{3}$$

The 't Hooft coupling in both cases is one and the same, $\lambda = 2Ng^2$, which entails, in turn, the equality of the dynamical scales, $\Lambda_P = \Lambda_D$. We will refer to the theories (2), (3) as parent/daughter. Having one and the same 't Hooft coupling, these theories are planar equivalent. This is as good as it gets, indeed.

Note that for the purpose of infrared regularization we will introduce a small gluino mass term $-(m/g^2)\mathrm{Tr}\lambda^2+$ h.c. The mass parameter m is assumed to be real and positive (a phase can be introduced through the θ term). The value of m must be the same in both theories since this parameter is physically observable.

Now, both theories are confining and have coincident spectra of composite bosons, up to $O(1/N^2)$ corrections. More exactly, we compare here excitation spectra over vacua which can be mapped one onto another, for instance, those characterized by real (and negative) gluino condensate $\langle \mathrm{Tr}\,\lambda^2\rangle$. One must be careful since the parent theory has $2N$ vacua while the daughter one N vacua. In fact, the boson spectra in adjacent vacua differ only by terms $O(m/N^2)$ (one should remember that at $m \neq 0$ the true vacuum is the one in which $\mathrm{Tr}\lambda^2$ is real and negative; others are quasistable, with an exponentially suppressed decay rate [7], $\Gamma \sim \exp(-N^4)$). Only for distant vacua, e.g. those with negative and positive $\mathrm{Re}\langle \mathrm{Tr}\,\lambda^2\rangle$, the spectra are shifted by $O(N^0)$. This fact is related to another similar statement. Switching on $\theta \neq 0$ changes bosonic spectra since at $m \neq 0$ particle masses are θ-dependent. However, changing θ from zero to $\delta\theta \sim 2\pi$ produces an impact on the spectrum suppressed by $1/N^2$. This can be readily seen e.g. from the SVZ sum rule [8] analysis at $\theta \neq 0$, for instance, for highly excited two-gluino bosons one can estimate

$$\delta m_n^2 \sim \frac{m\,\Lambda^{-2}}{g^2}\,N_c^{-2}\,\mathrm{Re}\,\langle \mathrm{Tr}\,\lambda^2\rangle\,, \tag{4}$$

where n is the number of the radial excitation, $n \gg 1$. CP odd quantities which might be generated at $\theta \neq 0$ are $O(1/N)$.

The impact of $\theta \neq 0$ becomes of order $O(N^0)$ only if $\delta\theta \sim 2\pi N$.

At the same time, even though the bosonic spectra are planar equivalent, the vacuum energy densities are not equal. The vacuum energy density [b]

$$\mathcal{E} = mg^{-2}\,\langle \mathrm{Tr}\,\lambda^2\rangle \sim N^2 \tag{5}$$

is sensitive to the overall number of the fundamental degrees of freedom. It is obvious that we cannot demand the equality of the vacuum energies in the parent/daughter theories. Equation (5) is fully compatible with the fact

[b] Tr is normalized in such a way that $\mathrm{Tr}\,(T^aT^b) = \delta^{ab}$ where T^a are color generators in the fundamental representation.

that the m dependence of the composite boson masses is identical in the parent and daughter theories. This can be seen from OPE for the two-point function

$$\left\langle g^{-2}\,\mathrm{Tr}\,\bar{\lambda}_{\dot{\alpha}}\lambda_{\alpha},\; g^{-2}\,\mathrm{Tr}\,\bar{\lambda}_{\dot{\beta}}\lambda_{\beta}\right\rangle \tag{6}$$

which scales as N^2. The mass correction to the above two-point function is given by (5). The relative weight of the leading (unit) operator and the mass correction term is N-independent.

Returning to the θ-dependence, the coincidence of the parent/daughter boson spectra can be maintained provided

$$\theta_D = \frac{1}{2}\,\theta_P\,. \tag{7}$$

As was mentioned, for $\theta = O(1)$ the vacuum angle effects in the spectra are irrelevant as they are suppressed by $1/N^2$. However, one can consider $\theta \sim CN2\pi$ where C is small numerically but not parametrically[c]. Then Eq. (7) follows from holomorphic dependence of appropriate quantities on complexified coupling constant which is dictated, in turn, by supersymmetry of the parent/daughter theories. I will further comment on this issue in Sect. 7.

The θ term has no impact whatsoever on perturbation theory. It is not seen at all in perturbation theory. Therefore, perturbative proofs of planar equivalence have nothing to say regarding this aspect. On the other hand, non-perturbative quantities, such as the gluino condensate, do carry a θ dependence which imposes the above identification of the parent/daughter vacuum angles. The Dashen points [9] in the parent/daughter theories do not match each other, as a consequence of a mismatch in the vacuum multiplicities. It is worth emphasizing that (i) in discussing the θ evolution we have to stick, at $\theta \geq \pi$, to a "wrong" (quasi-stable) vacuum which will ensure a smooth evolution; (ii) the Dashen phenomenon is then irrelevant in the leading in N approximation.

Besides particle excitations both theories have domain-wall excitations. The tensions of the elementary walls and their multiplicities scale as N and are, therefore, different in the parent/daughter theories. A common factor here is that all domain walls interpolating between vacua with distinct values of the gluino condensate are stable.

---

[c]If $C \sim 1$ one looses the vacuum quasi-stability.

4. A refinement of the proof of planar equivalence for orientifold daughter

A certain aspect in the previous analysis of non-perturbative planar equivalence between the SYM parent and orietifold daughter was treated at an intuitive level. This gap is closed in a refined proof [5] making use of the fermion loop expansion. The equivalence extends to θ effects, e.g. the topological susceptibility — a feature which is certainly lost in the case of the orbifold daughter. This is in one-to-one correspondence with the fact that the vacuum structure of the orientifold daughter at $N \to \infty$ is identical to that of the parent theory. In particular, there is an exact matching of the Dashen transitions.

5. The role of Z_2 invariance in the orbifold theory

The Lagrangian of the orbifold theory,

$$\mathcal{L} = -\frac{1}{4g^2}\left[\left(G^a_{\mu\nu}G^{\mu\nu\,,a}\right)_e + \left(G^a_{\mu\nu}G^{\mu\nu\,,a}\right)_m\right]$$

$$+\frac{1}{g^2}\left[\overline{\lambda^{i_e}_{j_m}}\,(i\,\not{D}\lambda)^{i_e}_{j_m} + \overline{\lambda^{i_m}_{j_e}}\,(i\,\not{D}\lambda)^{i_m}_{j_e}\right], \tag{8}$$

has an obvious discrete Z_2 symmetry with respect to the interchange $e \leftrightarrow m$. (Note that in Fig. 1 two Weyl spinors,

$$\lambda^{i_e}_{j_m} \quad \text{and} \quad \overline{\lambda^{j_m}_{i_e}}\,,$$

are combined in one Dirac spinor. For a while I will omit the subscript D in the gauge coupling. One should remember, however, that $g_D^2 = 2g_P^2$, see Eq. (1).)

A crucial physical question is whether or not this Z_2 symmetry is spontaneously broken. If it is dynamically broken, the number of vacua is doubled. As a manifestation of the discrete symmetry breaking, domain walls must emerge, which interpolate between the vacua related by the broken Z_2. The corresponding order parameters are Z_2 odd. For historical reasons, the Z_2 odd sector of the theory is referred to as a "twisted sector."

If the above Z_2 is not broken, the spectrum of the theory in each vacuum can be classified with regards to Z_2. For instance, Z_2 even particles do not mix with Z_2 odd, all domain walls of the unbroken theory are Z_2 symmetric, and so on.

The fate of nonperturbative planar equivalence between the orbifold theory and its supersymmetric parent is inseparable from the fate of Z_2.

As was shown in [10,11,12], if Z_2 is unbroken, perturbative planar equivalence extends to the nonperturbative level. In the opposite case of the dynamical Z_2 breaking, planar equivalence is not expected to survive at the nonperturbative level. A shift of the vacuum energy from zero ensues: the vacuum energy density is expected to become negative, see Sects. 6, 8 and 10. Other immediately observable consequences refer to the particle spectrum. Multiple (parity/spin) degeneracies which would be inherited from supersymmetric Yang-Mills under planar equivalence, will be lifted.

To see that this is indeed the case suffice it to note that if the twisted scalar field [d]

$$T \equiv \left(\mathrm{Tr}\, G_e^2 - \mathrm{Tr}\, G_m^2\right) \tag{9}$$

develops a Z_2-breaking vacuum expectation value, while its pseudoscalar counterpart

$$\tilde{T} \equiv \left(\mathrm{Tr}\, G_e \tilde{G}_e - \mathrm{Tr}\, G_m \tilde{G}_m\right) \tag{10}$$

does not, this will be transmitted to the untwisted sector e.g. through a term

$$\delta\mathcal{L} = \frac{1}{N}\, \tau^2\, \sigma\,, \tag{11}$$

where τ is a meson for which the interpolating field is T, while σ is the dilaton (the corresponding interpolating operator is $S = \mathrm{Tr}\, G_e^2 + \mathrm{Tr}\, G_m^2$). A vacuum expectation value $\langle\tau\rangle \sim N$ will entail a shift in $\langle\sigma\rangle \sim N$, which will lead, in turn, to a shift in the σ mass of order $O(N^0)$, not accompanied by a corresponding shift in the mass of the untwisted pseudoscalar meson.

Thus, understanding dynamics governing the Z_2 symmetry of the orbifold model is a key to solving the issue of nonperturbative planar equivalence in the case at hand. What does today's theory tell us on that?

6. The mode of Z_2 implementation

String theory prompts us [13,14] that in the non-supersymmetric (or $\mathcal{N} = 0$) orbifold daughter of $\mathcal{N} = 4$ SYM theory, the Z_2 symmetry is spontaneously

---

[d] Here and below the normalization of traces is such that

$$\mathrm{Tr}\, G^2 = \sum_{a=1}^{4N^2} G^a_{\mu\nu}\, G^{\mu\nu\ a}\,, \qquad \mathrm{Tr}\, (G^2)_e = \sum_{a=1}^{N^2} \left(G^a_{\mu\nu}\, G^{\mu\nu\ a}\right)_e\,,$$

and so on.

broken above a critical value of the 't Hooft coupling. The orbifold field theory under consideration can be described by a brane configuration of type-0 string theory [15]. Type-0 strings contain a closed-string tachyon mode in the twisted sector. The tachyon couples [14] to the twisted field (9) of the $SU_e(N)\times SU_m(N)$ gauge theory. The prediction of string theory [14] is that the perturbative vacuum at $\langle T\rangle = 0$ is unstable. In the *bona fide* vacua a condensate of the form

$$\left\langle \mathrm{Tr}\, G_e^2 - \mathrm{Tr}\, G_m^2 \right\rangle = \pm\Lambda^4 \tag{12}$$

must develop.

Of course, a long way lies between the above string construction and the orbifold field theory specified in Fig. 1 or Eq. (8) *per se.* Therefore, it is natural to address the issue of the spontaneous Z_2 breaking directly in field theory. In Ref. [16] (see also [11], v.1) low-energy theorems were suggested as a tool for proving nonequivalence of the orbifold daughter theories to the parent SYM theory. These theorems become instrumental under the assumption of exact coincidence between the corresponding vacuum condensates. However, as explained in Sect. 3, the vacuum condensate coincidence is not necessary, generally speaking. The above-mentioned low-energy theorems reflect not only the vacuum structure — they are potentially sensitive to the number of fundamental degrees of freedom. This aspect was pointed out in [12]. In passing from the orbifold theory to its parent the number of fundamental degrees of freedom doubles.

Relaxing the requirement of exact coincidence makes the low-energy theorems uninformative: allowing for unequal condensates one concludes that these theorems cannot prove or disprove the Z_2 symmetry breaking.

Another argument suggested in [11] is based on the domain wall dynamics. If Z_2 was unbroken and NPO conjecture valid, the domain walls in the orbifold theory that are inherited from SYM theory would be stable. Apparently, this is not the case. To discuss the issue in more detail I will have to briefly review what is known of the vacuum structure in the orbifold theory (Sect. 7).

Concluding this section, it is instructive to outline a possible scenario of the development of the tachyonic mode coupled to the twisted operator (9). Let us give a mass term m to the fermion field in (8),

$$\mathcal{L}_m = -m\, g^{-2}\, \bar{\Psi}\, \Psi\,. \tag{13}$$

This mass term is obviously Z_2 invariant. We will consider m as a free parameter, keeping the dynamical scale Λ fixed. Then, at $m/\Lambda \to \infty$ the fermion field can be integrated out leading to two disconnected SU(N) gauge theories, electric and magnetic. At finite but large values of m, there is a weak connection between the electric and magnetic theories which can be described by a (local) operator $m^{-4}\,\mathrm{Tr}\,G_e^2\,\mathrm{Tr}\,G_m^2$. The mass-squared matrix of the electric/magnetic scalar glueballs takes the form

$$\mathcal{M}^2 = \begin{pmatrix} \mu_e^2 & \alpha^2 \\ \alpha^2 & \mu_m^2 \end{pmatrix} \tag{14}$$

where $\mu_e^2 = \mu_m^2 = \mathrm{const}\,\Lambda^2$ and α is a small parameter proportional to m^{-2}. The Z_2 invariance of the theory manifests itself in the fact that $\mu_e^2 = \mu_m^2 \equiv \mu^2$. The eigenvalues of $\mathcal{M}^2$ are $\mu^2 \pm \alpha^2$. The corresponding eigenstates are built of mixtures of the electric and magnetic gluons, with Z_2 parity $+1$ and -1, respectively.

Now, let us diminish m moving towards Λ. This enhances interaction between the electric and magnetic sectors, which no longer can be described by a local operator. If at $m = 0$ the transition matrix element α^2 is larger than the diagonal ones μ^2, a negative eigenvalue in the twisted sector emerges. At a certain critical value of m,

$$m_* \sim \Lambda,$$

the Z_2-odd glueball becomes massless, while further decrease of m from m_* to zero makes the corresponding channel tachyonic causing condensation of the operator (9) and a radical vacuum restructuring signifying spontaneous breaking of the Z_2 invariance.

One can illustrate the very same statement in a slightly different language of effective Lagrangians. Indeed, if one approaches the critical value m_* from the large m side one can describe the vacuum structure by the effective Lagrangian of the type [17]

$$\mathcal{L} = S_e \ln \frac{S_e}{e} + S_m \ln \frac{S_m}{e} + \eta\, S_e\, S_m \tag{15}$$

where $S_{e,m} = \mathrm{Tr}\, G_{e,m}^2$, and I put $\Lambda = 1$. The above Lagrangian is explicitly Z_2 invariant. Of course, it is valid only at $\eta \ll 1$, where the vacuum solution is Z_2 invariant too, $S_e = S_m \approx 1$. Assume that, at a qualitative level, Eq. (15) can be extrapolated to $\eta \sim 1$. Then, at $\eta = e$ the vacuum solution is still Z_2 symmetric, $S_e = S_m = e^{-1}$, but the mass eigenvalue corresponding to $S_e - S_m$ vanishes. Further increase of η leads to Z_2-asymmetric vacuum solutions while the Z_2-symmetric extremum is no more minimum of the potential.

7. Vacuum structure of the orbifold daughter at a glance

The gauge group of the orbifold theory is a direct product of two SU(N)'s. Correspondingly, it has two vacuum angles conjugated to two distinct noncontractible cycles in the space of fields. We will introduce these two vacuum angles as follows:

$$\mathcal{L}_\theta = \frac{\theta_D}{32\pi^2}\left[\left(G^a_{\mu\nu}\tilde{G}^{\mu\nu\,,a}\right)_e + \left(G^a_{\mu\nu}\tilde{G}^{\mu\nu\,,a}\right)_m\right]$$

$$+\frac{\vartheta}{32\pi^2}\left[\left(G^a_{\mu\nu}\tilde{G}^{\mu\nu\,,a}\right)_e - \left(G^a_{\mu\nu}\tilde{G}^{\mu\nu\,,a}\right)_m\right]. \tag{16}$$

They refer to non-twisted and twisted sectors of the theory, respectively. Since the parent theory has no twisted sector, the NPO conjecture requires $\vartheta = 0$. Let us set $\vartheta = 0$ for the time being, and focus on θ_D.

The order parameter of the parent theory marking its $2N$ vacua, the gluino condensate, is mapped onto the fermion condensate $\bar{\Psi}\frac{1}{2}(1-\gamma_5)\Psi$ in the daughter theory. Note that this operator is Z_2 invariant; hence, its nonvanishing vacuum expectation value is insensitive to the spontaneous breaking of Z_2.

Following the standard line of reasoning, one can conclude that the fermion condensate does develop, and has N distinct values,

$$\left\langle \bar{\Psi}\frac{1}{2}(1-\gamma_5)\Psi\right\rangle = \text{const.}\, N\,\Lambda^3 \exp\left(i\,\frac{2\pi\, k + \theta_D}{N}\right), \qquad k = 1, 2, ..., N\,. \tag{17}$$

The N-valuedness of the fermion condensate is in one-to-one correspondence with the dependence on θ_D/N which, in turn, follows from the consideration of the chiral anomaly. Thus, the fermion condensate marks N distinct chirally asymmetric sectors.

In the absence of the fermion mass term, $m = 0$, the vacuum angle θ_D is physically unobservable. Indeed, at $m = 0$ the axial current is classically conserved. The chiral anomaly then allows one to completely rotate away the vacuum angle θ_D. No physically measurable quantity can depend on it. In particular, the vacuum energy is θ_D independent. Only if $m \neq 0$, the vacuum angle θ_D becomes observable.

Note that the dependence of the fermion condensate on θ_D indicated in Eq. (17) and the relation between the parent and daughter vacuum angles (7) are compatible with the θ dependence of the gluino condensate in SU($2N$) SYM theory,

$$\langle\lambda^a_\alpha\lambda^{a\,,\alpha}\rangle = -6(2N)\,\Lambda^3 \exp\left(i\,\frac{2\pi k + \theta_P}{2N}\right), \quad k = 1, ..., 2N\,. \tag{18}$$

The g_P^2/g_D^2 ratio, see Eq. (1), also matches. Thus, the fermion condensate of the orbifold theory could have been projected from the parent theory provided that the NPO conjecture was valid.

As an example, I depicted the chiral condensates of the parent/daughter theories at

$$\theta_D = \pi\,, \qquad \theta_P = 2\pi\,, \tag{19}$$

in Fig. 3. $P_{0,\pm 1}$ are the vacua of the SYM theory, while $D_{\pm 1}$ are the vacua of the orbifold theory. Since the vacua reflect the discrete chiral symmetry breaking, $2N$ vacua of SYM theory are degenerate, and so are N vacua of the orbifold theory. Whether or not they are degenerate between themselves, depends on the validity of planar equivalence.

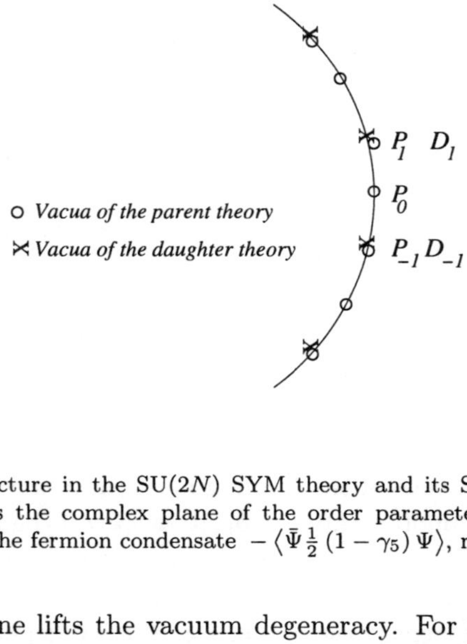

Figure 3. The vacuum structure in the SU(2N) SYM theory and its SU(N)×SU(N) orbifold daughter. Shown is the complex plane of the order parameters, the gluino condensate $-\langle\lambda^a_\alpha\lambda^{a\,,\alpha}\rangle$ and the fermion condensate $-\left\langle\bar\Psi\,\frac{1}{2}\,(1-\gamma_5)\,\Psi\right\rangle$, respectively.

Introducing $m \neq 0$ one lifts the vacuum degeneracy. For instance, for real and positive m the vacua $P_{\pm 1}$ are excited (quasistable) because

$$\mathcal{E}_{P_{\pm 1}} > \mathcal{E}_{P_0}\,.$$

At the same time, the daughter theory has two-fold degeneracy,

$$\mathcal{E}_{D_{+1}} = \mathcal{E}_{D_{-1}},$$

a phenomenon well-known at $\theta = \pi$, the so-called Dashen phenomenon [9]. As was explained in Sect. 3, θ-dependent effects are suppressed by $1/N$.

The fermion condensate (17) is a good order parameter for the chiral symmetry breaking. It cannot serve, however, as an order parameter for the Z_2 breaking. In the orbifold theory with the spontaneously broken Z_2 the fermion condensate does not differentiate those vacua which are connected to each other by Z_2 because it is Z_2-even. We must supplement (17) by a Z_2-odd expectation value of (9). This vacuum expectation value (VEV) is dichotomic. The fermion condensate (17) in conjunction with $\langle T \rangle = \pm\Lambda^4$ fully identifies each degenerate vacuum of the orbifold theory. If spontaneous breaking of the discrete chiral symmetry produces N vacua, this number is doubled in the process of Z_2 breaking. Somewhat symbolically, the corresponding vacuum structure is presented in Fig. 4. The angular coordinate represents the phase of (17), while the radial coordinate can take two distinct values representing the dichotomic parameter $\langle T \rangle$.

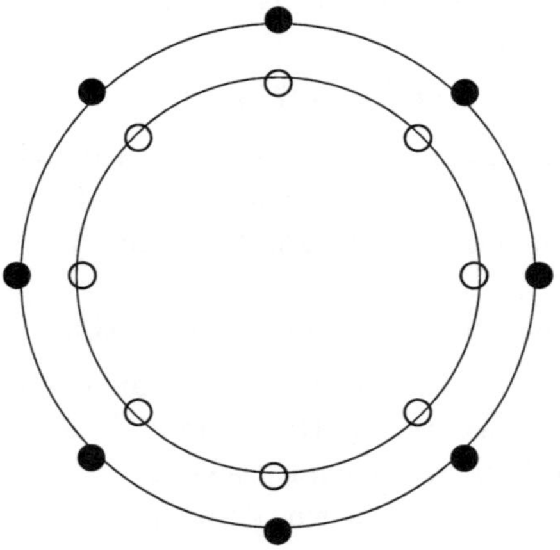

Figure 4. The vacuum structure of the SU(8)×SU(8) orbifold theory reflecting spontaneous breaking of the Z_2 symmetry.

The orbifold theory has a remarkable feature: because of its proven *perturbative* planar equivalence to SYM theory, the vacuum energy density — certainly a Z_2 symmetric parameter — can and does play the role of the order parameter for Z_2 breaking. The vacuum energy density is proportional to the vacuum expectation value of the Z_2-even gluon operator

$\langle \mathrm{Tr}\, G_e^2 + \mathrm{Tr}\, G_m^2 \rangle$. Indeed, this operator is related, in turn, to the total energy-momentum tensor of the theory,

$$\theta^\mu_\mu = -\frac{3N}{32\pi^2} \sum_{\ell=e,m} \left(G^a_{\mu\nu}\, G^a_{\mu\nu} \right)_\ell . \tag{20}$$

Ever since the gluon condensate was introduced in non-Abelian gauge theories [8] people tried to identify it as an order parameter. Nobody succeeded. The orbifold theory is the one where it is an order parameter, albeit in a special sense.

If Z_2 is unbroken, the orbifold theory is perfectly equivalent at $N \to \infty$ to SYM theory, and then $\langle \mathrm{Tr}\, G_e^2 + \mathrm{Tr}\, G_m^2 \rangle$ reduces to $\langle \mathrm{Tr}\, G^2_{\mathrm{SYM}} \rangle$. The latter condensate vanishes due to supersymmetry of the parent theory. Hence, the Z_2 symmetric vacua in the daughter theory would have vanishing vacuum energy density in the leading order in N.

When the Z_2-symmetric point becomes unstable, the Z_2-asymmetric vacua must have a negative energy density. Equation (20) implies then that in the genuine vacua

$$\langle \mathrm{Tr}\, G_e^2 + \mathrm{Tr}\, G_m^2 \rangle > 0 \tag{21}$$

at order $O(N^2)$. In this way the gluon condensate acquired the role of a Z_2 breaking order parameter, much in the same way as $\langle \mathrm{Tr}\, G^2_{\mathrm{SYM}} \rangle$ is the order parameter for supersymmetry breaking in SYM theory.

8. Domain-wall-based argument for Z_2 breaking

In this section we analyze the domain wall dynamics in the Z_2 orbifold theory. Since domain walls are "QCD D-branes" [18] a similarity between the wall dynamics and D-brane dynamics is clear.

Why domain walls? As well-known, domain walls are physical manifestations of spontaneously broken discrete symmetries. Since our consideration aims at exploring the Z_2 breaking in the orbifold daughter theory, addressing domain walls is an adequate maneuver.

To begin with, let me recapitulate the domain wall topic in the parent theory. SU($2N$) SYM theory has BPS domain walls [19] that carry both tension σ and charge Q (per unit area), with $\sigma = Q$. The expressions for the tension and charge can be written as follows [20]:

$$\sigma = \frac{3(2N)}{32\pi^2} \int_{\mathrm{wall}} dz\, \mathrm{Tr}\, G^2 , \tag{22}$$

$$Q = \frac{3(2N)}{32\pi^2} \int_{\mathrm{wall}} dz\, \mathrm{Tr}\, G\tilde{G} , \tag{23}$$

where z is the direction perpendicular to the wall plane. Equation (22) is a consequence of the scale anomaly. The walls interpolating between the adjacent vacua (e.g. P_0 and P_1 in Fig. 3) are called elementary, or 1-walls. One can consider bound states of the elementary walls too. These walls interpolate between the vacua i and $i+k$ with $k>1$ and are referred to as k-walls. For instance, the wall interpolating between P_{-1} and P_1 in Fig. 3 is a 2-wall. At $N=\infty$ it is marginally stable, since the tension of the 2-wall is twice the tension of the 1-wall. Although elementary walls do interact via the exchange of glueballs, there is an exact cancellation between the contribution of even- and odd-parity glueballs [20] at $N=\infty$. From the world-sheet theory standpoint, the no-force result is due to the Bose-Fermi degeneracy on the wall. I will return to the world-sheet theory shortly, after a brief remark regarding generalizations of Eqs. (22) and (23) in the orbifold daughter theory.

For the tension and charge of the orbifold theory domain walls one can write [e]

$$\sigma_D = \frac{3N}{32\pi^2}\int_{\rm wall} dz\,{\rm Tr}\,G_e^2 + \frac{3N}{32\pi^2}\int_{\rm wall} dz\,{\rm Tr}\,G_m^2\,, \tag{24}$$

$$Q_D = \frac{3N}{32\pi^2}\int_{\rm wall} dz\,{\rm Tr}\left(G\tilde{G}\right)_e + \frac{3N}{32\pi^2}\int_{\rm wall} dz\,{\rm Tr}\left(G\tilde{G}\right)_m\,. \tag{25}$$

It is suggestive to think of the domain walls in the orbifold theory as of marginally bound states of fractional "electric" and "magnetic" domain walls, with the following tensions and charges:

$$\sigma_e = \frac{3N}{32\pi^2}\int dz\,{\rm Tr}\,G_e^2\,, \qquad \sigma_m = \frac{3N}{32\pi^2}\int dz\,{\rm Tr}\,G_m^2\,,$$

$$Q_e = \frac{3N}{32\pi^2}\int dz\,{\rm Tr}\,(G\tilde{G})_e\,, \qquad Q_m = \frac{3N}{32\pi^2}\int dz\,{\rm Tr}\,(G\tilde{G})_m\,. \tag{26}$$

Assuming unbroken Z_2 s i.e. $\sigma_e = \sigma_m$, we would get

$$\sigma_{e,m} = \frac{1}{2}\left(\sigma_e + \sigma_m\right), \tag{27}$$

i.e., a fractional amount of tension, in full analogy with fractional D-branes. Then we would have to conclude that, say, at $k=2$ two parallel electric domain walls do not interact at $N=\infty$. The same would be valid for the

[e] In SYM theory such integrals are well-defined since $\langle G^2\rangle$ vanishes in any supersymmetric vacuum. In the orbifold theory this is not necessarily the case. The integrals in Eq. (24) must be properly regularized.

magnetic walls. Unfortunately, the world-sheet theory in the case at hand does not support this conclusion.

I will again start from SU($2N$) SYM theory. The world-sheet theory for k-walls in $\mathcal{N}=1$ gluodynamics was derived by Acharya and Vafa [21]. It was shown to be a (2+1)-dimensional $U(k)$ theory with level-$2N$ Chern–Simons term and it was shown to have (2+1)-dimensional $\mathcal{N}=1$ supersymmetry. The action of the theory is

$$S=\int d^3x\left\{\mathrm{Tr}\left(-\frac{1}{4e^2}F^2+\frac{2N}{16\pi}\epsilon^{ijk}A^iF^{jk}+\frac{1}{2}(D_i\Phi)^2\right)+\text{fermions}\right\}. \tag{28}$$

All fields in the action, including the fermion fields, transform in the adjoint representation of $U(k)$. For definiteness, we will consider a minimal case $k=2$.

In the orbifold daughter, the world-sheet theory becomes, by virtue of the orbifold procedure, a $\mathrm{U}_e(1)\times\mathrm{U}_m(1)$ gauge theory with a neutral scalar field and "bifundamental" fermions. The same conclusion on the world-sheet theory can be obtained directly through a consideration of type-0 string theory similar to that of Acharya and Vafa. In this case the world-sheet action is

$$S=\int d^3x\left\{\sum_{\ell=e,m}\left(-\frac{1}{4e^2}F_\ell^2+\frac{N}{16\pi}\epsilon^{ijk}A_\ell^i\,F_\ell^{jk}+\frac{1}{2}(\partial_i\Phi_\ell)^2\right)\right.$$
$$\left.+\;\bar{\Psi}\,(\Phi_e-\Phi_m)\,\Psi+\ldots\right\}. \tag{29}$$

The occurrence of the Yukawa coupling $\bar{\Psi}\,(\Phi_e-\Phi_m)\,\Psi$ in the daughter theory, with no counterpart in the parent one, is the fact of a special importance.

One can interpret the above expression as follows. The daughter wall is a sum of the electric and magnetic walls that interact with each other via the bifundamental fermions. The electric branes can be separated from the magnetic branes as is seen from the fact that the Yukawa term $\bar{\Psi}\,(\Phi_e-\Phi_m)\,\Psi$ in the action (29) can make the bifundamental fermion massive. The vacuum expectation values

$$\langle\Phi_e\rangle=v_e\,,\qquad\langle\Phi_m\rangle=v_m\,, \tag{30}$$

which can be chosen to be real are in one-to-one correspondence with the wall separation. If $v_e\neq v_m$ a mass μ for the world-sheet fermions is generated,

$$\mu=v_e-v_m\,. \tag{31}$$

At $\mu \to \infty$ the fermions decouple — we have two decoupled U(1) theories. The world-sheet theory on the separated electric (or magnetic) domain walls is just a bosonic U(1) gauge theory with a level-N Chern–Simons term. It is not supersymmetric. There is no reason for the wall tension non-renormalization and the no-force statement.

The above conclusion can be backed up by a calculation of the wall repulsion [11,22]. Needless to say, this repulsion is in contradiction with the NPO conjecture.

9. Back to the bulk theory

If the orbifold has Z_2-odd vacua, the tachyon field potential must have minima away from the origin, as shown in Fig. 5, cf. the last paragraph in Sect. 7. String theory gives us a hint that the point $T = 0$ is unstable. Field theory allows us to say that the potential $V(T)$ is bounded from below since the regime of large expectation values is fully controlled by semiclassical dynamics. From the field-theoretic standpoint it is clear that

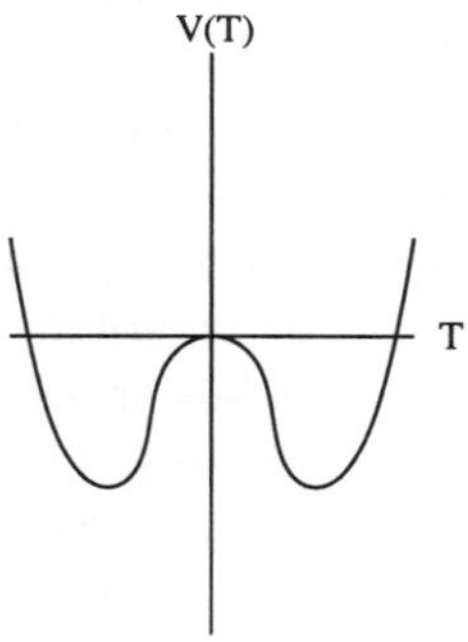

Figure 5. The tachyon field potential.

the only possibility open is that in the *bona fide* vacuum $\langle T \rangle \sim \Lambda^4$. Non-stabilization of tachyons would mean $\langle T \rangle \gg \Lambda^4$, which is ruled out.

In the parent SYM theory with the gauge group SU($2N$), there are $2N$ vacua, with the gaugino condensate as an order parameter, see Fig. 3. The domain walls interpolate between these $2N$ various vacua. In the daughter theory the situation is more complicated. Z_2 breaking implies that each vacuum of the N "false" perturbative vacua splits into two, see Fig. 4.

A scenario of the wall inheritance from the parent to daughter theory we have in mind is as follows. We first pretend that the daughter theory is planar equivalent to SYM, and that the Z_2 symmetry is unbroken. Start from a 2-wall in the parent theory. It will be inherited, as a minimal wall in the daughter theory. This is seen from Fig. 3. We may consider e.g. the wall connecting D_{-1} and D_1 in the daughter (this is a minimal wall in the daughter), versus the wall connecting P_{-1} and P_1 in the parent (this is a 2-wall in SYM theory).

In the parent theory two 1-walls comprising the 2-wall do not interact with each other (at $N = \infty$). If we consider them on top of each other, the world-volume theory has $U(2)$ gauge symmetry. However, nobody precludes us from introducing a separation. Then we will have U(1) on each 1-wall, U(1)×U(1) altogether. The tension of each 1-wall is 1/2 of the tension of the 2-wall, it is well-defined and receives no quantum corrections. The fact that the world-volume theory on each 1-wall is supersymmetric is in one-to-one correspondence with the absence of quantum corrections. In the daughter theory the minimal wall splits into one electric and one magnetic repelling each other. (The electric one connects D_{-1} with the would-be vacuum which is a counter-partner of P_0, the magnetic one connects the would-be vacuum which is a counter-partner of P_0 with D_1).

How can one visualize this situation?

In the parent theory we have degenerate minima at all points P_i. In the Z_2 broken orbifold theory these minima become saddle point (still critical points, but unstable). Near every second saddle point two minima develop. Of course, the walls that would be inherited from SYM are all unstable, with tachyonic modes. 1-walls are transformed into electric/magnetic walls of the orbifold theory, which are still unstable and, in fact, decay. Each of them separately could decay only into a "twisted wall" connecting white and adjacent black true vacua. The "untwisted" electric+magnetic wall can decay into a minimal stable wall of the daughter theory which connects two neighboring black vacua or two neighboring white vacua.

10. Why non-perturbative non-equivalence is natural?

I this section I will try to illustrate why a shift of the vacuum energy from zero is expected in the orbifold theory. Needless to say this can only happen if perturbative planar equivalence gives place to non-equivalence at the non-perturbative level. The issue to be discussed here is the vacuum angle dependence, see Eq. (16). In this section I will treat N as a fixed parameter assuming that transition to $N \to \infty$ is smooth, as is the case in pure Yang-Mills theory.

As was mentioned, physical quantities do not depend on θ_D, as this angle can be rotated away. A weak dependence appears if $m \neq 0$, but we will be interested in the limit $m \to 0$. For our present purposes θ_D is irrelevant and can be set at zero.

Unlike θ_D, the second vacuum angle, ϑ, cannot be rotated away: the only axial current of the theory is Z_2 even while the ϑ term in Eq. (16) is Z_2 odd. Thus, physics must be ϑ dependent even at $m \to 0$. Of course, at the end of the day we want to focus on the $\vartheta = 0$ sector. Nothing precludes us, however, from dealing with $\vartheta \neq 0$ sectors at intermediate stages of our consideration. Knowledge of pure YM theory and YM theory with massless quarks can be used as a reference frame and a guiding principle.

In pure YM theory the vacuum angle reflects a non-trivial topology in the space of fields and the possibility of tunneling [23], a nonperturbative effect which makes the vacuum energy θ dependent and decreases the vacuum energy at $\theta = 0$. Instantons exemplify the tunneling trajectories [24]. Massless quarks suppress instantons (and any other field configurations with nonvanishing topological charges), freeze tunneling and make physics (including the vacuum energy) θ independent. Likewise, in SYM theory instanton does not contribute to the vacuum energy because of the gluino zero modes (an instanton-antiinstanton configuration could contribute but it has a vanishing topological charge and is topologically unstable.) In the orbifold

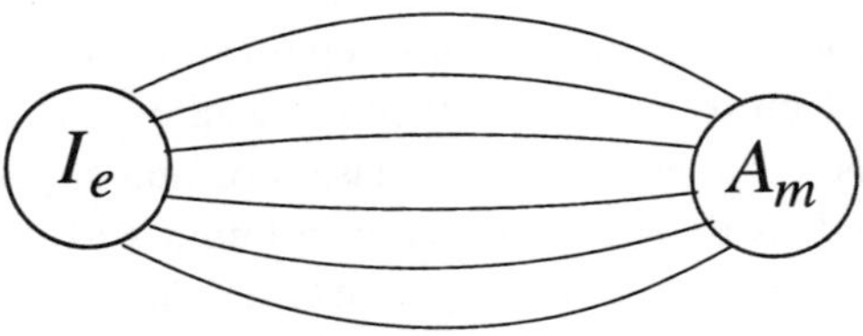

Figure 6. Topologically stable instanton-antiinstanton pairs in the orbifold theory. Instanton belongs to the electric SU(N) while antiinstanton to the magnetic SU(N).

theory we have two topological charges. Massless bifundamental fermions do suppress tunneling in the direction conjugate to θ_D. That's why physics cannot depend on this parameter. However, the orbifold theory exhibits a new phenomenon: topologically stable instanton-antiinstanton pairs, connected through fermion zero modes, see Fig. 6. The stability is due to the fact that they belong to distinct gauge factors. Therefore, although the overall topological charge (electric + magnetic) vanishes (all fermion zero modes are contracted), still instanton$_e$ cannot annihilate antiinstanton$_m$.

The "twisted" topological charge, conjugate to ϑ, is the difference between the electric topological charge and the magnetic one. Non-trivial topology and tunneling with regards to the twisted topological charge is not suppressed by massless fermions.

That's why physics does depend on ϑ. With regards to ϑ effects, the orbifold theory is expected to be similar to pure Yang–Mills, with no massless quarks. The instanton$_e$-antiinstanton$_m$ pair plays the role of the instanton in pure Yang–Mills. In particular, the vacuum energy $\mathcal{E}$ becomes a function of ϑ (more exactly, ϑ/N), and, if so, there is absolutely no reason for $\mathcal{E}(\vartheta) = 0$ at $\vartheta = 0$.

In fact, one is expected to find "vacuum families," of the type described by Witten [25] (see also [7]): a group of $\sim N$ quasistable "vacua" entangled in the process of ϑ evolution and interchanging their position each time ϑ reaches $k\pi$ where k is integer.[f] The issue of the dynamical Z_2 breaking in this language is formulated as follows: at $\vartheta = 0$ each vacuum family contains two degenerate stable vacua connected by a Z_2 transformation. At generic $\vartheta \neq 0$ the Z_2 symmetry of the action is explicitly broken by the ϑ term in Eq. (16).

11. Conclusions

Examples of cross-fertilization between string theories and gauge field theories are abundant. The topic of planar equivalence between supersymmetric and non-supersymmetric gauge theories emerged in this way. In the recent years it produced quite a few spectacular results and stimulated various activities in diverse directions. Two classes of non-supersymmetric models were identified as daughter theories: orbifold and orientifold. Planar equivalence is valid for both at the perturbative level.

In this talk I tried to summarize recent nonperturbative analyses of the orbifold theories. It was found, beyond reasonable doubt, that the Z_2 symmetry of the Z_2 orbifolds is the key to nonperturbative planar equivalence. If it is not dynamically broken, planar equivalence must extend to the nonperturbative level. The opposite is also true: spontaneous breaking of Z_2 entails a nonvanishing vacuum energy and a failure of planar equivalence. I discussed arguments in favor of nonperturbative nonequivalence such as domain wall dynamics and ϑ dependence. Unfortunately, there is no iron-clad proof of the statement. At a certain point, low-energy theorems seemed to

[f] This is in addition to N chiral sectors labeled by $\langle\bar{\Psi}\frac{1}{2}(1-\gamma_5)\Psi\rangle$. Note that the first crossover Dashen point is at $\vartheta = \pi/2$.

provide such a proof. It turned out, however, that they may or may not be relevant since they are sensitive not only to the vacuum structure of the parent/daughter theories, but also to the number of the fundamental degrees of freedom which is different in the parent/daughter theories.

In this sense, situation with the orientifold daughter theories is much more favorable. Nonperturbative planar equivalence certainly does hold for the orientifold theories. Why they are better than their orbifold cousins?

String theorists are familiar with this phenomenon. Type-II strings on orbifold singularities of the form C^3/Z_n , or type-0 strings always contain a tachyon in the twisted sector (and fractional branes).

For orientifold theories the situation is conceptually different. This non-supersymmetric gauge theory has no twisted sector and, in particular, it does not contain fractional domain walls; hence, it is guaranteed that the theory inherits its vacua from the SUSY parent. Similarly, the candidate for a string dual of the orientifold theory — Sagnotti's type-0$'$ model [26] — contains no tachyon since it was projected out by orientifolding.

The orientifold theory is closer to QCD. On the other hand, the orbifold theory has rich internal dynamics presenting, in a sense, a hybrid between QCD with massless quarks and pure Yang–Mills. Even though its planar equivalence to SYM theory is highly unlikely, it is an alluring target for future studies.

Acknowledgments

I am grateful to Adi Armoni and Sasha Gorsky for valuable comments on the manuscript. This work was supported in part by DOE grant DE-FG02-94ER408.

References

1. G. 't Hooft, *Nucl. Phys.* **B72**, 461 (1974).
2. M.J. Strassler, *On methods for extracting exact non-perturbative results in non-supersymmetric gauge theories,* hep-th/0104032. This work builds on Refs. [3].
3. S. Kachru and E. Silverstein, *Phys. Rev. Lett.* **80**, 4855 (1998) [hep-th/9802183]. A.E. Lawrence, N. Nekrasov and C. Vafa, *Nucl. Phys.* **B533**, 199 (1998) [hep-th/9803015]. M. Bershadsky, Z. Kakushadze and C. Vafa, *Nucl. Phys.* **B523**, 59 (1998) [hep-th/9803076]. M. Bershadsky and A. Johansen, *Nucl. Phys.* **B536**, 141 (1998) [hep-th/9803249]. Z. Kakushadze, *Nucl. Phys.* **B529**, 157 (1998) [hep-th/9803214]; *Phys. Rev.* **D58**, 106003 (1998) [hep-th/9804184]. M. Schmaltz, *Phys. Rev.* **D59**, 105018 (1999)

[hep-th/9805218]. A. Armoni and B. Kol, *JHEP* **9907**, 011 (1999) [hep-th/9906081].
4. A. Armoni, M. Shifman and G. Veneziano, *Nucl. Phys.* **B667**, 170 (2003) [hep-th/0302163]; *Phys. Rev. Lett.* **91**, 191601 (2003) [hep-th/0307097]; *Phys. Lett.* **B579**, 384 (2004) [hep-th/0309013].
5. A. Armoni, M. Shifman and G. Veneziano, *Phys. Rev.* **D71**, 045015 (2005) [hep-th/0412203].
6. A. Armoni, M. Shifman and G. Veneziano, *"From super-Yang–Mills theory to QCD: Planar equivalence and its implications,"* in *From Fields to Strings: Circumnavigating Theoretical Physics*, Eds. M. Shifman, J. Wheater and A. Vainshtein, (World Scientific, Singapore, 2004), Vol. 1, page 353 [hep-th/0403071].
7. M.A. Shifman, *Phys. Rev.* **D59**, 021501 (1999) [hep-th/9809184].
8. M.A. Shifman, A.I. Vainshtein and V.I. Zakharov, *Nucl. Phys.* **B147**, 385; 448 (1979).
9. R.F. Dashen, *Phys. Rev.* **D3**, 1879 (1971).
10. P. Kovtun, M. Unsal and L.G. Yaffe, *JHEP* **0507**, 008 (2005) [hep-th/0411177].
11. A. Armoni, A. Gorsky and M. Shifman, *Spontaneous Z_2 symmetry breaking in the orbifold daughter of $\mathcal{N}=1$ super-Yang-Mills theory, fractional domain walls and vacuum structure,* hep-th/0505022.
12. P. Kovtun, M. Unsal and L.G. Yaffe, *Can large N_c equivalence between supersymmetric Yang-Mills theory and its orbifold projections be valid?,* hep-th/0505075.
13. I.R. Klebanov and A.A. Tseytlin, *Nucl. Phys.* **B546**, 155 (1999).
14. I.R. Klebanov and A.A. Tseytlin, *JHEP* **9903**, 015 (1999) [hep-th/9901101]; I.R. Klebanov, *Phys. Lett.* **B466**, 166 (1999) [hep-th/9906220].
15. A. Armoni and B. Kol, *JHEP* **9907**, 011 (1999) [hep-th/9906081].
16. A. Gorsky and M. Shifman, *Phys. Rev.* **D67**, 022003 (2003) [hep-th/0208073].
17. A.A. Migdal and M.A. Shifman, *Phys. Lett.* **114**, 445 (1982).
18. E. Witten, *Nucl. Phys.* **B507**, 658 (1997) [hep-th/9706109].
19. G.R. Dvali and M.A. Shifman, *Phys. Lett.* **B396**, 64 (1997) (E) *ibid.* **B407**, 452 (1997) [hep-th/9612128].
20. A. Armoni and M. Shifman, *Nucl. Phys.* **B670**, 148 (2003) [hep-th/0303109].
21. B.S. Acharya and C. Vafa, *On domain walls of $\mathcal{N}=1$ supersymmetric Yang–Mills in four dimensions,* hep-th/0103011.
22. K. Zarembo, *Phys. Lett.* **B462**, 70 (1999);
A.A. Tseytlin and K. Zarembo, *Phys. Lett.* **B457**, 77 (1999).
23. R. Jackiw and C. Rebbi, *Phys. Rev. Lett.* **37**, 172 (1976);
C. G. . Callan, R. F. Dashen and D. J. Gross, *Phys. Lett.* **B63**, 334 (1976).
24. A.A. Belavin, A.M. Polyakov, A.S. Shvarts and Y.S. Tyupkin, *Phys. Lett.* **B59** (1975) 85.
25. E. Witten, *Phys. Rev. Lett.* **81**, 2862 (1998) [hep-th/9807109].
26. A. Sagnotti, *Some properties of open string theories,* hep-th/9509080; *Nucl. Phys. Proc. Suppl.* **56B**, 332 (1997) [hep-th/9702093].

MULTIPLE INTERACTIONS AND SATURATION IN HIGH ENERGY COLLISIONS

GÖSTA GUSTAFSON

Dept. of Theoretical Physics, Lund University
Sölvegatan 14A, S-22362 Lund, Sweden
E-mail: Gosta.Gustafson@thep.lu.se

Energy conservation is a large part of the NLO corrections to BFKL. A formalism is presented which includes effects of energy-momentum conservation in the dipole formalism in transverse coordinate space. Energy conservation gives a very strong suppression at small x, which implies that the effects of multi-pomeron exchange and saturation are pushed to very small x-values or very high energies. Perturbative QCD and PDFs from HERA can describe jets and $E_\perp$-flow in high energy pp collisions, but the multiplicity distributions are much more sensitive to non-perturbative effects. Analyses of minimum bias and underlying events at the Tevatron show that colours appear to combine to give strings which are as short as possible, in a way very different from e^+e^--annihilation. This leads to important questions about the properties of the pomeron, the relation between high multiplicity events and diffraction, and the applicability of the AGK cutting rules.

1. Introduction

In high energy collisions multiple interactions, saturation, and multi-pomeron exchange are essential problems, which still are largely unsolved. Constraints from unitarity imply that the observed power-like increase in the gluon density for small x cannot continue beyond limits. The gluon density has to saturate at sufficiently small x-values, but it is not yet clear whether saturation has already been observed in DIS at HERA. In high energy pp or p$\bar{\text{p}}$ collisions the inclusive cross section for minijet production is much larger than the total cross section. This implies that multiple hard subcollisions must be common, and this feature is clearly demonstrated in data from the Tevatron [1]. The more detailed properties of the underlying event and the relation to unitarity and diffraction is, however, still very unclear.

The leading order BFKL equation has solutions with asymptotic behavior $F \sim 1/x^\lambda$, with $\lambda = 4\ln 2 \cdot \bar{\alpha} \approx 0.5$, for $\bar{\alpha} \equiv 3\alpha_s/\pi = 0.2$. There are various corrections which reduce this increase at small x. Besides unitarity constraints and saturation, also the NLO corrections to the BFKL equation have a very large effect. Non-leading effects have been most easily studied in momentum space, while saturation effects are more easily treated in transverse coordinate space. Therefore it has been difficult to account for both effects in a unified formalism. It is well-known that a large fraction of the NLO corrections are related to energy conservation [2]. In Sec. 2 I want to discuss how these effects can be incorporated in the dipole formalism in transverse coordinate space, and be compared with the effects of multi-pomeron exchange and saturation. As an example some results for dipole-dipole scattering ($\gamma^*\gamma^*$-collisions) will be presented, and a more detailed discussion can be found in Ref. 3.

Perturbative QCD and PDFs from HERA can be used to calculate the production of transverse jets and $E_\perp$-flow in high energy pp collisions. At Tevatron energies the inclusive cross section for minijet production is about 300 mb, while the total inelastic non-diffractive cross section is only about 40 mb. Thus the average event must contain several minijets, which contribute to the total flow of transverse energy. The hadron multiplicity is, however, much more sensitive to non-perturbative effects than the jet cross section, and the effects of multiple subcollisions is here more difficult to determine. As a consequence models differ by factors 3–4 in their predictions for the multiplicity of hadrons at LHC [4].

These problems are related to the applicability of the Abramovskiĭ-Gribov-Kancheli [5] (AGK) cutting rules for multi-pomeron exchange diagrams. These were derived from Regge theory assuming the dominance of vertices with small momentum transfers. As discussed in the talk by Bartels [6], the color structure in QCD implies extra complications. The analyses of Tevatron data also shows that the situation is more complicated than the picture envisaged in the original paper by Gribov and coworkers. The interpretation of these results will be discussed in more detail in Sec. 3.

2. NLO Corrections and Saturation in DIS

2.1. *Formalism*

Because the transverse coordinates are unchanged in a high energy collision, unitarity constraints are generally more easy to take into account in a formalism based on the transverse coordinate space representation, and several suggestions for how to include saturation effects in such a formalism have been proposed. Golec–Biernat and Wüsthoff [7] formulated a dipole model, in which a virtual photon is treated as a $q\bar{q}$ or $q\bar{q}g$ system impinging on a proton, and this approach has been further developed by several authors (see e.g. Refs. 8 and 9). Mueller [10] has formulated a dipole cascade model in transverse coordinate space, which reproduces the BFKL equation, and in which it is also possible to account for multiple sub-collisions. Within this formalism Balitsky and Kovchegov [11] have derived a non-linear evolution equation, which also takes into account these saturation effects from multi-pomeron exchange.

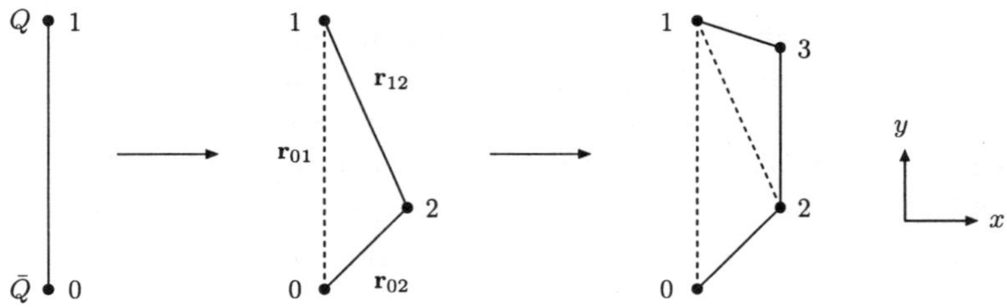

Figure 1. A quark-antiquark dipole in transverse coordinate space is split into successively more dipoles via gluon emission.

Study the process $\gamma^* \to Q\bar{Q} \to Qg\bar{Q} \to Qgg\bar{Q} \to \ldots$. Here a virtual gluon is split into a $Q\bar{Q}$ colour dipole, which is first split into two dipoles by the emission of a gluon, then into three dipoles by a second gluon, etc. The process is illustrated in transverse coordinate space in Fig. 1. The probability for such a dipole splitting is given by the expression (for notation see Fig. 1)

$$\frac{dP}{dy} = \frac{\bar{\alpha}}{2\pi} d^2\mathbf{r}_2 \frac{r_{01}^2}{r_{02}^2\, r_{12}^2} \cdot S; \quad S = \exp\left[-\frac{\bar{\alpha}}{2\pi}\int dy \int d^2\mathbf{r}_2 \frac{r_{01}^2}{r_{02}^2\, r_{12}^2}\right]. \tag{1}$$

Here S denotes a Sudakov form factor. We note that the integral over $d^2\mathbf{r}_2$ in the exponent diverges for small values of r_{02} and r_{12}. Therefore Mueller introduced a cutoff ρ, such that the integration region satisfies $r_{02} > \rho$ and $r_{12} > \rho$. A small cutoff value ρ will here imply that we get very many dipoles with small r-values.

If a dipole size, $\mathbf{r}$, is small, it means that the gluons are well localized, which must imply that transverse momenta are correspondingly large. This implies that not only the new gluon gets a large $k_\perp \sim 1/r$, also the original gluon, which is close in coordinate space, gets a corresponding recoil. Let us study the example in Fig. 2. For the emissions of the gluons marked 2, 3, and 4 the dipole sizes become smaller and smaller, $a >> b >> c >> d$, in each step of the evolution. The corresponding $k_\perp$ therefore become larger and larger in each step. After the minimum dipole, with size d, the subsequent emissions, 5, and 6, give again larger dipoles with correspondingly smaller $k_\perp$ values. The probability for this chain is proportional to

$$\frac{d^2\mathbf{r}_2\, a^2}{a^2\, b^2} \cdot \frac{d^2\mathbf{r}_3\, b^2}{b^2\, c^2} \cdot \frac{d^2\mathbf{r}_4\, c^2}{c^2\, d^2} \cdot \frac{d^2\mathbf{r}_5\, d^2}{e^2\, e^2} \cdot \frac{d^2\mathbf{r}_6\, e^2}{f^2\, f^2} \tag{2}$$

For the first emissions, 2 and 3, we in this expression recognize the product of factors $d^2\mathbf{r}_i/r_i^2 \sim \prod d^2\mathbf{k}_i/k_i^2$, just as is expected from a "DGLAP evolution" of a chain with monotonically increasing $k_\perp$. Emission number 4 corresponds to the minimum dipole size, d, and we here note that the factors of d cancel in Eq. (2). We therefore get the weight $d^2\mathbf{r}_4 \sim d^2\mathbf{k}_{\max}/k^4_{\max}$, which corresponds to a hard gluon-gluon collision. When the dipole sizes get larger again, this gives factors corresponding to a "DGLAP chain" from the other end of the chain, up to the central hard subcollision.

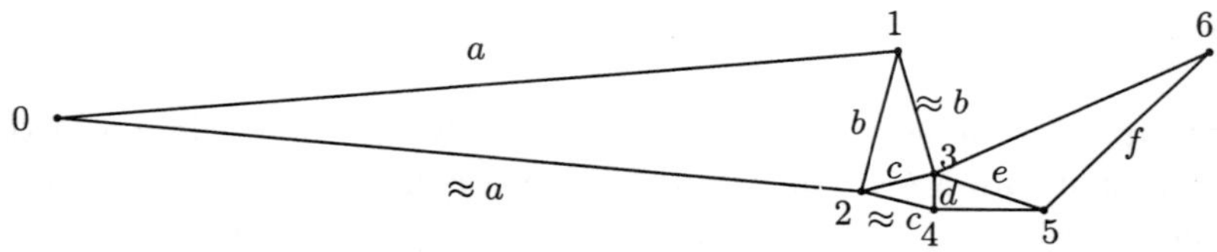

Figure 2. A dipole cascade, where a chain of smaller and smaller dipoles is followed by a set of dipoles with increasing sizes. This is interpreted as one $k_\perp$-ordered cascade from the left and one from the right, up to a central hard subcollision, which is represented by the dipole with minimum size and therefore maximum $k_\perp$.

It is also easy to see that for a chain with increasing dipole sizes up to a maximum value, $r_{\max}$, which thus corresponds to a minimum transverse momentum, $k_{\perp\min}$, we get the weight $d^2\mathbf{r}_{\max}/r^4_{\max} \sim d^2\mathbf{k}_{\min}$. Therefore there is no singularity for the minimum $k_\perp$-value. This result agrees exactly with the result in the Linked Dipole Chain model, LDC [12], which is a reformulation of the CCFM model [13], interpolating between DGLAP and BFKL for non-$k_\perp$-ordered chains.

To study $\gamma^*\gamma^*$ scattering we imagine that the two virtual photons split up into quark-antiquark pairs, which develop into dipole cascades as schematically illustrated in Fig. 3. When the two central dipoles collide, it implies a recoupling, as indicated by the arrow, and the probability for this

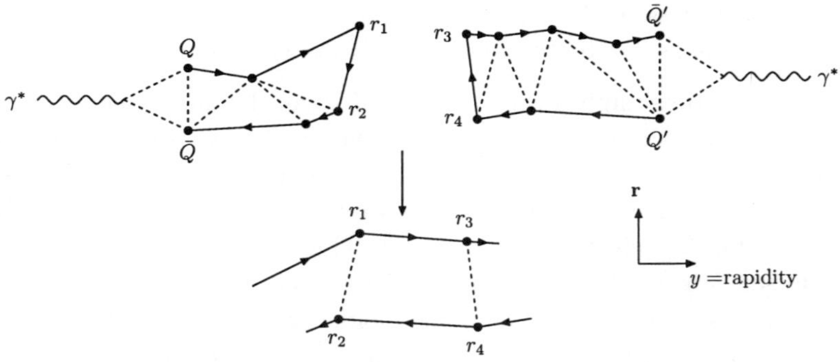

Figure 3. A symbolic picture of a $\gamma^*\gamma^*$ collision in rapidity-$\mathbf{r}_\perp$-space. The two dipole chains interact and recouple with probability f given by Eq. (3).

is given by the expression [14]

$$f = \frac{\alpha_s^2}{2}\left\{\ln\left[\frac{|\mathbf{r}_1 - \mathbf{r}_3|\cdot|\mathbf{r}_2 - \mathbf{r}_4|}{|\mathbf{r}_1 - \mathbf{r}_4|\cdot|\mathbf{r}_2 - \mathbf{r}_3|}\right]\right\}^2. \tag{3}$$

As the dipole cascades from the two virtual photons branch out, it is also possible to have *multiple interactions* with dipoles from the left and from the right. The total cross section is then given by

$$\sigma \sim \int d^2b(1 - e^{-\sum f_{ij}}), \tag{4}$$

where b denotes the impact parameter.

With a small cutoff ρ ($r > \rho$) we get, as mentioned above, very many small dipoles. If these are interpreted as real emissions, it would imply a violation of energy-momentum conservation. The emission of these small dipoles must be compensated by virtual emissions. Thus the result in Eq. (4) will describe the inclusive cross section, but the many dipoles produced in all the branching chains will not correspond to the production of exclusive final states. To be able to describe final state properties we will make the following conjecture:

In the cascade an emission with a (transverse) dipole size r corresponds to transverse momentum $k_\perp = 2/r$, as discussed above. Emissions satisfying energy-momentum conservation correspond to real emissions. Keeping only these emissions corresponds to keeping what is called "primary gluons" in Ref. 12 and "backbone gluons" in Ref. 15. These emissions determine both the total cross section and the structure of exclusive final states, and all softer emissions can be treated as final state emissions.

This conjecture implies that a chain of small dipoles is regarded as a single "effective" dipole. Keeping only the energy conserving emissions implies a *dynamical cutoff*, $\rho(\Delta y)$, which is large for small steps in rapidity, Δy, but gets smaller for larger Δy. (Alternatively it could be described as a cutoff for Δy which depends on ρ.) This is similar to the approach by Andersen-Stirling [16] and Orr-Stirling [17] in transverse momentum space. Conserving also the negative lightcone momentum, p_-, implies that we in a similar way also get a maximum value for r in each emission.

This conjecture is very easy to implement in a MC simulation. The result is that the number of dipoles grows much more slowly with energy. It is straightforward to calculate cross sections and to study saturation effects, by comparing the unitarised expression $\int d^2b(1-e^{-\sum f_{ij}})$ in Eq. (4) with $\int d^2b \sum f_{ij}$ representing single pomeron exchange. (The large numerical complications in MCs without energy conservation, discussed in Ref. 14, are not present.) Some results are presented in the next section.

2.2. *Application to Dipole-Dipole Scattering*

We here show some results obtained for a fixed coupling $\bar{\alpha} = 0.2$ [3]. Let us study the scattering of two dipoles with sizes r_1 and r_2. With a fixed coupling the scaled cross section, σ/r_2^2, depends only on the ratio r_1/r_2. We can imagine a target with size $r_2 \sim 1/M$, and a varying projectile size $r_1 \sim 1/\sqrt{Q^2}$. The results in Fig. 4 show that the cross section grows faster with the total rapidity range, $Y \sim \ln s$, for smaller r_1 (larger Q^2). This corresponds to a larger effective power λ_{eff} for larger virtuality Q^2, in a way qualitatively similar to the behavior of the proton structure function.

The effect of energy conservation is demonstrated in Fig. 5 by the results obtained for the case $r_1 = r_2$, with a constant cutoff, $\rho = 0.02\, r_i$. We see that energy conservation has a very strong effect, reducing σ by almost an order of magnitude for $Y \sim 13$. In Fig. 5 we also see that without energy-momentum conservation the effect of multiple pomeron exchange (saturation) is about a factor 2 for $r_1 = r_2$ and $Y = 13$. The much smaller cross section obtained with energy-momentum conservation implies that the saturation effect here is much less important, being only $\approx 20\%$ for the same parameter values.

Reference 3 also presents results for collisions between a dipole and a toy model nucleus or proton. As expected, the effect of unitarisation

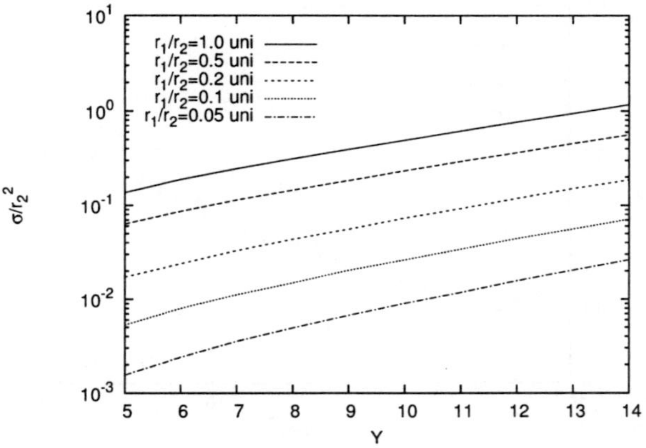

Figure 4. The scaled cross section, σ/r_2^2, for scattering of two dipoles with sizes r_1 and r_2 respectively, as a function of $Y = \ln s$ for different ratios r_1/r_2.

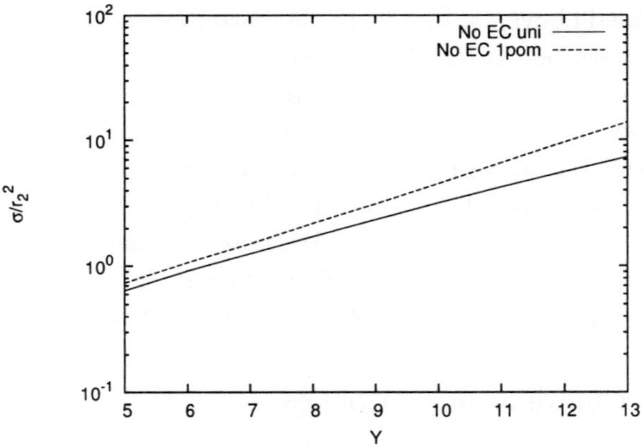

Figure 5. The scaled cross section obtained without energy conservation for $r_1 = r_2$ (solid line). The dashed line shows the one-pomeron exchange contribution.

grows with nuclear size and with the size of the projectile. For a small projectile of size 0.1 GeV^{-1} one can see the effect of colour transparency, as the cross sections for the unitarised and the one pomeron calculations are almost identical, while for a larger projectile there is a clear effect from unitarisation. However, even for $r_{\text{proj}} = 1\ \text{GeV}^{-1}$ and a nucleus with $A = 200$ this reduction due to multi-pomeron exchange is only about 20% in the rapidity interval $10 - 14$.

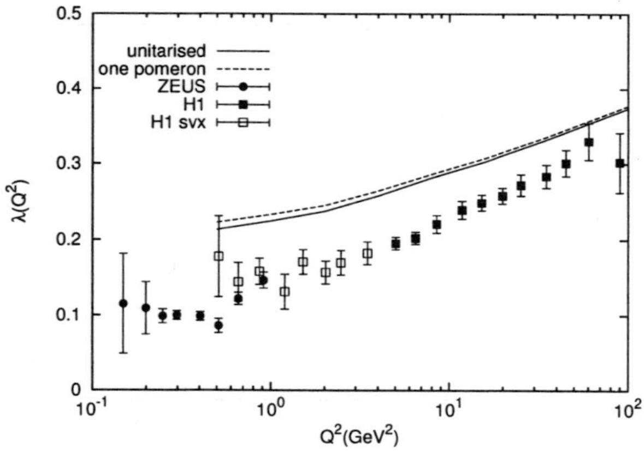

Figure 6. The effective slope λ measured at different Q^2 compared with data from ZEUS [18], filled squares, and H1 [19], open squares. The full line is the result of the toy model including unitarisation, and the dashed line is the single pomeron contribution.

As mentioned above, for dipole-dipole scattering the logarithmic slope λ is increasing when one of the dipoles gets smaller, corresponding to larger Q^2. This is also seen when we apply our toy model to a single proton. Figure 6 shows the effective slope as a function of Q^2 compared to data from ZEUS and H1, and we can see that there is a qualitative agreement between the model and the data. The figure shows also non-unitarised results from one-pomeron exchange. We note also here that the effect of unitarisation is not large, and the dependence on Q^2 is here a kinematic effect from energy conservation rather than an effect of saturation.

3. Multiple Collisions in pp Collisions

In pp collisions the multiplicity of final state hadrons depends sensitively on the colour connections between the produced partons. This implies that the result depends on soft non-perturbative effects. Multiple interactions are related to multiple pomeron exchange, which is expected to obey the Abramovskiĭ-Gribov-Kancheli cutting rules [5]. These rules were derived from Regge theory in a multiperipheral model, but such a multiperipheral chain has important similarities with a gluonic chain in QCD. An essential feature is the dominance of small momentum transfers at each vertex. The colour structure of QCD gives, however, some extra complications as discussed by J. Bartels at this workshop [6].

The pomeron is identified with two-gluon exchange, and multiple chains correspond to multi-pomeron exchange. For the example of two-pomeron exchange, the AGK rules give the relative weights $1 : -4 : 2$ for cutting 0, 1 or 2 pomerons. These ratios imply that the two-pomeron diagram contributes to the multiplicity *fluctuations*, but has no effect on the number of produced particles, determined by $\sum n\sigma_n$. This result can also be generalized to the exchange of more pomerons.

Similar cutting rules apply to a diagram with two pomerons attached to one of the colliding protons and one pomeron to the other, connected by a central triple-pomeron coupling. In Ref. 5 this and similar diagrams are, however, expected to give smaller contributions.

A hard gg → gg subcollision will imply that the two proton remnants carry colour octet charges. This is expected to give two colour triplet strings, or two cluster chains, connecting the two remnants and the two final state gluons. In the string model the strings are stretched between the remnants, with the gluons acting as kinks on the strings. As indicated in Fig. 7, these kinks can either be on different strings or both on one of the strings depending on the colour flow in the gluon-gluon scattering subprocess [20] [a]. Including initial state radiation will give extra kinks, which due to colour coherence will be connected so as to result in minimal extra string length, as schematically indicated in Fig. 8a.

Multiple collisions with two independent gg → gg scatterings would be expected to correspond to two cut pomerons, with four triplet strings stretched between the proton remnants as indicated in Fig. 8b. This would give approximately a doubled multiplicity, in accordance with the AGK cutting rules. However, the CDF data show that this is *far from reality.*

Rick Field's successful tune A [1] is a fit using an early PYTHIA version based on a model by Sjöstrand and van Zijl [21]. Already in their early analysis it was realized that four strings as in Fig. 8b would give too high multiplicity. Therefore, in this PYTHIA version there are three possible string connections for a secondary hard subcollision. 1) An extra closed string loop between the two final state gluons. 2) A single string between the scattered partons, which are then treated as a $q\bar{q}$ system. 3) The new hard gluons are inserted as extra kinks among the initial state radiations, in a way which corresponds to minimum extra string length. In the successful tune A the last possibility is chosen in 90% of the cases, which corresponds

[a] There are also interference terms for which the colour connections cannot be determined from perturbative QCD. These terms are, however, suppressed by factors $\sim 1/N_c^2$ [20].

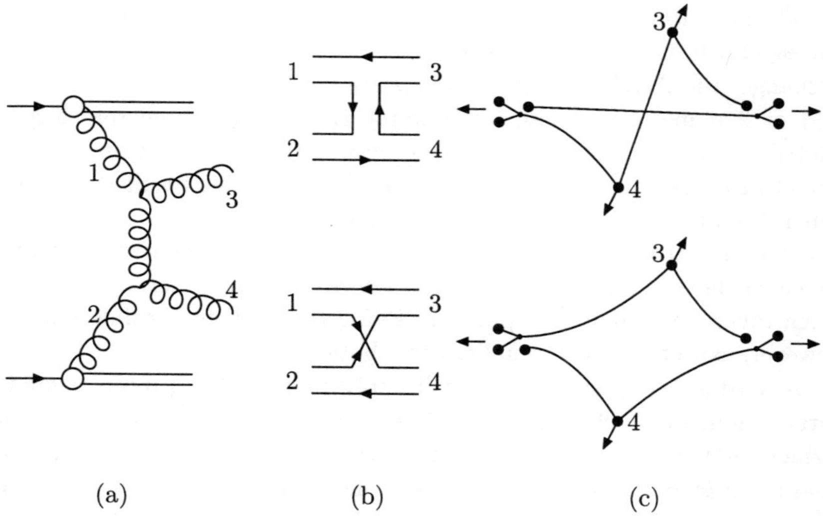

Figure 7. For the gluon-gluon subcollision in Fig. (a), the different colour flow patterns give four different colour connecting string configurations. Two possibilities are shown in Figs. (b) and (c), and the other two are obtained by interchanging colour and anticolour.

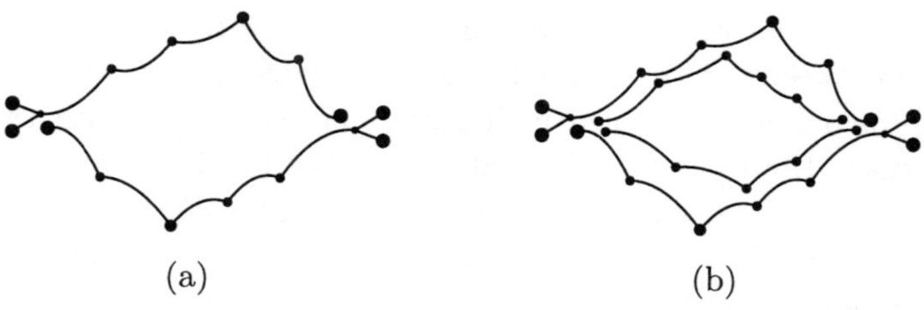

Figure 8. (a) Initial state radiation corresponds to more kinks on the strings. Due to colour coherence these kinks are connected so that the total string length is minimal. (b) Two cut pomerons with two hard subcollisions would naively be expected to result in four strings and approximately doubled multiplicity.

to *minimal extra multiplicity.* The default PYTHIA tune, which contained equal probabilities for the three cases, does not give a good fit. A more advanced treatment of pp collisions presented in Ref. 22 is implemented in a new PYTHIA version (PYTHIA 6.3 [23]). This model does, however, not work as well as Field's tune A of the older model.

Thus we have the following conclusions from the CDF data:

i) Multiple interactions are definitely needed.

ii) The data are fitted if colours rearrange so that secondary hard scatterings give minimum extra string length, i.e. minimum extra multiplicity.

Consequently, two independent hard collisions do not correspond to two cut pomeron ladders stretched between the proton remnants. It also does not correspond to a cut pomeron loop in the centre. Instead it looks like a single ladder, with a higher density of gluon rungs in the central region.

How can this be understood? It raises a set of important questions: What does it imply for the AGK rules and the diffractive gap survival probability? Do rescattering and unitarity constraints (and AGK) work in the initial perturbative phase? If so, does this correspond to an initial hard collision inside a confining bag, with the final state partons colour connected in a later non-perturbative phase?

We can compare with the situation in e^+e^--annihilation. If two gluons are emitted from the quark or antiquark legs, these gluons form a colour singlet with probability $\sim 1/N_c^2$. They could then hadronize as a separate system. Analyses of data from LEP indicate that such isolated systems are suppressed, possibly even more than by a factor $\sim 1/N_c^2$.

Thus we have three important questions:

– Why do the strings make the shortest connections in $\approx 100\%$ in pp and almost never in e^+e^-?

– How do multiplicity fluctuations and the relation *diffraction* $\leftrightarrow$ *high multiplicity events* reflect features of AGK in ep, γp, and pp?

– Or is the pomeron a much more complicated phenomenon than the simple ladder envisaged by Abramovskiĭ-Gribov-Kancheli?

– How could this be tested experimentally?

– If these questions can be answered at HERA and the Tevatron, it would be of great value for the analyses of both new and old physics at the LHC.

4. Summary

NLO corrections and saturation at small x

Energy conservation is a large part of the NLO corrections to BFKL. A formalism is presented which includes effects of energy-momentum conservation in the dipole formalism in transverse coordinate space. Energy conservation gives a very strong suppression at small x, reducing the cross section for dipole-dipole scattering by an order of magnitude for $\ln(1/x) \sim 13$.

This implies that the effects of multi-pomeron exchange and saturation are pushed to very small x-values or very high energies. It also has the result that the effective logarithmic growth, λ, increases with larger Q^2, in qualitative agreement with results from HERA. For dipole-nucleus scattering the effects of saturation is important only for large nuclei and relatively small Q^2.

Multiple scattering in pp collisions

Perturbative QCD and PDFs from HERA can describe jets and $E_\perp$-flow in high energy pp collisions. The multiplicity distributions are, however, much more sensitive to non-perturbative effects. PYTHIA tune A reproduces the essential features of minimum bias and underlying events at the Tevatron. This shows that colours appear to combine to give strings which are as short as possible, in a way very different from e^+e^--annihilation.

This leads to a set of questions: What is the relation between high multiplicity events and diffraction? Do unitarity effects and AGK cutting rules work as expected in an initial perturbative phase, followed by colour recombination in a later soft phase? Or is the pomeron a much more complicated phenomenon than the simple ladder envisaged by Abramovskiĭ-Gribov-Kancheli? Can analyses of data from HERA and the Tevatron improve predictions for, and analyses of, results at the LHC?

References

1. R. Field, *Proceedings of XXXIV Int. Symp. on Multiparticle Dynamics*, July 26 – Aug. 1, 2004, Sonoma County, USA, *Acta Phys. Polon.* **B36**, 167 (2005).
2. See e.g. G.P. Salam, *Acta Phys. Polon.* **B30**, 3679 (1999).
3. E. Avsar, G. Gustafson and L. Lönnblad, *JHEP* **0507**, 62 (2005).
4. A. Moraes, *Contribution to the workshop HERA and the LHC*, CERN Oct. 11 – 13, 2004.
5. V.A. Abramovskiĭ, V.N. Gribov and O.V. Kancheli, *Yad. Fiz.* **18**, 595 (1973).
6. J. Bartels, contribution to these proceedings;
   J. Bartels, M. Salvadore and G.P. Vacca, *Eur. Phys. J.* **C42**, 53 (2005).
7. K. Golec-Biernat and M. Wüsthoff, *Phys. Rev.* **D59**, 014017 (1999).
8. J.R. Forshaw, G. Kerley and G. Shaw, *Phys. Rev.* **D60**, 074012 (1999).
9. J. Bartels, K. Golec-Biernat and H. Kowalski, *Phys. Rev.* **D66**, 014001 (2002).
10. A.H. Mueller, *Nucl. Phys.* **B415**, 373 (1994);
    A.H. Mueller and B. Patel, *Nucl. Phys.* **B425**, 471 (1994);
    A.H. Mueller, *Nucl. Phys.* **B437**, 107 (1995).
11. I.I. Balitsky, *Nucl. Phys.* **B463**, 99 (1996);
    V. Kovchegov, *Phys. Rev.* **D60**, 034008 (1999).
12. B. Andersson, G. Gustafson and J. Samuelsson, *Nucl. Phys.* **B467**, 443

(1996);
B. Andersson, G. Gustafson and H. Kharraziha, *Phys. Rev.* **D57**, 5543 (1998);
H. Kharraziha and L. Lönnblad, *JHEP* **9803**, 006 (1998).
13. M. Ciafaloni, *Nucl. Phys.* **B269**, 49 (1988);
S. Catani, F. Fiorani and G. Marchesini, *Nucl. Phys.* **B336**, 18 (1990).
14. G.P. Salam, *Nucl. Phys.* **B461**, 512 (1996);
A.H. Mueller and G.P. Salam, *Nucl. Phys.* **B475**, 293 (1996).
15. G.P. Salam, *JHEP* **9903**, 009 (1999).
16. J.R. Andersen and W.J. Stirling, *JHEP* **0302**, 018 (2003).
17. L.H. Orr and W.J. Stirling, *Phys. Rev.* **D56**, 97 (1997).
18. J. Breitweg *et al.* (ZEUS Coll.), *Phys. Lett.* **B487**, 53 (2000).
19. C. Althoff *et al.* (H1 Coll.), *Eur. Phys. J.* **C21**, 33 (2001).
20. G. Gustafson, *Z. Phys.* **C15**, 155 (1982).
21. T. Sjöstrand and M. van Zijl, *Phys. Rev.* **D36**, 2019 (1987).
22. T. Sjöstrand and P. Skands, *JHEP* **0403**, 053 (2004); *Eur. Phys. J.* **C39**,129 (2005).
23. T. Sjöstrand, L. Lönnblad, S. Mrenna and P. Skands, *"PYTHIA 6.3: Physics and manual"*, hep-ph/0308153 (2003).

FROM QUANTUM BLACK HOLES TO RELATIVISTIC HEAVY IONS*

D. KHARZEEV
Nuclear Theory Group,
Physics Department,
Brookhaven National Laboratory,
Upton, NY 11973, USA
E-mail: kharzeev@bnl.gov

The quantum vacuum becomes unstable in the presence of an event horizon; a famous example of that is quantum evaporation of black holes. In this talk I speculate that the universal and apparently thermal character of hadron production in various processes may originate in the decay of the QCD vacuum induced by a pulse of strong chromo–electric field.

1. Introduction

V.N. Gribov realized early on that quantum mechanics should make the black holes unstable. This appeared to have far–reaching implications, and the topic remains in the focus of attention today, sparking lively controversies and calling for a new theoretical break-through in reconciling quantum theory with general relativity. Moreover, the phenomenon of vacuum decay induced by a strong field occurs not only in the vicinity of black holes. As I will discuss in this talk, the familiar process of pair creation in an electric field may be interpreted in this way. The analogous phenomenon in QCD may be responsible for the universal thermal character of hadron production in various processes.

Indeed, there is something puzzling in the data on hadron production in various processes, from e^+e^- annihilation and deep–inelastic scattering to heavy ion collisions: the relative abundances of different hadron species appear to follow the statistical distribution with a surprising accuracy (for reviews, see e.g. [1,2]). Moreover, at small transverse momenta the spectra of

*This work is supported by the U.S. Department of Energy under Contract No. DE-AC02-98CH10886.

the produced hadrons also look approximately thermal. While in heavy ion collisions it is possible to expect the emergence of statistical distributions as a result of intense re–interactions between the produced particles, this seems very implausible in e^+e^- annihilation at high energies $\sqrt{s}$, where the process of hadronization is stretched in space over a long distance $\sim \sqrt{s}/2\mu^2$ and the density of produced hadrons is small ($\mu \sim \Lambda_{\rm QCD}$ is an infrared cut-off describing the hadronization scale).

Statistical distributions may emerge as a result of the saddle–point approximation to the multi–particle phase space, when the dynamics is inessential and the production mechanism is "phase space dominated" (see, for example, [3] and references therein). However, in e^+e^- annihilation this mechanism can hardly be expected to work: the jet structure, the angular distributions of the produced hadrons, and inter–jet correlations point to the all–important role of QCD dynamics of gluon radiation (for a recent review, see [4]), and thus the "phase space dominance" cannot be invoked. It thus looks natural to think that the emergence of statistical hadron abundances has something to do with the process of hadronization – in other words, with the way in which the QCD vacuum responds to the external color fields, as was advocated by Dokshitzer [4].

This attractive and deep idea however does not yet allow one to understand why the vacuum response to the external carriers of color field leads to the emergence of statistical distributions. To do this, we are forced to discuss the structure of the QCD vacuum. Little is known about it, but we do know that QCD vacuum is populated by the fluctuations of color fields, some of which are of semi–classical nature [5]. It may therefore be useful to examine what is known at present about the interactions of charged quanta with background classical fields.

This logic will first lead us to the discussion of quantum fluctuations in the background of the gravitational field of a black hole, where the quantum radiation appears to be thermal as shown by Hawking [6]. It was demonstrated by Unruh [7] that the Hawking phenomenon should be present in any non–inertial frame; indeed, this is required by Einstein's Equivalence Principle. Then we will proceed to the process of electron–positron pair production in the background electric field, analyzed in QED by Schwinger [8]. We will see that the Schwinger formula for the rate of e^+e^- pair production in a constant electric field allows for a simple statistical interpretation; moreover, for the case of a time–dependent field pulse, the spectra of the produced particles become thermal.

These two examples inspire the picture in which the thermal character of hadron production emerges from the interactions with the vacuum chromo–electric field, the strength of which in the color flux model is parameterized in terms of the string tension, or the slope of Regge trajectories b. Indeed, we will see how the well–known formula for the Hagedorn temperature

$$T_{Hagedorn} = \frac{\sqrt{6}}{4\pi} \frac{1}{\sqrt{b}} \tag{1}$$

can be derived in this way.

What happens if we create semi–classical color fields of strength exceeding the strength of the vacuum fields? This can be achieved in the collisions of heavy ions at high energies, which are accompanied by a short pulse of the chromo–electric field $E \sim Q_s^2/g$ of duration $\tau \sim Q_s^{-1}$; here Q_s is the saturation scale [9,10], and g is the strong coupling. I will argue that such a strong color field induces the creation of partons with a distribution which is isotropic and thermal in a co–moving local frame, with an effective temperature $T_{in} \simeq Q_s/2\pi$. For high enough energies and heavy enough nuclei, the value of T_{in} exceeds the Hagedorn temperature; the produced thermal matter is thus in the deconfined phase. The phase transition in this case can also be understood in a geometrical picture, in which the acceleration $a \simeq Q_s$ in the chromo–electric field determines the curvature of space in the non–inertial Rindler frame. This is mathematically analogous to the phase transition induced by the presence of a massive black hole. This talk is based on the paper with Kirill Tuchin [11], to which I refer for details and references.

2. Quantum Fluctuations in the Classical Background

2.1. *Black hole evaporation*

In 1974 Hawking [6] demonstrated that the black holes emit radiation with a thermal spectrum, of an effective temperature

$$T_H = \frac{\kappa}{2\pi}, \tag{2}$$

where $\kappa = 4M$ is the acceleration of gravity at the surface of a black hole of mass M. The thermal character of the black hole radiation stems from the presence of the event horizon, which hides the interior of the black hole from an outside observer. The rate of pair production in the gravitational background of a black hole can be evaluated by considering the tunneling through the event horizon. Parikh and Wilczek [12] showed that the imaginary part of the action for this classically forbidden process corresponds to the exponent of the Boltzmann factor describing the thermal emission[a].

Unruh [7] has found that a similar effect arises in a uniformly accelerated frame, where an observer detects an apparent thermal radiation with the temperature

$$T_U = \frac{a}{2\pi}; \tag{3}$$

(a is the acceleration). The event horizon in this case emerges due to the existence of causally disconnected regions of space–time, conveniently described by using the Rindler coordinates.

2.2. *Pair production in a constant electric field*

The effects associated with a heat bath of temperature (3) usually are not easy to detect because of the smallness of the acceleration a in realistic experimental conditions. For example, for the acceleration of gravity on the surface of Earth $g \simeq 9.8$ m s^{-2} the corresponding temperature is only $T \simeq 4 \times 10^{-20}$ K. (The wavelength of the thermal radiation in this case is about 1 parsec - so it would have to be detected on a cosmic scale!)

Much larger accelerations can be achieved in electromagnetic fields, and Bell and Leinaas [13] considered the possible manifestations of the Hawking–Unruh effect in particle accelerators. They argued that the presence of an apparent heat bath can cause beam depolarization. Indeed, if the energies of spin–up $E_\uparrow$ and spin–down $E_\downarrow$ states of a particle in the magnetic field of an accelerator differ by $\Delta E = E_\uparrow - E_\downarrow$, the Hawking–Unruh effect would lead to the thermal ratio of the occupation probabilities

$$\frac{N_\uparrow}{N_\downarrow} \simeq \exp\left(-\frac{\Delta E}{T_U}\right), \tag{4}$$

where the Unruh temperature (3) is determined by the particle acceleration. According to (4), a pure polarization state of a particle in the accelerator is inevitably diluted by the acceleration.

If the energy spectrum of an accelerated observer is continuous, as is the case for a particle of mass m with a transverse (with respect to the

---

[a]Conservation laws also imply a non-thermal correction to the emission rate [12], possibly causing a leakage of information from the black hole.

direction of acceleration) momentum p_T, a straightforward extension of (4) leads to a thermal distribution in the "transverse mass" $m_T = \sqrt{m^2 + p_T^2}$:

$$W_m(p_T) \sim \exp\left(-\frac{m_T}{T_U}\right). \tag{5}$$

An important example is provided by the dynamics of charged particles in external electric fields. Consider a particle of mass m and charge e in an external electric field of strength E. Under the influence of the Lorentz force, it moves with an acceleration $a = eE/m$, and the corresponding temperature is $T_U = a/2\pi$. The Boltzman factor $\exp(-m/T_U)$ entering the particle creation rate in this case is

$$W_m^E \sim \exp\left(-\frac{2\,\pi m^2}{eE}\right). \tag{6}$$

This expression looks familiar – in fact, (6) differs from the classic Schwinger result for the rate of particle production in a constant electric field only by a factor of two in the exponent.

Is this a coincidence? To answer this question, let us have a closer look at the corresponding action:

$$S = \int (-m\,ds - e\,\varphi\,dt\,), \tag{7}$$

where φ is the electric potential, which for a constant electric field aligned along the x axis is $\varphi = -Ex$ modulo an additive constant; the invariant interval is given by $ds^2 = (1 - v^2(t))\,dt^2$. Using the equations of motion, we can evaluate the action to find

$$\begin{aligned} S(\tau) &= \int^{\tau} dt\,(-m\sqrt{1 - v(t)^2} + e\,E\,x(t)) \\ &= -\frac{m}{a}\,\text{arcsinh}(a\,\tau) + \frac{e\,E}{2\,a^2}\left(a\,\tau\,(\sqrt{1 + a^2\,\tau^2} - 2) + \text{arcsinh}(a\,\tau)\right). \end{aligned} \tag{8}$$

In classical mechanics the equations of motion completely specify the trajectory of a uniformly accelerating particle moving under the influence of a constant force $\vec{F} = -e\nabla\varphi$. In contrast, in quantum theory the particle has a finite probability to be found under the potential barrier $V(x) = eEx$ in the classically forbidden region. Mathematically, it comes about since the action (8) being an analytic function of τ has an imaginary part in the Euclidean space

$$\text{Im}\,S(\tau) = \frac{m\,\pi}{a} - \frac{e\,E\,\pi}{2\,a^2} = \frac{\pi\,m^2}{2\,e\,E}. \tag{9}$$

It is well known that a quasi-classical exponent describing the decay of a metastable state is given by the imaginary part of the action. The rate of tunneling under the potential barrier in the quasi-classical approximation is thus given by

$$W_m^E \sim \exp(-2\,\mathrm{Im}S) = \exp\left(-\frac{\pi\, m^2}{e\,E}\right). \tag{10}$$

Equation (10) gives the probability to produce a particle and its antiparticle (each of mass m) out of the vacuum by a constant electric field E; note that the incorrect factor of 2 in the exponent of (6) has now disappeared due to the contribution of the field term in the action (7).

The ratio of the probabilities to produce states of masses M and m is then

$$\frac{W_M^E}{W_m^E} = \exp\left(-\frac{\pi\,(M^2 - m^2)}{e\,E}\right). \tag{11}$$

The relation (11) allows a dual interpretation in terms of both Unruh and Schwinger effects (see e.g. [11,14,15,16] and references therein). First, consider a detector with quantum levels m and M moving in a constant electric field. Each level is accelerated differently, however if the splitting is not large, $M - m \ll m$ we can introduce the average acceleration of the detector

$$\bar{a} = \frac{2\,e\,E}{M + m}. \tag{12}$$

Substituting (12) into (11) we arrive at

$$\frac{W_M^E}{W_m} = \exp\left(\frac{2\,\pi\,(M - m)}{\bar{a}}\right). \tag{13}$$

This expression is reminiscent of the Boltzmann weight in a heat bath with an effective temperature (3): $T = \bar{a}/2\pi$. It implies that the detector is effectively immersed in a photon heat bath at temperature $T \approx eE/\pi m$. This is the renown Unruh effect [7].

It is important to remember that for a constant electric field the momentum distribution of the produced charged particles allows a statistical interpretation only in the transverse to the field direction. However for a short pulse of an electric field of duration $\tau \leq m/(eE)$ the distribution becomes thermal in all three directions (see [11] and references therein) – physically, this happens because the field has not enough time to perform work on curving the momenta of the produced particles.

Let us now discuss (10) and (11) from the viewpoint of a field-theoretical derivation done by Schwinger [8]. It is clear that (10) cannot be expanded neither in powers of the coupling e, nor in powers of the field E, and so cannot be reproduced in any finite order of perturbation theory. Schwinger considered the one–loop QED action describing the electron–positron fluctuations in the background of the external electric field E. He has found that the series in the number of external field insertions indeed diverges (is not Borel summable). As a result the action ceases to be real, and develops an imaginary part, similarly to the simple semi–classical example considered above.

The Hawking–Unruh interpretation therefore appears to capture an essentially non–perturbative dynamics. Indeed, a more rigorous treatment (for a review, see e.g. [17]) allows to establish that the Bogoliubov transformations relating the particle creation and annihilation operators in Minkowski and Rindler spaces describe a rearrangement of the vacuum structure which cannot be captured by perturbative series.

3. Event horizon and thermalization in high energy hadronic interactions

3.1. *Unruh effect, Hagedorn temperature, and parton saturation*

We are now ready to address the case of hadronic interactions at high energies, which is the main subject of this talk. Consider a high–energy hadron of mass m and momentum P which interacts with an external field (e.g., another hadron) and transforms into a hadronic final state of invariant mass $M \gg m$. This transformation is accompanied by a change in the longitudinal momentum

$$q_L = \sqrt{E^2 - m^2} - \sqrt{E^2 - M^2} \simeq \frac{M^2 - m^2}{2P}, \tag{14}$$

and therefore by a deceleration; we assumed that the particle m is relativistic, with energy $E \simeq p$.

The probability for a transition to a state with an invariant mass M is given by

$$P(M \leftarrow m) = 2\pi |T(M \leftarrow m)|^2 \; \rho(M), \tag{15}$$

where $T(M \leftarrow m)$ is the transition amplitude, and $\rho(M)$ is the density of hadronic final states. According to the results of the previous section, we expect that under the influence of deceleration a which accompanies the change of momentum (14), the probability $|T|^2$ will be determined by the

Unruh effect and will be given by

$$|T(M \leftarrow m)|^2 \sim \exp(-2\pi M/a) \tag{16}$$

in the absence of any dynamical correlations; we assume $M \gg m$.

To evaluate the density of states $\rho(M)$, let us first use the dual resonance model (see e.g. [18,19]), in which

$$\rho(M) \sim \exp\left(\frac{4\pi}{\sqrt{6}}\, \sqrt{b}\, M\right), \tag{17}$$

where b is the universal slope of the Regge trajectories, related to the string tension σ by the relation $\sigma = 1/(2\pi b)$.

The unitarity dictates that the sum of the probabilities (15) over all finite states M should be finite. Therefore, by converting the sum into integral over M one can see that the eqs (16) and (17) impose the following bound on the value of acceleration a:

$$\frac{a}{2\pi} \equiv T \leq \frac{\sqrt{6}}{4\pi}\, \frac{1}{\sqrt{b}} \equiv T_{Hag}\,. \tag{18}$$

The quantity on the r.h.s. of (18) is known as the Hagedorn temperature [20] – the "limiting temperature of hadronic matter" derived traditionally from hadron thermodynamics. In our case it stems from the existence of a "limiting acceleration" a_0:

$$a_0 = \sqrt{\frac{3}{2}}\, b^{-1/2}\,. \tag{19}$$

The meaning of the "limiting temperature" in hadron thermodynamics is well-known: above it, hadronic matter undergoes a phase transition into the deconfined phase, in which the quarks and gluons become the dynamical degrees of freedom. To establish the meaning of the limiting acceleration (19), let us consider a dissociation of the incident hadron into a large number $n \gg 1$ of partons. In this case the phase space density (17) can be evaluated by the saddle point method ("statistical approximation"), with the result (see e.g. [21])

$$\rho(M) \sim \exp(\beta M)\,; \tag{20}$$

where β^{-1} is determined by a typical parton momentum in the center-of-mass frame of the partonic configuration. When interpreted in partonic language, eq(17) thus implies a constant value of mean transverse momentum $\bar{p}_T \sim \beta^{-1} \sim b^{-1/2}$. On the other hand, in the parton saturation picture, the mean transverse momentum has to be associated with the "saturation scale" Q_s determined by the density of partons in the transverse plane within the wave function of the incident hadron (or a nucleus). This leads to the phase space density $\rho(M) \sim \exp(M/Q_s)$. The unitarity condition and the formulae (16), (15) thus lead us to the acceleration $a = Q_s$, which can exceed (19), and to the conclusion that the final partonic states are described by a thermal distribution with the temperature

$$T = \frac{Q_s}{2\pi}. \tag{21}$$

The same result can be obtained by considering the acceleration $a = gE/m$ of a parton with off-shellness $m \equiv \sqrt{p^2} \simeq Q_s$ in an external color field $gE \simeq Q_s^2$. It is interesting to note that to exceed the limiting acceleration (19), and thus the limiting Hagedorn temperature (18) for the produced hadronic matter, one has to build up strong color fields, exceeding $gE_0 \sim 1/b$. This is achieved by parton saturation in the Color Glass Condensate, when $gE \simeq Q_s^2 > gE_0$ at sufficiently high energies and/or large mass numbers of the colliding nuclei. Parton saturation in the initial wave functions thus seems to be a necessary pre-requisite for the emergence of a thermal deconfined partonic matter in the final state.

The thermal distribution is built over the time period of

$$t_{\text{therm}} \sim T^{-1} = \frac{\pi}{Q_s}. \tag{22}$$

As discussed above, this apparent thermalization originates from the presence of the event horizon in an accelerating frame: the incident hadron decelerates in an external color field, which causes the emergence of the causal horizon. Quantum tunneling through this event horizon then produces a thermal final state of partons, in complete analogy with the thermal character of quantum radiation from black holes.

3.2. *The space–time picture of relativistic heavy ion collisions*

In the conventional space–time picture of a relativistic heavy ion collision, the colliding heavy ions approach the interaction region along the light cone from $x = t = -\infty$ and $x = -t = \infty$. The partons inside the nuclei in the spirit of the collinear factorization approach are assumed to have a vanishing transverse momentum k_T, have a zero virtuality $k^2 = -k_T^2 = 0$, and thus are also localized on the light cone at $\pm x = t$. The collision at $x = t = 0$ produces the final state particles with transverse momenta p_T which according to the uncertainty principle approach their mass shell at a proper time $\tau = (t^2 - x^2)^{1/2} = \tau_0 \sim 1/p_T$. For further discussion it is convenient to introduce the Rindler coordinates

$$\rho^2 = x^2 - t^2; \qquad \eta = \frac{1}{2} \ln \left| \frac{t+x}{t-x} \right|; \tag{23}$$

The surface of a fixed proper time which is a hyperbola in Minkowski space thus represents a line at $\rho^2 = x^2 - t^2 = \rho_0^2 < 0$ in Rindler space. The Rindler coordinate η in high energy physics is often called a space–time rapidity.

Consider now the case when partons in the wave functions of the colliding nuclei have non-vanishing transverse momenta, as in the Color Glass Condensate picture where their transverse momenta are on the order of the saturation scale Q_s. In this case the partons are space–like $k^2 = -k_T^2$ and are located off the light cone. As the colliding nuclei approach each other, these partons begin to interact; note that since they are space–like, their interactions are acausal, and are responsible for the breakdown of factorization for the parton modes with $k_T \leq Q_s$. The interactions of partons with the color fields $gE \simeq Q_s^2$ of another nucleus decelerate them, with a typical acceleration $|a| \simeq Q_s$. The space–time trajectories of the partons are thus given by hyperbolae in Minkowski space, or by the lines with a fixed value of $\rho^2 = -\rho_0^2 = Q_s^2 > 0$ in Rindler space. For partons moving in the left and right sectors of space–time with an acceleration $|a| = \rho^{-1}$ the light cone surfaces $\rho^2 = 0$, $\eta = \pm\infty$ represent the event horizon of the future. The information from the future is hidden from them. However, as discussed above, the future sector can be reached from the left and right sectors by quantum tunneling.

3.3. *QCD vacuum decay in the parton language*

It is important to mention that this picture does not imply any "bouncing" of the colliding nuclei – the on–shell particles, as well as the high–momentum valence partons from the colliding nuclei are transparent for each other. The soft components of the parton wave functions however do interact strongly. Nevertheless, the whole picture at first glance looks completely orthogonal to the conventional parton model. Indeed, in parton model the Weizsäcker–Williams gluon fields surrounding the valence quarks are transverse, with $\vec{E} \perp \vec{H} \perp \vec{p}$ ($\vec{p}$ is the momentum of the quark). The corresponding "equivalent gluons" are almost on mass shell, and no longitudinal chromo–electric field is present: the gluon field tensor is flat in the longitudinal direction: $F_{+-} = 0$ ("+" and "−" refer to the light–cone components).

However, a more careful analysis reveals that this picture is not complete: the configuration of the produced gluon field is characterized by $F_{+-} \neq 0$, with a substantial longitudinal chromo–electric field E_z. In the Color Glass Condensate picture, the strength of the field is $E_z \sim Q_s^2/g$, and the duration of the pulse is $\sim Q_s^{-1}$. Note that in the conventional string model picture, the produced chromo–electric field is purely longitudinal, with the strength proportional to the string tension. This exhibits a possible continuity between the string and parton approaches to multi–particle production, and suggests the existence of the minimal allowed value of the saturation momentum $Q_{s\ min}$. Basing on the arguments given above and (18), one is led to the conclusion that

$$Q_{s\ min} = 2\pi T_{Hagedorn} \simeq 1 \text{ GeV} . \tag{24}$$

A quantitative analysis of parton production in the longitudinal chromo–electric field performed in the framework of the Color Glass Condensate approach is underway [22].

4. Summary

In this talk I have argued that the statistical features of multi–particle production may emerge as a consequence of the QCD analog of the black hole evaporation. The acceleration, and the emergence of the corresponding event horizon for partons, is caused by the pulse of chromo–electric field which accompanies inelastic interactions at high energies. For hadron interactions at moderate energies, the effective temperature appears equal to the Hagedorn temperature (18). Once the strength of the chromo–electric field $E \sim Q_s^2/g$ exceeds a critical value determined by (24), the partons are produced with an effective temperature $T > T_{Hagedorn}$, i.e. in a deconfined state.

The ideas of V.N. Gribov penetrate almost all of the contemporary high energy and nuclear physics. It is quite likely that his influence will become even more profound in the future.

I am very grateful to J. Nyiri, Yu. Dokshitzer and P. Levai for the invitation to this Workshop. I thank E. Levin and K. Tuchin for the enjoyable collaboration on the topics presented above.

References

1. F. Becattini, *J. Phys. Conf. Ser.* **5**, 175 (2005) [arXiv:hep-ph/0410403].
2. P. Braun-Munzinger, K. Redlich and J. Stachel, arXiv:nucl-th/0304013.
3. W. Deinet and D.H. Rischke, arXiv:nucl-th/0507076.
4. Yu.L. Dokshitzer, *Acta Phys. Polon.* **B36**, 361 (2005) [arXiv:hep-ph/0411227]; *Nucl. Phys.* **A755**, 229 (2005).
5. A.A. Belavin, A.M. Polyakov, A.S. Shvarts and Y.S. Tyupkin, *Phys. Lett.* **B59**, 85 (1975).
6. S.W. Hawking, *Commun. Math. Phys.* **43**, 199 (1975).
7. W.G. Unruh, *Phys. Rev.* **D14**, 870 (1976).
8. J.S. Schwinger, *Phys. Rev.* **82**, 664 (1951).
9. L.V. Gribov, E.M. Levin and M.G. Ryskin, *Phys. Rep.* **100**, 1 (1983); A.H. Mueller and J.W. Qiu, *Nucl. Phys.* **B268**, 427 (1986).
10. L.D. McLerran and R. Venugopalan, *Phys. Rev.* **D49**, 2233 (1994) [arXiv:hep-ph/9309289], *Phys. Rev.* **D49**, 3352 (1994) [arXiv:hep-ph/9311205].
11. D. Kharzeev and K. Tuchin, *Nucl. Phys.* **A753**, 316 (2005) [arXiv:hep-ph/0501234].
12. M.K. Parikh and F. Wilczek, *Phys. Rev. Lett.* **85**, 5042 (2000) [arXiv:hep-th/9907001].
13. J.S. Bell and J.M. Leinaas, *Nucl. Phys.* **B212**, 131 (1983); *Nucl. Phys.* **B284**, 488 (1987).
14. R. Parentani and S. Massar, *Phys. Rev.* **D55**, 3603 (1997) [arXiv:hep-th/9603057].
15. C. Gabriel and P. Spindel, *Annals Phys.* **284**, 263 (2000) [arXiv:gr-qc/9912016].
16. N.B. Narozhny, V.D. Mur and A.M. Fedotov, *Phys. Lett.* **A315**, 169 (2003) [arXiv:hep-th/0304010].
17. N.D. Birrell and P.C.W. Davies, "*Quantum Fields In Curved Space*," Cambridge, UK: University Press, (1982) 340p.
18. V. De Alfaro, S. Fubini, G. Furlan and C. Rosetti, "*Currents in Hadron Physics*", North Holland Publishing Co,. 1973, Amsterdam–London.
19. H. Satz, *Phys. Lett.* **B44**, 373 (1973).
20. R. Hagedorn, *Nuovo Cim. Suppl.* **3**, 147 (1965).
21. E. Byckling and K. Kajantie, "*Particle kinematics*" John Wiley and Sons, 1973, London–New York–Sydney–Toronto.
22. D. Kharzeev, E. Levin and K. Tuchin, *to appear.*

HIGH ENERGY SCATTERING IN QCD: DIPOLE APPROACH WITH POMERON LOOPS

EUGENE LEVIN *

Department of Particle Physics, School of Physics and Astronomy, Raymond and Beverly Sackler Faculty of Exact Science Tel Aviv University, Tel Aviv, 69978, Israel

E-mail: leving@post.tau.ac.il, levin@mail.desy.de, elevin@quark.phy.bnl.gov

In this talk we discuss the BFKL Pomeron Calculus and its interrelation with the colour dipole approach. The two key problems we consider are the probabilistic interpretation of the FKL Pomeron Calculus and the possible scenario for the asymptotic behaviour of the scattering amplitude at high energy in QCD.

1. Introduction

The simplest approach that we can propose for high energy interaction is based [1,2] on the BFKL Pomeron [3] and reggeon-like diagram technique for the BFKL Pomeron interactions [4,5,6,7]. This technique which is a generalization of the Gribov Reggeon Calculus [8], can be written in the elegant form of the functional integral (see Ref. [5] and the next section); and it is a challenge to solve this theory in QCD finding the high energy asymptotic behaviour. However, even this simple approach has not been solved during three decades of attempts by the high energy community. This failure stimulates a search for deeper understanding of physics which is behind the BFKL Pomeron Calculus. In particular, at the end of the Reggeon era it was understood [9,10,11] that the Reggeon Calculus can be reduced to a Markov process [12] for the probability of finding a given number of Pomerons at fixed rapidity Y. Such an interpretation, if it would be reasonable in QCD, can be useful, since it allows us to use powerful methods of statistical physics in our search for the solution.

The goal of this talk is to consider two main problems: the probabilistic

*Work partially supported by the Israel Science Foundation, founded by the Israeli Academy of Science and Humanities.

interpretation of the BFKL Pomeron Calculus based on the idea that colour dipoles are the correct degrees of freedom at high energy QCD [13] and the possible solution for the scattering amplitude at high energy. Colourless dipoles play two different roles in our approach. First, they are partons ('wee' partons) for the BFKL Pomeron. This role is not related to the large N_c approximation and, in principle, we can always speak about probability to find a definite number of dipoles instead of defining the probability to find a number of the BFKL Pomerons. The second role of the colour dipoles is that at high energies we can interprete the vertices of Pomeron merging and splitting in terms of the probability for two dipoles to annihilate in one dipole and of the probability for the decay of one dipole into two. It was shown in Ref. [13] that $I\!P \to 2I\!P$ splitting can be described as the process of the dipole decay into two dipoles. However, the relation between the Pomeron merging ($2I\!P \to I\!P$) and the process of annihilation of two dipoles into one dipole is not so obvious and it will be discussed here.

Our presentation is based on Refs. [14,15,16,17,18] and we would like to thank Michael Lublinsky and Alex Prygarin for their contributions and for a great pleasure to work with them.

The outline of the talk looks as follows. In the next section we will discuss the BFKL Pomeron Calculus in the elegant form of the functional integral, suggested by M. Braun about five years ago [5]. We will show that the intensive recent work on this subject [19,20,16] confirms the BFKL Pomeron Calculus in spite of the fact that these attempts were based on slightly different but not more general assumptions.

In the third section the general approach based on the generating functional will be discussed. The set of equations for the amplitude of n-dipole interaction with the target will be obtained and the interrelations between vertices of the Pomeron interactions and the microscopic dipole processes will be considered.

The fourth section is devoted to the toy model which simplifies the QCD interaction and allows us to see the main properties of high energy amplitude. In particular, we are going to discuss the solution of the equations for the asymptotic behaviour of the scattering amplitude at high energies.

In the fifth section we will consider a general case of QCD and will discuss the probabilistic interpretation as well as the scenario for the solution using the experience with the toy model.

In conclusion we are going to compare our approach with other approaches on the market.

2. The BFKL Pomeron Calculus

The main ingredient of this calculus is the Green function of the BFKL Pomeron describing the propagation of a pair of gluons from rapidity y' and points x'_1 and x'_2 to rapidity y and points x_1 and x_2 [a]. Since the Pomeron does not carry colour in the t-channel we can treat the initial and final coordinates as coordinates of quark and antiquark in a colourless dipole. This Green function is well known [21] and has a form:

$$G(x_1, x_2; y|x'_1, x'_2; y') = \Theta(y - y') \cdot$$
$$\cdot \sum_{n=-\infty}^{\infty} \int \frac{d\nu}{2\pi i} d^2 x_0 \, e^{\omega(n,\nu)(y-y')} \, \lambda(n,\nu) \, E(x_1, x_2; x_0|\nu) \, E^*(x'_1, x'_2; x_0|\nu), \quad (1)$$

where vertices E are given by

$$E(x_1, x_2; x_0|\nu) = \left(\frac{x_{12}}{x_{10}\, x_{20}}\right)^h \left(\frac{x^*_{12}}{x^*_{10}\, x^*_{20}}\right)^{\tilde{h}}. \quad (2)$$

Here $x_{ik} = x_i - x_k$, $x_i = x_{i,x} + i x_{i,y}$ [b], $x^*_i = x_{i,x} + i x_{i,y}$; $h = (1-n)/2 + i\nu$ and $\tilde{h} = 1 - h^*$. The energy levels $\omega(n,\nu)$ are the BFKL eigenvalues

$$\omega(n,\nu) = \bar{\alpha}_S \left(\psi(1) - Re\psi(\frac{|n|+1}{2} + i\nu)\right), \quad (3)$$

where $\Psi(z) = d \ln \Gamma(z)/dz$ and $\Gamma(z)$ is the Euler gamma function, $\bar{\alpha}_S = \alpha_S N_c/\pi$ and finally

$$\lambda(n,\nu) = \frac{1}{[(n+1)^2 + 4\nu^2][(n-1)^2 + 4\nu^2]}. \quad (4)$$

The interaction between Pomerons is described by the triple Pomeron vertex which can be written in the coordinate representation [5] for the following process: two gluons with coordinates x_1 and x_3 at rapidity y_1 decay into two gluons pairs with coordinates $x"_1$ and $x"_2$ at rapidity y_2 and $x"_3$ and $x"_4$ at rapidity y_3 due to the Pomeron splitting at rapidity y. It looks like

$$\frac{\pi \, \bar{\alpha}_S^2}{N_c} \int \frac{d^2 x'_1 \, d^2 x_2 \, d^2 x_3}{x^2_{12}\, x^2_{23}\, x^2_{13}} \left(G(x_1, x_3; y_1|x'_1, x'_3; y) \, \overleftarrow{L}_{1',3'}\right) \cdot$$
$$\cdot \, G(x'_1, x'_2; y|x"_1, x"_2; y_2) \, G(x'_2, x'_3; y|x"_3, x"_4; y_2), \quad (5)$$

[a] Coordinates x_i here are two dimensional vectors and, strictly speaking, should be denoted as $\vec{x}_i$ or $\mathbf{x_i}$. However, we will use the notation x_i hoping that it will not cause difficulties in understanding.

[b] $x_{i,x}$ and $x_{i,y}$ are components of the two dimensional vector x_i on the x and y axis.

where

$$\overleftarrow{L}_{1',3'} \;=\; r'^4_{13}\, p^2_{1'}\, p^2_{3'} \;\text{ with } p^2 \;=\; -\,\nabla^2 \tag{6}$$

and the arrow shows the direction of action of the operator L. For the inverse process of merging of two Pomerons into one we have

$$\frac{\pi\,\bar{\alpha}_S^2}{N_c}\int \frac{d^2x'_1\,d^2\,x_2\,d^2x_3}{x^2_{12}\,x^2_{23}\,x^2_{13}}\,G(x"_1,x"_2;y_2|x'_1,x'_2;y)\,\cdot$$
$$\cdot\, G(x"_3,x"_4;y_2|x'_2,x'_3;y)\cdot\left(\vec{L}_{1',3'}\,G(x'_1,x'_3;y|x_1,x_3;y_1)\right)\ . \tag{7}$$

The theory with the interaction given by Eq. (5) and Eq. (7) can be written through the functional integral [5]

$$Z[\Phi,\Phi^+] \;=\; \int D\Phi\, D\Phi^+\, e^S \;\text{ with}\, S \;=\; S_0 \,+\, S_I \,+\, S_E\ , \tag{8}$$

where S_0 describes free Pomerons, S_I corresponds to their mutual interaction while S_E relates to the interaction with the external sources (target and projectile). From Eq. (5) and Eq. (7) it is clear that

$$S_0 \;=\; \int dy\,dy'\,d^2x_1\,d^2x_2\,d^2x'_1\,d^2x'_2$$
$$\Phi^+(x_1,x_2;y)\,G^{-1}(x_1,x_2;y|x'_1,x'_2;y')\,\Phi(x'_1,x'_2;y') \tag{9}$$
$$S_I \;=\; \frac{\pi\tilde{\alpha}_S^2}{N_c}\int dy\int\frac{d^2x_1\,d^2x_2\,d^2x_3}{x^2_{12}\,x^2_{23}\,x^2_{13}}\,\cdot$$
$$\cdot\{\left(\vec{L}_{1,3}\,\Phi(x_1,x_3;y)\right)\,\cdot\,\Phi^+(x_1,x_2;y)\,\Phi^+(x_2,x_3)\;+\;h.c.\}\,. \tag{10}$$

For S_E we have local interaction both in rapidity and in coordinates, namely,

$$S_E \;=\; -\int dy\,d^2x_1\,d^2x_2$$
$$\cdot\{\Phi(x_1,x_2;y)\,\tau_{pr}(x_1,x_2;y)\;+\;\Phi^+(x_1,x_2;y)\,\tau_{tar}(x_1,x_2;y)\}\ , \tag{11}$$

where τ_{pr} and τ_{tar} stands for the projectile and target, respectively. The form of function τ depends on the non-perturbative input in our problem; and for the case of a nucleus target they are written in Ref. [5].

For the case of the projectile being a dipole that scatters off a nucleus the scattering amplitude has the form

$$T(Y,x_1,x_2) \;=\; -\,\frac{4\,\pi^2\,\bar{\alpha}_S}{N_c}\,\frac{\int\, D\Phi\,D\Phi^+\,\Phi(x_1,x_2;Y)\,e^S}{\int\, D\Phi\,D\Phi^+\;e^S|_{S_E=0}}\ , \tag{12}$$

where the extra α_S comes from our normalization.

For further presentation we need some properties of the BFKL Green function [21]:

1. Generally,

$$G^{-1}(x_1,x_2;y|x'_1,x'_2;y') \;=\; p_1^2\,p_2^2\left(\frac{\partial}{\partial y}+H\right) \;=\; \left(\frac{\partial}{\partial y}+H^+\right)p_1^2\,p_2^2\,; \tag{13}$$

$$H f(x_1,x_2;y) \;=\; \frac{\bar{\alpha}_S}{2\pi}\int\frac{d^2x_3\,x_{12}^2}{x_{23}^2\,x_{13}^2}\left(f(x_1,x_2;y)\;-\;f(x_1,x_3;y)\;-\;f(x_3,x_2;y)\right)\,; \tag{14}$$

2. The initial Green function (G_0) is equal to

$$G_0(x_1,x_2;y|x'_1,x'_2;y') \;=\; \frac{\pi^2}{2}\ln\frac{x_{1,1'}^2\,x_{2,2'}^2}{x_{1,2'}^2\,x_{1',2}^2}\,\ln\frac{x_{1,1'}^2\,x_{2,2'}^2}{x_{1,2}^2\,x_{1',2'}^2}\,; \tag{15}$$

3. It should be stressed that

$$\nabla_1^2\,\nabla_2^2\,G_0(x_1,x_2;y|x'_1,x'_2;y') \;=\; \delta^{(2)}(x_1-x'_1)\,\delta^{(2)}(x_2-x'_2)\;+\;\delta^{(2)}(x_1-x'_2)\,\delta^{(2)}(x_2-x'_1)\,; \tag{16}$$

4. In the sum of Eq. (1) only the term with $n=0$ is essential for the high energy asymptotic behaviour since all $\omega(n,\nu)$ with $n\geq 1$ are negative and, therefore, lead to contributions that decrease with energy. Taking into account only the first term one can see that G is the eigenfunction of operator L_{13}, namely

$$\begin{aligned} L_{13}\,G_0(x_1,x_2;y|x'_1,x'_2;y') \;&=\; \frac{1}{\lambda(0,\nu)}\,G_0(x_1,x_2;y|x'_1,x'_2;y') \\ &\approx\; G_0(x_1,x_2;y|x'_1,x'_2;y')\,. \end{aligned} \tag{17}$$

The last equation holds only approximately in the region where $\nu \ll 1$, but this is the most interesting region which is responsible for the high energy asymptotic behaviour of the scattering amplitude.

Using Eq. (5), Eq. (7) and Eq. (14) we can easily obtain the chain equation for the multi-dipole amplitude $T^{(n)}$ noticing that every dipole interacts only with one Pomeron (see Eq. (12)). The equation is shown in

Fig. (1) and it has the following form

$$\frac{\partial T^{(1)}(y;x_1,x_2)}{\partial y} \;=\; \frac{\bar{\alpha}_S}{2\,\pi}\int d^z\,K(x_1,x_2;z)$$
$$\Big(T^{(1)}(y;x_1,z)\;+\;T^{(1)}(y;z,x_2)\;-\;T^{(1)}(y;x_1,x_2\;-\;T^{(2)}(y;x_1,z;z,x_2)\Big) \tag{18}$$

$$\frac{\partial T^{(2)}(y;x_1,x_2;x_3,x_4)}{\partial y} \;=\; \frac{\bar{\alpha}_S}{2\,\pi}\int d^z\,K(x_1,x_2;z)$$
$$\Big(T^{(2)}(y;x_1,z;x_3,x_4)\;+\;T^{(2)}(y;z,x_2;x_3,x_4)\;-\;T^{(2)}(y;x_1,x_2;x_3,x_4)\,-$$
$$-\,T^{(3)}(y;x_1,z;z,x_2;x_3,x_4)\;+\;\{x_1\to x_3,x_2\to x_4\}\Big)\;+$$
$$+\,(\frac{4\pi^2\bar{\alpha}_S}{N_c})^2\frac{\bar{\alpha}_S}{2\,\pi}\int\Gamma_{2\to 1}(x_1,x_2;x_3,x_4|x',x")\,\nabla^2_{x'}\nabla^2_{x"}T^{(1)}(y;x',x")\;. \tag{19}$$

Deriving Eq. (18) and Eq. (19) we use Eq. (15) and Eq. (16) as well as the normalization condition (see Eq. (12)) for the scattering amplitude. These two equations are the same as in Ref. [22]. This shows that in the papers [19,22], actually, the same approach as the BFKL Pomeron Calculus has been developed (much later, of course), in spite of the fact that the authors think that they are doing something more general.

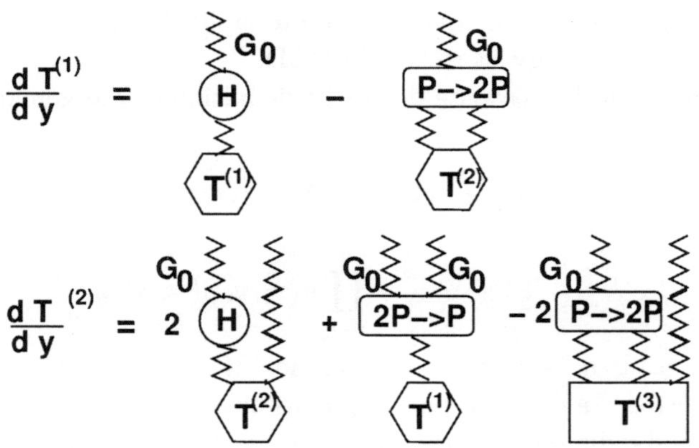

Figure 1. The graphic form of equations for the multi dipole amplitude.

Kernel $K(x,y|z)$ is defined as

$$K(x,y|z) \;=\; \frac{(x-y)^2}{(x-z)^2\,(z-y)^2}\;. \tag{20}$$

Assuming $T^{(2)} = T^{(1)}T^{(1)}$, we obtain the Balitsky-Kovchegov equation [23,24]. We can do this only if we can argue why the Pomeron splitting is more important than the Pomeron merging. For example this assumption is reasonable for the scattering of the dipole with the nucleus target. Generally speaking, the splitting and merging have the same order in α_S (see Eq. (5) and Eq. (7)). In Eq. (18) and Eq. (19) these two processes look like having a different order of magnitude in α_S but this fact does not interrelate with any physics and reflect only our normalization. However, we will see that for a probabilistic interpretation the correct normalization is very important.

Function $\Gamma_{2\to 1}$ is equal to

$$\Gamma_{2\to 1}(x_1, y_1; x_2, y_2|x, y) = \\ = \int d^2 z K(x, y; z)\, G_0(x_1, y_1|x, z)\, G_0(x_1, y_1|z, y)\,, \qquad (21)$$

where G_0 is given by Eq. (15) and we use the notation (x, y) for a colourless pair of gluons (colour dipole).

3. Generating functional and probabilistic interpretation

In this section we discuss the main equations of the BFKL Pomeron Calculus in the formalism of the generating functional, which we consider as the most appropriate technique for the probabilistic interpretation of this approach to high energy scattering in QCD.

To begin with, let us write down the definition of the generating functional [13]

$$Z(Y - Y_0;\, [u]) \equiv \sum_{n=1} \int P_n\,(Y - Y_0;\, x_1, y_1; \ldots; x_i, y_i; \ldots; x_n, y_n) \cdot \\ \cdot \prod_{i=1}^{n} u(x_i, y_i)\, d^2 x_i\, d^2 y_i\,, \qquad (22)$$

where $u(x_i, y_i) \equiv u_i$ is an arbitrary function of x_i and y_i. The coordinates (x_i, y_i) describe the colorless pair of gluons or a dipole. P_n is a probability density to find n dipoles with the size $x_i - y_i$, and with the impact parameter $(x_i + y_i)/2$. It follows from the physical meaning of P_n and the definition in Eq. (22) directly that the functional in Eq. (22) obeys the condition

$$Z(Y - Y_0;\, [u = 1]) = 1\,. \qquad (23)$$

The physical meaning of this equation is that the sum over all probabilities is equal to unity.

Introducing vertices for the dipole process: $1 \to 2$ ($V_{1\to 2}(x, y \to x_1, y_1 + x_2, y_2)$), $2 \to 1$ ($V_{2\to 1}(x_1, y_1 + x_2, y_2 \to x, y)$) and $2 \to 3$ ($V_{2\to 3}(x_1, y_1 + x_2, y_2 \to x'_1, y'_1 + x'_2, y'_2 + x'_3, y'_3)$ we can write a typical birth-death equation in the form

$$\frac{\partial P_n(\ldots, x_i, y_i; \ldots)}{\partial Y} =$$

$$= \sum_i V_{1\to 2} \bigotimes (P_{n-1}(\ldots, x_i, y_i; \ldots) - P_n(\ldots, x_i, y_i; \ldots))$$

$$+ \sum_{i>k} V_{2\to 1} \bigotimes (P_{n+1}(; x_i, y_i; \ldots; x_k, y_k;) - P_n(; x_i, y_i;))$$

$$+ \sum_{i>k} V_{2\to 3} \bigotimes (P_{n-1}(; x_i, y_i;; x_k, y_k;) - P_n(; x_i, y_i;; x_k, y_k;)) \quad (24)$$

Multiplying this equation by the product $\prod_{i=1}^{n} u_i$ and integrating over all x_i and y_i, we obtain the following linear equation for the generating functional:

$$\frac{\partial Z[u]}{\partial Y} = \chi[u] \, Z[u] \quad (25)$$

with

$$\chi[u] = - \int d^4 q d^4 q_1 \, d^4 q_2$$

$$\Big(V_{1\to 2}(q \to q_1 + q_2) \, (-u(q) + u(q_1) \, u(q_2)) \, \frac{\delta}{\delta u(q)} -$$

$$-V_{2\to 1}(q_1 + q_2 \to q) \, (u(q_1) \, u(q_2) - u(q)) \, \frac{1}{2} \, \frac{\delta^2}{\delta u(q_1) \, \delta u(q_2)} \Big) \,. \quad (26)$$

Trying to make our presentation more transparent we omitted in Eq. (26) the term corresponding to the $2 \to 3$ transition (see Ref. [16] for full presentation).

Eq. (25) is a typical diffusion equation or Fokker-Planck equation [12], with a diffusion coefficient depending on u. This is the master equation of our approach, and the goal of this talk is to find the correspondence of this equation with the BFKL Pomeron Calculus and the asymptotic solution to this equation. In spite of the fact that this is a functional equation we intuitively feel, that this equation could be useful since we can develop a direct method for its solution and, on the other hand, there exist many studies of such an equation in the literature (see for example Ref. [12]) as well as some physical realizations in statistical physics. The intimate relation between the Fokker-Planck equation and the high energy asymptotic was first pointed out by Weigert [25] in the JIMWLK approach [27], and has been discussed in Refs. [26,19,20].

The scattering amplitude can be defined as a functional [24,15]

$$\begin{aligned} N\left(Y;[\gamma_i]\right) &= \\ &= -\sum_{n=1}^{\infty}\int \gamma_n(x_1,y_1;\ldots;x_n,y_n;Y_0)\prod_{i=1}^{n}\frac{\delta}{\delta u_i}Z\left(Y,[u_i]\right)|_{u_i=1}\,d^2x_i\,d^2y_i \\ &= -\sum_{n=1}^{\infty}(-1)^n\int \gamma_n(x_1,y_1;\ldots;x_n,y_n;Y_0)\rho(x_1,y_1;;x_n,y_n;Y-Y_0) \end{aligned} \tag{27}$$

The physical meaning of functions γ_n is the imaginary part of the amplitude of the interaction of n dipoles with the target at low energies. All these functions should be taken from the non-perturbative QCD input. However, in Refs. [14,15,16] it was shown that we can introduce the amplitude of interaction of n dipoles $(\gamma_n(x_1,y_1;\ldots;x_n,y_n;Y)$ at high energies (large values of rapidity Y) and Eq. (22), Eq. (25) and Eq. (27) can be rewritten as a chain set of equation for $\gamma_n(x_1,y_1;\ldots;x_n,y_n;Y)$. The equation has the form[c]

$$\begin{aligned} &\frac{\partial\,\gamma_n\left(r_1,b_1,\ldots,r_n,b_n\right)}{\partial Y} \quad = \\ &2\sum_{i=1}^{n}\int d^4q'\,d^4q\,V_{1\,\rightarrow\,2}(q_i;\,q\,q')\,\gamma_n\left(\ldots q'\ldots\right) \\ &-\sum_{i=1}^{n}\int d^4q_1'\,d^4q_2'\,V_{1\,\rightarrow\,2}(q_i;\,q_1'\,q_2')\,\gamma_n\left(\ldots,q_i\ldots\right) \\ &-\sum_{i=1}^{n-1}\int d^4q\,d^4q' V_{1\,\rightarrow\,2}(q_i;\,q\,q')\,\gamma_{n+1}\left(\ldots q\ldots q'\right) \\ &-\sum_{i>j}^{n}\int d^4\,q\,V_{2\,\rightarrow\,1}\left(q_i\,q_j;\,q\right)\cdot\gamma_{n-1}\left(q_i\ldots q_j\ldots q\right) \\ &+2\sum_{i=1}^{n}\int d^4q\,d^4q' V_{2\rightarrow 1}\left(qq_i;q'\right)\,\gamma_{n-1}\left(\ldots q_i\ldots q\right)\,+ \\ &+\sum_{i,k,i>k}^{n}\int d^4q\,V_{2\rightarrow 1}\left(q_iq_k;q\right)\,\gamma_n\left(\ldots q_i\ldots q_k\ldots\right) \end{aligned} \tag{28}$$

[c]This equation is Eq. (2.19) of Ref. [16] but, hopefully, without misprints, part of which has been noticed in Ref. [22].

Comparing this equation for $\gamma_1 \equiv T^{(1)}$ and $\gamma_2 \equiv T^{(2)}$ one can see that

$$V_{1\to 2} = \frac{\bar{\alpha}_S}{2\,\pi}\Gamma_{1\to 2} \;=\; \frac{\bar{\alpha}_S}{2\,\pi}\,K\,(x,y;z)\;; \tag{29}$$

$$\begin{aligned} V_{2\to 1} \;=\; & \left(\frac{4\pi^2\bar{\alpha}_S}{N_c}\right)^2 \frac{\bar{\alpha}_S}{2\,\pi}\,(-\,\Gamma_{2\to 1}(x_1,y_1+x_2,y_2\to x,y)+ \\ & +\;\int d^2x\,d^2y\,\Gamma_{2\to 1}(x_1,y_1+x_2,y_2\to x,y)\,\cdot \\ & \cdot\;\left(\delta^{(2)}(x_1-x)\delta^{(2)}(y_1-y)\,+\,\delta^{(2)}(x_2-x)\delta^{(2)}(y_2-y)\right)) \end{aligned} \tag{30}$$

with $\Gamma_{2\to 1}$ is given by Eq. (21).

4. A toy model: Pomeron interaction and probabilistic interpretation

In this section we consider the simple toy model in which the probabilities to find n-dipoles do not depend on the size of dipoles [13,14,16,17]. In this model the master equation (25) has a simple form

$$\frac{\partial Z}{\partial Y} \;=\; -\Gamma(1\to 2)\,u(1-u)\,\frac{\partial Z}{\partial u} + \Gamma(2\to 2)\,u(1-u)\,\frac{\partial^2 Z}{(\partial u)^2} \tag{31}$$

and this equation generates the Pomeron splitting $G_{I\!P\to 2I\!P} = -\Gamma(1\to 2)$, Pomerons merging $G_{2I\!P\to I\!P} = \Gamma(2\to 1)$ and also the two Pomeron scattering $G_{2I\!P\to 2I\!P} = -\Gamma(2\to 1)$. It is easy to see that neglecting the $u^2\partial^2 Z/(\partial u)^2$ term in Eq. (31), we cannot provide a correct sign for Pomerons merging $G_{2I\!P\to I\!P}$.

Eq. (31) is the diffusion with the u dependence in the diffusion coefficient. For $u < 1$ the diffusion coefficient is positive and the equation has a reasonable solution. If $u > 1$, the sign of this coefficient changes and the equation gives a solution which increases with Y and $Z(Y)$ and it cannot be treated as the generating function for the probabilities to find n dipoles (Pomerons) (see Refs. [9,11,17] for details). We can see the same features in the asymptotic solution that is the solution to Eq. (31) with the l.h.s. equal to zero. It is easy to see that this solution has the form

$$Z(u;Y\to\infty) \;=\; e^{-\,\kappa\,(1-u)}\,,\;\text{ with }\kappa = \frac{\Gamma(1\to 2)}{\Gamma(2\to 1)} \;=\; \frac{2\,N_c^2}{\bar{\alpha}_S^2} \gg 1\;. \tag{32}$$

One can see that for negative κ this solution leads to $Z > 1$ for $u < 1$. This shows that we cannot give a probabilistic interpretation for such a solution.

Using the asymptotic solution and finding the typical u at high energies from the unitarity constraints [17,28,29] we can determine the asymptotic limit for the high energy amplitude. It turns out that

$$N(Y \to \infty) \;=\; 1 \;-\; e^{-\kappa} \tag{33}$$

It was shown in Ref. [17] that we can search for the correction to the solution of Eq. (33) in the form

$$Z(u;Y) \;=\; Z(u;Y \to \infty) \;+\; \Delta Z(u;Y) \tag{34}$$

Assuming $\Delta Z(u;Y)(u;Y \;\to\; \infty) \;\gg\; Z(u;Y \;\to\; \infty)$ we can write for $\Delta Z(u;Y)$ the linear equation [17] and the solution of this equation decreases with energy.

It is interesting to notice that this asymptotic amplitude does not show a black disc behaviour at high energies. The behaviour of the scattering amplitude, given by Eq. (33), corresponds to the so called gray disc behaviour.

5. Probabilistic interpretation in QCD

Eq. (29) has a very simple physical meaning describing the Pomeron splitting as the decay process of one dipole into two dipoles. It turns out that the vertices for $2I\!P \to 3I\!P$ and $2I\!P \to 2I\!P$ can be easily understood as a dipole 'swing'. What we mean is that with some probability two quarks of a pair of dipoles can exchange their antiquarks to form another pair of dipoles [16]. Naturally, this process has an $1/N_c^2$ suppression and it correctly reproduces the splitting and rescattering of two Pomerons that has been explicitly calculated from the diagrams [7].

Eq. (30)[d] is more difficult to view as the vertex for the transition of two dipoles into one. Indeed, the integral over coordinates of the produced dipole is positive, namely $\int V(2 \to 1) d^2x d^2y = \int \Gamma(2 \to 1) d^2x d^2y$. However, Eq. (30) leads, generally speaking, to a negative vertex in some regions of the phase space. The detailed study of this vertex will be published soon [18]. Here we want to point out that the key problem is not in the probabilistic interpretation of the microscopic process of two dipoles to one dipole transition but the fact that a negative vertex $V(2 \to 1)$ means that

[d]This equation is quite different from the equation which is obtained in Ref. [16] (see also Ref. [22]). The main difference stems from the correct use of the BFKL Pomeron Calculus for determining this vertex while in Refs. [16,22] the Born diagram was used which does not and cannot give a correct expression.

in some kinematic region we have a negative diffusion coefficient which results in a solution that increases at large values of rapidity Y. To save such a theory, we have to introduce other Pomeron interactions like $2I\!P \to 3I\!P$ and/or $2I\!P \to 2I\!P$ transitions.

In Ref. [16,17] attempts were made to deal with such Pomeron interactions. It turns out that at large values of Y the $2 \to 1$ process contribute in a very limited part of the kinematic region with a very specific function $u(x, y)$. In this particular region the vertex $V(2 \to 1)$ is positive [18]. Therefore, we could use the probabilistic interpretation but we need to study this process better and deeper to obtain the final result.

In the toy model it has been shown [9,11] that we can generate the $2I\!P \to 3I\!P$ vertex without the process of two dipoles to one dipole transition. The transition of two dipoles to two or more dipoles also leads to this vertex. Similar ideas are developed in Ref. [31] in QCD. However, we need to pay a price: the contribution to Eq. (25) resulting in the correct sign for $2I\!P \to I\!P$ interaction will be negative. In other words, we can add to Eq. (25) the contribution

$$\int \prod_{i=1}^{2} d^2 x_i \, d^2 y_i \prod_{i=1}^{2} d^2 x'_i \, d^2 y'_i \, V_{2\to 2}(x_1, y_1 + x_2, y_2 \to x'_1, y'_1 + x'_2, y'_2) \cdot$$
$$\cdot \, u(x'_1, y'_1) \, u(x'_2, y'_2) \, \frac{1}{2} \frac{\partial}{\partial \, u(x_1, y_1)} \, \frac{\partial}{\partial \, u(x_2, y_2)} \,, \qquad (35)$$

which will give the $2I\!P \to I\!P$ vertex in the form

$$\Gamma(2I\!P \to I\!P)(x_1, y_1 + x_2, y_2 \to x'_1, y'_1) =$$
$$- \int d^2 x'_2 \, d^2 y'_2 \; V_{2\to 2}(x_1, y_1 + x_2, y_2 \to x'_1, y'_1 + x'_2, y'_2) \,. \quad (36)$$

Therefore, we have to assume that $V_{2\to 2} < 0$. As we have discussed, generally speaking, it means that our problem has no solution. However, in QCD the situation is much better. Indeed, to generate a correct $2I\!P \to I\!P$ vertex we need to introduce $V_{2\to 2} \propto \bar{\alpha}_S^3$. The Feynman diagrams for this transition are obvious (see Fig. (2-b)). However, the main contribution stems from the diagrams of Fig. (2-a)–type which are of the order of $\bar{\alpha}_S / N_c^2$. Therefore, diagrams of Fig. (2-b)–type are small corrections to the main contribution and could be negative, in spite of the fact that the diagrams shown in Fig. (2-b) actually give a positive contribution. We will discuss this scenario in all details in our paper [18].

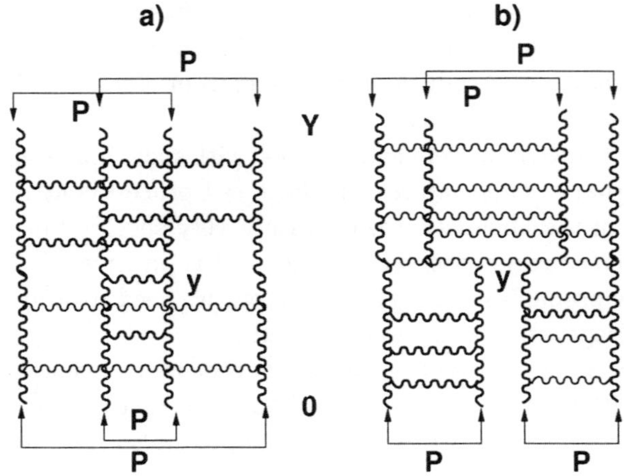

Figure 2. The diagrams for $2I\!P \to 2I\!P$ process. Fig. (2)-a shows the diagram which is of the order $\bar{\alpha}_S/N_c^2$ while Fig. (2)-a leads to a contribution $\propto \bar{\alpha}_S^3$.

6. Conclusions

Being elegant and beautiful, the BFKL Pomeron Calculus has a clear disadvantage: it is not a theoretically closed theory. Indeed, we need to add to the formalism of the BFKL Pomeron Calculus the theoretical ideas what kind of Pomeron interactions we should take into account and why. Of course, the Feynman diagrams in leading log $1/x$ approximation of perturbative QCD allow us, in principle, to calculate all possible Pomeron interactions but, practically, it is a very hard job. Even if we will calculate these vertices we need to understand what set of vertices we should take into account for the calculation of the scattering amplitude. This is the reason why we have to develop a more general formalism. Fortunately, such a formalism has been built and it is known under the abbreviation JIMWLK-Balitsky approach [23,27,30]. In this approach we are able to calculate all vertices for Pomeron interactions as it was demonstrated in Ref. [31] and it solves the first part of the problem: the determination of all possible Pomeron interactions. However, we need to understand what vertices we should take into account for the calculation of the scattering amplitude. We hope that a further progress in going beyond of the BFKL Pomeron Calculus (see Refs. [31,32]) will lead to such a development of the BFKL Pomeron Calculus that we will have a consistent theoretical approach. Hopefully this approach will be simpler than the Lipatov's effective action [33] which is not easier to solve than the full QCD Lagrangian.

The probabilistic interpretation gives a practical method for creating a Monte Carlo code in the spirit of the approach suggested in Ref. [34]. This code will allow us to find a numerical solution to the problem and to consider the inclusive observables. This extention is very desirable since most of the experimental data exist just for these observables.

This year, my teacher, Prof. Gribov, would have been 75 years old. I am happy to report him that his ideas are still alive and still give the simplest and the most elegant approach to high energy interaction. The high energy behaviour of the scattering amplitude is still an open theoretical question. We have learned a lot and we are able to foresee a lot of difficulties ahead. Therefore, it is a good problem to be solved.

References

1. V.N. Gribov, E.M. Levin and M.G. Ryskin, *Phys. Rep.* **100**, 1 (1983).
2. A.H. Mueller and J. Qiu, *Nucl. Phys.* **B268**, 427 (1986).
3. E.A. Kuraev, L.N. Lipatov, and F.S. Fadin, *Sov. Phys. JETP* **45**, 199 (1977); Ya.Ya. Balitsky and L.N. Lipatov, *Sov. J. Nucl. Phys.* **28**, 22 (1978).
4. J. Bartels, M. Braun and G.P. Vacca, *Eur. Phys. J.* **C40**, 419 (2005) [arXiv:hep-ph/0412218]; J. Bartels and C. Ewerz, *JHEP* **9909**, 026 (1999) [arXiv:hep-ph/9908454]; J. Bartels and M. Wusthoff, *Z. Phys.* **C66**, 157 (1995); A.H. Mueller and B. Patel, *Nucl. Phys.* **B425**, 471 (1994) [arXiv:hep-ph/9403256]; J. Bartels, Z. Phys. **C60**, 471 (1993).
5. M. Braun, arXiv:hep-ph/0504002; *Eur. Phys. J.* **C16**, 337 (2000) [arXiv:hep-ph/0001268]; M. Braun, *Eur. Phys. J.* **C6**, 321 (1999) [arXiv:hep-ph/9706373]; M.A. Braun and G.P. Vacca, *Eur. Phys. J.* **C6**, 147 (1999) [arXiv:hep-ph/9711486].
6. H. Navelet and R. Peschanski, *Nucl. Phys.* **B634**, 291 (2002) [arXiv:hep-ph/0201285]; *Phys. Rev. Lett.* **82**, 137 (1999) [arXiv:hep-ph/9809474]; *Nucl. Phys.* **B507**, 353 (1997) [arXiv:hep-ph/9703238].
7. J. Bartels, L.N. Lipatov and G.P. Vacca, *Nucl. Phys.* **B706**, 391 (2005) [arXiv:hep-ph/0404110].
8. V.N. Gribov, *Sov. Phys. JETP* **26**, 414 (1968) [*Zh. Eksp. Teor. Fiz.* **53**, 654 (1967)].
9. P. Grassberger and K. Sundermeyer, *Phys. Lett.* **B77**, 220 (1978).
10. E. Levin, *Phys. Rev.* **D49**, 4469 (1994).
11. K.G. Boreskov, *"Probabilistic model of Reggeon field theory,"* arXiv:hep-ph/0112325 and reference therein.
12. C.W. Gardiner, *"Handbook of Stochastic Methods for Physics, Chemistry and the Natural Science"*, Springer-Verlag, Berlin, Heidelberg 1985.
13. A.H. Mueller, *Nucl. Phys.* **B415**, 373 (1994); *ibid* **B437**, 107 (1995).
14. E. Levin and M. Lublinsky, *Nucl. Phys.* **A730**, 191 (2004) [arXiv:hep-ph/0308279].
15. E. Levin and M. Lublinsky, *Phys. Lett.* **B607**, 131 (2005).

16. E. Levin and M. Lublinsky, *Nucl. Phys.* **A763**, 172 (2005).
17. E. Levin, *Nucl. Phys.* **A763**, 140 (2005).
18. E. Levin and A. Prygarin, (in preparation).
19. E. Iancu and D.N. Triantafyllopoulos, *Nucl. Phys.* **A756**, 419 (2005); *Phys. Lett.* **B610**, 253 (2005).
20. A.H. Mueller, A.I. Shoshi and S.M.H. Wong, *Nucl. Phys.* **B715**, 440 (2005) [arXiv:hep-ph/0501088].
21. L.N. Lipatov, *Phys. Rept.* **286**, 131 (1997) [arXiv:hep-ph/9610276]; Sov. Phys. JETP **63**, 904 (1986) and references therein.
22. E. Iancu, G. Soyez and D.N. Triantafyllopoulos, arXiv:hep-ph/0510094.
23. I. Balitsky, [arXiv:hep-ph/9509348]; *Phys. Rev.* **D60**, 014020 (1999) [arXiv:hep-ph/9812311].
24. Yu.V. Kovchegov, *Phys. Rev.* **D60**, 034008 (1999) [arXiv:hep-ph/9901281].
25. H. Weigert, *Prog. Part. Nucl. Phys.* **55**, 461 (2005) [arXiv:hep-ph/0501087] and references therein.
26. J.P. Blaizot, E. Iancu and H. Weigert, *Nucl. Phys.* **A713**, 441 (2003) [arXiv:hep-ph/0206279].
27. J. Jalilian-Marian, A. Kovner, A. Leonidov and H. Weigert, *Phys. Rev.* **D59**, 014014 (1999) [arXiv:hep-ph/9706377]; *Nucl. Phys.* **B504**, 415 (1997) [arXiv:hep-ph/9701284]; J. Jalilian-Marian, A. Kovner and H. Weigert, *Phys. Rev.* **D59**, 014015 (1999) [arXiv:hep-ph/9709432]; A. Kovner, J.G. Milhano and H. Weigert, *Phys. Rev.* **D62**, 114005 (2000) [arXiv:hep-ph/0004014] ; E. Iancu, A. Leonidov and L.D. McLerran, *Phys. Lett.* **B510**, 133 (2001) [arXiv:hep-ph/0102009]; *Nucl. Phys.* **A692**, 583 (2001) [arXiv:hep-ph/0011241]; E. Ferreiro, E. Iancu, A. Leonidov and L. McLerran, *Nucl. Phys.* **A703**, 489 (2002) [arXiv:hep-ph/0109115]; H. Weigert, *Nucl. Phys.* **A703**, 823 (2002), [arXiv:hep-ph/0004044].
28. E. Iancu and A.H. Mueller, *Nucl. Phys.* **A730**, 460, 494 (2004) [arXiv:hep-ph/0308315],[arXiv:hep-ph/0309276].
29. M. Kozlov and E. Levin, *Nucl. Phys.* **A739**, 291 (2004) [arXiv:hep-ph/0401118].
30. L. McLerran and R. Venugopalan, *Phys. Rev.* **D 49**, 2233, 3352 (1994); **D 50**, 2225 (1994); **D 53**, 458 (1996); **D 59**, 09400 (1999).
31. A. Kovner and M. Lublinsky, ArXiv:hep-ph/0510047; arXiv:hep-ph/0503155; *Phys. Rev. Lett.* **94**, 181603 (2005) [arXiv:hep-ph/0502119]; *JHEP* **0503**, 001 (2005) [arXiv:hep-ph/0502071]; *Phys. Rev.* **D71**, 085004 (2005) [arXiv:hep-ph/0501198].
32. Y. Hatta, E. Iancu, L. McLerran and A. Stasto, *Nucl. Phys.* **A762**, 272 (2005) [arXiv:hep-ph/0505235]; arXiv:hep-ph/0504182.
33. L.N. Lipatov, *Nucl. Phys.* **B452**, 69 (1995).
34. A.H. Mueller and G.P. Salam, *Nucl. Phys.* **B475**, 293 (1996) [arXiv:hep-ph/9605302]; G.P. Salam, *Nucl. Phys.* **B461**, 512 (1996).

GRIBOV APPROACH TO NUCLEAR SHADOWING AND COLLISIONS OF HEAVY IONS *

A.B. KAIDALOV

Institute of Theoretical and Experimental Physics,
B. Cheremushkinskaya 25, 117259, Moscow, Russia
E-mail: kaidalov@itep.ru

The approach to interaction of hadrons and virtual photons with nuclei, proposed by Gribov, is reviewed and applied to description of nuclear structure functions in the small x region. The problem of saturation of parton densities in deep inelastic scattering (DIS) in the limit $x \to 0$ is discussed and its relation to heavy ion collisions at very high energies is investigated in the framework of reggeon theory. It is emphasized that the shadowing effects are important in a definite kinematic region. It is shown how the prediction of the Glauber model for the density of hadrons produced in the central rapidity region in nucleus–nucleus interactions are modified due to shadowing of small-x-partons. Calculations show that shadowing effects are important at RHIC energies, but the "saturation" at these energies is not achieved yet. Production of particles with large transverse momenta and jets both in $AuAu$ and dAu collisions is discussed.

1. Introduction

The modification of parton densities inside nuclei in the small x region is a very interesting phenomenon [1]. It is related to an important problem of "saturation" of parton densities in the limit $x \to 0$. The "saturation" of partons plays an important role in heavy ion collisions at very high energies. In this paper I shall use the approach to nuclear interactions, proposed by Gribov [2], for investigation of nuclear shadowing in the small x region and for relation of these effects to particle production in heavy ion collisions. In the following I shall use the reggeon approach to high-energy interactions of hadrons and nuclei. V.N. Gribov has made very essential contributions to this approach. He has introduced a leading pole in the complex angular momentum plane [3], which determines asymptotics of diffractive processes

*This work is supported in part by the grants: CRDF RUP2-MO-03, RFBR 04-02-17263, SS-1774.2003.2.

(nowadays called the Pomeron), investigated the main properties of Regge–poles [4] and Regge–cuts [5] and developed the reggeon diagram technique [6].

An important contribution to the theory of multiparticle production has been made by V.N. Gribov together with V.A. Abramovsky and O.V. Kancheli [7] and is usually referred to as the AGK–cutting rules. These rules are widely used for construction of QCD–based models for high–energy interactions (for reviews see [8,9]). In this paper I shall extensively use AGK–cutting rules and Gribov reggeon diagram technique to calculate inclusive particle spectra and particle densities in the central rapidity region for heavy ion collisions. The values of particle and energy densities are very important for the problem of creation of the quark–gluon plasma in heavy ion collisions.

In the first part of the paper (Section 2) I shall shortly review Gribov's approach to high–energy interactions of hadrons with nuclei [2] and note the difference in the space–time picture of interactions at low and high energies. The difference between the Glauber approximation [10] and the general Gribov theory will be emphasized.

Investigation of the deep inelastic scattering (DIS) in the region of small x-Bjorken gives an important information on the behavior of partonic densities. Experimental data on small-x DIS obtained at HERA [11] show that distributions of quarks and gluons have a fast increase as x decreases up to values $\sim 10^{-4}$. In reggeon theory this increase is related to an intercept of the leading Pomeron singularity $\alpha_{I\!P}(0)$: $xq(x)$ or $xg(x) \sim \frac{1}{x^{\Delta}}$, where $\Delta \equiv \alpha_{I\!P}(0) - 1$ (see for example [12]). There are good reasons to believe (for reviews on this subject see [13]) that this fast increase will be modified for even smaller values of x due to influence of unitarity effects. In the limit $x \to 0$ partonic distributions $xq(x)(xg(x))$ reach the limit of "saturation" and increase only as powers of $ln\frac{1}{x}$. The problem of "saturation" is even more important in heavy ion collisions, where densities of partons in colliding nuclei are very large. This problem has been studied in QCD perturbation theory [14] and partonic configurations, which include nonlinear effects in QCD evolution, were called "color glass condensate" [15]. However the pattern of "saturation" is much more general.

In Section 3 I shall review theoretical approaches to the origin of Pomeron in QCD. In Section 4 the problem of small x behavior in DIS will be considered and the model [16] based on reggeon theory and dipole picture of interaction of virtual photons with nucleons will be outlined. This model gives a self consistent description of structure functions of a proton and diffraction dissociation of a photon in a broad region of virtualities Q^2.

I shall discuss the pattern of "saturation" in this model.

It will be shown in Section 5 that Gribov's approach to interactions with nuclei allows to calculate shadowing effects in the small x region for structure functions of nuclei. Production of hadrons and jets in heavy ion collisions is described in Section 6. It is emphasized that the coherence condition strongly limits the kinematic region where the shadowing effects for inclusive cross sections are important. In the same approach, which leads to description of nuclear structure functions, it is shown that the shadowing effect in heavy ion collisions leads to a sizable reduction of inclusive particle spectra in comparison with the Glauber model already at RHIC energies. Theoretical predictions agree with the existing experimental data. It is emphasized, however, that at RHIC (and even LHC) energies the situation is far from the "saturation" limit. The effects, which lead to a suppression of production for particles with large $p_\perp$ at RHIC, are discussed.

2. Gribov theory of high-energy nuclear interactions

In classical papers of Gribov [2] it was shown how to incorporate the Glauber model [10] for interactions of hadrons with nuclei into a general framework of relativistic quantum theory. Consider an amplitude of elastic scattering for high–energy hadron–nucleus interactions. In the Glauber model it is described by successive rescatterings of initial hadron on nucleons of the nucleus. However, as it was emphasized by Gribov [2,17], the space–time picture of the interaction at high–energy $E > m_h \mu R_A$ (μ is a characteristic hadronic scale ~ 1 GeV and R_A is the radius of the nucleus) is completely different from this simple picture. It corresponds to coherent interactions of a fluctuation of the initial hadron, which is "prepared" long before its interaction with the nucleus. A very important result of Gribov [2] is that nevertheless the elastic hA–amplitude can be written as a sum of the diagrams with elastic rescatterings, which give the same result as the Glauber model, plus all possible diffractive excitations of the initial hadron. At not too high energies $E_L \sim 10^2$ GeV these terms lead to corrections to the Glauber approximation of $10-20\%$ for the total hA cross section [18,19].

We will show below that at very high energies and for inclusive cross sections this modification of the Glauber approximation is very important. The difference between the Glauber model and Gribov's theory is essential for understanding shadowing corrections for structure functions of nuclei related to hard processes on nuclei and for many aspects of multiparticle production on nuclei [20].

An important consequence of the space–time structure of interactions of hadrons with nuclei is the theorem, based on AGK–cutting rules [7], that for inclusive cross sections all rescatterings cancel with each other and these cross sections are determined by the diagrams of impulse approximation. Note, however, that this result, valid asymptotically in the central rapidity region, only applies to the diagrams of the Glauber–type, i.e. when masses of diffractive intermediate states are limited and do not increase with energy. As a result, the inclusive cross section for the production of a hadron a is expressed, for a given impact parameter b, in terms of inclusive cross section for hN interactions

$$E\frac{d^3\sigma^a_{hA}(b)}{d^3p} = T_A(b)E\frac{d^3\sigma^a_{hN}}{d^3p} \ , \tag{1}$$

where $T_A(b)$ is the nuclear profile function $T_A(b) = \int dz \rho_A(r)$, and $\rho_A(r)$ is the nuclear density, $(\int d^2b \ T_A(b) = A)$. After integration over b we get

$$E\frac{d^3\sigma^a_{hA}}{d^3p} = A \ E\frac{d^3\sigma^a_{hN}}{d^3p} \ . \tag{2}$$

For inclusive cross sections in nucleus-nucleus A_1A_2–collisions the result of the Glauber approximation is very simple to formulate due to the AGK cancellation theorem. It is possible to prove, for an arbitrary number of interactions of nucleons of both nuclei, that all rescatterings cancel in the same way as for hA–interactions. Thus a natural generalization of eq. (1) for inclusive spectra of hadrons produced in the central rapidity region in nucleus–nucleus interactions takes place in the Glauber approximation

$$E\frac{d^3\sigma^a_{A_1A_2}(b)}{d^3p} = T_{A_1A_2}(b) \ E\frac{d^3\sigma^a_{NN}}{d^3p} \ , \tag{3}$$

where $T_{A_1A_2}(b) = \int d^2s T_{A_1}(\vec{s})T_{A_2}(\vec{b} - \vec{s})$.

After integration over b eq. (3) reads

$$E\frac{d^3\sigma^a_{A_1A_2}}{d^3p} = A_1A_2 \ E\frac{d^3\sigma^a_{NN}}{d^3p} \ . \tag{4}$$

Note that eqs.(3),(4) are valid for an arbitrary set of Glauber type diagrams.

The densities of charged particles can be obtained from eqs.(3), (4) by dividing them by the total inelastic cross section of nucleus–nucleus interaction. For example

$$\frac{dn^{ch}_{A_1A_2}(b)}{dy} = \frac{T_{A_1A_2}(b)}{\sigma^{in}_{A_1A_2}}\frac{d\sigma^{ch}_{NN}}{dy} = n_{A_1A_2}(b)\frac{d\sigma^{ch}_{NN}}{dy} \tag{5}$$

and

$$\frac{dn^{ch}_{A_1A_2}}{dy} = \frac{A_1A_2}{\sigma^{in}_{A_1A_2}}\frac{d\sigma^{ch}_{NN}}{dy} . \tag{6}$$

In the following we shall use these results to calculate particle densities in the central rapidity region at high energies.

3. Pomeron and small-x physics

Scattering amplitudes of high energy hadronic interactions, as well as of scattering of a virtual photon on a proton at fixed virtuality Q^2 and very high c.m. energy W are described in reggeon theory by exchange of the Pomeron pole and multi-Pomeron cuts. The structure function $F_2(x, Q^2)$ in pole approximation for $x \approx \frac{Q^2}{W^2} \to 0$ behaves as $\sim \frac{1}{x^\Delta}$, where $\Delta \equiv \alpha_{I\!P}(0)-1$. At large Q^2 $F_2(x,Q^2) \sim \sum_i e_i^2 x(q_i(x,Q^2)+\overline{q_i}(x,Q^2))$. Here $q_i(x, Q^2)$ is the distribution of quarks of flavor i. Thus for $\Delta > 0$ density of partons increases as $x \to 0$. The value of Δ was calculated in QCD perturbation theory. Summation of the leading $ln\frac{1}{x}$ leads to the BFKL [21] Pomeron with the value of $\Delta = \frac{12ln2}{\pi}\alpha_s \approx 0.5$, which corresponds to a very fast increase of parton densities. In this leading approximation the Pomeron corresponds to a cut in j-plane. In the next-to-leading approximation the leading singularity is the Regge pole [a] and the value of Δ substantially decreases to the values $\Delta = 0.15 \div 0.25$ [22]. This leading pole strongly depends on the nonperturbative region of small momentum transfer [23]. A connection of the Pomeron with spectrum of glueballs and the role of nonpertubative effects were studied in Ref. [24], using the method of vacuum correlators [25]. It was shown [24] that confinement effects and mixing of gluonic and $q\overline{q}$-Regge-trajectories are important for the value of the intercept of the Pomeron trajectory. The Pomeron pole with $\Delta > 0$ leads for $s \equiv W^2 \to \infty$ to a violation of s-channel unitarity. It is well known (see for example [12]) that the unitarity is restored if the multi Pomeron exchanges in the t-channel are taken into account. In the following I will use for this intercept the value $\Delta \approx 0.2$, found from the analysis of experimental data on high-energy hadronic interactions with an account of multi Pomeron contributions [26].

[a]There is a sequence of poles in j-plane concentrated at point j=1.

4. Models for small-x DIS

Experimental data at HERA [11] show that distributions of quarks and gluons have a fast rise as x decreases. For parametrization of $F_2(x, Q^2)$ at small $x \sim \frac{1}{x^\Delta}$ data demonstrate that Δ increases with Q^2 and is close to 0.2 at $Q^2 \sim 10$ GeV^2. This was interpreted in Ref. [27] as a consequence of decrease of multi Pomeron contributions as Q^2 increases. A similar interpretation in terms of "saturation" of parton densities has been proposed by Golec-Biernat and Wusthoff (GBW) [28]. In the pomeron theory the multi Pomeron contributions are related to diffractive production proceses and for self consistency it is important to describe simultaneously total cross sections (structure functions) and diffractive processes. At high energies virtual photons dissociate first to $q\overline{q}$-pairs, which interact with a target. For small relative transverse distance $r_\perp$ between q and $\overline{q}$ such a dipole has a small ($\sim r_\perp^2$) total interaction cross section and small diffraction dissociation cross section. On the other hand for pairs with large size these cross sections are not small. As a result the role of multi Pomeron rescatterings decreases as Q^2 increases. Explicit models, which take these effects into account, have been constructed in the framework of the reggeon theory [16]. Contribution of small size configurations to $\sigma_s(tot)$ has been described in QCD perturbation theory in the form similar to the one used in Ref. [28]:

$$\sigma_s^{(tot)T(L)}(s, Q^2) = 4 \int d^2b \int_0^{r_0} d^2r \int_0^1 dz |\Psi^{T(L)}(r, z, Q^2)|^2 \cdot \sigma_s(b, r, s, Q^2) , \tag{7}$$

where $\Psi^{T,L}(r, z, Q^2)$ are wave functions of $q\overline{q}$-pair for transverse (longitudinal) photons. Integration on transverse size r was limited by the value $r_0 \approx 0.2$ fm.

Contributions of large distance dipoles were described in the same framework as hadronic interactions. An important difference from the GBW-model is that the model [16] corresponds to summation of a certain class of reggeon diagrams and formulas for cross sections $\sigma(b, r, s, Q^2)$ are given in the impact parameter representation. Also the fan-type diagrams with triple Pomeron interaction are included [16]. This leads to a qualitative difference with GBW-model in the pattern of "saturation" for very small $x (x \leq 10^{-5})$ [b]. In particular, in the "saturation" region cross sections of $q\overline{q}$-pairs with a target does not tend to a constant (as in GBW-model),

[b] Models of Ref. [16] as well as GBW-model [28] give a good description of HERA data on structure functions F_2 and diffractive production.

but have the Froissart type increase $\sim ln^2\frac{1}{x}$. This is due to a logarithmic increase of the interaction radius and is out of scope of QCD perturbation theory. It is worth to emphasize that the model of Ref. [16] predicts that for the function F_2 (and thus for distributions of quarks and antiquarks) an approach to the unitarity limit or "saturation" regime is very slow and for large Q^2 ($Q^2 \sim 10^2$ GeV^2) it is not achieved even for $x \sim 10^{-10}$. On the other hand for distribution of gluons the unitarity effects should be more pronounced and should lead to a modification of x–dependence already at $x \sim 10^{-4}$, see Ref. [29].

5. Structure functions of nuclei in the shadowing region

Study of nuclear structure functions in the small-x region provides an important information on unitarity effects for dense partonic systems. The total cross section of a virtual photon-nucleus interaction can be written as a series in number of rescatterings on nucleons of a nucleus

$$\sigma^{(tot)}_{\gamma^* A} = \sigma^{(1)}_{\gamma * A} + \sigma^{(2)}_{\gamma * A} + ... \tag{8}$$

The first term corresponds to a sum of incoherent interactions and is equal to[c]

$$\sigma^{(1)}_{\gamma * A} = A\sigma_{\gamma^* N} \quad . \tag{9}$$

The second term describes the shadowing and according to the Gribov theory [2] can be expressed in terms of cross section of diffraction dissociation of a photon on a nucleon target

$$\sigma^{(2)} = -4\pi \int d^2b\, T_A^2(b) \int dM^2 \frac{d\sigma^{DD}_{\gamma^* N}(t=0)}{dM^2 dt} f_A(t_{min}) \,. \tag{10}$$

The minimal value of momentum transfer t_{min} to nucleons depends on the longitudinal fraction of momentum transfer $x_{I\!P} = \frac{Q^2+M^2}{s} = \frac{x}{\beta}$, where $\beta = \frac{Q^2}{Q^2+M^2}$:

$$t_{min} = -x_{I\!P}^2 m_N^2 \,. \tag{11}$$

[c] In the small x region it is possible to neglect by a difference between interaction of γ^* with p and n.

I would like to emphasize that the shadowing effects are different from zero only in the region of very small $x_{I\!P}$ or x, where $t_{min} << \frac{1}{R_A^2}$ (R_A is the nuclear radius). The same condition of coherence corresponds to lifetimes of the initial hadronic fluctuation $\tau_h \sim \frac{1}{m_N x}$ much larger than radius of a nucleus. These well known conditions for the existence of shadowing are important also for heavy ion collisions (see below). Higher order rescatterings are model dependent. In Refs. [30,31] they were calculated in the Schwimmer model [32], where

$$F_{2A}/AF_{2N} = \int d^2b \frac{T_A(b)}{1+F(x,Q^2)T_A(b)} , \tag{12}$$

with

$$F(x,Q^2) = 4\pi \int dM^2 \left(\frac{d\sigma^{DD}_{\gamma^* N}(t=0)}{dM^2 dt} \right) \cdot \left(\frac{f_A(t_{min})}{\sigma_{\gamma^* N}(x,Q^2)} \right) . \tag{13}$$

Using an information on diffraction dissociation in $\gamma^* N$-collisions it is possible to calculate nuclear structure functions in a good agreement with experimental data [30,31] (see also [33]). These calculations show that the shadowing effects increase as x decreases in the region $x < 0.1$. The shadowing correction $\sigma^{(2)}/\sigma^{(1)}$ increases with atomic number $\sim A^{1/3}$, but in the region of $x \sim 10^{-2}$ relevant for RHIC energies (see below) and for $Q^2 \sim 1$ GeV^2 a decrease of F_{2A} due to shadowing is less than 25% even for heavy nuclei (Au).

Thus the shadowing effects for distributions of quarks (which are directly measured in DIS) can be well described theoretically and prediction for smaller x, relevant for LHC energies, look reliable. Information on shadowing effects for gluons is more limited. This information can be obtained from the data on "gluonic content" of the Pomeron [34], using the same method as for quarks. Recent calculations [35] show that qualitatively shadowing for gluons has similar behavior as the one for quarks, but the function $F(x,Q^2)$ in eq.(12) is about 50% larger for gluons.

6. Heavy ion collisions

In heavy ion collisions the problem of shadowing of soft partons of colliding nuclei is very important. This phenomenon takes place only in the kinematic region where longitudinal momentum fractions of soft partons satisfy to the condition $x_i << 1/m_N R_A (i = 1,2)$ discussed in the previous Section. Values of x_i can be determined from the kinematic conditions

$$x_1 \cdot x_2 = \frac{M_\perp^2}{s} , \qquad x_1 - x_2 = x_F , \tag{14}$$

where $M_\perp$ is the transverse mass of the system produced in the collision and x_F is its Feynman x.

For example at RHIC energy $\sqrt{s} = 200$ GeV in the central rapidity region ($x_F \approx 0$) the shadowing for $AuAu$ collisions is possible only for transverse momenta less than several GeV. In the forward region ($x_2 << x_1 \approx x_F$) the shadowing is absent for partons of the nucleus 1 and extends to $p \perp \sim 10$ GeV for partons of nucleus 2.

It is possible to calculate the shadowing effects for inclusive densities of produced particles using the Gribov approach [2] to nuclear interactions. For the models of the Glauber-type, which do not take into account interactions between Pomerons, the AGK theorem [7] takes place. As a result simple equations (3)-(5) are valid.

It was emphasized in the Section 2, that eqs. (3)-(5) are valid in the Glauber model for $x_F \approx 0$ in the limit $s \to \infty$. For finite energies there are corrections due to energy-momentum conservation effects [36].

In reggeon theory the shadowing effects discussed above are related to diagrams with interactions between Pomerons. They lead to a substantial decrease of particle densities at high energies. In string models interactions between Pomerons can be considered as interactions of strings, which can lead to a percolation in the string system [37].

The same model, which has been used for description of shadowing effects for nuclear structure functions, was used for prediction of inclusive particle densities in heavy ion collisions [36]. In this approach the expression (5) for densities of particles in A_1A_2 collisions is modified as follows [36]

$$\frac{dn_{A_1A_2}}{dy} = n_{A_1A_2}(b)\frac{dn_{NN}}{dy}\gamma_{A_1}\gamma_{A_2} , \tag{15}$$

where

$$\gamma_A = \int d^2b \frac{T_A(b)}{1 + F(x, Q^2)T_A(b)} . \tag{16}$$

For LHC eqs.(15),(16) predict a decrease of particle densities by a factor ~ 4 compared to the Glauber model [36], while for RHIC the suppression is ≈ 2. Detailed predictions for shadowing suppressions are given in Ref. [31]. This approach agrees [12] with the results of experiments at RHIC [38,39]. Dependence of number of produced particles on number of participant nucleons is also in agreement with data [40].

The kinematic borders for shadowing discussed above correspond also to an essential change in the space–time picture of nuclear interactions and change [41] in AGK cutting rules. For example for $x_{A_2} << \frac{1}{m_N R_{A_2}}$ the lifetime for the initial state configuration of a nucleus A_1 is larger than the size of a nucleus A_2. In this case conditions for coherence are satisfied and usual AGK cutting rules [7] are valid. For $x_{A_2} \geq \frac{1}{m_N R_A}$ space–time picture is close to final state reinteractions of produced particles with formation time [41]. The last situation corresponds to a picture used in most of the existing models of heavy ion collisions. Note that it is valid only in the non-shadowing region. This situation is very general. For J/ψ production in hadron-nucleus interactions it was discussed in details in Ref. [42]. It has important implications for production of particles and jets with large $p_\perp$ at RHIC. For $x_F = 0$ and $p_\perp \geq 1 \div 2$ GeV the shadowing effects at $\sqrt{s} = 200$ GeV are small as was emphasized above and is demonstrated by the data on dAu collisions [43]. However in this region of $x_{A_i} \geq \frac{1}{mR_A}$ the final state interaction effects are important. In particular an account of Cronin effect and interaction of a particle (jet) with comovers allowed to describe [44] a suppression of particle production at large $p_\perp$ observed at RHIC [39] for $AuAu$ collisions. The data in dAu collisions [43] show that in the fragmentaion region of d there is a suppression for production of particles with large $p_\perp$. This can be explained in present approach: due to small values of $x_{Au} \sim 10^{-3}$ in this region the shadowing of partons becomes important. It was shown recently [35], that shadowing of gluons, calculated in Gribov's approach, allows to describe experimental data on dAu collisions in the forward rapidity region.

7. Conclusions

The Gribov theory of high-energy interactions is able to describe a large class of physical processes including small-x DIS and nuclear interactions. In this theory the Glauber approximation to nuclear dynamics is valid in the region of not too high energies and is modified already starting from RHIC energies.

The problem of "saturation" of parton densities is related in the Gribov theory to interactions between Pomerons. These effects are especially important for heavy ion collisions. The shadowing effects are clearly seen in the RHIC data, but the "saturation" is not achieved at RHIC energies in the central rapidity region.

References

1. M. Arneodo, *Phys. Rept.* **240**, 301 (1993).
2. V.N. Gribov, *JETP* **56**, 892 (1969);
   V.N. Gribov, *JETP* **57**, 1306 (1969).
3. V.N. Gribov, *JETP* **41**, 667 (1961) (*Sov. Phys. JETP* **14**, 47 (1962)).
4. V.N. Gribov, *JETP* **41**, 1962 (1961) (*Sov. Phys. JETP* **14**, 1395 (1962));
   V.N. Gribov, I.Ya. Pomeranchuk, *Phys. Rev. Lett.* **2**, 232 (1962); *JETP* **43**, 1556 (1962).
5. V.N. Gribov, I.Ya. Pomeranchuk and K.A. Ter-Martirosyan, *Phys. Rev.* **B139**, 184 (1965).
6. V.N. Gribov, *JETP* **57**, 654 (1967).
7. V.A. Abramovsky, V.N. Gribov and O.V. Kancheli, *Sov. J. Nucl. Phys.* **18**, 308 (1974).
8. A. Kaidalov, in *"QCD at 200 TeV"*, edited by L. Cifarelli and Yu. Dokshitzer, Plenum Press (1992) p.1 ; *Yad. Fiz.* **66**, 2044 (2003).
9. A. Capella, U. Sukhatme, C.I. Tan and J. Tran Thanh Van, *Phys. Rep.* **236**, 225 (1994).
10. R.J. Glauber, *Lectures in Theoretical Physics*, Ed. W.E. Britten, New York.: Int. Publ. 1959, v.1, p.315.
11. T. Ahmed *et al.* (H1 Collaboration), *Phys. Lett.* **299B**, 374 (1992);
    C. Adloff *et al.* (H1 Collaboration), *Nucl. Phys.* **B497**, 3 (1997);
    M. Derrick *et al.* (ZEUS Collaboration), *Phys. Lett.* **293B**, 465 (1992);
    J. Breitweg *et al.* (ZEUS Collaboration), *Phys. Lett.* **407B**, 432 (1997).
12. A.B. Kaidalov, *Surveys in High Energy Phys.* **9**, 143 (1996); *ibid.* **16**, 267 (2001).
13. L.V. Gribov, E.M. Levin and M.G. Ryskin, *Phys. Rep.* **100**, 1 (1983);
    E. Laenen and E. Levin, *Ann. Rev. Nucl. Part.* **44**, 199 (1994);
    A.H. Mueller, hep-ph/9911289.
14. L. McLerran and R. Venugopalan, *Phys. Rev.* **D49**, 2233 (1994); **D50**, 2225 (1994); **D53**, 458 (1996);
    J. Jalilian-Marian *et al.*, *Phys. Rev.* **D59**, 014014, 034007 (1999);
    A. Kovner, L. McLerran and H. Weigert, *Phys. Rev.* **D52**, 3809, 6231 (1995);
    Yu.V. Kovchegov and A.H. Mueller, *Nucl. Phys.* **B529**, 451 (1998);
    Yu.V. Kovchegov, A.H. Mueller and S. Wallon, *Nucl. Phys.* **B507**, 367 (1997);
    A.H. Mueller, *Nucl. Phys.* **B558**, 285 (1999).
15. L. McLerran, *Nucl. Phys.* **A702**, 49 (2002).
16. A. Capella, E. Ferreiro, A.B. Kaidalov and C.A. Salgado, *Nucl. Phys.* **B593**, 336 (2001); *Phys. Rev.* **D63**, 054010 (2001).
17. V.N. Gribov, *Lectures at VII. LIYAF Winter School of Physics*, 1973, v.II, p.5.
18. A.B. Kaidalov and L.A. Kondratyuk, *JETP Letters* **15**, 170 (1972); Nucl. Phys. **B56**, 90 (1973).
19. V.V. Anisovich, L.G. Dakhno and P.E. Volkovitsky, *Yad.Fiz.* **15**, 168 (1972).
20. A.B. Kaidalov, *Proc Quark Matter'90*, *Nucl. Phys.* **A552**, 39c (1991).

21. E.A. Kuraev, L.N. Lipatov and V.S. Fadin, *Sov. Phys. JETP* **44**, 433 (1976); **45**, 199 (1977); Ya.Ya. Balitsky and L.N. Lipatov, *Sov. J. Nucl. Phys.* **28**, 822 (1979).
22. V.S. Fadin and L.N. Lipatov, *Phys. Lett.* **429B**, 127 (1998);
G. Camici and M. Ciafaloni, *Phys. Lett.* **430B**, 349 (1998);
S.J. Brodsky *et al.*, *JETP Lett.* **70**, 155 (1999).
23. L.P.A. Haakman, O.V. Kancheli and J.H. Koch, *Nucl. Phys.*, **B518**, 275 (1998).
24. A.B. Kaidalov and Yu.A. Simonov, *Phys. Lett.* **477B**, 163 (2000);
A.B. Kaidalov and Yu.A. Simonov, *Yad. Fiz.* **63**, 1507 (2000).
25. Yu.A. Simonov, *Nucl. Phys.* **B307**, 512 (1988).
26. A.B. Kaidalov, L.A. Ponomarev and K.A. Ter-Martirosyan, *Sov. J. Nucl. Phys.* **44**, 468 (1986).
27. A. Capella, A. Kaidalov, C. Merino and J. Tran Thanh Van, *Phys. Lett.* **337B**, 358 (1994).
28. K. Golec-Biernat and M. Wüsthoff, *Phys. Rev.* **D59**, 014017 (1999); **D60**, 114023 (1999).
29. A.B. Kaidalov, V.A. Khoze, A.D. Martin and M.G. Ryskin, *Phys. Lett.* **B567**, 61 (2003).
30. A. Capella, A. Kaidalov, C. Merino, D. Pertermann and J. Tran Than Van, *Eur. Phys. J.* **C5**, 111 (1998).
31. J.L. Albacete, N. Armesto, A. Capella, A.B. Kaidalov and C. Salgado, *Eur. Phys. J.* **C29**, 531 (2003).
32. A. Schwimmer, *Nucl. Phys.*, **B94**, 445 (1975).
33. L. Frankfurt, V. Guzey, M.M. McDermott and M. Strikman, *JHEP* **0202**, 027 (2002); L. Frankfurt, V. Guzey and M. Strikman, hep-ph/0303022.
34. C. Adloff *et al.* (H1 Collaboration), *Abstract 980, Presented at ICHEP02*, Amsterdam, July 2002.
35. I.C. Arsene, A.B. Kaidalov and K. Tywoniuk, *paper submitted to the Conference QM2005.*
36. A. Capella, A. Kaidalov and J. Tran Thanh Van, *Heavy Ion Phys.* **9**, 169 (1999).
37. M.A. Braun, F. del Morel and C. Pajares, *Nucl. Phys.*, **A715**, 791 (2003).
38. B.B. Back *et al.* (PHOBOS Collaboration), *Phys. Rev. Lett.* **87**, 102303 (2001).
39. S.S. Adler *et al.* (PHENIX Collaboration), *Phys. Rev. Lett.* **91**, 072301 (2003).
40. A. Capella and D. Sousa, *Phys. Lett.* **511B**, 185 (2001). .
41. K.G. Boreskov *et al.*, *Yad. Fiz.* **53**, 569 (1991).
42. K.G. Boreskov, A. Capella, A. Kaidalov and J. Tran Tranh Van, *Phys. Rev.* **D47**, 919 (1993).
43. I. Arsene *et al.* (BRAHMS Collaboration), *Phys. Rev. Lett.* **93**, 242303 (2004).
44. A. Capella, E.G. Ferreiro, A.B. Kaidalov and D. Sousa, *Eur. Phys. J.* **C40**, 129 (2005).

GRIBOV REGGEON CALCULUS, DGLAP EQUATION AND INTEGRABILITY*

L.N. LIPATOV

Petersburg Nuclear Physics Institute,
Gatchina, St.Petersburg, 188300, Russia
E-mail: lipatov@thd.pnpi.spb.ru

We review the Regge description of high energy hadron interactions. It is shown, that gluon in QCD belongs to a Regge family. Pomeron and Odderon are composite states of reggeized gluons. The BFKL equation for the Pomeron wave function is discussed. In the multi-colour QCD the BKP equations for composite states of several gluons turn out to be integrable. To construct the Gribov reggeon calculus we formulate an effective lagrangian for the interaction of QCD partons with reggeized gluons. The effective vertices satisfy recurrent relations and are constructed explicitly in the momentum representation. We discuss the BFKL and DGLAP equations in N=4 SUSY. The intercept of the Pomeron at large couplings is calculated in the framework of the AdS/QFT correspondence.

1. Introduction

Let us consider the hadron-hadron scattering

$$A + B \to A' + B' \tag{1}$$

in the Regge kinematics

$$s \gg -t \simeq m^2\,,\; t \leq 0\,, \tag{2}$$

where the Mandelstam variables are defined below

$$s = (p_A + p_B)^2\,,\; t = (p_A - p_{A'})^2\,,\; u = (p_A - p_{B'})^2\,. \tag{3}$$

It is assumed usually, that the scattering amplitude in this kinematics has the Regge form

$$M(s,t) = \sum_k \xi_k(t)\, g_k(t)\, \left(\frac{s}{m^2}\right)^{j_k(t)} g_k(t) \tag{4}$$

*This work is supported by RFBR grant 04-02-17094, RSGSS-1124.2003.2 .

with the Regge trajectory

$$j_k(t) = j_k^{(0)} + j'_k(0)\, t + ... \tag{5}$$

and the signature factor

$$\xi_k(t) = \frac{1 - P_k\, e^{-i\pi j(t)}}{\sin\left(\pi j(t)\right)}\,, \;\; P_k = \pm 1\,. \tag{6}$$

The spins and masses of hadrons belonging to the trajectory are related as follows

$$S = j_k(M^2)\,. \tag{7}$$

The total cross-section can be calculated in terms of the scattering amplitude with the use of the optical theorem

$$\sigma_t = \frac{Im\, M(s,0)}{s}\,. \tag{8}$$

The most important contribution to the total cross-section is caused by the Pomeron exchange in the t-channel

$$\sigma_t \sim (s/m^2)^{\Delta}\,, \;\; P_p = 1\,. \tag{9}$$

Its Regge trajectory is parametrized by the intercept Δ and slope j'_P

$$j_P(t) = 1 + \Delta + j'_P\, t\, + ...\,, \;\; \Delta \ll 1\,. \tag{10}$$

The Pomeron provides the fulfillment of the Pomeranchuck theorem

$$\frac{\sigma_t^{pp}}{\sigma_t^{p\overline{p}}} \to 1\,. \tag{11}$$

Around 40 years ago V.N. Gribov derived for the pomeron interactions the Reggeon Calculus being a quantum field theory, in which the complex Reggeon field $\phi(\overrightarrow{\rho}, y)$ depends on the impact parameter coordinates $\overrightarrow{\rho}$ and the rapidity

$$y = \ln \frac{E + |\overrightarrow{p}|}{E - |\overrightarrow{p}|}\,.$$

Below we shall consider the similar approach in QCD where the Pomeron is a glueball state.

2. BFKL Pomeron

In the leading logarithmic approximation (LLA) the scattering amplitude in QCD has the form

$$M_{AB}^{A'B'}(s,t) = s \sum_{n=0}^{\infty} (\alpha_s \ln s)^n \, a_n(t) \,, \; \alpha_s = \frac{g^2}{4\pi} \to 0 \,. \tag{12}$$

In the Born approximation it is factorized

$$M_{AB}^{A'B'}(s,t)|_{Born} = g \, T^c_{A'A} \, \delta_{\lambda_{A'}\lambda_A} \, \frac{2s}{t} \, g \, T^c_{B'B} \, \delta_{\lambda_{B'}\lambda_B} \,. \tag{13}$$

After summing radiative corrections we obtain for the amplitude the Regge-type expression [1]

$$M_{AB}^{A'B'}(s,t) = M_{AB}^{A'B'}(s,t)|_{Born} \, s^{\omega(t)} \,, \tag{14}$$

where the gluon Regge trajectory is

$$\omega(-|q|^2) = -\frac{\alpha_c}{4\pi^2} \, N_c \int d^2k \, \frac{|q|^2}{|k|^2|q-k|^2} \approx -\frac{\alpha_c}{2\pi} \, \ln \frac{|q^2|}{\lambda^2} \,. \tag{15}$$

The particles at high energies for the process $AB \to A'B'd_1 ... d_{n-1}$ in LLA are produced in the multi-Regge kinematics

$$s \gg s_r = (k_{r-1} + k_r)^2 \gg |q_r|^2 \,, \qquad\qquad k_r = q_r - q_{r+1} \,. \tag{16}$$

The gluon production amplitude in this region has the multi-Regge form [1]

$$M_{2\to 1+n} \sim \frac{s_1^{\omega_1}}{|q_1|^2} \, g T^{d_1}_{c_2 c_1} C(q_2, q_1) \, \frac{s_2^{\omega_2}}{|q_2|^2} ... C(q_n, q_{n-1}) \, \frac{s_n^{\omega_n}}{|q_n|^2} \,, \tag{17}$$

where the Reggeon-Reggeon-gluon vertex for the produced gluon with a definite helicity is

$$C(q_2, q_1) = \frac{q_2 \, q_1^*}{q_2 - q_1} \,. \tag{18}$$

It is convenient to introduce the complex variables for the gluon transverse coordinates and momenta

$$\rho_k = x_k + i y_k \,, \; \rho_k^* = x_k - i y_k \,, \; p_k = i \frac{\partial}{\partial \rho_k} \,, \; p_k^* = i \frac{\partial}{\partial \rho_k^*} \,. \tag{19}$$

Then in the coordinate representation the Balitsky-Fadin-Kuraev-Lipatov (BFKL) equation is [1]

$$E \, \Psi(\vec{\rho}_1, \vec{\rho}_2) = H_{12} \, \Psi(\vec{\rho}_1, \vec{\rho}_2) \,, \qquad\qquad \Delta = -\frac{\alpha_s N_c}{2\pi} \, E \,. \tag{20}$$

One can write the BFKL Hamiltonian in the form [2]

$$H_{12} = \ln\,|p_1 p_2|^2 + \frac{1}{p_1 p_2^*}\ln\,|\rho_{12}|^2 p_1 p_2^* + \frac{1}{p_1^* p_2}\ln\,|\rho_{12}|^2 p_1^* p_2 - 4\psi(1)\,, \quad (21)$$

where $\rho_{12} = \rho_1 - \rho_2$. The kinetic energy is proportional to the gluon Regge trajectories and the potential energy $\sim \ln\,|\rho_{12}|^2$ is related to the Fourie transformation of the product of two vertices $C(q_2, q_1)$.

The BFKL Hamiltonian is invariant under the Moebius transformation [3]

$$\rho_k \to \frac{a\rho_k + b}{c\rho_k + d_k}\,. \quad (22)$$

For the classification of its eigenstates one can introduce the Casimir operators of this group

$$M^2 = \left(\sum_{r=1}^{2} \vec{M}^{(r)}\right)^2 = \rho_{12}^2\, p_1\, p_2\,,\; M^{*2} = (M^2)^*\,. \quad (23)$$

Their eigenvalue equations

$$M^2 f_{m,\widetilde{m}} = m(m-1) f_{m,\widetilde{m}}\,, \qquad M^{*2} f_{m,\widetilde{m}} = \widetilde{m}(\widetilde{m}-1) f_{m,\widetilde{m}} \quad (24)$$

define the conformal weights having the form

$$m = 1/2 + i\nu + n/2\,,\; \widetilde{m} = 1/2 + i\nu - n/2 \quad (25)$$

for the principal series of unitary representations.

The Hamiltonian has the property of the holomorphic separability [4]

$$H_{12} = h_{12} + h_{12}^*\,, \quad (26)$$

where the holomorphic hamiltonian h_{12} equals

$$h_{12} = \ln(p_1 p_2) + \frac{1}{p_1}\ln\rho_{12}\, p_1 + \frac{1}{p_2}\ln\rho_{12}\, p_2 - 2\psi(1)\,. \quad (27)$$

3. Integrability of the BFKL dynamics

One can write the Bartels-Kwiecinskii-Praszalowicz equation [5] for the n-gluon state as follows

$$E\Psi(\vec{\rho}_1, ..., \vec{\rho}_n) = \sum_{k<l} \frac{T_k^a T_l^a}{(-N_c)} H_{k,l}\, \Psi(\vec{\rho}_1, ..., \vec{\rho}_n)\,, \quad (28)$$

where

$$[T_k^a, T_l^b] = i f_{abc} T_k^c\, \delta_{kl}\,,\; T_{bc}^a = -i f_{abc}\,. \quad (29)$$

This equation is especially simple in the multi-colour QCD. In particular, for the eigenfunction of H we have the holomorphic factorization [4]

$$\Psi(\vec{\rho}_1,\vec{\rho}_2,...,\vec{\rho}_n)=\sum_{r,s}a_{r,s}\,\Psi_r(\rho_1,...,\rho_n)\,\Psi_s(\rho_1^*,...,\rho_n^*)\,, \tag{30}$$

$$H=\frac{h+h^*}{2}\,,\qquad h=\sum_{r=1}^{n}h_{r,r+1}\,,\qquad [h,h^*]=0\,, \tag{31}$$

$$h_{12}=\sum_{r=1}^{2}\left[\ln\,p_r+\frac{1}{p_r}\,\ln\left(\rho_{12}\right)\,p_r-\psi(1)\right] \tag{32}$$

and the duality symmetry [6]

$$\rho_{r,r+1}\to p_r\to\rho_{r-1,r}\,. \tag{33}$$

The holomorphic hamiltonian h has the integrals of motion [7]

$$q_r=\sum_{k_1<k_2<...<k_r}\rho_{k_1k_2}\rho_{k_2k_3}...\rho_{k_{r-1}k_r}\,p_{k_1}p_{k_2}...p_{k_r}\,,\qquad [q_r,h]=0\,, \tag{34}$$

in an accordance with the fact, that the obtained reggeon theory at large N_c coincides with the Heisenberg spin model

$$h=\sum_{r=1}^{n}h(\vec{M}_r\vec{M}_{r+1})\,, \tag{35}$$

where $\vec{M}_r$ are the generators of the Moebius group

$$M_k^z=\rho_k\partial_k\,,\ M_k^+=\partial_k\,,\ M_k^-=-\rho_k^2\partial_k\,. \tag{36}$$

In particular for the Pomeron ($n=2$) we obtain the following wave function [3]

$$f_{m,\widetilde{m}}(\overrightarrow{\rho_1},\overrightarrow{\rho_2};\overrightarrow{\rho_0})=\left(\frac{\rho_{12}}{\rho_{10}\,\rho_{20}}\right)^m\left(\frac{\rho_{12}^*}{\rho_{10}^*\,\rho_{20}^*}\right)^{\widetilde{m}} \tag{37}$$

with the corresponding energy having the holomorphic separability property

$$E_{m,\widetilde{m}}=\varepsilon_m+\varepsilon_{\widetilde{m}}\,,\qquad \varepsilon_m=\psi(m)+\psi(1-m)-2\psi(1)\,. \tag{38}$$

Thus, the intercept of the BFKL pomeron is [1]

$$\Delta=4\,\frac{\alpha_s}{\pi}N_c\,\ln 2 \tag{39}$$

and, as a result, one obtains the violation of the Froissart bound

$$\sigma\sim s^{\Delta}>c\ln^2 s\,. \tag{40}$$

Therefore there is a problem of restoration of the s-channel unitarity for scattering amplitudes. The consistent way to solve this problem is to use an effective field theory for reggeized gluons similar to the Gribov reggeon calculus for Pomerons. In the next section for the interaction of usual partons and reggeized gluons we construct the effective action local in a certain interval of their rapidity.

4. Effective field theory for high energies

In the next-to-leading approximation one should consider more complicated processes - the production of particle clusters in the multi-Regge kinematics

$$P_A + P_B = Q_1 + Q_2 + ... + Q_n \ , \ s \gg s_i = 2Q_iQ_{i+1} \gg |q_i^2| \ , \tag{41}$$

$$Q_i^2 = M_i^2 \ , \quad Q_k = \sum_j p_j^{(k)} \ , \quad k = 1, 2, ..., n \, . \tag{42}$$

We shall write below the effective action for interactions of particles incide each cluster. To begin with, we fix their rapidities y

$$y = \frac{1}{2} \ln \frac{\epsilon_k + |k|}{\epsilon_k - |k|} \tag{43}$$

in the interval $|y - y_0| < \eta$, $\eta << \ln s$. The corresponding gluon and quark fields are

$$v_\mu(x) = -iT^a v_\mu^a(x) \, , \quad \psi(x) \, , \quad \bar{\psi}(x) \, , \quad [T^a, T^b] = i f_{abc} T^c \, . \tag{44}$$

Let us introduce also the fields describing the production and annihilation of reggeized gluons

$$A_\pm(x) = -iT^a A_\pm^a(x) \, . \tag{45}$$

Under the global colour group rotations the fields are transformed in the standard way

$$\delta v_\mu(x) = [v_\mu(x), \chi], \ \delta\psi(x) = -\chi \, \psi(x), \ \delta A(x) = [A(x), \chi] \, , \tag{46}$$

but under the local gauge transformations we have

$$\delta v_\mu(x) = \frac{1}{g}[D_\mu, \chi(x)], \ \delta\psi(x) = -\chi(x) \, \psi(x), \ \delta A_\pm(x) = 0 \, . \tag{47}$$

In the quasi-multi-Regge kinematics one obtains the kinematical constraints on the reggeon fields

$$\partial_\mp A_\pm(x) = 0 \, . \tag{48}$$

We can construct an effective action for the multi-Regge processes [8], but it is difficult to write this action in a gauge-invariant form. For this purpose one should consider the more general quasi-multi-Regge kinematics and introduce new effective vertices with an arbitrary number of gluons. The corresponding effective action local in the rapidity y has the form [9]

$$S = \int d^4x \, (L_0 + L_{ind}) \,, \tag{49}$$

where the usual QCD Lagrangian is

$$L_0 = i\bar{\psi}\hat{D}\psi + \frac{1}{2}Tr\, G_{\mu\nu}^2 \,, \quad D_\mu = \partial_\mu + g v_\mu \,, \quad G_{\mu\nu} = \frac{1}{g}[D_\mu, D_\nu] \tag{50}$$

and the induced contribution is given below

$$L_{ind} = Tr\,(L_{ind}^k + L_{ind}^{GR}) \,, \quad L_{ind}^k = -\partial_\mu A_+^a \partial_\mu A_-^a \,. \tag{51}$$

Here the reggeon-gluon interaction can be presented in two different forms

$$\begin{aligned} L_{ind}^{GR} = & -\frac{1}{g}\partial_+ \, P\exp\left(-\frac{1}{2}\int_{-\infty}^{x^+} v_+(x')d(x')^+\right) \partial_\sigma^2 A_- \\ & -\frac{1}{g}\partial_- \, P\exp\left(-\frac{1}{2}\int_{-\infty}^{x^-} v_-(x')d(x')^+\right) \partial_\sigma^2 A_+ \\ = & \; Tr\left(v_+ - g v_+ \frac{1}{\partial_+} v_+ + g^2 v_+ \frac{1}{\partial_+} v_+ \frac{1}{\partial_+} v_+ - \ldots\right) \partial_\sigma^2 A_- \\ & + \left(v_- - g v_- \frac{1}{\partial_-} v_- + g^2 v_- \frac{1}{\partial_-} v_- \frac{1}{\partial_-} v_- - \ldots\right) \partial_+^2 A_+ \,. \end{aligned} \tag{52}$$

5. Effective vertices

We can formulate the Feynman rules in the momentum space [10]. For this purpose it is needed to take into account the momentum conservation for induced vertices

$$k_0^+ + k_1^+ + \ldots + k_m^+ = 0 \,. \tag{53}$$

The simple examples of the vertices are given below

$$\Delta_{a_0 a_1 c}^{\nu_0 \nu_1 +} = \vec{q}_\perp^{\,2}\, T_{a_1 a_0}^c \, (n^+)^{\nu_1} \frac{1}{k_1^+} (n^+)^{\nu_0} \,, \tag{54}$$

$$\Delta_{a_0 a_1 a_2 c}^{\nu_0 \nu_1 \nu_2 +} = \vec{q}_\perp^{\,2}\, (n^+)^{\nu_0} (n^+)^{\nu_1} (n^+)^{\nu_2} \left(\frac{T_{a_2 a_0}^a \, T_{a_1 a}^c}{k_1^+ k_2^+} + \frac{T_{a_2 a_1}^a \, T_{a_0 a}^c}{k_0^+ k_2^+}\right) \,. \tag{55}$$

We have their factorization in a general case

$$\Delta^{\nu_0\nu_1...\nu_r+}_{a_0a_1...a_rc} = -\vec{q}_\perp^{\,2} \prod_{s=0}^{r} (n^+)^{\nu_s}\, 2\,Tr\,(T^c G_{a_0a_1...a_r}) \tag{56}$$

with the following representation for $G_{a_0a_1...a_r}$ [10]

$$G_{a_0a_1...a_r} = \sum_{\{i_0,i_1,...,i_r\}} \frac{T^{a_{i_0}}T^{a_{i_1}}T^{a_{i_2}}...T^{a_{i_r}}}{k^+_{i_0}(k^+_{i_0}+k^+_{i_1})...(k^+_{i_0}+k^+_{i_1}+...+k^+_{i_{r-1}})}\,, \tag{57}$$

These vertices satisfy recurrent relations (Ward identities) [9]

$$k_r^+\,\Delta^{\nu_0\nu_1...\nu_r+}_{a_0a_1...a_r}(k_0^+,...,k_r^+) =$$
$$\sum_{i=0}^{r-1} if_{aa_ma_i}\Delta^{\nu_0...\nu_r+}_{...a_{i-1}aa_{i+1}...a_r}(k_0^+,...,k^+_{i-1},k^+_i+k^+_r,k^+_{i+1},...)\,. \tag{58}$$

The simplest Reggeon-particle vertices appearing in LLA are

$$\frac{\Gamma^{\nu\nu'+}(P_A,q)}{2} = -\frac{q^2(n^+)^\nu(n^+)^{\nu'}}{2P_A^+} + P_A^+ g^{\nu\nu'} - (n^+)^\nu P_A^{\nu'} - (n^+)^{\nu'} P^\nu_{A'}\,, \tag{59}$$

$$\frac{C^\mu(q_1,q_2)}{2} = (n^-)^\mu(q_1^+ + \frac{q_1^2}{q_2^-}) - (n^+)^\mu(q_2^- + \frac{q_2^2}{q_1^+}) + (q_2-q_1)^\mu\,. \tag{60}$$

The more complicated Reggeon-Reggeon-gluon-gluon vertex is given below

$$\frac{\Gamma^{-\nu_1\nu_2+}_{ca_1a_2d}(q_1;p_1,p_2;q_2)}{ig^2} = T_1((n^-)^{\nu_1}(n^+)^{\nu_2} - (n^-)^{\nu_2}(n^+)^{\nu_1})$$
$$-2(T_2-T_3)g^{\nu_1\nu_2} + T_2(n^-)^{\nu_1}(n^+)^{\nu_2} - T_3(n^-)^{\nu_2}(n^+)^{\nu_1}$$
$$+\frac{T_1}{p_{12}^2}C^\eta(q_1,q_2)\gamma^{\nu_1\nu_2\eta}(-p_1,-p_2,p_{12})$$
$$+\frac{T_3}{(p_2-q_2)^2}\Gamma^{\eta\nu_1-}(q_1,p_1-q_1)\Gamma^{\eta\nu_2+}(p_2-q_2,q_2)$$
$$-\frac{T_2}{(p_1-q_2)^2}\Gamma^{\eta\nu_2-}(q_1,p_2-q_1)\Gamma^{\eta\nu_1+}(p_1-q_2,q_2)$$
$$-\Delta^{\rho\nu_1\nu_2+}_{ca_1a_2d}(q_1,p_1,p_2,q_2)(n^-)_\rho + \Delta^{-\nu_1\nu_2\eta}_{ca_1a_2d}(q_1,p_1,p_2,q_2)(n^+)_\eta\,, \tag{61}$$

where

$$T_1 = f_{a_1a_2r}f_{cdr},\; T_2 = f_{a_2cr}f_{a_1dr},\; T_3 = f_{ca_1r}f_{a_2dr}\,. \tag{62}$$

One can calculate also the amplitude for the production of three gluons on the mass shell in the collision of two reggeized gluons [10]. This amplitude is needed for the construction of the BFKL equation in the next-to-next-to-leading approximation.

We can derive from this effective action also the Balitsky-Kovchegov equation

$$\frac{\partial N_{\vec{\rho}_1,\vec{\rho}_2}}{\partial Y} = \bar{\alpha}_s \int \frac{d^2\rho_0}{2\pi} \left| \frac{\rho_{12}}{\rho_{10}\rho_{20}} \right|^2 (N_{\vec{\rho}_1,\vec{\rho}_0} + N_{\vec{\rho}_2,\vec{\rho}_0} - N_{\vec{\rho}_1,\vec{\rho}_2} - N_{\vec{\rho}_1,\vec{\rho}_0} N_{\vec{\rho}_2,\vec{\rho}_0}) \,. \tag{63}$$

6. Anomalous dimensions of twist-2 operators

Usual parton distributions are expressed in terms of the corresponding unintegrated quantities

$$f_a(x, Q^2) \;=\; \int_{k_\perp^2 < Q^2} dk_\perp^2 \, \varphi_a(x, k_\perp^2) \,, \quad a = g, q, s \,. \tag{64}$$

Momenta of parton distributions

$$f_a(j, Q^2) = \int_0^1 dx \, x^{j-1} \, f_a(x, Q^2) \tag{65}$$

allow one to simlify the DGLAP equation using the anomalous dimensions matrix

$$\frac{d}{d \ln Q^2} f_a(j, Q^2) = \sum_b \gamma_{ab}(j) f_b(j, Q^2) \,. \tag{66}$$

These momenta are proportional to the light-cone components of the local twist-2 operators

$$O^a = \tilde{n}^{\mu_1} ... \tilde{n}^{\mu_j} \, O^a_{\mu_1,...,\mu_j} \,, \quad \tilde{O}^a = \tilde{n}^{\mu_1} ... \tilde{n}^{\mu_j} \tilde{O}^a_{\mu_1,...,\mu_j} \,. \tag{67}$$

Their anomalous dimensions are the same also for the mixed projections of these tensors

$$\tilde{n}^{\mu_1} ... \tilde{n}^{\mu_{1+\omega}} \, O^a_{\mu_1,...,\mu_{1+\omega},\sigma_1,...,\sigma_{|n|}} \, l_\perp^{\sigma_1} ... l_\perp^{\sigma_{|n|}} \,. \tag{68}$$

The solution of the BFKL equation due to its Möbius invariance depends on the anomalous dimension $\gamma = \frac{1}{2} + i\nu$ and the conformal spin $|n|$ which coincides with the number of indices of the tensor O^a multiplied by the transverse vector $l_\perp$.

7. BFKL and DGLAP equations in $N = 4$ SUSY

In the next-to-leading approximation the eigenvalue of the BFKL kernel is

$$\omega = \omega_1(n,\,\gamma) + 4\;\hat{a}^2\,\Delta(n,\gamma)\,,\;\;\hat{a} = g^2 N_c/(16\pi^2) \tag{69}$$

In QCD it is non-analytic as a function of the conformal spin $|n|$ [11]

$$\Delta_{QCD}(n,\gamma) = c_0\delta_{n,0} + c_2\delta_{n,2} + \textit{analytic terms}\,, \tag{70}$$

but in $N = 4$ SUSY non-analytic terms are cancelled [12].

Moreover, in this model we have the Hermitian separability

$$\Delta(n,\gamma) = \phi(M) + \phi(M^*) - \frac{\rho(M) + \rho(M^*)}{2\hat{a}/\omega}\,, \quad M = \gamma + \frac{|n|}{2}\,, \tag{71}$$

$$\rho(M) = \beta'(M) + \frac{1}{2}\zeta(2)\,,\;\beta'(z) = \frac{1}{4}\left[\Psi'\Big(\frac{z+1}{2}\Big) - \Psi'\Big(\frac{z}{2}\Big)\right] \tag{72}$$

Note, that in the last expression all special functions have the property of maximal complexity [12].

$$\phi(M) = 3\zeta(3) + \Psi''(M) - 2\Phi(M) + 2\beta'(M)\Big(\Psi(1) - \Psi(M)\Big)\,, \tag{73}$$

$$\Phi(M) = \sum_{k=0}^{\infty} \frac{\beta'(k+1)}{k+M} + \sum_{k=0}^{\infty} \frac{(-1)^k}{k+M} \left(\Psi'(k+1) \;-\; \frac{\Psi(k+1) - \Psi(1)}{k+M}\right) \tag{74}$$

For one loop anomalous dimension matrix in the case $N = 4$ the calculations were done in Ref. [13]. In this case one can introduce the following twist-2 Wilson operators

$$\mathcal{O}^g_{\mu_1,...,\mu_j} = \hat{S} G^a_{\rho\mu_1} D_{\mu_2} D_{\mu_3} ... D_{\mu_{j-1}} G^a_{\rho\mu_j}\,, \tag{75}$$

$$\tilde{\mathcal{O}}^g_{\mu_1,...,\mu_j} = \hat{S} G^a_{\rho\mu_1} D_{\mu_2} D_{\mu_3} ... D_{\mu_{j-1}} \tilde{G}^a_{\rho\mu_j}\,, \tag{76}$$

$$\mathcal{O}^q_{\mu_1,...,\mu_j} = \hat{S} \bar{\Psi}^a \gamma_{\mu_1} D_{\mu_2} ... D_{\mu_j} \Psi^a\,, \tag{77}$$

$$\tilde{\mathcal{O}}^q_{\mu_1,...,\mu_j} = \hat{S} \bar{\Psi}^a \gamma_5 \gamma_{\mu_1} D_{\mu_2} ... D_{\mu_j} \Psi^a\,, \tag{78}$$

$$\mathcal{O}^\varphi_{\mu_1,...,\mu_j} = \hat{S} \bar{\Phi}^a D_{\mu_1} D_{\mu_2} ... D_{\mu_j} \Phi^a\,. \tag{79}$$

Their anomalous dimension matrix is [13]

$$\gamma_{gg} = \frac{4}{j-1} - \frac{4}{j} + \frac{4}{j+1} - \frac{4}{j+2} - 4S_1(j)\,, \quad \gamma_{q\varphi} = \frac{8}{j}\,, \tag{80}$$

$$\gamma_{gq} = \frac{4}{j-1} - \frac{4}{j} + \frac{2}{j+1}\,, \quad \gamma_{g\varphi} = \frac{4}{j-1} - \frac{4}{j}\,, \tag{81}$$

$$\gamma_{qg} = \frac{8}{j} - \frac{16}{j+1} + \frac{16}{j+2}\,, \quad \gamma_{qq} = \frac{8}{j} - \frac{8}{j+1} - 4S_1(j), \tag{82}$$

$$\gamma_{\varphi g} = \frac{12}{j+1} - \frac{12}{j+2}\,, \quad \gamma_{\varphi q} = \frac{6}{j+1}\,, \quad \gamma_{\varphi\varphi} = -4S_1(j), \tag{83}$$

$$\widetilde{\gamma}_{gg} = \frac{8}{j} - \frac{8}{j+1} - 4S_1(j)\,, \quad \widetilde{\gamma}_{gq} = \frac{4}{j} - \frac{2}{j+1}\,, \tag{84}$$

$$\widetilde{\gamma}_{qg} = -\frac{8}{j} + \frac{16}{j+1}\,, \quad \widetilde{\gamma}_{qq} = -\frac{4}{j} + \frac{4}{j+1} - 4S_1(j) \tag{85}$$

and its eigenvalues after the diagonalization $U\gamma U^+$ are

$$\begin{vmatrix} -4S_1(j-2) & 0 & 0 \\ 0 & -4S_1(j) & 0 \\ 0 & 0 & -4S_1(j+2) \end{vmatrix}, \quad \begin{vmatrix} -4S_1(j-1) & 0 \\ & \\ 0 & -4S_1(j+1) \end{vmatrix} \tag{86}$$

proportional to the universal function (with schifted arguments)

$$U\gamma U^+ = \gamma_{uni}(j)\,, \quad \gamma_{uni}^{(0)}(j) = -4S_1(j-2)\,, \quad S_r(j) = \sum_{i=1}^{j} \frac{1}{i^r} \tag{87}$$

which leads to an integrability of the evolution equations for the matrix elements of quasi-partonic operators in $N = 4$ SUSY [13].

8. Two- and three-loop universal anomalous dimension in $N = 4$

In an accordance with the fact, that the eigenvalue of the BFKL equation is expressed in terms of the most complicated harmonic sums and using the hypothesis [12], that the anomalous dimension in $N = 4$ theory can be obtained from the BFKL equation by the analytic continuation of its kernel, one can assume, that the perturbative expansion of the universal anomalous dimension

$$\gamma_{uni}(j) = \hat{a}\gamma_{uni}^{(0)}(j) + \hat{a}^2\gamma_{uni}^{(1)}(j) + \hat{a}^3\gamma_{uni}^{(2)}(j) + ... \tag{88}$$

contains only the most complicated functions in each order of the perturbation theory. With such assumption we can obtain [12]

$$\frac{1}{8}\,\gamma_{uni}^{(1)}(j+2) = 2S_1(j)\left(S_2(j) + S_{-2}(j)\right) - 2S_{-2,1}(j) + S_3(j) + S_{-3}(j)\,, \tag{89}$$

where

$$S_{-r}(j)=\sum_{i=1}^{j}\frac{(-1)^i}{i^r}\,,\quad S_{-2,1}=\sum_{m=1}^{j}\frac{(-1)^m}{m^2}S_1(m)\,. \tag{90}$$

This result was verified by the direct calculation of the anomalous dimension matrix in two loops [14].

On the other hand, in 2004 the three-loop anomalous dimension matrix for QCD was calculated [15]. It gave us a possibility to find the universal anomalous dimension in three loops for $N=4$ SUSY using the above hypothesis concerning the most complicated harmonic sums appearing in the perturbation theory [16]

$$\begin{aligned}\frac{1}{32}\,\gamma^{(2)}_{uni}(j+2) = {} & 24\,S_{-2,1,1,1}-12\,(S_{-3,1,1}+S_{-2,1,2}+S_{-2,2,1})\\ & +6\,(S_{-4,1}+S_{-3,2}+S_{-2,3})-3\,S_{-5}-2\,S_3\,S_{-2}-S_5\\ & -2\,S_1^2\,(3\,S_{-3}+S_3-2\,S_{-2,1})-S_2\,(S_{-3}+S_3-2\,S_{-2,1})\\ & -S_1\left(8\bar{S}_{-4}+\bar{S}_{-2}^2+4S_2\bar{S}_{-2}+2S_2^2\right)\\ & -S_1\left(3S_4-12\bar{S}_{-3,1}-10\bar{S}_{-2,2}+16\bar{S}_{-2,1,1}\right)\,,\end{aligned} \tag{91}$$

where

$$S_a(j)=\sum_{m=1}^{j}\frac{1}{m^a}\,,\qquad S_{a,b,c,\cdots}(j)=\sum_{m=1}^{j}\frac{1}{m^a}\,S_{b,c,\cdots}(m)\,, \tag{92}$$

$$S_{-a}(j)=\sum_{m=1}^{j}\frac{(-1)^m}{m^a}\,,\quad S_{-a,b,\cdots}(j)=\sum_{m=1}^{j}\frac{(-1)^m}{m^a}\,S_{b,\cdots}(m)\,, \tag{93}$$

$$\overline{S}_{-a,b,c\cdots}(j)=(-1)^j S_{-a,b,\ldots}(j)+S_{-a,b,\cdots}(\infty)\Big(1-(-1)^j\Big)\,. \tag{94}$$

9. Comparison with other approaches

The three-loop anomalous dimension for $N=4$ SUSY at $j=1+\omega\to 1$

$$\gamma^{N=4}_{uni}(j)=\hat{a}\,\frac{4}{\omega}-32\zeta_3\,\hat{a}^2+32\zeta_3\,\hat{a}^3\,\frac{1}{\omega}+\ldots \tag{95}$$

is in an agreement with the predictions of the BFKL equation [12].

Near the negative even points $j+2r=\omega\to 0$ one can verify, that the anomalous dimension satisfy the equation

$$\gamma_{uni}=4\,\frac{\hat{a}}{\omega}+\frac{\gamma^2_{uni}}{\omega} \tag{96}$$

corresponding to the ressummation of the double logarithmic terms $\sim\alpha/\omega^2$.

Further, one can find the universal anomalous dimension at large j

$$\gamma_{uni} = a(z)\,\ln j\,, \quad z = \frac{\alpha N_c}{\pi} = 4\hat{a} \tag{97}$$

in the perturbation theory from our above results

$$a(z) = -z + \frac{\pi^2}{12}\,z^2 - \frac{11}{720}\pi^4 z^3 + \ldots. \tag{98}$$

On the other hand using the well-known AdS/CFT correspondence between the superstring model on the anti-de-Sitter space and the $N = 4$ supersymmetric Yang-Mills theory, A. Polyakov with collaborators obtained the strong coupling prediction for this quantity

$$\lim_{z\to\infty} a = -z^{1/2} + \frac{3\ln 2}{4\pi} + \ldots. \tag{99}$$

In Ref. [14] the resummation of the perturbation theory was suggested in the form

$$\widetilde{a} = -z + \frac{\pi^2}{12}\,. \tag{100}$$

The prediction of this equation for three loops is in a rather good agreement with the exact result in the perturbation theory

$$\widetilde{a}^2 = -z + \frac{\pi^2}{12}\,z^2 - \frac{1}{72}\pi^4 z^3 + \ldots \tag{101}$$

and for large $\widetilde{a}$.

10. BFKL Pomeron and graviton in N=4 SUSY

In this section the Pomeron intercept in the $N = 4$ supersymmetric gauge theory will be calculated at large coupling constants [17]. To begin with, we can simplify the eigenvalue equation for the BFKL Pomeron in the diffusion approximation as follows

$$j = 2 - \Delta - D\nu^2\,, \tag{102}$$

where the parameters Δ and D are some functions of the coupling constant. For large coupling constants it is natural to expect, that the Pomeron coincides with the graviton and therefore $\Delta << 1$. The anomalous dimensions of twist-2 operators can be expressed in terms of ν as follows

$$\gamma = 1 + \frac{j-2}{2} + i\nu\,. \tag{103}$$

Due to the energy-momentum conservation we have $\gamma|_{j=2} = 0$ and therefore γ can be expressed in terms of one parameter Δ

$$\gamma = (j-2)\left(\frac{1}{2} - \frac{1/\Delta}{1+\sqrt{1+(j-2)/\Delta}}\right). \tag{104}$$

Using the AdS/CFT correspondence the eigenvalue equation for the BFKL pomeron wave function in the diffusion approximation for small Δ can be written as the graviton Regge trajectry

$$j = 2 + \frac{\alpha'}{2}\,t\,, \quad t = E^2/R^2\,, \quad \alpha' = \frac{R^2}{2}\,\Delta\,. \tag{105}$$

On the other hand, the above expression for γ at large coupling constants and $j \to \infty$ can be compared with the Polyakov predicton from the AdS/CFT correspondence

$$\gamma_{|z\to\infty} = -\sqrt{j-2}\,\Delta^{-1/2}_{|j\to\infty} = \sqrt{\pi j}\,z^{1/4} \tag{106}$$

Therefore we obtain the following Pomeron intercept at large α

$$j = 2 - \Delta\,, \quad \Delta = \frac{1}{\pi}\,z^{-1/2} \tag{107}$$

On the other hand we obtain for the slope of the anomalous dimension at $j = 2$ in the perturbation theory

$$\gamma'(2) = \frac{1}{2} - \frac{1}{2\Delta} = b = -\frac{\pi^2}{6}z + \frac{\pi^4}{72}\,z^2 - \frac{\pi^6}{540} + \ldots\,. \tag{108}$$

Similar to the case of the case $j \to \infty$ we use the following resummation procedure [14]

$$\frac{\pi^2}{6}z = -\widetilde{b} + \frac{1}{2}\widetilde{b}^2\,. \tag{109}$$

The weak and strong coupling asymptotics of the solution of this equation is given below

$$\widetilde{b} = -\frac{\pi^2}{6}z + \frac{\pi^4}{72}\,z^2 - \frac{\pi^6}{432} + \ldots\,, \qquad \lim_{z\to\infty}\widetilde{\Delta} = \frac{\sqrt{3}}{2\pi}z^{-1/2}\,, \tag{110}$$

which is in an approximate agreement with our above results.

An additional information concerning the intercept of the BFKL Pomeron in the $N = 4$ SUSY can be obtained from the next-to-leading approximation to the BFKL equation

$$\omega_0 = 1 - \Delta = 4\ln 2\,z - a_1\,z^2 + \ldots\,, \tag{111}$$

$$D = 14\,\zeta(3)\,z - b_1 z^2 + ...\,, \qquad \zeta(3) = \sum_{n=1}^{\infty} \frac{1}{n^3}\,, \tag{112}$$

where the coefficients a_1 and b_1 are known. We can perform the resummation of the perturbation theory results in an agreement with its agreement with their asymptotic behaviour at large coupling constants

$$4\ln 2\, z = \frac{(1-\Delta) - \left(2 - a_1\,(4\ln 2)^{-2}\right)(1-\Delta)^2}{\Delta^2}\,, \tag{113}$$

$$14\,\zeta(3)\, z = \frac{1 - \sqrt{1 - 2\,D^3 + cD^4}}{D^2}\,. \tag{114}$$

11. Discussion

It is important, that in QCD the gluons and quarks are reggeized. In the leading and next-to-leading approximations the Pomeron is a composite state of two gluons. The bare Pomeron intercept is above unity and we have the unitarization problem of the BFKL dynamics. For the purposes of finding radiative corrections to the BFKL equation and for its unitarization it is possible to use the effective action for the particle interactions in the quasi-multi-Regge kinematics. Above we constructed the Feynman rules for reggeon vertices contained in this action. They satisfy the Ward identities and were presented in an explicit form in the momentum representation. Thus, the Reggeon calculus in the form of a 2+1 field theory can be derived from QCD. In the framework of this approach the t-channel unitarity is automatically fulfilled. The s-channel unitaririty is incorporated in the reggeon theory through the bootstrap equations (see Ref. [1]) and various relations among reggeon vertices obtained from the effective action. The next-to-leading corrections to the eigenvalue of the BFKL kernel in QCD contain non-analytic terms in the conformal spin $|n|$. In the extended $N = 4$ SUSY this eigenvalue is an analytic function of $|n|$ and has the property of a hermitian separability. Moreover, it contains only the harmonic sums having the maximal complexity. Using the hypothesis that the universal anomalous dimensions of the twist-2 operators should contain also the most complicated harmonic sums, we calculated it up to the third order. We suggested a resummation procedure and verified the strong coupling constant predictions obtained from AdS/CFT correspondence. In particular, the intercept of the BFKL pomeron in $N = 4$ SUSY at strong couplings was calculated in this regime.

References

1. V.S. Fadin, E.A. Kuraev and L.N. Lipatov, *Phys. Lett.* **B60**, 50 (1975); Ya.Ya. Balitsky, *Yad. Fiz.* **28**, 1597 (1978).
2. L.N. Lipatov, *Phys. Lett.* **B309**, 394 (1993).
3. L.N. Lipatov, *JETP* **90**, 1536 (1986).
4. L.N. Lipatov, *Phys. Lett.* **B251**, 284 (1990).
5. J. Bartels, *Nucl. Phys* **B175**, 365 (1980);
   J. Kwiecinski and M. Praszalowicz, *Phys. Lett.* **B94**, 413 (1980).
6. L.N. Lipatov, *Nucl. Phys.* **B548**, 328 (1999).
7. L.N. Lipatov, *High energy asymptotics of multi-colour QCD and exactly solvable lattice models*, Padova preprint DFPD/93/TH/70, hep-th/9311037, unpublished.
8. L.N. Lipatov, *Nucl. Phys.* **B365**, 614 (1991).
9. L.N. Lipatov, *Nucl. Phys.* **B452**, 369 (1995).
10. E. Antonov, I. Cherednikov, E. Kuraev and L.N. Lipatov, *Nucl. Phys.* **B721**, 111 (2005).
11. V. Fadin and L.N. Lipatov, *Phys. Lett* **B429**, 127 (1998); M. Ciafaloni and G. Camici, *Phys. Lett* **B430**, 349 (1998); A. Kotikov and L.N. Lipatov, *Nucl. Phys.* **B582**, 19 (2000).
12. A. Kotikov and L.N. Lipatov, *Nucl. Phys.* **B661**, 19 (2003).
13. L.N. Lipatov, *talk at "Perspectives in Hadronic Physics", Proc. of Conf. ICTP.* Triest, Italy, May 1997.
14. A. Kotikov, L.N. Lipatov and V. Velizhanin, *Phys. Lett.* **B557**, 114 (2003).
15. S. Moch, J.A.M. Vermaseren and A. Vogt, *Nucl. Phys.* **B688**, 101 (2004).
16. A. Kotikov, L.N. Lipatov, A. Onishchenko and V. Velizhanin, *Phys. Lett.* **B595**, 521 (2004).
17. A. Kotikov, L.N. Lipatov, A. Onishchenko and V. Velizhanin, *Phys. Lett.* **B** *in print.*

NONLINEAR $k_\perp$-FACTORIZATION: A NEW PARADIGM FOR AN IN-NUCLEUS HARD QCD *

N. N. NIKOLAEV

Institut f. Kernphysik, Forschungszentrum Jülich, D-52425 Jülich, Germany and L.D.Landau Institute for Theoretical Physics, Chernogolovka, Russia
E-mail: N.Nikolaev@fz-juelich.de

W. SCHÄFER

Institut f. Kernphysik, Forschungszentrum Jülich, D-52425 Jülich, Germany
E-mail: Wo.Schaefer@fz-juelich.e

B. G. ZAKHAROV

L.D.Landau Institute for Theoretical Physics, Chernogolovka, Russia
E-mail: B.Zakharov@fz-juelich.de

V. R. ZOLLER

Institute for Theoretical and Experimental Physics, 117218, Russia E-mail: Zoller@hereon.itep.ru

We review the origin, and salient features, of the breaking of the conventional linear $k_\perp$-factorization for an in-nucleus hard pQCD processes. A realization of the nonlinear $k_\perp$-factorization which emerges instead is shown to depend on color properties of the underlying pQCD subprocesses. We discuss the emerging universality classes and extend nonlinear $k_\perp$-factorization to AGK unitarity rules for the excitation of the target nucleus.

1. Introduction

An extension of factorization theorems to nuclear targets is of burning urgency - the notion of nuclear gluon densities can be made meaningful only if they furnish a unified description of the whole variety of nuclear hard processes. A large thickness of a target nucleus introduces a new

*This work is supported partly by grants DFG 436 RUS 17/101/04 and DFG RUS 17/138/05.

scale – the so-called saturation scale Q_A^2 –, which breaks the familiar linear $k_\perp$-factorization theorems for hard scattering in a nuclear environment. Here we review the recent work by the ITEP-Jülich-Landau collaboration in which a new concept of the nonlinear $k_\perp$-factorization has been introduced and elaborated [1,2,3,4,5,6]. To this end, we recall the 1975 observation [7], based on the insight from Gribov [8] and Kancheli [9], that the Lorentz contraction of ultrarelativistic nucleus entails a spatial overlap/fusion/screening of partons from nucleons at the same impact parameter if

$$x \lesssim x_A = 1/R_A m_N, \tag{1}$$

where R_A is the radius of a nucleus of mass number A.

Within the fusion reinterpretation of multiple gluon exchanges, a nuclear glue will be a nonlinear functional of the free nucleon glue: the same sea and glue will be shared by many nucleons. Specifically, we derived the collective glue of j-overlapping nucleons. It is a basis for the definition of the collective nuclear unintegrated glue $\phi(b, \boldsymbol{\kappa})$, per unit area in the impact parameter plane, which describes the coherent diffractive dijet production off nuclei [10,1], for instance, in πA collisions or DIS. We found that the so-defined collective nuclear glue provides the linear $k_\perp$-factorization for the nuclear structure function and the leading single-quark spectrum, but this is rather an exception due to the Abelian nature of these observables. All other single-jet and dijet cross sections prove to be highly nonlinear functionals – quadratures – of the collective nuclear glue [2,3,4,5,6]. Any application of linear $k_\perp$–factorization to nuclei is entirely unwarranted. Here we review how the concrete form of the nonlinearity depends on the relevant pQCD subprocesses and define the universality classes of nonlinear $k_\perp$-factorization for production of hard dijets. We also report the nonlinear $k_\perp$-factorization solution [11] to the Abramovsky-Gribov-Kancheli (AGK) unitarity problem [12].

At the heart of our approach is the equivalence between the parton fusion description of the shadowing introduced in 1975 [7] and the unitarization of the color dipole–nucleus interaction [13]. The major technical problem in the unitarization program is the non-Abelian evolution of color dipoles in a nuclear environment, and we present a closed-form solution based on the multiple-scattering theory for color dipoles [1,14]. A very rich pattern of nonlinear $k_\perp$-factorization relations emerges already for the lowest order pQCD subprocesses considered here.

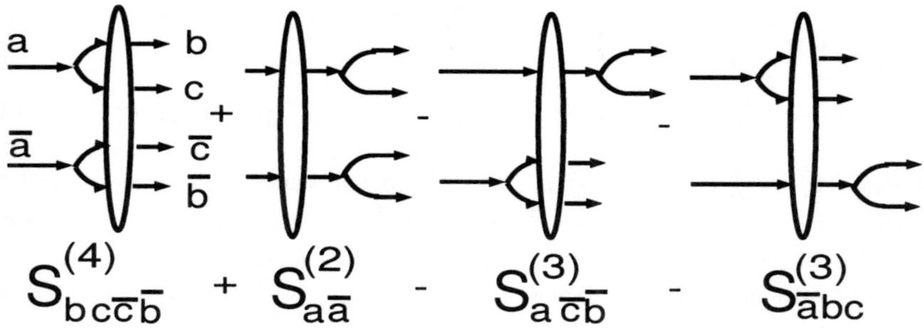

Figure 1. The S-matrix structure of the two-body density matrix for excitation $a \to bc$.

2. The master formula for nuclear dijets

In the laboratory frame, dijet production in the parton-nucleus collisions can be viewed as an excitation $a \to bc$, where $a = \gamma^*, q, g,\ b, c = q, \bar{q}g$. The parton fusion condition (1) amounts to the coherency of excitation over the whole diameter of the nucleus. Excitation of the perturbative $|bc\rangle$ Fock state of the physical projectile $|a\rangle$ by one-gluon exchange, $ag \to bc$, leaves the target nucleon debris in the color excited state. For nuclear targets one has to deal with multiple gluon exchanges which are enhanced by a large target thickness.

According to the general rules of quantum mechanics, the two-particle spectrum is the Fourier transform of the two-body density matrix. The derivation of the master formula for the dijet spectrum, based on the technique developed in [1,14] is found in [3]:

$$\frac{d\sigma(a^* \to bc)}{dz_b d^2\boldsymbol{p}_b d^2\boldsymbol{p}_c} = \frac{1}{(2\pi)^4} \int d^2\boldsymbol{b}_b d^2\boldsymbol{b}_c d^2\boldsymbol{b}'_b d^2\boldsymbol{b}'_c \tag{2}$$
$$\times \exp[-i\boldsymbol{p}_b(\boldsymbol{b}_b - \boldsymbol{b}'_b) - i\boldsymbol{p}_c(\boldsymbol{b}_c - \boldsymbol{b}'_c)]\Psi(z_b, \boldsymbol{b}_b - \boldsymbol{b}_c)\Psi^*(z_b, \boldsymbol{b}'_b - \boldsymbol{b}'_c)$$
$$\times \left\{ S^{(4)}_{\bar{b}\bar{c}cb}(\boldsymbol{b}'_b, \boldsymbol{b}'_c, \boldsymbol{b}_b, \boldsymbol{b}_c) + S^{(2)}_{\bar{a}a}(\boldsymbol{b}', \boldsymbol{b}) - S^{(3)}_{\bar{b}\bar{c}a}(\boldsymbol{b}, \boldsymbol{b}'_b, \boldsymbol{b}'_c) - S^{(3)}_{\bar{a}bc}(\boldsymbol{b}', \boldsymbol{b}_b, \boldsymbol{b}_c) \right\}.$$

If $\boldsymbol{b}_a = \boldsymbol{b}$ is the projectile's impact parameter, then $\boldsymbol{b}_b = \boldsymbol{b} + z_c\boldsymbol{r}$, $\boldsymbol{b}_c = \boldsymbol{b} - z_b\boldsymbol{r}$, where $z_{b,c}$ stand for the fraction of the lightcone momentum of the projectile a carried by partons b and c, $\Psi(z, \boldsymbol{r})$ stands for the lightcone wave function of the $|bc\rangle$ Fock state of the projectile, its connection to the parton-splitting functions is found in [3]. All $S^{(n)}$ describe a scattering of color-singlet systems of n partons, as indicated in Fig. 1. This is the crucial point - in the course of our derivation of the dijet spectra and single-jet spectra we only deal with infrared-safe observables. $S^{(2)}$ and $S^{(3)}$ are readily calculated in terms of the 2-parton and 3-parton dipole cross sections [13,15,14]. For the dilute-gas nucleus one uses the Glauber-Gribov multiple-scattering theory formula [16,17]

$$\mathsf{S}^{(n)}(\boldsymbol{b}_c', \boldsymbol{b}_b', \boldsymbol{b}_c, \boldsymbol{b}_b) = \mathsf{S}[\boldsymbol{b}, \Sigma_n] = \exp\{-\frac{1}{2}\Sigma^{(n)}(\boldsymbol{b}_c', \boldsymbol{b}_b', \boldsymbol{b}_c, \boldsymbol{b}_b)T(\boldsymbol{b})\}, \quad (3)$$

where $T(\boldsymbol{b}) = \int dr_z n_A(r_z, \boldsymbol{b})$ is the optical thickness of a nucleus at the impact parameter $\boldsymbol{b}$ and $n_A(r_z, \boldsymbol{b})$ is the nuclear matter density. The major nontrivial task is a calculation of the coupled-channel operator Σ_4 [1,5,6]. The relevant basis of color states depends on the pQCD subprocess,

$$\begin{aligned}
\gamma^* &\to q\bar{q} : 1_1 + 8_{N_c^2}, \\
g &\to q\bar{q} : 1_1 + 8_{N_c^2},, \\
q &\to qg : 3_{N_c} + \{6 + 15\}_{N_c^3}, \\
g &\to gg : 1_1 + \{8_A + 8_S\}_{N_c^2} + \{10 + \overline{10} + 27 + R_7\}_{N_c^4},
\end{aligned}$$

where the subscripts indicate the N_c-dependence of the size of the relevant color multiplets. For the description of the two-gluon multiplet R_7 which exists for $N_c > 3$ only, see [6]. These size-dependent subspaces of multiplets prove very useful in the diagonalization of the non-Abelian evolution of color dipoles. In the case of nuclear targets one must distinguish truly inelastic processes, which leave the target nucleus in the color excited state, and coherent diffraction $aA \to (bc)A$ with retention of the target nucleus in the ground state.

3. The $k_\perp$-factorization for DIS off free nucleons

The unintegrated gluon density in the target nucleon,

$$\mathsf{F}(x, \kappa^2) = \frac{\partial G(x, \kappa^2)}{\partial \log \kappa^2}, \quad (4)$$

furnishes a universal description of the proton structure function $F_{2p}(x, Q^2)$ and of the final states. For instance, the linear $k_\perp$-factorization for forward quark-antiquark dijets reads (for applications, see [18] and references therein)

$$\frac{2(2\pi)^2 d\sigma_N(\gamma^* \to Q\overline{Q})}{dz d^2\boldsymbol{p} d^2\boldsymbol{\Delta}} = f(x, \boldsymbol{\Delta}) \left|\Psi(z, \boldsymbol{p}) - \Psi(z, \boldsymbol{p} - \boldsymbol{\Delta})\right|^2, \quad (5)$$

where $\boldsymbol{\Delta} = \boldsymbol{p}_q + \boldsymbol{p}_{\bar{q}}$ is the jet-jet decorrelation (acoplanarity) momentum, $\boldsymbol{p} \equiv \boldsymbol{p}_{\bar{q}}$, $z \equiv z_{\bar{q}}$ refer to the $\bar{q}$-jet, and

$$f(x, \boldsymbol{\kappa}) = \frac{4\pi\alpha_S}{N_c} \cdot \frac{1}{\kappa^4} \cdot \mathsf{F}(x, \boldsymbol{\kappa}^2). \quad (6)$$

From the unitarity point of view, Eq. (5) corresponds to the unitarity cuts of diagrams Figs. 2a,b for the forward Compton scattering amplitude. Notice, that the dijet cross section is a direct probe of the free-nucleon unintegrated glue $f(x, \boldsymbol{\Delta})$; the point that the jet-jet decorrelation momentum $\boldsymbol{\Delta}$ comes from the transverse momentum of the exchanged gluon is obvious form Figs. 2a,b for the forward Compton scattering amplitude. Notice, that the dijet cross section is a direct probe of the free-nucleon unintegrated glue $f(x, \boldsymbol{\Delta})$; the point that the jet-jet decorrelation momentum $\boldsymbol{\Delta}$ comes from the transverse momentum of the exchanged gluon is obvious.

The focus of the further discussion is on how linear $k_\perp$-factorization (5) is modified by multiple gluon exchanges predominant in hard production off nuclear targets.

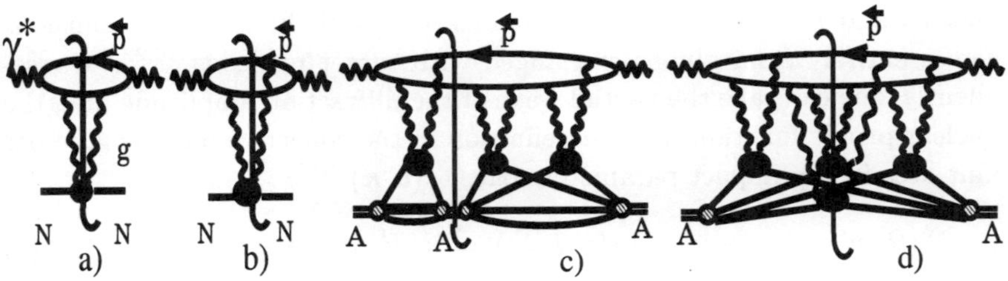

Figure 2. The typical unitarity cuts and dijet final states in DIS : (a),(b) - free-nucleon target, (c) - coherent diffractive DIS off a nucleus, (d) - truly inelastic DIS with multiple color excitation of the nucleus.

4. What is the collective unintegrated nuclear glue?

While seeking the factorization properties of an in-nucleus hard QCD, one must define the collective glue in terms of a certain observable such that the so-defined glue enters in a universal manner a description of all other nuclear processes. Taking DIS as an example, the same nuclear glue must describe the total DIS cross section, its coherent diffractive unitarity cuts, Fig. 2c, and multiple color excitation unitarity cuts, Fig. 2d, of the nuclear forward Compton scattering amplitude.

DIS off a nucleus at $x \lesssim x_A$ can be described in terms of the color dipole-nucleus cross-section [13]

$$\sigma_A(\boldsymbol{r}) = 2\int d^2\boldsymbol{b}\Big\{1 - \exp[-\frac{1}{2}\sigma(x,\boldsymbol{r})T(\boldsymbol{b})]\Big\}, \tag{7}$$

where

$$\sigma(x,\boldsymbol{r}) = \int d^2\boldsymbol{\kappa} f(x,\boldsymbol{\kappa})[1-\exp(i\boldsymbol{\kappa}\boldsymbol{r})] \tag{8}$$

is the $q\bar{q}$ dipole cross section [13]. The x-dependence of $\sigma(x,\boldsymbol{r})$ is governed by the color-dipole evolution equation [15,19]; for the sake of brevity, here-below we suppress the x-dependence. The nuclear dipole cross section (7) sums in the compact form the Glauber-Gribov multiple-scattering diagrams and is a basis for the quantitative description of nuclear shadowing in DIS [20].

It is remarkable, that although a deposition of dozen MeV energy will break any heavy nucleus, at $x \lesssim x_A$ such a coherent diffraction makes $\approx 50\%$ of the total cross section of DIS off heavy nucleus [21]. To the lowest order in pQCD, the coherent diffractive final state consists of the back-to-back dijet with vanishing transverse momentum transfer $\boldsymbol{\Delta}$ to the target nucleus and the large transverse momentum $\boldsymbol{p}$ of dijets in πA collisions comes entirely from gluons exchanged with target nucleons [10,22]. Consequently, one can take the partial wave of the diffraction amplitude, i.e., the nuclear profile function, for the definition of the collective nuclear glue per unit area in the impact parameter space, $\phi(\boldsymbol{b},\boldsymbol{\kappa})$ [10,1]:

$$\Gamma_A(\boldsymbol{b},\boldsymbol{r}) = [1-\exp(-\frac{1}{2}\sigma(\boldsymbol{r})T(\boldsymbol{b}))] = \int d^2\boldsymbol{\kappa}\phi(\boldsymbol{b},\boldsymbol{\kappa})\{1-\exp[i\boldsymbol{\kappa}\boldsymbol{r}]\}\,. \tag{9}$$

In can be expanded as

$$\phi(\boldsymbol{b},\boldsymbol{\kappa}) = \frac{1}{\sigma_0}\sum_{j=1}^{\infty} w_j(\boldsymbol{b}) f^{(j)}(\boldsymbol{\kappa}), \tag{10}$$

where the collective glue $f^{(j)}(\boldsymbol{\kappa})$ of j overlapping nucleons of the Lorentz-contracted nucleus satisfies the convolution representation

$$f^{(j)}(\boldsymbol{\kappa}) = \frac{1}{\sigma_0}(f^{(j-1)}\otimes f)(\boldsymbol{\kappa}), \tag{11}$$

and a probability to find j overlapping nucleons equals

$$w_j(\boldsymbol{b}) = \frac{\nu_A^j(\boldsymbol{b})}{j!}\exp\left[-\nu_A(\boldsymbol{b})\right],\;\; \nu_A(\boldsymbol{b}) = \frac{1}{2}\sigma_0 T(\boldsymbol{b}), \tag{12}$$

$\sigma_0 = \int d^2\boldsymbol{\kappa} f(\boldsymbol{\kappa})$ is the dipole cross section $\sigma(\boldsymbol{r})$ for large dipoles $\boldsymbol{r}$.

Subsequently we will use also $\Phi(\boldsymbol{b},\boldsymbol{\kappa}) = \phi(\boldsymbol{b},\boldsymbol{\kappa}) + w_0(\boldsymbol{b})\delta(\boldsymbol{\kappa})$, which is the Fourier transform of the S-matrix for an elementary dipole,

$$\mathsf{S}[\boldsymbol{b},\sigma(\boldsymbol{r})] = \int d^2\boldsymbol{\kappa}\Phi(\boldsymbol{b},\boldsymbol{\kappa})\exp[i\boldsymbol{\kappa}\boldsymbol{r}]\,. \tag{13}$$

The antishadowing properties [1,10] of the so-defined collective glue $\phi(\boldsymbol{b},\boldsymbol{\kappa})$ are responsible for the familiar Cronin effect [3]. It furnishes the linear $k_\perp$-factorization for the nuclear structure function $F_{2A}(x,Q^2)$ and the forward single jets in DIS off nuclei, precisely the same as for the free-nucleon target in terms of $f(\boldsymbol{\kappa})$. This linear $k_\perp$-factorization is rather an exception because of their special Abelian features. Dijets in DIS, and dijet and single-jet spectra in hA collisions, however, prove to be highly nonlinear functionals of $\Phi(\boldsymbol{b},\boldsymbol{\kappa})$ [1,2,3,4,5,6]. We dubbed this property the nonlinear $k_\perp$-factorization.

An important point is that color dipole cross section depends on the $SU(N_c)$ representation the partons belong to. For this reason the collective nuclear glue can not be described by one scalar function, it is rather a density matrix in the space of color representations, for examples see [3,6].

5. The origin of nonlinear $k_\perp$-factorization

The fundamental point is that the calculation of the single-jet and dijet cross sections can be reduced to an interaction with the nucleus of color-singlet multiparton systems [14], as exemplified by our master formula (3). The interaction cross-section for such systems of n partons is a (matrix) superposition of elementary dipole cross-sections. Upon the diagonalization of the non-Abelian evolution problem, the matrix elements of nuclear S-matrices $\mathsf{S}^{(n)}$, $n \geq 3$, in (3) will be a (matrix) products of nuclear S-matrices for elementary dipoles. In the calculation of the Fourier transform of the two-body density matrix, nuclear S-matrices for elementary dipoles will be replaced by their Fourier transforms (13). Consequently, the Fourier transforms of $\mathsf{S}^{(n)}$ would give rise to multiple convolutions of $\Phi(\boldsymbol{b},\boldsymbol{\kappa})$, and the nuclear dijet spectra will be highly nonlinear quadratures of collective nuclear glue. The specific expansion for eigenvalues of $\mathsf{S}^{(n)}$ in term of elementary dipole cross sections, and the specific form of the emerging nonlinear $k_\perp$-factorization thereof, depends on the color representations of the initial-state parton and final-state dijet, still there emerges a well-defined pattern of the universality classes.

6. Why collective glue for slices of a nucleus is a must?

In the realm of pQCD with partons in the fundamental and adjoint multiplets, one finds an exact product representations for nuclear S-matrices for $n = 2, 3$ in terms of the S-matrices for elementary dipoles. The case of $n = 4$ is much more tricky. It is remarkable how the forbidding three-channel for qg dijets, and seven-channel for gg dijets, non-Abelian intranuclear evolution can be diagonalized exactly for large N_c due the block-diagonal structure of the four-parton cross-section operator $\hat{\Sigma}^{(4)}$. We explicitly represent it in the form

$$\hat{\Sigma}^{(4)} = \hat{\Sigma}^{(0)} + \hat{\omega}\,, \tag{14}$$

where $\hat{\Sigma}^{(0)}$ is the diagonal operator, while the off-diagonal $\hat{\omega}$, proportional to the transition cross section $\Omega(\mathbf{s}, \boldsymbol{r}, \boldsymbol{r}')$ introduced in [1,5], describes the $1/N_c$ suppressed transitions between the subspaces formed by the states having equal dimension at large N_c as expounded in Sec. 2. The expansion (14) is a a basis for the systematic large-N_c perturbation theory. One can address it, for instance, making use of the Sylvester expansion [1], which has a generic form (we show the result for the case of gg-dijets, the detailed description of eigenstates $|e_i\rangle$ and eigenvalues Σ_i is found in [6])

$$\begin{aligned} \mathsf{S}[\boldsymbol{b}, \hat{\sigma}^{(4)}(\mathbf{s}, \boldsymbol{r}, \boldsymbol{r}')] &= |e_2\rangle\langle e_2| \exp[-\frac{1}{2}\Sigma_2 T(\boldsymbol{b})] \\ &+|e_4\rangle\langle e_2| \frac{\Omega(\mathbf{s}, \boldsymbol{r}, \boldsymbol{r}')}{\sqrt{2}N_c(\Sigma_2 - \Sigma_4)} \left\{ \exp[-\frac{1}{2}\Sigma_4 T(\boldsymbol{b})] - \exp[-\frac{1}{2}\Sigma_2 T(\boldsymbol{b})] \right\}. \end{aligned} \tag{15}$$

Here $\Sigma_{2,4}$ – the relevant eigenvalues of $\hat{\Sigma}^{(0)}$ – are certain superpositions of the elementary dipole cross sections, While the Fourier transform of the diagonal term in (16) will be a multiple convolution of the collective nuclear glue, for the off-diagonal transitions, which excite gg-dijets in higher color multiplets, $10, \overline{10}, 27, R_7$, this is the dead end, though. The denominator $(\Sigma_2 - \Sigma_4)$ in the Sylvester expansion (16) blocks further analytic calculations. For instance, in their closely related study, Kovchegov and Jalilian-Marian [23] stop at precisely this point; in order to proceed further one has to rely upon the brute force numerical Fourier transform, which masks important distinction between different production mechanisms and salient features of the production dynamics.

The way to the explicit analytic quadratures is paved by our integral representation [1], which to the first order in the off-diagonal perturbation $\hat{\omega}$ takes the form (similar representation is found for higher orders of large-N_c perturbation theory [1])

$$\mathsf{S}[\boldsymbol{b}, \hat{\Sigma}^{(0)} + \hat{\omega}] - \mathsf{S}[\boldsymbol{b}, \hat{\Sigma}^{(0)}] = -\frac{1}{2}T(\boldsymbol{b}) \int_0^1 d\beta\, \mathsf{S}[\boldsymbol{b}, (1-\beta)\hat{\Sigma}^{(0)}]\, \hat{\omega}\, \mathsf{S}[\boldsymbol{b}, \beta\hat{\Sigma}^{(0)}]. \tag{16}$$

This integral representation is much more than a technical trick to achieve the convolution representation for the four-body S-matrix. Specifically, $\hat{\omega}$

describes the hard excitation of dijets in higher color multiplets at the depth β from the front face of the nucleus, while $\mathsf{S}[\boldsymbol{b}, \beta\hat{\Sigma}^{(0)}]$ and $\mathsf{S}[\boldsymbol{b}, (1-\beta)\hat{\Sigma}^{(0)}]$ describe the Initial (ISI) and Final State Interaction (FSI) in slices $[0, \beta]$ and $[\beta, 1]$ of the nucleus, respectively. Now, the color multiplet of the dijet is different from that of the incident parton, and the resulting distinction between ISI and FSI is an integral part of the dijet production dynamics, which can not be described entirely in terms of the classical gluon field of a whole nucleus [24]. The emergence of $\mathsf{S}[\boldsymbol{b}, \beta\hat{\Sigma}^{(0)}]$ and $\mathsf{S}[\boldsymbol{b}, (1-\beta)\hat{\Sigma}^{(0)}]$ calls upon the nuclear glue defined for the slice $[0, \beta]$ of the nucleus ($0 \leq \beta \leq 1$)

$$\exp\left[-\frac{1}{2}\beta\sigma(\boldsymbol{r})T(b)\right] = \int d^2\boldsymbol{\kappa}\Phi(\beta; \boldsymbol{b}, \boldsymbol{\kappa}) \exp(i\boldsymbol{\kappa}\boldsymbol{r}), \tag{17}$$

and the corresponding overlap probabilities $w_j(\beta; \boldsymbol{b})$. We will need also the wave function of bc Fock states coherently distorted in this slice,

$$\Psi(\beta; z, \boldsymbol{p}) \equiv \int d^2\boldsymbol{\kappa}\Phi(\beta; \boldsymbol{b}, \boldsymbol{\kappa})\Psi(z, \boldsymbol{p}+\boldsymbol{\kappa}). \tag{18}$$

Now we are in the position to present our explicit results on nonlinear $k_\perp$-factorization. We start with the short digression into the single-jet problem.

7. The fate of $k_\perp$-factorization for single-jet spectra in pA collisions

The integration over the transverse momentum of the unobserved parton in the master formula (3) is straightforward. The unobserved parton and its antiparton will enter at the same impact parameter and multiparton color singlet states simplify to the two-parton ones. Still, the non-Abelian features of QCD manifest themselves in the breaking of linear $k_\perp$-factorization. Here we show the the large-N_c result for the radiation of gluons from quarks, $q^* \to qg$ [3], which is directly relevant to jet production in the proton hemisphere of pA collisions at RHIC [25,26]. The linear $k_\perp$-factorization holds for the free-nucleon target,

$$\frac{(2\pi)^2 d\sigma_A(q^* \to gq)}{dz_g d^2\boldsymbol{p}_g} = \frac{1}{2}\int d^2\boldsymbol{\kappa} f(\boldsymbol{\kappa}) \tag{19}$$
$$\times\left\{\left|\Psi(z_g, \boldsymbol{p}_g) - \Psi(z_g, \boldsymbol{p}_g+\boldsymbol{\kappa})\right|^2 + \left|\Psi(z_g, \boldsymbol{p}_g+\boldsymbol{\kappa}) - \Psi(z_g, \boldsymbol{p}_g+z_g\boldsymbol{\kappa})\right|^2\right\},$$

while the same spectrum for the nuclear target is of the two-component form

$$\frac{(2\pi)^2 d\sigma_A(q^* \to gq)}{dz_g d^2\boldsymbol{p}_g d^2\boldsymbol{b}} = S[\boldsymbol{b},\sigma_0] \int d^2\boldsymbol{\kappa}\phi(\boldsymbol{b},\boldsymbol{\kappa}) \tag{20}$$

$$\times\Big\{|\Psi(z_g,\boldsymbol{p}_g) - \Psi(z_g,\boldsymbol{p}_g+\boldsymbol{\kappa})|^2 + |\Psi(z_g,\boldsymbol{p}_g+\boldsymbol{\kappa}) - \Psi(z_g,\boldsymbol{p}_g+z_g\boldsymbol{\kappa})|^2\Big\} +$$

$$\int d^2\boldsymbol{\kappa}_1 d^2\boldsymbol{\kappa}_2 \phi(\boldsymbol{b},\boldsymbol{\kappa}_1)\phi(\boldsymbol{b},\boldsymbol{\kappa}_2)|\Psi(z_g,\boldsymbol{p}_g+z_g\boldsymbol{\kappa}_1) - \Psi(z_g,\boldsymbol{p}_g+\boldsymbol{\kappa}_1+\boldsymbol{\kappa}_2)|^2 .$$

The first component is an exact counterpart of the free-nucleon spectrum: it is linear $k_\perp$-factorizable, but is suppressed by the nuclear absorption factor $S[\boldsymbol{b},\sigma_0(x)]$. For central interactions of the main experimental interest, the gluon spectrum is entirely dominated by the second component which is a non-linear – quadratic – functional of the collective nuclear glue. A full compendium of nonlinear $k_\perp$-factorization results for single jets from all possible pQCD subprocesses is found in [3].

There is a conspicuous difference between the z_g-dependence of the free-nucleon and nuclear spectra. This amounts to the $\boldsymbol{p}_g$-dependence of the Landau-Pomeranchuk-Migdal effect; the same applies to the spectrum of leading quarks and nuclear quenching of forward jets in pA collisions [3].

8. Nonlinear $k_\perp$-factorization for dijets: the case of DIS

Here we report closed-form analytic results for quark-antiquark dijets in DIS to the leading order in $1/N_c$, higher orders can be derived following Ref. [1]. In DIS off nuclei, dipoles propagate as color-singlets until at depth β from the front face of a nucleus they excite into the octet state; the non-Abelian evolution in the slice $[\beta, 1]$ consists of color rotations within the octet state. Combining the truly inelastic and diffractive components of DIS,

$$\frac{(2\pi)^2 d\sigma_A(\gamma^* \to Q\overline{Q})}{d^2\boldsymbol{b}dzd^2\boldsymbol{p}d^2\boldsymbol{\Delta}} = \frac{1}{2}T(\boldsymbol{b})\int_0^1 d\beta \int d^2\boldsymbol{\kappa}_1 d^2\boldsymbol{\kappa}$$

$$\times f(\boldsymbol{\kappa})\Phi(1-\beta,\boldsymbol{b},\boldsymbol{\Delta}-\boldsymbol{\kappa}_1-\boldsymbol{\kappa})\Phi(1-\beta,\boldsymbol{b},\boldsymbol{\kappa}_1)$$

$$\times\Big|\Psi(\beta;z,\boldsymbol{p}-\boldsymbol{\kappa}_1) - \Psi(\beta;z,\boldsymbol{p}-\boldsymbol{\kappa}_1-\boldsymbol{\kappa})\Big|^2$$

$$+\delta^{(2)}(\boldsymbol{\Delta})\Big|\Psi(1;z,\boldsymbol{p}) - \Psi(z,\boldsymbol{p})\Big|^2 . \tag{21}$$

The diffractive component of Eq. (21), $\propto \delta^{(2)}(\boldsymbol{\Delta})$, gives exactly back-to-back dijets (for the $\boldsymbol{\Delta}$-dependence for finite-size nuclei see Ref. [10]). It is a quadratic functional of the collective nuclear glue. The first component in

(21) – truly inelastic DIS with excitation of color-octet dijets – is of fifth order in gluon field densities: a linear one of the free-nucleon glue $f(\boldsymbol{\kappa})$ which describes the hard singlet-to-octet transition and a quartic functional of the collective nuclear glue for the two slices of a nucleus: the coherent ISI's of color-singlet dipoles in the slice $[0, \beta]$ and incoherent FSI's of color-octet dipoles in the slice $[\beta, 1]$.

9. Nonlinear $k_\perp$-factorization for dijets: universality classes

We start with the universality class of coherent diffraction. The origin of coherent diffractive DIS is a difference between the in-vacuum and intranuclear-distorted wave function of the $q\bar{q}$ state of the photon:

$$\frac{(2\pi)^2 d\sigma_A(\gamma^* \to q\bar{q})}{d^2 b dz d^2\boldsymbol{p} d^2\boldsymbol{\Delta}} = \delta^{(2)}(\boldsymbol{\Delta})\Big|\Psi(1; z, \boldsymbol{p}) - \Psi(z, \boldsymbol{p})\Big|^2 . \tag{22}$$

In the case of $q \to qg$ the incident partons are colored, and intranuclear attenuation of the incident quark wave entails the suppression of coherent diffraction by the factor $w_0^2(b)$, similar, but stronger intranuclear attenuation is found for $g \to gg$ [6]. This nuclear suppression of coherent diffraction can be identified with Bjorken's diffractive gap survival probability [27], its process dependence makes clear that the concept of hard factorization for the nuclear pomeron makes no sense. In all above cases coherent diffraction is of leading order in $1/N_c$, while coherent diffraction $g \to q\bar{q}$ is suppressed, which adds further case against the hard factorization for nuclear pomerons.

The universality class of dijets in higher color miltiplets excited from partons in lower multiplet is the most interesting one. As an example we cite the large-N_c result [5]

$$\begin{aligned}
\frac{d\sigma\big(q^* \to qg(6+15)\big)}{d^2 b dz d^2\boldsymbol{\Delta} d^2\boldsymbol{p}} &= \frac{1}{2(2\pi)^2} T(b) \int_0^1 d\beta \\
&\times \int d^2\boldsymbol{\kappa} d^2\boldsymbol{\kappa}_1 d^2\boldsymbol{\kappa}_2 d^2\boldsymbol{\kappa}_3 \delta(\boldsymbol{\kappa} + \boldsymbol{\kappa}_1 + \boldsymbol{\kappa}_2 + \boldsymbol{\kappa}_3 - \boldsymbol{\Delta}) \\
&\times \underbrace{\Phi(\beta; \boldsymbol{b}, \boldsymbol{\kappa}_3)}_{Quark\ ISI} \underbrace{f(\boldsymbol{\kappa}) \left|\Psi(\beta; z, \boldsymbol{p} - \boldsymbol{\kappa}_2 - \boldsymbol{\kappa}_3) - \Psi(\beta; z, \boldsymbol{p} - \boldsymbol{\kappa}_2 - \boldsymbol{\kappa}_3 - \boldsymbol{\kappa})\right|^2}_{Hard\ Excitation} \\
&\times \underbrace{\Phi(1-\beta; \boldsymbol{b}, \boldsymbol{\kappa}_1)}_{Quark\ FSI} \underbrace{\Phi(\frac{C_A}{C_F}(1-\beta); \boldsymbol{b}, \boldsymbol{\kappa}_2)}_{Gluon\ FSI},
\end{aligned} \tag{23}$$

where we indicated the rôle of different factors in the integrand. Notice the sixth order nonlinearity of the dijet spectrum (23) in nuclear and free-nucleon glue. The most nontrivial point here is how incoherent distortions of the incident quark wave come along with the coherent distortions of the qg wave function in the same slice $[0, \beta]$ of the nucleus.

The ratio of Casimirs C_A/C_F in $\Phi(\frac{C_A}{C_F}(1-\beta), \boldsymbol{b}, \boldsymbol{\kappa}_2)$ for FSI of gluons is a reminder that the collective nuclear glue derives from a density matrix in color space [1,3]. In DIS, $\gamma^* \to q\bar{q}(8)$, ISI of the photon vanishes and $\Phi(\beta; \boldsymbol{b}, \boldsymbol{\kappa}_3) = \delta^{(2)}(\boldsymbol{\kappa}_3)$, for the incident gluons, $g \to gg(10 + \overline{10} + 27 + R_7)$, the quark FSI factor must be swapped for the gluon FSI factor, and there will be a slight modification of incoherent ISI distortions - at large N_c the gluon behaves like the uncorrelated quark-antiquark pair [3,6]. The striking distinction between ISI and FSI requires the collective nuclear glue for slices of the nucleus – nonlinear $k_\perp$-factorization can not be described by the classical gluon field of the whole nucleus. The nonlinear $k_\perp$-factorization entails a nuclear enhancement of the decorrelation of dijets [1,5,6].

Universality class of dijets in the same lower color multiplet as the beam parton has its own unique features. For instance, in the case of color-triplet qg dijets [5]

$$\frac{d\sigma\big(q^*A \to qg(3)\big)}{d^2b dz d^2\boldsymbol{\Delta} d^2\boldsymbol{p}} = \frac{1}{(2\pi)^2}\phi(b, \boldsymbol{\Delta}) \left|\Psi(1; z, \boldsymbol{p} - \boldsymbol{\Delta}) - \Psi(z, \boldsymbol{p} - z\boldsymbol{\Delta})\right|^2 . \tag{24}$$

The free-nucleon cross section has exactly the same form in terms of $f(\boldsymbol{\Delta})$ and the in-vacuum wave functions. Recalling that $\Psi(z, \boldsymbol{p} - z\boldsymbol{\Delta})$, which has a collinear singularity, is the probability amplitude to find the qg state in the physical quark, we reinterpret this result as a fragmentation of the quark, scattered quasi-elastically with the differential cross section $\propto \phi(b, \boldsymbol{\Delta})$, see also below. The change from the free-nucleon to a nuclear target,

$$|\Psi(z, \boldsymbol{p} - \boldsymbol{\Delta}) - \Psi(z, \boldsymbol{p} - z\boldsymbol{\Delta})|^2 \Longrightarrow |\Psi(1; z, \boldsymbol{p} - \boldsymbol{\Delta}) - \Psi(z, \boldsymbol{p} - z\boldsymbol{\Delta})|^2,$$

must be interpreted as a nuclear modification of the hard fragmentation function of the quark.

Apart from the fact that ISI of incident gluons looks like an ISI of the uncorrelated quark-antiquark pair, the nonlinear $k_\perp$-factorization for $g \to q\bar{q}(8)$, $g \to gg(8_A + 8_S)$, $g \to gg(8_S)$ is very similar to that of $q \to qg(3)$ [6]. In $g \to gg(8_A + 8_S)$ too there emerges the nuclear modified hard fragmentation function of the gluon.

10. Nonlinear $k_\perp$-factorization for dijets: unitarity cuts and AGK rules for color excitation of the target nucleus

Our technique can readily be extended [11] to partial cross sections $d\sigma_j$ for final states with j color-excited nucleons of the target nucleus – in the language of the so-called AGK unitarity rules that corresponds to j cut pomerons [12]. This multiplicity j controls the hadronic multiproduction in the nucleus hemisphere, i.e., the collision centrality, as well as the nonperturbative energy-loss contribution to the quenching of forward jets. The simplest case is the AGK rule for the universality class of coherent diffraction:

$$d\sigma_j = \delta_{j0} d\sigma_D. \tag{25}$$

Now we notice that the differential cross section of quark-nucleon quasi-elastic scattering $qN \to q'N^*$ equals

$$\frac{d\sigma_{qN}}{d^2\boldsymbol{\kappa}} = \frac{1}{2} f(\boldsymbol{\kappa}). \tag{26}$$

Here the target debris N^* is in the color-excited state. Then, the convolution property (11) of the collective glue $f^{(j)}(\boldsymbol{\kappa})$ relates it to the differential cross section of j-fold incoherent quasi-elastic scattering,

$$f^{(j)}(\boldsymbol{\kappa}) \propto \frac{d\sigma^{(j)}}{d^2\boldsymbol{\kappa}}. \tag{27}$$

Evidently, the latter is simply a j-fold convolution of single scattering cross sections. This observation paves a way to a simple interpretation of $f^{(j)}(\boldsymbol{\kappa})$ in terms of j cut pomerons.

We illustrate the emerging AGK rule for the universality class of dijets in the same lower color representation as the beam parton on an example of the reaction $q \to qg(3)$ with j color-excited nucleons of the target nucleus:

$$\begin{aligned} &\frac{d\sigma_j\left(q^*A \to qg(3)\right)}{d^2b dz d^2\boldsymbol{\Delta} d^2\boldsymbol{p}} = \\ &= \frac{1}{(2\pi)^2\sigma_0} w_j(b) f^{(j)}(\boldsymbol{\Delta}) \left|\Psi(1; z, \boldsymbol{p}-\boldsymbol{\Delta}) - \Psi(z, \boldsymbol{p} - z\boldsymbol{\Delta})\right|^2. \end{aligned} \tag{28}$$

It makes the interpretation of this final state as a result of fragmentation of the quasi-elastically scattered quark an obvious one - the nuclear-distorted fragmentation function, its collinear pole included, does not depend on the multiplicity of cut pomerons j. Notice the uncut pomerons which enter the intranuclear distortion of $\Psi(1; z, \boldsymbol{p} - \boldsymbol{\Delta})$.

As an example of AGK rules from the universality class of dijets in higher color multiplet excited from partons in lower multiplet we show the large-N_c result for $q \to qg(6+15)$ with j color-excited nucleons of the target nucleus:

$$
\begin{aligned}
&\frac{d\sigma_j(q^* \to qg(6+15))}{d^2bdzd^2\boldsymbol{\Delta}d^2\boldsymbol{p}} = \frac{1}{(2\pi)^2}T(b)\int_0^1 d\beta \\
&\times \int d^2\boldsymbol{\kappa}d^2\boldsymbol{\kappa}_1 d^2\boldsymbol{\kappa}_2 d^2\boldsymbol{\kappa}_3 \delta(\boldsymbol{\kappa}+\boldsymbol{\kappa}_1+\boldsymbol{\kappa}_2+\boldsymbol{\kappa}_3-\boldsymbol{\Delta}) \\
&\times \sum_{n,k,m} \delta(j-n-k-m-1) \underbrace{w_m(\beta;\boldsymbol{b})\frac{f^{(m)}(\boldsymbol{\kappa}_3)}{\sigma_0}}_{Quark\ \ ISI} \\
&\times \underbrace{f(\boldsymbol{\kappa})\left|\Psi(\beta;z,\boldsymbol{p}-\boldsymbol{\kappa}_3-\boldsymbol{\kappa}_2)-\Psi(\beta;z,\boldsymbol{p}-\boldsymbol{\kappa}_3-\boldsymbol{\kappa}_2-\boldsymbol{\kappa})\right|^2}_{Hard\ \ excitation} \\
&\times \underbrace{w_k\Big(\frac{C_A}{C_F}(1-\beta);\boldsymbol{b}\Big)\frac{f^{(k)}(\boldsymbol{\kappa}_2)}{\sigma_0}}_{Gluon\ \ FSI} \times \underbrace{w_n\Big((1-\beta);\boldsymbol{b}\Big)\frac{f^{(n)}(\boldsymbol{\kappa}_1)}{\sigma_0}}_{Quark\ \ FSI} \qquad (29)
\end{aligned}
$$

It corresponds to the following counting of cut pomerons: one cut pomeron for hard excitation at the depth β; m cut pomerons for m-fold quasi-elastic intranuclear scattering of the incident quark in the slice $[0,\beta]$; k cut pomerons for k-fold quasi-elastic intranuclear scattering of the final-state gluon in the slice $[\beta,1]$ and n cut pomerons for n-fold quasi-elastic intranuclear scattering of the final-state quark in the slice $[\beta,1]$ of the nucleus. There are more uncut pomerons which describe the coherent distortion of the qg wave function in the slice $[0,\beta]$. The functional form of the hard excitation factor, which can also be re-interpreted as the (non-local) gluon radiation vertex, is preserved. However, upon the convolution with partial collective glue for initial-state and final-state slices of the nucleus, its numerical value will depend on the multiplicity of cut pomerons in the intimal-state and final-state channels.

New features of our AGK rules for an in-nucleus hard QCD are noteworthy: (i) we established simple relationship between cut pomerons, quasi-elastic scattering of partons and collective glue for overlapping nucleons of the Lorentz-contracted nucleus, (ii) the concrete realization of the AGK rules depends on the universality class the specific pQCD subprocess belongs to, (iii) the coherent nuclear distortions described by uncut pomerons persist in all universality classes, (iv) the gluon radiation vertex depends on the multiplicity of cut pomerons.

Hard processes in a nuclear environment must be described by nonlinear $k_\perp$-factorization in terms of the collective nuclear glue defined through the coherent diffractive dijet production. Any application of the familiar linear $k_\perp$-factorization would be entirely erroneous. We reported explicit quadratures for single-jet to dijet spectra from all pQCD subprocesses. Indispensable virtue of nonlinear $k_\perp$-factorization is a nontrivial interplay of incoherent rescatterings and coherent distortions of the dijet wave function. Still another important virtue is that it requires collective glue for slices of a nucleus - nonlinear $k_\perp$-factorization can not be described by a classical gluon field of the whole nucleus. Our connection between the collective glue for overlapping nucleons and the cross section of multiple quasi-elastic scattering of partons gives a simple interpretation of the AGK unitarity rules for excitation of the target nucleus. These unitarity rules can be applied to evaluation of the energy loss and quenching of forward jets.

Acknowledgments

This presentation is is a humble tribute to the memory of a great theorist Vladimir N. Gribov, alias BH, whose technique we used heavily in the reported work. We hope BH would have liked the results. NNN is grateful to J. Nyiri, Yu. Dokshitzer and P. Levai for the invitation to this Workshop.

References

1. N.N. Nikolaev, W. Schäfer, B.G. Zakharov and V.R. Zoller, *J. Exp. Theor. Phys.* **97**, 441 (2003).
2. N.N. Nikolaev, W. Schäfer, B.G. Zakharov and V.R. Zoller, *Phys. Atom. Nucl.* **68**, 661 (2005).
3. N.N. Nikolaev and W. Schäfer, *Phys. Rev.* **D71**, 14023 (2005).
4. N.N. Nikolaev, W. Schäfer and B.G. Zakharov, *Phys. Rev. Lett.* **95**, 221803 (2005).
5. N.N. Nikolaev, W. Schäfer, B.G. Zakharov and V.R. Zoller, *Phys. Rev.* **D72**, 034033 (2005).
6. N.N. Nikolaev, W. Schäfer and B.G. Zakharov, arXiv:hep-ph/0508310.
7. N.N. Nikolaev and V.I. Zakharov, *Sov. J. Nucl. Phys.* **21**, 227 (1975); *Phys. Lett.* **B55**, 397 (1975).
8. V.N. Gribov, *Moscow 1 ITEP school, v.1 'Elementary particles'*, p.65 (1973), e-Print Archive: hep-ph/0006158.
9. O. V. Kancheli, *JETP Lett.* **18**, 274 (1973) [*Pisma Zh. Eksp. Teor. Fiz.* **18**, 465 (1973)].
10. N.N. Nikolaev, W. Schäfer and G. Schwiete, *Phys. Rev.* **D63**, 014020 (2001); *JETP Lett.* **72**, 405 (2000).

11. N.N. Nikolaev, W. Schafer and B.G. Zakharov, paper in preparation.
12. V.A. Abramovsky, V.N. Gribov and O.V. Kancheli, *Yad. Fiz.* **18**, 595 (1973) [*Sov. J. Nucl. Phys.* **18**, 308 (1974)].
13. N.N. Nikolaev and B.G. Zakharov, *Z. Phys.* **C49**, 607 (1991);
14. N.N. Nikolaev, G. Piller and B.G. Zakharov, *J. Exp. Theor. Phys.* **81**, 851 (1995); *Z. Phys.* **A354**, 99 (1996).
15. N.N. Nikolaev and B.G. Zakharov, *J. Exp. Theor. Phys.* **78**, 598 (1994); [*Zh. Eksp. Teor. Fiz.* **105**, 1117 (1994)]; N.N. Nikolaev and B.G. Zakharov, *Z. Phys.* **C64**, 631 (1994).
16. R.J. Glauber, in *Lectures in Theoretical Physics,* edited by W. E. Brittin *et al.*, (Interscience Publishers, Inc., New York, 1959), Vol. 1, p. 315.
17. V.N. Gribov, *Sov. Phys. JETP* **29**, 483 (1969); [*Zh. Eksp. Teor. Fiz.* **56**, 892 (1969)].
18. A. Szczurek, N.N. Nikolaev, W. Schäfer and J. Speth, *Phys. Lett.* **B500**, 254 (2001).
19. N.N. Nikolaev, B.G. Zakharov and V.R. Zoller, *JETP Lett.* **59**, 6 (1994).
20. V. Barone, M. Genovese, N.N. Nikolaev, E. Predazzi and B.G. Zakharov, *Z. Phys.* **C58**, 541 (1993).
21. N.N. Nikolaev, B.G. Zakharov and V.R. Zoller, *Z. Phys.* **A351**, 435 (1995).
22. N.N. Nikolaev and B.G. Zakharov, *Phys. Lett.* **B332**, 177 (1994).
23. J. Jalilian-Marian and Y.V. Kovchegov, *Phys. Rev.* **D70**, 114017 (2004).
24. L.D. McLerran and R. Venugopalan, *Phys. Rev.* **D49**, 2233 (1994); R. Venugopalan, arXiv: hep-ph/0412396. and references therein.
25. C. Adler *et al.* [STAR Collaboration], *Phys. Rev. Lett.* **90**, 082302 (2003).
26. I. Arsene *et al.* [BRAHMS Collaboration], *Phys. Rev. Lett.* **93**, 242303 (2004).
27. J.D. Bjorken, *Phys. Rev.* **D47**, 101 (1993).

RECOMBINATION MODEL FOR HADRON PRODUCTION IN pp AND pA INTERACTIONS

PETER PREŠINSKÝ, NEVA PIŠÚTOVÁ, AND JÁN PIŠÚT

Department of Theoretical Physics,
Faculty of Mathematics, Physics and Informatics,
Comenius University, SK-84248 Bratislava, Slovakia

We reanalyze an old recombination model of hadron production in proton - proton interactions. We start with comparing the results with data on hadron production in pp interactions. The model shows reasonable agreement with data on the production of non - strange hadrons. The model also reasonably describes the production of $s\bar{s}$ pairs, but it has problems with describing the ratio of strange baryons to strange mesons. We also make first attempts to extend the model to proton - nucleus collisions.

1. Introduction

In the seventies we had in Bratislava a small group meeting weekly at working seminars, reading, presenting and discussing papers we found interesting and occasionally trying to write up something. In that time one of our most important source were lectures presented at the ITEP and Leningrad schools of Theoretical physics and Stanford seminars. I still remember in detail how we studied the Gribov lectures on space-time evolution of hadron collisions within the parton model [1]. These were very impressive and influential lectures. Later on when I had the opportunity to meet Vladimir Gribov in persona I realized that he himself and his work was much, much more than these lectures on the foundations of the parton model.

Some of the papers which emerged from the work of our small group concerned the Monte Carlo model of particle production in hadron-hadron collisions [2]. Similar models were conceived elsewhere and some of them earlier. There was the work done by the Leningrad group [3], continued and developed later as described in the monography [4], Budapest work on quarkochemistry [5] and later the ALCOR model [6] and an early work done at Stanford [7]. The most successful, although with a bit different scheme of hadronization was the well known LUND model [8].

We have almost forgotten our old recombination model, but Vlado Černý wished to make some comparison of the model with the NA49 data and the model was put into C++ language and used by his diploma student M. Stankoviansky [9]. A year later I wished to have a look on the strange particle production in our old model and to make an attempt to extend it to pA collisions. The work was done by my diploma student Peter Prešinský and the results are contained in his diploma Thesis [10]. Finally, in order to finish this work in a reasonable time I asked my wife, who used the model in past, for some assistance.

Another reason for playing with this historical model was the enormous increase in the available computer power. It is easy to check that the spead of a standard laptop in 2003 was about as high as the spead of the mainframe computer used in the mid-seventies at CERN. To check that one has just to use any program from the computer library and published also in the journal Comp. Phys. Communications. In that journal it is indicated how much CPU time the code requires. We did that with the Jadach's generator [11] which we have used to generate the longitudinal phase space distributions. In the 70s we had in Bratislava at our disposal a computer which was much slower than the CERN mainframe and we were happy to generate a few hundred events, nowadays one can generate milions of events in a reasonable time on a laptop and this was another reason why to have a look on this old stuff.

The last reason to come back to our old model was the hope to have much more accurate data. Unfortunately this is only partly true, but the data may improve significantly in the future, when the final NA49 data will be published.

In what follows I will briefly mention the assumptions of our old recombination model, present some results obtained by Peter Prešinský and discuss the problems the model apparently encounters in comparison with more recent data.

2. A simple Monte Carlo recombination model of of hadron production in *pp* interactions

The model was based on the following assumptions (for more details see [2]). The two colliding hadrons are supposed to be coherent superpositions of valence quarks, quark-antiquark pairs from the "sea" and gluons. The collision starts by the interaction of wee partons. This destroys the coherence of partons within hadrons and leads to the formation of a compound state.

During this process gluons are converted to $Q\bar{Q}$ pairs. Hadrons are formed by the recombination of quarks and antiquarks from the intermediate state - after the conversion of gluons. The recombination is supposed to be of short range in rapidity and to lead to particles from the octet of pseudoscalar mesons and nonet of vector mesons as well as to baryons from the octet and decuplet and their antiparticles. Local conservation of colour is assumed. In the next step the unstable particles decay, producing particles detected in the experiment.

In our approach we do not try to construct a dynamical model for the first stage of the *pp* collision but we start directly with generating quarks and antiquarks just after the conversion of gluons to $Q\bar{Q}$ pairs and before the recombination of Q's and $\bar{Q}$'s to hadrons. To each configuration is assigned the weight given by the cylindrical phase space multiplied by the appropriate power of a free parameter g, referred to as "coupling constant". This parameter regulates the multiplicity distribution of final state hadrons. The weight of the event contains also a factor pushing the valence quarks to the higher values of Feynman x.

The probability of finding six valence quarks (in the case of *pp* collision) with rapidities and transverse momenta $y_1, \vec{p}_{T1}, ..., y_6, \vec{p}_{T6}$, n quarks with $y_i, \vec{p}_{Ti}$, (i=7, ... n+6), and n antiquarks with $y_j, \vec{p}_{Tj}$, j=n+7, ..., 2n+6 is given by the expression (N=2n+6)

$$dP_N(y_1, \vec{p}_{T1}, ...y_N, \vec{p}_{TN} = KG^{2N}\left(\prod_1^6 |x_i|^{1/2}\right) \cdot$$

$$\cdot\, exp(-\sum_1^N p_{Ti}^2/R^2)\delta(\sum \vec{p}_{Ti})\delta(\sum_1^N p_{Ti})\delta(E - \sum_1^N E_i)\prod_1^N dy_i d^2 p_{Ti} \quad (1)$$

Here K is a normalization constant, $|x_i|^{1/2}$ assigns larger probabilities to larger values of the Feynman x_i and factors $exp(-p_{Ti}^2/R^2)$ provide the transverse momentum cut-off.

After the coupling constant g, another free parameter is the ratio of the production of $s\bar{s}$ quarks by the qluon conversion.

$$\lambda = \frac{< s\bar{s} >}{< u\bar{u} > + < d\bar{d} >} \quad (2)$$

There are still three free parameters to be specified. The former is the ratio of vector to pseudoscalar meson formation out of a particular $Q\bar{Q}$ pair and the later is the ratio of a member of baryon decuplet and baryon octet formation out of a particular triplet of three quarks. Based on our older experience and on the other models, like Ref.[8] we fixed both these ratios at the value of 1. The last free parameter is the transverse momentum cut-off R in Eq. (1). We took the same value of R as in our older work in Ref. [2] so we were left with two free parameters G and λ.

3. Multiplicities, strangeness production and charged particle distributions

We shall now compare the results following from our old model with the data and fix the parameters G and λ. The comparison with charged particle multiplicities in *pp* interactions at 200 GeV/c is shown in Fig. 1.

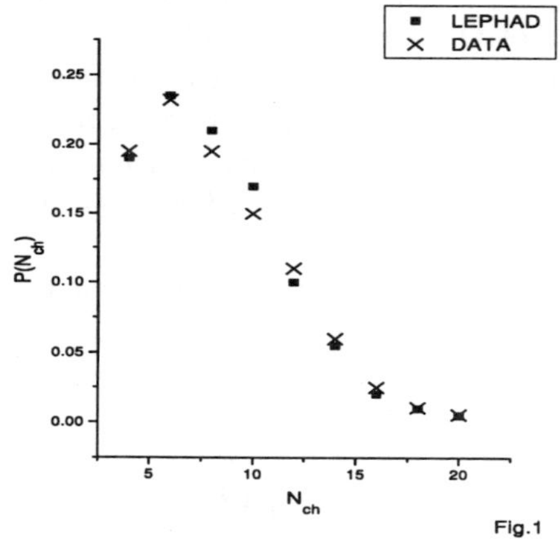

Figure 1. Charged particle multiplicities in *pp* interactions at 200 GeV/c from Ref.[12] compared with our model. Results of the model are denoted as "LEPHAD".

A reasonable agreement is reached with $G \approx 0.45$. In order to determine the value of the parameter λ we have compared data on strange quark production with the model. The average number of $s\bar{s}$ pairs produced is given by the expression [14] :

$$< s\bar{s} > \approx \frac{1}{2}\left(< \Lambda > + < \Sigma > + < \bar{\Lambda} > + < \bar{\Sigma} >\right) + \\ + \frac{1}{2}\left(< K^0 > + < \bar{K}^0 > ++ < K^+ > + < K^- >\right) + \frac{2}{3} < \eta > \quad (3)$$

Comparison with data in Refs.[12,13] leads to $\lambda \approx 0.11$.

Having fixed the parameters we can now proceed to making more detailed comparison of results of the model with data.

4. Rapidity distributions of mesons and baryons in *pp* interactions

In Fig. 2 and 3. we show the comparison of rapidity distribution of positive and negative pions with the data.

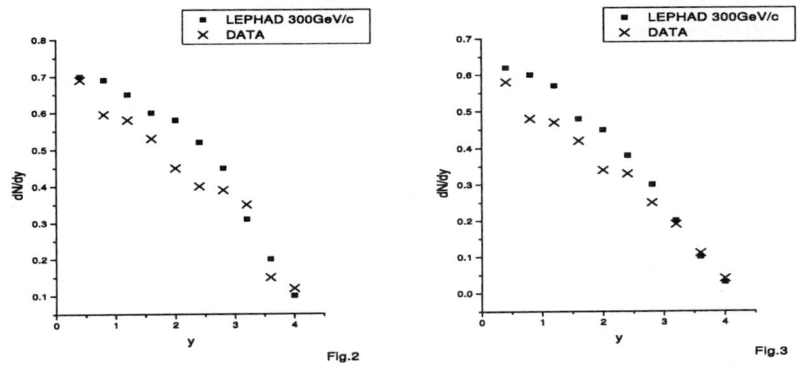

Figure 2-3. Rapidity distribution of positive pions (left hand side) and negative pions (right hand side) in *pp* collisions at 300 GeV/c. Results of the model are denoted as "LEPHAD". Data are from Refs. [12,13].

In Fig. 4 and 5 we compare the results on proton and antiproton rapidity distributions with the data.

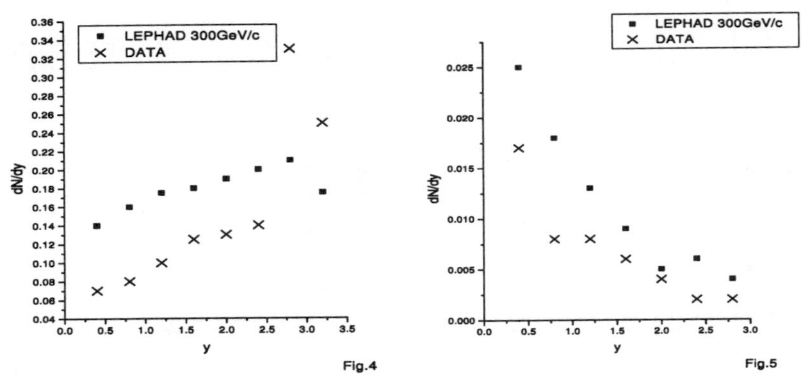

Figure 4-5. Rapidity distribution of protons (left hand side) and antiprotons (right hand side) in *pp* collisions at 300 GeV/c. Results of the model are denoted as "LEPHAD". Data are from Refs. [12,13].

Looking at Figs. 2.-5. one can see, that general trends of rapidity dependence of π^+, π^-, p and $\bar{p}$ are reasonably reproduced by the model, although the model predictions are somewhat above the data in all cases. The data on p and $\bar{p}$ production seem also to be more overshot that the data on π^+ and π^-. The disagreement is not huge and one could think that by changing a bit the value of the "coupling constant" G one could improve the agreement with the data.

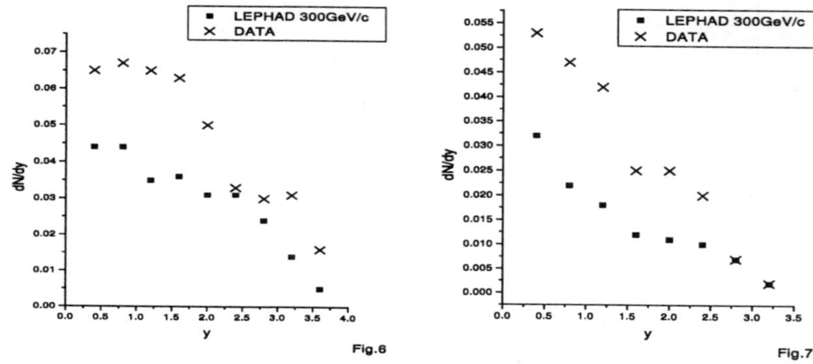

Figure 6-7. Rapidity distribution of K^+ (left hand side) and K^- (right hand side) in pp collisions at 300 GeV/c. Results of the model are denoted as "LEPHAD". Data are from Refs. [12,13].

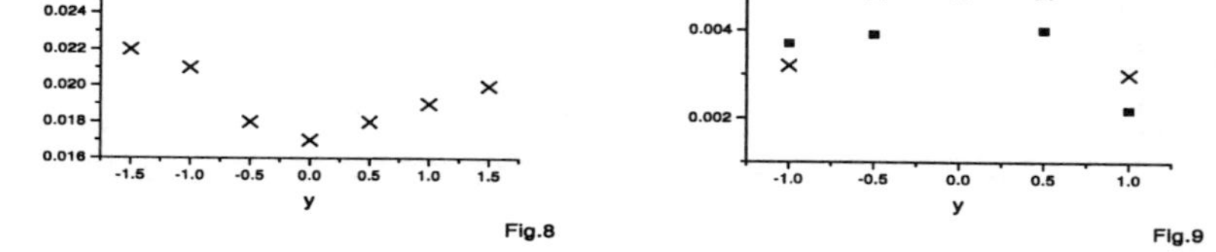

Figure 8-9. Rapidity distribution of Λs (left hand side) and $\bar{\Lambda}$s (right hand side) in pp collisions at 300 GeV/c. Results of the model are denoted as "LEPHAD". Data are from Refs. [12,13].

The disagreement becomes worse in the case of strange particle production. In Figs. 6. and 7. we show the rapidity dependence of K^+ and K^- production. In both cases the model prediction undershoots the data. The rapidity dependence of data on Λ and $\bar{\Lambda}$ production is compared with the model in Figs. 8 and 9. The disagreement is worse, since for the K^+ and K^- the model is bellow the data, whereas for Λ and $\bar{\Lambda}$ the model is above the data, in both cases roughly by a factor of two. This problem cannot be solved by changing the value of λ.

The consequences of the disagreement may lead to two conclusions. Either the normalization of the preliminary data of the CERN NA49 Collaboration as presented in Ref. [13] will be modified in the final publications, or the model is wrong. In the latter case the problem lies probably in the oversimplified recombination scheme and some other effects should be included, like the presence of diquarks, used e.g. in Ref. [8]. Before making definite conclusions it is advisable to wait until definite data are published.

5. Extension of the model to *pA* interactions

The simplest view of the pA collision is probably the one in which one takes it as a series of subsequent collisions of the proton with a tube of nucleons in A. The average number of subsequent collisions is

$$<n>= l \cdot \rho \cdot \sigma \tag{4}$$

Here l is the length of the path of proton in the nucleus at a given value of the impact parameter, ρ is the nuclear density and σ is the non-diffractive nucleon-nucleon cross-section. Eq. (4) is valid only in the approximation of homogeneous nuclear density. More accurate expression corresponding to Woods-Saxon density distribution is a bit more complicated but the principle remains the same.

In the simplest picture one also assumes that in each collision the incident proton looses some fraction of its energy, but its quantum numbers (the valence quarks) are present up to the last collision.

This picture indicates a simple generalization of our old model to pA interactions. The simplest picture of subsequent energy lost is obtained when we assume that in each of subsequent collisions the rapidity of the proton is decreased by a constant value. Although the exact value of the rapidity loss in one collision is exactly unknown, the experimental data of CERN NA49 suggest something like $\Delta y \approx 0.5$. When taking this value it is easy to calculate the energy proton looses in subsequent collisions. Working in the laboratory frame and taking the incident proton of 200 GeV/c we obtain the following energy losses.

- First collision, 78,7 GeV/c
- Second collision, 47,73 GeV/c
- Third collision, 29,15 GeV/c
- Fourth collision, 17,56 GeV/c
- Fifth collision, 10,65 GeV/c
- etc.

Since valence quarks of the incident proton are preserved up to the last collision we have first a few collisions (all except the last one) with only three valence quarks moving left and no valence quarks moving right. In the last collision there are three valence quarks moving left and three valence quarks moving right. For each of the collisions one generates the quarks and antiquarks with weights given by the Eq. (1). In the next step one merges the systems formed in subsequent collisions to one system of quarks and antiquarks and then the whole system is recombined to hadrons according to the same scheme as in the case of *pp* collision.

The Eq. (1) is valid only in the cms of the proton-nucleon collision but this leads only to technical problems. One has to shift for each of subsequent collisions to and back from the laboratory system to the c.m. system and vice versa.

The preliminary results of the collisions of a proton with a few nucleons look quite reasonably. To continue in this direction makes however sense only if the model will not be excluded by the normalization of the final NA49 data.

Acknowledgements

One of the authors (J.P.) is indebted to Julia Nyiri and Peter Lévai for the invitation to the seminar organized on the occassion of the 75th birthday of the great man and great physicist Vladimir N. Gribov. I am glad that I have had also the occasion to meet at this seminar friends some of whom I have not met for years. I would also like to thank to Vlado Černý and Peter Lichard for many discussions and collaboration on our old Monte Carlo model.

References

1. V.N. Gribov, *Elementary particles, Proceedings of the first ITEP School on Theoretical Physics*, Atomizdat, Moscow, 1973, Vol. I., p.65.
2. V. Černý, P. Lichard and J. Pišút, *Phys. Rev.* **D16**, 2822 (1977); *Phys. Rev.* **D18**, 2409 (1978); *Phys. Rev.* **D18**, 4052 (1978).
3. V.V. Anisovich and V.M. Shekhter, *Nucl. Phys.* **B55**, 455 (1973); V.V. Anisovich *et al.*, *Nucl. Phys.* **B55**, 474 (1973).
4. V.V. Anisovich, M.N. Kobrinsky, J. Nyiri and Yu.M. Shabelski, *Quark Model and High Energy Collisions*, World Scientific, Singapore, 1985.
5. I. Montvay and J. Zimányi, *Nucl. Phys.* **A316**, 490 (1979).
6. T.S. Biró, P. Lévai and J. Zimányi, *Phys. Lett.* **B347**, 6 (1995); P. Lévai and J. Zimányi, *Phys. Lett.* **B304**, 203 (1993).
7. J.D. Bjorken and G.R. Farrar, *Phys. Rev.* **D9**, 1449 (1974).
8. B. Andersson, G. Gustafson and B. Nilsson-Almquist, *Nucl. Phys.* **B281**, 289 (1987); B. Andersson, G. Gustafson, G. Ingelman and T. Sjöstrand, *Phys. Rep.* **97**, 31 (1983).
9. M. Stankoviansky, Diploma Thesis, 2003, Department of Theoretical Physics, Faculty of Mathematics and Physics, Comenius University, Bratislava.
10. P. Prešinský, *Production of strange particles in pA collisions in a simple recombination model.* Diploma Thesis 2005, Comenius University, Bratislava.
11. S. Jadach, *Comput. Phys. Commun.* **9**, 297 (1975).
12. M. Jacob, *Phys. Rep.* **315**, 7 (1999).
13. D. Barna, *Strange particle production in nuclear collisions at CERN NA49*, Thesis, Central Research Institute for Physics, Academy of Science, Hungary, Budapest, 2003.
14. R. Lietava and J. Pišút, *Eur. J. Phys.* **C5**, 135 (1998).

BOSE-EINSTEIN OR HBT CORRELATION SIGNATURE OF A SECOND ORDER QCD PHASE TRANSITION

T. CSÖRGŐ

MTA KFKI RMKI, P.O. Box 49, Budapest 114, 1525, Hungary
e-mail: csorgo@sunserv.kfki.hu

S. HEGYI

MTA KFKI RMKI, P.O. Box 49, Budapest 114, 1525, Hungary
e-mail: hegyi@rmki.kfki.hu

T. NOVÁK

Radboud University, 1 Toernooiveld, Nijmegen, 6525 ED, The Netherlands,
e-mail: novakt@hef.ru.nl

W. A. ZAJC

Dept. Physics, Columbia University, 538 W 120th Street, New York, NY 10027, USA
e-mail: zajc@nevis.columbia.edu

For particles emerging from a second order QCD phase transition, we show that a recently introduced shape parameter of the Bose-Einstein correlation function, the Lévy index of stability equals to the correlation exponent - one of the critical exponents that characterize the behavior of the matter in the vicinity of the second order phase transition point. Hence the shape of the Bose-Einstein / HBT correlation functions, when measured as a function of bombarding energy and centrality in various heavy ion reactions, can be utilized to locate experimentally the second order phase transition and the critical end point of the first order phase transition line in QCD.

1. Introduction

The study fractal phenomena was initiated in high energy particle and nuclear physics by Bialas and Peschanski in ref. [1], with the motivation of searching for a second order phase transition by studying intermittency or the the power-law behavior of moments of the multiplicity distribution in narrowing bins of the momentum space, see refs. [2,3] for excellent reviews on this topic.

The mathematical properties of Bose-Einstein correlation functions for Lévy stable sources were written up by three of us in refs. [4,5], and are recapitulated in the next section. In ref. [6] we have added a physical interpretation and showed, that in case of jet physics, the fractal properties of QCD cascades can naturally be measured by the Lévy index of stability of the Bose-Einstein correlation functions. Our analytic results were similar in spirit to the numerical investigations of Wilk and collaborators in ref. [7]. Note that these correlations are frequently referred to as Hanbury Brown - Twiss or HBT correlations in the literature of heavy ion physics.

Bialas realized, that Bose-Einstein correlations and intermittency might be deeply connected [8], and considered a distribution of Gaussians where the radius parameter of the Gaussian has a power-law distribution, thus giving a way to the study fractals in coordinate space with the help of Bose-Einstein correlations. Brax and Peschanski were the first to introduce Lévy distributions, in momentum space, to multiparticle production in high energy physics [9]. They have suggested to use the measured value of the Lévy index of stability to signal quark gluon plasma production in heavy ion physics. Here we reconnect these seemingly different topics, and show how the excitation function of the shape parameter of the correlation function can be utilized to locate experimentally the critical end-point of QCD.

2. Bose-Einstein correlations & Lévy stable sources

The two-particle Bose-Einstein correlation function is defined with the help of the two-particle and single-particle invariant momentum distributions as:

$$C_2(\mathbf{k}_1,\mathbf{k}_2) = \frac{N_2(\mathbf{k}_1,\mathbf{k}_2)}{N_1(\mathbf{k}_1)N_1(\mathbf{k}_2)}. \tag{1}$$

If long-range correlations can be neglected or corrected for, and if the short-range correlations are dominated by Bose-Einstein correlations, this two-particle Bose-Einstein correlation function is related to the Fourier-transformed source distribution. For clarity, let us consider the case of a one-dimensional, factorized source,

$$S(x,k) = f(x)\,g(k). \tag{2}$$

In this case [4,5], the Bose-Einstein correlation function is

$$C_2(k_1,k_2) = 1 + |\tilde{f}(q)|^2, \tag{3}$$

where the Fourier transformed source density (often referred to as the *characteristic function*) and the relative momentum are defined as

$$\tilde{f}(q) = \int \mathrm{d}x \exp(iqx)\, f(x), \qquad q = k_1 - k_2. \tag{4}$$

For the case of the jets decaying to jets to jets and so on, as well as at a second order phase transition, where fluctuations appear on all possible scales with a power-law tailed distribution, the final position of a particle is given by a large number of position shifts, hence the distribution of the final position x is obtained as a convolution,

$$x = \sum_{i=1}^{n} x_i, \qquad f(x) = \int \Pi_{i=1}^{n} dx_i \Pi_{j=1}^{n} f_j(x_j)\, \delta(x - \sum_{k=1}^{n} x_k). \qquad (5)$$

In the case of particle emission from QCD jets, that the fractal defining the particle emission is infrared stable: adding one more, very soft gluon does not change the resulting source distributions. A similar property holds for systems at a second order phase transition: the system becomes invariant under a renormalization group transformation. Bose-Einstein correlation functions for such particle emitting sources were evaluated recently by three of us, which we summarize here following refs. [4,5].

Various forms of the Central Limit Theorem state, that under certain conditions, the distribution of the sum of large number of random variables converges (for $n \to \infty$) to a limit distribution. In case of "normal" elementary processes, that have finite means and variances, the limit distribution of their sum is a Gaussian. In case of random motion in a thermal medium, such position distribution corresponds to normal diffusion. However, near a second order phase transition point, fluctuations appear on all scales and the variance of the elementary process diverges, corresponding to the so-called anomalous diffusion. In this case, the system still may be invariant under convolution, and the shape of the limit distribution becomes independent from the number of elementary steps.

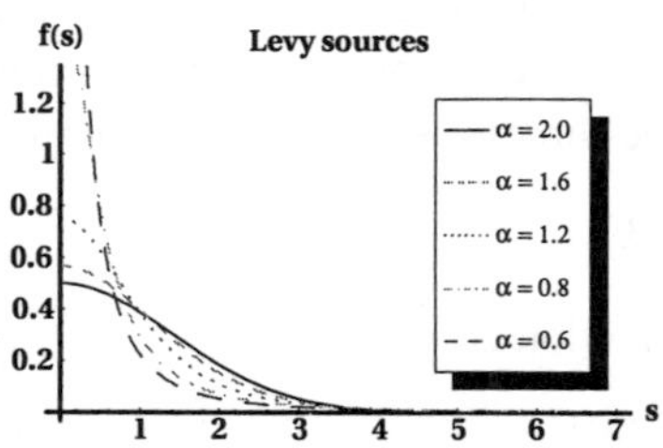

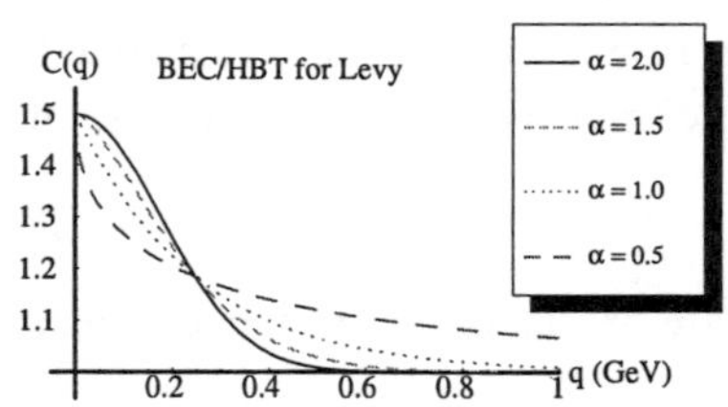

Figure 1. (left) Source functions for univariate symmetric Lévy laws, as a function of the dimensionless variable $s = r/R$, for various values of the Lévy index of stability, α. (right) Bose-Einstein correlation (or HBT) correlation functions for univariate symmetric Lévy laws, for a fixed scale parameter of $R = 0.8$ fm and various values of the Lévy index of stability, α.

Stable distributions are precisely those limit distributions that can occur in Generalized Central Limit theorems. Their study was begun by the mathematician Paul Lévy in the 1920's. The stable distributions can be given in terms of their characteristic functions, as the Fourier transform of a convolution is a product of the Fourier-transforms,

$$\tilde{f}(q) = \prod_{i=1}^{n} \tilde{f}_i(q), \tag{6}$$

and limit distributions appear when the convolution of one more elementary process does not change the shape of the limit distribution, but it results only in a modification of the parameters of the limit distribution. The characteristic function of univariate and symmetric stable distributions is

$$\tilde{f}(q) = \exp\left(iq\delta - |\gamma q|^{\alpha}\right), \tag{7}$$

where the support of the density function $f(x)$ is $(-\infty, \infty)$. Deep mathematical results imply that the index of stability, α, satisfies the inequality $0 < \alpha \leq 2$, so that the source distribution be always positive. These Lévy distributions are indeed stable under convolutions, in the sense of the following relations:

$$\tilde{f}_i(q) = \exp\left(iq\delta_i - |\gamma_i q|^{\alpha}\right), \qquad \prod_{i=1}^{n} \tilde{f}_i(q) = \exp\left(iq\delta - |\gamma q|^{\alpha}\right), \tag{8}$$

$$\gamma^{\alpha} = \sum_{i=1}^{n} \gamma_i^{\alpha}, \qquad \delta = \sum_{i=1}^{n} \delta_i. \tag{9}$$

Thus the Bose-Einstein correlation functions for uni-variate, symmetric stable distributions (after a core-halo correction, and a re-scaling) read as

$$C(q;\alpha) = 1 + \lambda \exp(-|qR|^{\alpha}). \tag{10}$$

Refs. [4,5] discuss further examples and details and generalize these results to three dimensional, hydrodynamically expanding, core-halo type sources, as well as to three-particle correlations.

3. The anomalous dimension of QCD jets and BE/HBT

In QCD, jets emit jets that emit additional jets and so on. The resulting fractal structure of QCD jets was related to intermittency and power-law dependence of multiplicity moments on the bin-size in momentum space with the help of a beautiful geometric interpretation of the color dipole picture in refs. [10,11,12], and an infrared stable measure on the parton states, related to the hadronic multiplicity distribution. These ideas were developed further in refs. [13,14,15,16] as well as in refs. [17,18].

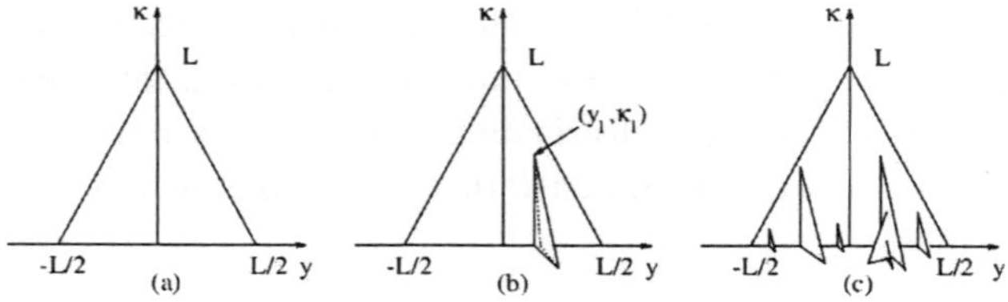

Figure 2. The phase-space of QCD jets in the (y,κ) plane, where $\kappa = \log(k_t^2)$. (a) The phase space available for a gluon emitted by a high energy $q\bar{q}$ system is a triangular region in the (y,κ) plane. (b) If one gluon is emitted at (y_1,κ_1), the phase space for a second (softer) gluon is given by the area of this folded surface. (c) The total gluonic phase space can be described by this multifaceted fractal surface[10,11,12].

A high energy $q\bar{q}$ system radiates gluons according to the dipole formula

$$dn = \frac{3\alpha_s}{4\pi^2}\frac{dk_\perp^2}{k_\perp^2}dy\,d\phi, \tag{11}$$

hence the phase-space for the emission of a gluon is given by the relation

$$|y| \leq \frac{1}{2}\ln(s/k_\perp^2), \tag{12}$$

which corresponds to the triangular region in a $(y, \ln k_\perp^2)$ diagram as shown in Fig. 2 (a). If two gluons are emitted, then the distribution of the hardest gluon is described by eq. (11). The distribution of the second, softer, gluon corresponds to two dipoles, the first is stretched between the quark and the first gluon, and the second between the first gluon and the anti-quark. The phase-space available for the second gluon corresponds to the folded surface in Fig. 2 (b), with the constraint $k_{\perp,2}^2 < k_{\perp,1}^2$, as the first gluon is assumed to be the hardest one. This procedure can be generalized so that the emission of a third, still softer gluon corresponds to radiation from three color dipoles, with n gluons emitted already the emission of the $n+1$-th gluon is given by a chain of $n+1$ dipoles. Thus, with many gluons, the gluonic phase space can be represented by a multi-faceted surface as illustrated in Fig. 2 (c). Each gluon adds a fold to the surface, which increases the phase-space for softer gluons. (Note, that in this process the recoils are neglected, as is normal in leading log approximation). Due to its iterative nature, the process generates a Koch-type fractal curve at the base-line. The length of this base-line of the partonic structure on Figure 2 (c) is proportional to the particle multiplicity. This curve is longer, when studied with higher resolution: it is a fractal curve, embedded into the four-dimensional energy-momentum space, characterized by the fractal dimension

$$d_f = 1 + \sqrt{\frac{3\alpha_s}{2\pi}}, \tag{13}$$

or one plus the anomalous dimension of QCD [10,11,12]. With the help of the Lund string fragmentation picture, this fractal in momentum space is mapped into a fractal in coordinate space, and the constant of conversion is the hadronic string tension, $\kappa \approx 1$ GeV/fm. This mapping does not change the fractal properties of the curve.

A walk, where the length of the steps is given by a Lévy distribution, and the direction of the steps is random, corresponds to a fractal curve, in physical terms it can be interpreted as the path of a test particle performing a generalized Brownian motion. This motion is referred to as anomalous diffusion and the probability that the test particle diffuses to distances r greater than a certain value of $|s|$ is given by $P(r > |s|) \propto |s|^{-\alpha}$. This relation is valid for anomalous diffusion in any dimensions. Thus the Lévy index of stability α is the fractal dimension of the trajectory of the corresponding anomalous diffusion [19]. When we apply this result to QCD, there are two key considerations.

First, if gluon radiation is neglected, the $q\bar{q}$ system hadronizes as a 1+1 dimensional hadronic string, which has no fractal structure. If the gluon emission is switched on, the emission of gluon n from one of the n dipoles corresponds to a step of an anomalous diffusion in the plane transverse to the given dipole. Hence the anomalous dimension of QCD equals to the Lévy index of stability of this anomalous diffusion,

$$\sqrt{\frac{3\alpha_s}{2\pi}} = \alpha_{\text{Lévy}}. \tag{14}$$

Second, data on Bose-Einstein correlations are often determined in terms of the invariant momentum difference $Q_{\text{inv}} = \sqrt{-(p_1 - p_2)^2}$. Bose-Einstein correlation functions that depend on this invariant momentum difference can be obtained within the framework of the so-called τ-model. This model assumes a broad proper-time distribution, $H(\tau)$ and very strong correlations between coordinate and momentum in all directions, $x^\mu/\tau \propto p^\mu/m_{(t)}$. Hence $(x_1 - x_2)(p_1 - p_2) \propto \tau Q_{\text{inv}}^2$, see refs. [20,21] for details. In this case, the Bose-Einstein correlation function measures the Fourier-transformed proper-time distribution $\tilde{H}$ in the following, unusual manner:

$$C_2(Q_{\text{inv}}) \simeq 1 + \lambda Re \tilde{H}^2 \left(\frac{Q_{\text{inv}}^2}{2m_{(t)}} \right), \tag{15}$$

where $m_{(t)}$ stands for the (transverse) mass of the pair for (two)- or more jet events. From this relation it follows, that $\alpha_{\text{BEC}} = 2\alpha_{\text{Lévy}}$. Thus we find the following relationship between the running QCD coupling constant α_s and the exponent of an invariant relative momentum dependent Bose-Einstein correlation function α_{BEC}:

$$\alpha_s = \frac{\pi}{6}\alpha_{\text{BEC}}^2. \tag{16}$$

In ref. [6] we have compared this leading log result to NA22 and UA1 correlation data of refs. [22,23] and found a reasonable agreement with these data.

4. Bose-Einstein correlations at a second order QCD phase transition

The main motivation behind the experimental and theoretical program of high energy physics is to study the phase diagram of hot and dense hadronic matter. According to recent lattice QCD calculations at finite temperature and baryon density, there exist a line of first order phase transitions that separates the hadronic and the quark-gluon plasma (QGP) state. This line of the first order phase transitions ends at the critical end-point (CEP), where the transition from hadron gas to QGP becomes a second order phase transition. Recent lattice QCD calculations located [24] this CEP at $T_E = 162 \pm 2$ MeV and $\mu_E = 360 \pm 40$ MeV. Below these baryochemical values, the transition from a hadron gas to a QGP becomes a cross-over, and at vanishing net baryon density the critical temperature becomes $T_c = 164 \pm 2$ MeV (the errors are statistical only). In this calculation, the quark masses were already at the physical value, but the continuum extrapolation was missing. S. Katz presented improvements at the Quark Matter 2005 conference [25], using physical quark masses and working towards the continuum extrapolation. He reported $T_c = 189 \pm 8$ MeV for the critical temperature at $\mu_B = 0$.

At the CEP, the second order phase transition is characterized by the fixed point of the renormalization group transformations. In a quark-gluon plasma, the vacuum expectation value of the quark condensate $c = \langle \overline{q}q \rangle$ vanishes , while in the hadronic phase, this vacuum expectation value becomes non-zero. The correlation function of the order parameter is defined as $\rho(R) = \langle c(r+R)c(r) \rangle - \langle c \rangle^2$ and measures the spatial correlation between the pions. At the CEP, this correlation function decays as

$$\rho(R) \propto R^{-(d-2+\eta)}, \tag{17}$$

a power-law. The parameter η is called as the exponent of the correlation function.

For Lévy stable sources, corresponding to an anomalous diffusion with large fluctuations in coordinate space, the correlation between the initial and actual positions decays also as a power-law, where the exponent is given by the Lévy index of stability α as

$$\rho(R) \propto R^{-(1+\alpha)}. \tag{18}$$

As we are considering a QCD phase transition in a $d = 3$ three-dimensional coordinate space we find that the correlation exponent equals to the Lévy index of stability, $\alpha = \eta$. Stephanov, Rajagopal and Shuryak pointed out [26], that the universality class of the second order QCD phase transition is that of the 3d Ising model. For this universality class, the correlation exponent has been determined by Rieger [27] as

$$\alpha(\text{Lévy}) = \eta(\text{3d Ising}) = 0.50 \pm 0.05. \tag{19}$$

Fig. 1 indicates that the change in the shape of the correlation function is rather significant, if α decreases from its Gaussian value of 2 to 0.5, its characteristic value at the 2nd order QCD phase transition point. Fig. 3 illustrates how this shape parameter of the Bose-Einstein correlation function may depend on the relative temperature near the critical point. Hence studying the *shape* parameter of the two-particle Bose-Einstein or HBT correlation functions as a function of the bombarding energy or the centrality of the heavy ion collisions, a previously unknown tool is obtained to determine if the pions are emitted from the neighborhood of the critical end point of the QCD phase diagram.

Furthermore, based on an universality class argument, we have determined that the second order QCD phase transition at the critical end point will be signaled with the value of $\alpha = 0.5$, a very spiky Bose-Einstein correlation function indeed.

5. Conclusions

We have recapitulated earlier results that indicate, that the general shape of the Bose-Einstein or HBT correlation functions is a stretched exponential or Lévy stable form, where the Lévy index of stability becomes a new shape parameter of the correlation function with $0 < \alpha \leq 2$ and the popular Gaussian parameterization corresponds to the $\alpha = 2$ particular, special case. Then we have studied two physically interesting examples.

In case of particle emission from jets, we have recapitulated the connection between the stability index of the Bose-Einstein/HBT correlation functions and the running coupling constant of QCD.

We have also considered a scenario, when the power-law tail of a Lévy distribution of the particle emission *in the coordinate space* appears due to a second-order QCD phase transition. In this case, the Lévy index of stability of the Bose-Einstein or HBT correlation function was shown to be equal to the correlation exponent of QCD. This value is known to be 0.5 ± 0.05 from universality class

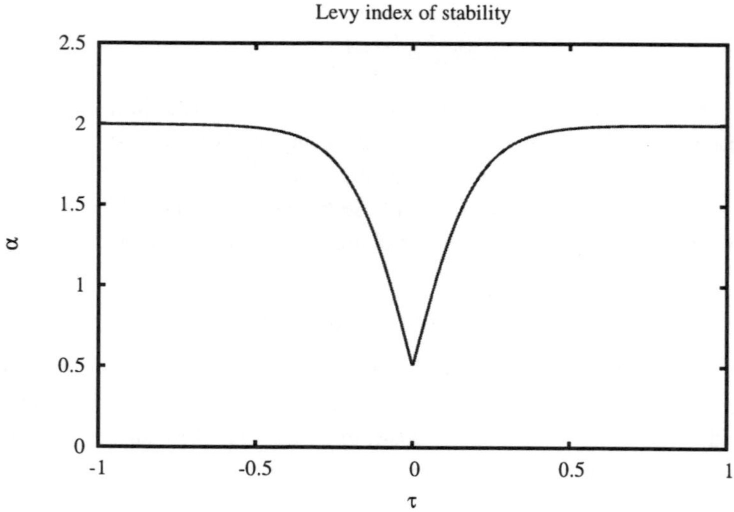

Figure 3. Illustration of the behavior of the Lévy index of stability of Bose-Einstein correlations as a function of the dimensionless temperature variable $\tau = (T - T_c)/T_c$ in the neighborhood of the critical endpoint of the 1st order phase transition line in QCD. At the critical endpoint, the phase transition becomes 2nd order and the Lévy index of stability decreases to the correlation exponent of QCD. As this transition has the same universality class as that of the 3d Ising model, one expects a decrease from the $\alpha \approx 2$ values that are characteristic to a Boltzmann gas and normal diffusion to $\alpha = 0.5$, corresponding to the correlation exponent of QCD at the critical endpoint. As shown in Fig. 1, such a change in the shape parameter makes the Bose-Einstein correlation functions much sharper than a simple Gaussian, so the spiky structure of the correlation function could be used to search for this point experimentally.

considerations. Hence by measuring the excitation function of the Lévy index of stability (the shape parameter of the two-particle Bose-Einstein or HBT correlation functions), one can experimentally determine the bombarding energy and centrality range where a heavy ion collision hits the critical end point of QCD. Clearly, more work is necessary to check to what extent this interesting effect can be masked by the decays of various resonances, hydrodynamic expansion, and by the time evolution of the particle emitting source between the second order phase transition point and the freeze-out temperature.

Acknowledgments

It is our pleasure to acknowledge the inspiring discussions with A. Bialas, R. Glauber, W. Kittel and R. Peschanski. This research was supported by the NATO Collaborative Linkage Grant PST.CLG.980086, by the Hungarian - US MTA OTKA NSF grant INT0089462 and by the OTKA grants T038406 and T049466.

Dedicated to the memory of V. N. Gribov

References

1. A. Bialas and R. Peschanski, *Nucl. Phys.* **B273**, 703 (1986).
2. A. Bialas, *Nucl. Phys.* **A525**, 345 (1991).
3. E.A. De Wolf, I.M. Dremin and W. Kittel, *Phys. Rep.* **270**, 1 (1996).
4. T. Csörgő, S. Hegyi and W.A. Zajc, *Eur. Phys. J.* **C36**, 67 (2004).
5. T. Csörgő, S. Hegyi and W.A. Zajc, arXiv:nucl-th/0402035.
6. T. Csörgő, S. Hegyi, T. Novák and W.A. Zajc, *Acta Phys. Polon.* **B36**, 329 (2005).
7. O.V. Utyuzh, G. Wilk and Z. Wlodarczyk, *Phys. Rev.* **D61**, 034007 (2000).
8. A. Bialas, *Acta Phys. Polon.* **B23**, 561 (1992).
9. P. Brax and R. Peschanski, *Phys. Lett.* **B253**, 225 (1991).
10. P. Dahlqvist, B. Andersson and G. Gustafson, *Nucl. Phys.* **B328**, 76 (1989).
11. G. Gustafson and A. Nilsson, *Nucl. Phys.* **B355**, 106 (1991).
12. G. Gustafson, *Proc Int. Workshop on Correlations and Multiparticle Production, Marburg, West Germany, 1990*, World Scientific, Singapore, 1991, eds. M. Plümer, S. Raha and R.M. Weiner.
13. G. Gustafson and A. Nilsson, *Z. Phys.* **C52**, 533 (1991).
14. G. Gustafson, *Nucl. Phys.* **B392**, 251 (1993).
15. B. Andersson, G. Gustafson and J. Samuelsson, *Nucl. Phys.* **B463**, 217 (1996).
16. Y.L. Dokshitzer and I.M. Dremin, *Nucl. Phys.* **B402**, 139 (1993).
17. W. Ochs and J. Wosiek, *Phys. Lett.* **B305**, 144 (1993).
18. W. Ochs and J. Wosiek, *Z. Phys.* **C68**, 269 (1995).
19. V. Seshardi and B.J. West, *Proc. Nat. Acad. Sci. USA* **79**, 4501 (1982).
20. T. Csörgő and J. Zimányi, *Nucl. Phys.* **A517**, 588 (1990).
21. T. Csörgő, in *Proc. 32nd International Symposium on Multiparticle Dynamics (ISMD 2002)*, Alushta, Ukraine, Sep 2002, in **Alushta 2002, Multiparticle dynamics**, p.38 [arXiv:hep-ph/0301164].
22. N.M. Agababian *et al.* [EHS/NA22 Coll.], *Z. Phys.* **C59**, 405 (1993).
23. N. Neumeister *et al.* [UA1-Min. Bias-Coll.], *Z. Phys.* **C60**, 633 (1993).
24. Z. Fodor and S.D. Katz, JHEP **0404**, 050 (2004) [arXiv:hep-lat/0402006].
25. S.D. Katz, arXiv:hep-ph/0511166, see `www.kfki.hu/events/hun/qm2005/plenary/Katz/0930_aug5GlobeKatzt.pdf`
26. M.A. Stephanov, K. Rajagopal and E.V. Shuryak, *Phys. Rev. Lett.* **81**, 4816 (1998).
27. H. Rieger *Phys. Rev.* **B52**, 6659 (1995).

PROBING THE EQUATION OF STATE OF NEUTRON-RICH MATTER WITH HEAVY-ION COLLISIONS

BAO-AN LI

Department of Chemistry and Physics, P.O. Box 419,
Arkansas State University, State University, AR 72467-0419, USA
E-mail: Bali@astate.edu

The equation of state (EOS) of neutron-rich matter plays many important roles in both nuclear physics and astrophysics. However, it is still very poorly known, especially its isospin dependent part, namely the symmetry energy. Nuclear reactions with neutron-rich nuclei, stable and/or radioactive, provide a great opportunity to pin down the EOS of isospin asymmetric nuclear matter. In this talk, we report some recent theoretical progress in this field within an isospin and momentum dependent transport model (IBUU04). Using in-medium nucleon-nucleon cross sections consistent with the nuclear mean field used in the transport model, a symmetry energy of $E_{sym}(\rho) \approx 31.6(\rho/\rho_0)^{0.69}$ was found most acceptable compared with both the MSU isospin diffusion data and the presently acceptable neutron-skin thickness in ^{208}Pb. The isospin dependent part $K_{asy}(\rho_0)$ of isobaric nuclear incompressibility was further narrowed down to -500 ± 50 MeV. Predictions on several observables sensitive to the density dependence of the symmetry energy at supranormal densities accessible at GSI and the planned Rare Isotope Accelerator (RIA) are also made.

1. Introduction

The Equation of State (EOS) of isospin asymmetric nuclear matter can be written within the well-known parabolic approximation, which has been verified by all many-body theories, as

$$E(\rho, \delta) = E(\rho, \delta = 0) + E_{sym}(\rho)\delta^2 + \mathcal{O}(\delta^4), \quad (1)$$

where $\delta \equiv (\rho_n - \rho_p)/(\rho_p + \rho_n)$ is the isospin asymmetry and $E_{sym}(\rho)$ is the density-dependent nuclear symmetry energy. The latter is very important for many interesting astrophysical problems [1], the structure of radioactive nuclei [2,3] and heavy-ion reactions [4,5,6,7]. Unfortunately, the density dependence of symmetry energy $E_{sym}(\rho)$, especially at supranormal densities, is still poorly known. Predictions based on various many-body theories diverge widely at both low and high densities. In fact, even the sign of the symmetry energy above $3\rho_0$ remains uncertain [8]. Fortunately, heavy-ion reactions, especially those induced by radioactive beams, provide a unique opportunity to pin down the density dependence of nuclear symmetry energy in terrestrial laboratories. Significant progress in determining the symmetry energy at subnormal densities has been made recently both experimentally and theoretically [9,10,11,12,13]. High energy radioactive beams to be available at GSI and RIA will allow us to determine the symmetry energy at supranormal densities.

To extract information about the EOS of neutron-rich matter from nuclear reactions induced by radioactive beams, one needs reliable theoretical tools. Transport models are especially useful for this purpose. Especially for central, energetic reactions at RIA and GSI, transport models are the most useful tool for understanding the role of isospin degree of freedom in the reaction dynamics and extract information about the EOS of neutron-rich matter. Significant progress has been made recently in improving semi-classical transport models for nuclear reactions induced by neutron-rich nuclei. While developing practically implementable quantum transport theories is a long term goal, applications of the semi-classical transport models have enabled us to learn a great deal of interesting physics about neutron-rich matter from heavy-ion reactions. In the following we outline the major ingredients of an isospin and momentum dependent transport model applicable for heavy-ion reactions induced by both stable and radioactive beams [14]. This model has been found very useful in understanding a number of new phenomena associated with the isospin degree of freedom in heavy-ion reactions. Based on applications of this model, we highlight here the most recent progress in determining the symmetry energy at subnormal densities and present our predictions on several most sensitive probes of the symmetry energy at supranormal densities.

2. Major ingredients of an isospin and momentum dependent transport model for nuclear reactions induced by neutron-rich nuclei

Crucial to the extraction of critical information about the $E_{\rm sym}(\rho)$ is to compare experimental data with transport model calculations. We outline here the major ingredients of the version IBUU04 of an isospin- and momentum-dependent transport model for nuclear reactions induced by radioactive beams [14]. The single nucleon potential is one of the most important inputs to all transport models. Both the isovector (symmetry poten-

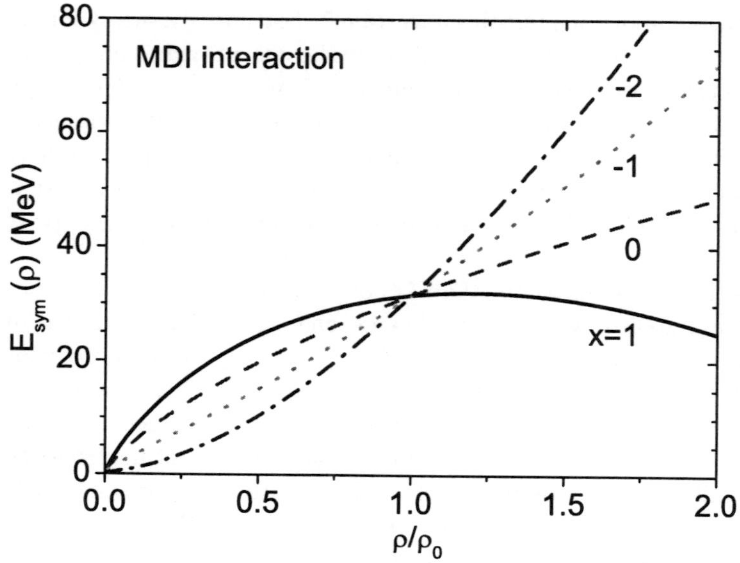

Figure 1. Density dependence of the symmetry energy for four x parameters.

tial) and isoscalar parts of this potential should be momentum dependent due to the non-locality of strong interactions and the Pauli exchange effects in many-fermion systems [15]. In the IBUU04, we use a single nucleon potential derived from the Hartree-Fock approximation using a modified Gogny effective interaction (MDI) [16], i.e.,

$$\begin{aligned} U(\rho,\delta,\vec{p},\tau,x) &= A_u(x)\frac{\rho_{\tau'}}{\rho_0} + A_l(x)\frac{\rho_\tau}{\rho_0} \\ &+ B(\frac{\rho}{\rho_0})^\sigma(1-x\delta^2) - 8\tau x\frac{B}{\sigma+1}\frac{\rho^{\sigma-1}}{\rho_0^\sigma}\delta\rho_{\tau'} \\ &+ \frac{2C_{\tau,\tau}}{\rho_0}\int d^3p'\frac{f_\tau(\vec{r},\vec{p'})}{1+(\vec{p}-\vec{p'})^2/\Lambda^2} \\ &+ \frac{2C_{\tau,\tau'}}{\rho_0}\int d^3p'\frac{f_{\tau'}(\vec{r},\vec{p'})}{1+(\vec{p}-\vec{p'})^2/\Lambda^2}. \end{aligned} \quad (2)$$

In the above $\tau = 1/2$ $(-1/2)$ for neutrons (protons) and $\tau \neq \tau'$; $\sigma = 4/3$; $f_\tau(\vec{r},\vec{p})$ is the phase space distribution function at coordinate $\vec{r}$ and momentum $\vec{p}$. The parameters $A_u(x), A_l(x), B, C_{\tau,\tau}, C_{\tau,\tau'}$ and Λ were obtained by fitting the momentum-dependence of the $U(\rho,\delta,\vec{p},\tau,x)$ to that predicted by the Gogny Hartree-Fock and/or the Brueckner-Hartree-Fock calculations, the saturation properties of symmetric nuclear matter and the symmetry energy of 30 MeV at normal nuclear matter density $\rho_0 = 0.16$ fm^{-3} [16]. The incompressibility K_0 of symmetric nuclear matter at ρ_0 is set to be 211 MeV. The parameters $A_u(x)$ and $A_l(x)$ depend on the x parameter according to

$$A_u(x) = -95.98 - x\frac{2B}{\sigma+1}, \quad A_l(x) = -120.57 + x\frac{2B}{\sigma+1}. \tag{3}$$

The parameter x can be adjusted to mimic predictions on the $E_{sym}(\rho)$ by microscopic and/or phenomenological many-body theories. The last two terms contain the momentum-dependence of the single-particle potential. The momentum dependence of the symmetry potential stems from the different interaction strength parameters $C_{\tau,\tau'}$ and $C_{\tau,\tau}$ for a nucleon of isospin τ interacting, respectively, with unlike and like nucleons in the background fields. More specifically, we use $C_{unlike} = -103.4$ MeV and $C_{like} = -11.7$ MeV. As an example, shown in Fig. 1 is the density dependence of the symmetry energy for $x = -2, -1, 0$ and 1.

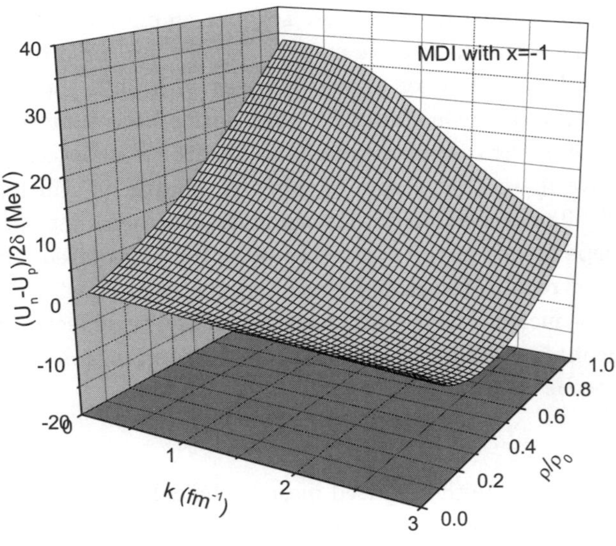

Figure 2. Symmetry potential as a function of momentum and density for MDI interaction with $x = -1$.

Systematic analyses of a large number of nucleon-nucleus and (p,n) charge exchange scattering experiments at beam energies below about 100

MeV indicate undoubtedly that the symmetry potential at ρ_0, i.e., the Lane potential, decreases approximately linearly with increasing beam energy E_{kin} according to $U_{Lane} = a - bE_{kin}$ where $a \simeq 22 - 34$ MeV and $b \simeq 0.1 - 0.2$ [17,18]. This provides a stringent constraint on the symmetry potential. The potential in eq.2 meets this requirement very well as seen in Fig. 2 where the symmetry potential $(U_n - U_p)/2\delta$ as a function of momentum and density for the parameter $x = -1$ is displayed.

One characteristic feature of the momentum dependence of the symmetry potential is the different effective masses for neutrons and protons in isospin asymmetric nuclear matter, i.e.,

$$\frac{m_\tau^*}{m_\tau} = \left\{1 + \frac{m_\tau}{\hbar^2 k}\frac{dU_\tau}{dk}\right\}^{-1}_{k=k_\tau^F}, \tag{4}$$

where k_τ^F is the nucleon Fermi wave number. With the potential in eq. 2, since the momentum-dependent part of the nuclear potential is independent of the parameter x, the nucleon effective masses are independent of the x parameter too. Shown in Fig. 3 are the nucleon effective masses as a function of density (upper window) and isospin asymmetry (lower window). It is seen that the neutron effective mass is higher than the proton effective mass and the splitting between them increases with both the density and isospin asymmetry of the medium [14]. We notice here that the momentum dependence of the symmetry potential and the associated splitting of nucleon effective masses in isospin asymmetric matter is still highly contoversial [19,20]. The experimental determination of both the density and momentum dependence of the symmetry potential is required. Since both the incoming current in the initial state and the level density of the final state in nucleon-nucleon (NN) scatterings depend on the effective masses of colliding nucleons in medium, the in-medium nucleon-nucleon cross sections are expected to be reduced by a factor

$$\sigma_{NN}^{medium}/\sigma_{NN} = (\mu_{NN}^*/\mu_{NN})^2 \tag{5}$$

where μ_{NN} and μ_{NN}^* are the reduced mass of the colliding nucleon pairs in free-space and in medium, respectively [21,22,23]. At least at low momenta, this scaling model is consistent with predictions based on more microscopic many-body theories [24]. Thus, because of the reduced in-medium nucleon effective masses and their dependence on the density and isospin asymmetry of the medium, the in-medium NN cross sections are not only reduced compared to their free-space values, the nn and pp cross sections are split and the difference between them grows in more asymmetric matter. Other

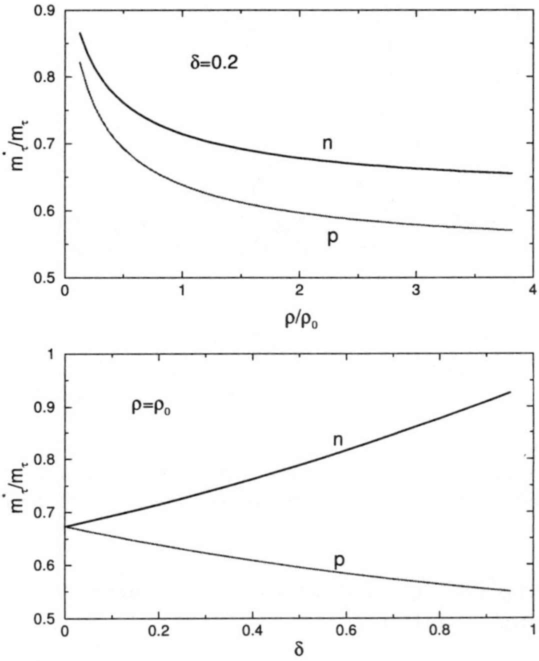

Figure 3. Nucleon effective masses in asymmetric matter as a function of density (upper window) and isospin asymmetry (lower window).

details, such as the initialization of colliding nuclei in phase space, Pauli blocking, etc can be found in our earlier publications [4,5,14].

3. Application of the model

The model outlined above have many applications in heavy-ion reactions induced by both stable and radioactive beams. In this section, we illustrate several examples of studying the role of isospin degree of freedom in the reaction dynamics and extracting the density dependence of nuclear symmetry energy. Besides probes of symmetry energy at subnormal densities, several observables sensitive to the high density behavior of the symmetry energy have been also proposed largely based on transport model simulations.

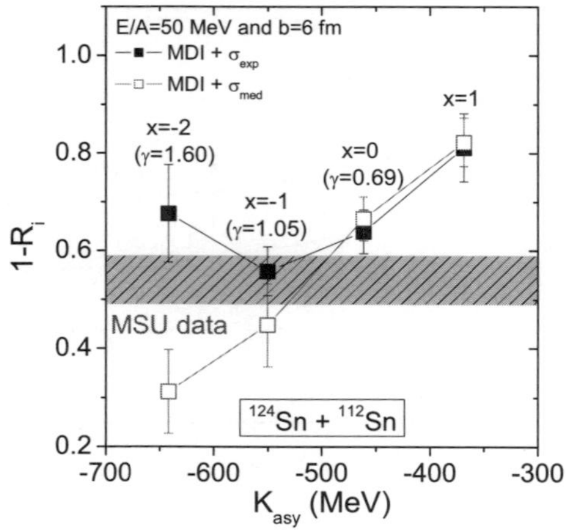

Figure 4. (Color online) The degree of isospin transport as a function of $K_{\rm asy}(\rho_0)$ with the free (filled squares) and in-medium (open squares) nucleon-nucleon cross sections.

3.1. *Symmetry energy of neutron-rich mnatter at subnormal densities*

Tsang et al. [9] recently studied the degree of isospin diffusion in the reaction ^{124}Sn + ^{112}Sn by measuring [25]

$$R_i = \frac{2X_{^{124}Sn+^{112}Sn} - X_{^{124}Sn+^{124}Sn} - X_{^{112}Sn+^{112}Sn}}{X_{^{124}Sn+^{124}Sn} - X_{^{112}Sn+^{112}Sn}} \tag{6}$$

where X is the average isospin asymmetry $\langle\delta\rangle$ of the ^{124}Sn-like residue. Shown in Fig. 4 is a comparison of the averaged strength of isospin transport $1-R_i$ obtained with the free and in-medium NN cross sections, respectively, as a function of the asymmetric part of the isobaric incompressibility of nuclear matter at ρ_0 [10]

$$K_{\rm asy}(\rho_0) \equiv 9\rho_0^2 \left(d^2E_{\rm sym}/d\rho^2\right)_{\rho_0} - 18\rho_0 \left(dE_{\rm sym}/d\rho\right)_{\rho_0}. \tag{7}$$

First, it is interesting to note that with the in-medium NN cross sections the strength of isospin transport $1 - R_i$ decreases monotonically with the decreasing value of x. With the free-space NN cross sections, however, there appears to be a minimum at around $x = -1$. Moreover, this minimum is the point closest to the experimental data. This allowed us to extract the value of $K_{asy}(\rho_0) = -550 \pm 100$ MeV. With the in-medium NN cross sections we can now further narrow down the $K_{asy}(\rho_0)$ to be about -500 ± 50 MeV.

Shown also in the figure are the γ values used in fitting the symmetry energy with $E_{\rm sym}(\rho) = 31.6(\rho/\rho_0)^\gamma$. The results with the in-medium NN cross sections constrain the γ parameter to be between 0.69 and 1.05. The lower value is close to what is extracted from studying giant resonances [26]. Moreover, the same equations of state have also been used to calculate the thickness of neutron-skin in ^{208}Pb [11,12,13]. It was found that with $x = 0$ the resulting thickness of neutron-skin in ^{208}Pb is close to the upper limit of the currently experimentally acceptable range. Thus combining both the isospin diffusion and neutron-skin studies, we are able to conclude that the most acceptable symmetry energy is $E_{\rm sym}(\rho) = 31.6(\rho/\rho_0)^{0.69}$. The value of $\gamma = 1.05$ extracted earlier using the free-space NN cross sections sets an upper limit [10]. The overall effect of the in-medium NN cross sections on isospin transport is a result of a complicated combination of both the mean field and the NN scatterings. Generally speaking, the symmetry potential effects on the isospin transport will become weaker when the NN cross sections are larger while the symmetry potential effects will show up more clearly if smaller NN cross sections are used.

3.2. *Isospin asymmetry of dense matter formed in high energy heavy-ion reactions*

What are the maximum baryon density and isospin asymmetry that can be achieved in central heavy-ion collisions at the highest beam energy expected at RIA? This is an interesting question relevant to the study of the EOS of asymmetric nuclear matter. To answer this question we show in Fig. 5 the central baryon density (upper window) and the average $(n/p)_{\rho\geq\rho_0}$ ratio (lower window) of all regions with baryon densities higher than ρ_0 in the reaction of $^{132}Sn+^{124}Sn$ at a beam energy of 400 MeV/nucleon and an impact parameter of 1 fm. It is seen that the maximum baryon density is about 2 times the normal nuclear matter density. Moreover, the compression is rather insensitive to the symmetry energy because the latter is relatively small compared to the EOS of symmetric matter around this density. The high density phase lasts for about 15 fm/c from 5 to 20 fm/c for this reaction. It is interesting to see that the isospin asymmetry of the high density region is quite sensitive to the symmetry energy. The soft (e.g., $x = 1$) symmetry energy leads to a significantly higher value of $(n/p)_{\rho\geq\rho_0}$ than the stiff one (e.g., $x = -2$). This is consistent with the well-known isospin fractionation phenomenon. Because of the $E_{sym}(\rho)\delta^2$ term in the EOS of asymmetric nuclear matter, it is energetically more favorable to have a higher isospin asymmetry δ in the high density region with a softer symmetry energy functional $E_{sym}(\rho)$. In the supranormal density region, as shown in Fig. 1, the symmetry energy changes from being soft to stiff when the parameter x varies from 1 to -2. Thus the value of $(n/p)_{\rho\geq\rho_0}$ becomes lower as the parameter x changes from 1 to -2. It is worth mentioning that the initial value of the quantity $(n/p)_{\rho\geq\rho_0}$ is about 1.4 which is less than the average n/p ratio of 1.56 of the reaction system. This is because of the neutron-skins of the colliding nuclei, especially that of the projectile ^{132}Sn. In the neutron-rich nuclei, the n/p ratio on the low-density surface is much higher than that in their interior. It is clearly seen that the dense region can become either neutron-richer or neutron-poorer with respect to the initial state depending on the symmetry energy functional $E_{sym}(\rho)$ used.

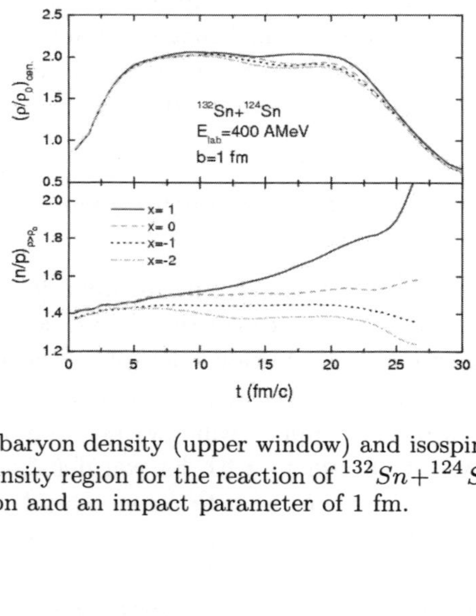

Figure 5. Central baryon density (upper window) and isospin asymmetry (lower window) of high density region for the reaction of $^{132}Sn+^{124}Sn$ at a beam energy of 400 MeV/nucleon and an impact parameter of 1 fm.

3.3. *Pions yields and π^-/π^+ ratio as a probe of the symmetry energy at supranormal densities*

At the highest beam energy at RIA, pion production is significant. Pions may thus carry interesting information about the EOS of dense neutron-rich matter [27,28]. Shown in Fig.6 are the π^- and π^+ yields as a function of the x parameter. It is interesting to see that the π^- multiplicity depends more sensitively on the symmetry energy. The π^- multiplicity increases by about 20% while the π^+ multiplicity remains about the same when the x

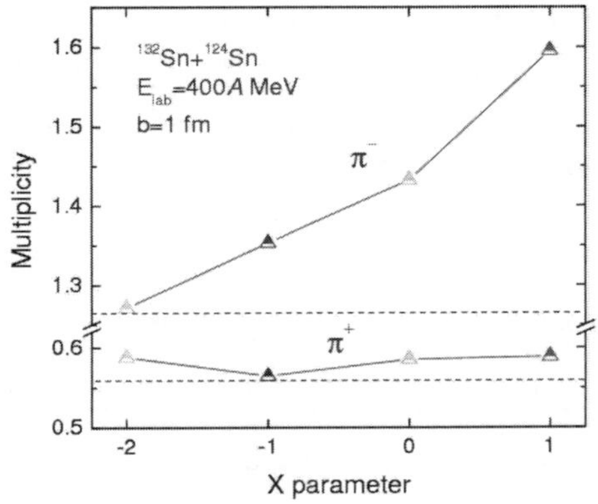

Figure 6. The π^- and π^+ yields as functions of the x parameter.

parameter is changed from -2 to 1. The multiplicity of π^- is about 2 to 3 times that of π^+. This is because the π^- mesons are mostly produced from neutron-neutron collisions. Moreover, with the softer symmetry energy the high density region is more neutron-rich due to isospin fractionation [28]. The π^- mesons are thus more sensitive to the isospin asymmetry of the reaction system and the symmetry energy. However, one should notice that it is well known that pion yields are also sensitive to the symmetric part of the nuclear EOS. It is thus hard to get reliable information about the symmetry energy from π^- yields alone. Fortunately, the π^-/π^+ ratio is a better probe since statistically this ratio is only sensitive to the difference in the chemical potentials for neutrons and protons [29]. This expectation is well demonstrated in Fig. 7. It is seen that the pion ratio is quite sensitive to the symmetry energy, especially at low transverse momenta. Thus, it is promising that the high density behavior of nuclear symmetry energy $E_{sym}(\rho)$ can be probed using the π^-/π^+ ratio.

3.4. *Isospin fractionation and n-p differential flow at RIA and GSI*

The degree of isospin equilibration or translucency can be measured by the rapidity distribution of nucleon isospin asymmetry $\delta_{free} \equiv (N_n - N_p)/(N_n + N_p)$ where N_n (N_p) is the multiplicity of free neutrons (protons) [30]. Although it might be difficult to measure directly δ_{free} because it requires the detection of neutrons, similar information can be extracted from ratios of light clusters, such as, $^3H/^3He$, as demonstrated recently within a

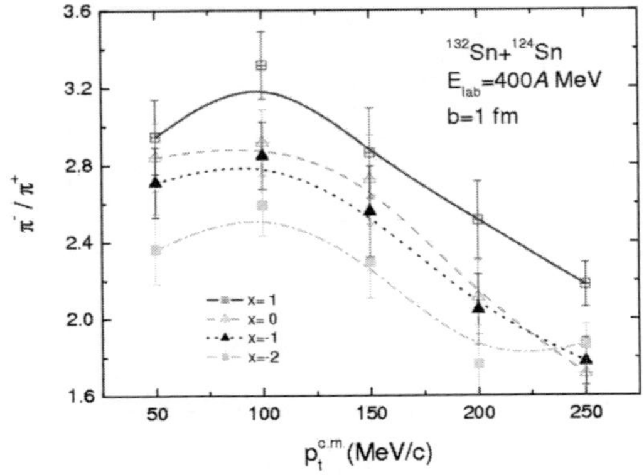

Figure 7. The π^-/π^+ ratio as a function of transverse momentum.

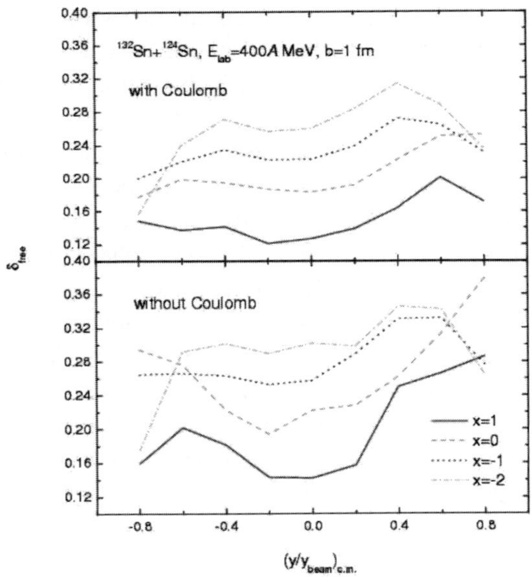

Figure 8. Isospin asymmetry of free nucleons with and without the Coulomb force.

coalescence model [31,32]. Shown in Fig. 8 are the rapidity distributions of δ_{free} with (upper window) and without (lower window) the Coulomb potential. It is interesting to see that the δ_{free} at midrapidity is particularly sensitive to the symmetry energy. As the parameter x increases from -2

to 1 the δ_{free} at midrapidity decreases by about a factor of 2. Moreover, the forward-backward asymmetric rapidity distributions of δ_{free} with all four x parameters indicates the apparent nuclear translucency during the reaction [33].

Another observable that is sensitive to the high density behavior of symmetry energy is the neutron-proton differential flow [34]

$$F^x_{n-p}(y) \equiv \sum_{i=1}^{N(y)} (p^x_i w_i)/N(y), \tag{8}$$

where $w_i = 1(-1)$ for neutrons (protons) and $N(y)$ is the total number of free nucleons at rapidity y. The differential flow combines constructively effects of the symmetry potential on the isospin fractionation and the collective flow. It has the advantage of maximizing the effects of the symmetry potential while minimizing those of the isoscalar potential. Shown in Fig. 9 is the n-p differential flow for the reaction of $^{132}Sn +^{124} Sn$ at a beam energy of 400 MeV/nucleon and an impact parameter of 5 fm. Effects of the symmetry energy are clearly revealed by changing the parameter x.

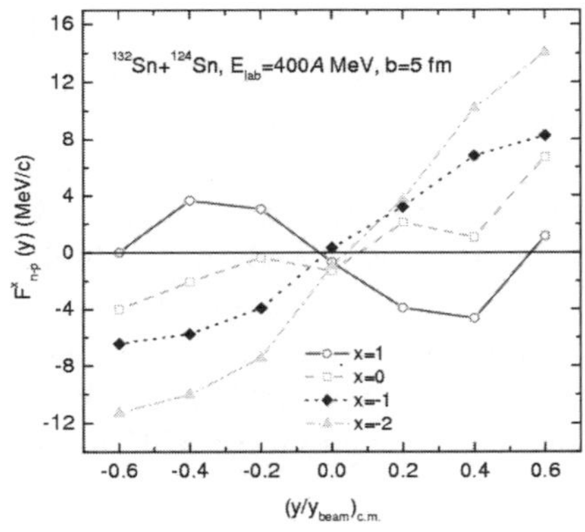

Figure 9. Neutron-proton differential flow at RIA and GSI energies

4. Conclusions

The EOS of neutron-rich matter has been a fundamentally important topic in both nuclear physics and astrophysics. Nuclear reactions induced by neutron-rich nuclei provide a great opportunity to pin down the EOS of neutron-rich matter. Transport models are powerful tools for investigating especially central heavy-ion reactions induced by neutron-rich nuclei. Applications of these models will help us to understand the isospin dependence of in-medium nuclear effective interactions. Comparing with the experimental data, we can extract the isospin dependence of thermal, mechanical and transport properties of asymmetric nuclear matter playing important roles in nuclei, neutron stars and supernovae. Currently, the most important issue is the density dependence of the nuclear symmetry energy. The latter is very important for both nuclear physics and astrophysics. Significant progress has been made recently by the heavy-ion community in determining the density dependence of the nuclear symmetry energy. Based on transport model calculations, a number of sensitive probes of the symmetry energy have been found. The momentum dependence in both the isoscalar and isovector parts of the nuclear potential was found to play an important role in extracting accurately the density dependence of the symmetry energy. Comparing with the experimental data on isospin diffusion and the neutron-skin thickness in ^{208}Pb, we have extracted a symmetry energy of $E_{sym}(\rho) \approx 31.6(\rho/\rho_0)^{0.69}$ at subnormal densities. It would be interesting to compare this conclusion with those extracted from studying other observables. More experimental data including neutrons with neutron-rich beams in a broad energy range are needed. Looking forward to experiments at RIA and GSI with high energy radioactive beams, we hope to pin down the symmetry energy at supranormal densities in the near future.

I would like to thank Lie-Wen Chen, Campak B. Das, Charles Gale, Subal Das Gupta, Che Ming Ko, Andrew W. Steiner, Gao-Chan Yong and Wei Zuo for their productive collaborations with me on the work reported here. This work was supported in part by the US National Science Foundation under grant No. PHYS0354572, PHYS0456890 and the NASA-Arkansas Space Grants Consortium award ASU15154.

References

1. J.M. Lattimer and M. Prakash, *Phys. Rep.* **333**, 121 (2000); *Astr. Phys. Jour.* **550**, 426 (2001); *Science Vol.* **304**, 536 (2004);
   A. W. Steiner, M. Prakash, J.M. Lattimer and P.J. Ellis, *Phys. Rep.* **411**, 325 (2005).
2. B.A. Brown, *Phys. Rev. Let.* **85**, 5296 (2000).
3. J.R. Stone *et al.*, *Phys. Rev.* **C68**, 034324 (2004).
4. B.A. Li, C.M. Ko and W. Bauer, *Int. J. Mod. Phys.* **E7**, 147 (1998).
5. *Isospin Physics in Heavy-Ion Collisions at Intermediate Energies*, Eds. B. A. Li and U. Schröder, Nova Science Publishers, Inc, New York, 2001.

6. P. Danielewicz, R. Lacey and W.G. Lynch, *Science* **298**, 1592 (2002).
7. V. Baran, M. Colonna, V. Greco and M. Di Toro, *Phys. Rep.* **410**, 335 (2005).
8. I. Bombaci, Chapter 2 in ref. [5].
9. M.B. Tsang *et al.*, *Phys. Rev. Lett.* **92**, 062701 (2004).
10. L.W. Chen, C.M. Ko and B.A. Li, *Phys. Rev. Lett.* **94**, 32701 (2005).
11. A.W. Steiner and B.A. Li, *Phys. Rev.* **C72**, 041601 (2005).
12. B.A. Li and L.W. Chen, ArXiv:nucl-th/0508025.
13. L.W. Chen, C.M. Ko and B.A. Li, ArXiv:nucl-th/0509009.
14. B.A. Li, C.B. Das, S. Das Gupta and C. Gale, *Phys. Rev.* **C69**, 011603 (2004); *Nucl. Phys.* **A735**, 563 (2004).
15. I. Bombaci and U. Lombardo, *Phys. Rev.* **C44**, 1892 (1991);
W. Zuo, L.G. Gao, B.A. Li, U. Lombardo, and C.W. Shen, *Phys. Rev.* **C72**, 014005 (2005).
16. C.B. Das, S. Das Gupta, C. Gale and B.A. Li, *Phys. Rev.* **C67**, 034611 (2003).
17. P.E. Hodgson, *The Nucleon Optical Model*, 1994, World Scientific.
18. G.W. Hoffmann and W.R. Coker, *Phys. Rev. Lett.* **29**, 227 (1972).
19. J. Rizzo, M. Colonna, M. Di Toro and V. Greco, *Nucl. Phys.* **AA732**, 202 (2004).
20. B.A. Li, *Phys. Rev.* **C69**, 064602 (2004).
21. V.R. Pandharipande and S.C. Pieper, *Phys. Rev.* **C45**, 791 (1991).
22. M. Kohno, M. Higashi, Y. Watanabe and M. Kawai, *Phys. Rev.* **C57**, 3495 (1998).
23. D. Persram and C. Gale, *Phys. Rev.* **C65**, 064611 (2002).
24. F. Sammarruca, ArXiv:nucl-th/0506081;
F. Sammarruca and P. Krastev, ArXiv:nucl-th/0509011.
25. F. Rami *et al.*, *Phys. Rev. Lett.* **84**, 1120 (2000).
26. J. Piekarewicz and S.P. Weppner, ArXiv:nucl-th/0509019.
27. B.A. Li, *Phys. Rev. Lett.* **88**, 192701 (2002); *Nucl. Phys.* **A708**, 365 (2003); *Phys. Rev.* **C67**, 017601 (2003).
28. B.A. Li, G.C. Yong and W. Zuo, *Phys. Rev.* **C71**, 014608 (2005).
29. G.F. Bertsch, *Nature* **283**, 280 (1980);
A. Bonasera and G.F. Bertsch, *Phys. Lett.* **B195**, 521 (1987).
30. B.A. Li, C.M. Ko and Z.Z. Ren, *Phys. Rev. Lett.* **78**, 1644 (1997);
B.A. Li, *Phys. Rev.* **C69**, 034614 (2004).
31. L.W. Chen, C.M. Ko and B.A. Li, *Phys. Rev.* **C68**, 017601 (2003); *Nucl. Phys.* **A729**, 809 (2003).
32. L.W. Chen, C.M. Ko and B.A. Li, *Phys. Rev.* **C69**, 054606 (2004).
33. B.A. Li, G.C. Yong and W. Zuo, *Phys. Rev.* **C71**, 014608 (2005); *ibid.*, **C71**, 044604 (2005).
34. B.A. Li, *Phys. Rev. Lett.* **85**, 4221 (2000).

SENSITIVITY OF THE CHIRAL PHASE TRANSITION OF QCD TO THE SCALAR MESON SECTOR *

T. HERPAY

Department of Physics of Complex Systems,
Eötvös University, H-1117 Budapest, Hungary
E-mail: herpay@complex.elte.hu

A. PATKÓS†

Department of Atomic Physics, Eötvös University,
H-1117 Budapest, Hungary
E-mail:patkos@ludens.elte.hu

ZS. SZÉP

Research Group for Statistical Physics of the Hungarian Academy of Sciences,
H-1117 Budapest, Hungary
E-mail:szepzs@achilles.elte.hu

Gribov's theory of light quark confinement implies the existence of two kinds of scalar bound states. The phase diagram of the three-flavor QCD is mapped out in the $(m_\pi - m_K)$–plane with help of the $SU_L(3) \times SU_R(3)$ linear sigma model supplemented with the assumption that the masses of the so-called superbound scalars do not change under the variation of the pion and kaon mass. The phase boundary along the $m_\pi = m_K$ line is found in the interval 15 MeV $< m_{\rm crit} <$ 25 MeV, irrespective which $f_0 - \sigma$ linear combination is identified with the pure superbound state.

1. Introduction

Volodia Gribov repeatedly has expressed in the early nineteen-eighties his view that deconfinement in QCD is going to turn out to be a complete analogue of atomic ionisation, which one would not call a phase transition. There is increasing evidence that the temperature driven transition from the hadronic phase to quark-gluon plasma is a smooth crossover indeed [1].

*This work is supported by Hungarian Science Fund (OTKA) under contract No. 046129.
†Speaker

He based his confinement theory also on an analogy with a phenomenon of atomic physics, e.g. supercritical binding. This would occur for light quarks irrespective to the accurate value of their mass [2]. The role of chiral symmetry is not evidently important in this theory, and the consequences on its spontaneous breakdown were not yet fully clarified [3].

In this contribution to the commemoration of the 75th anniversary of V.N. Gribov, we exploit semi-quantitatively his ideas on the nature of the scalar meson sector when discussing the chiral symmetry restoring quark-hadron transition in different points of the $m_\pi - m_K$-plane with help of a variant of the linear sigma model ($L\sigma M$). In a recent publication [4] we constructed a continuation of the parameters of this model from the physical point to an arbitrary mass-point m_π, m_K requiring agreement with results of Chiral Perturbation Theory (ChPT) for the decay constants f_π, f_K and the trace of the squared mass matrix in the $\eta - \eta'$ sector, M_η^2, at the tree level. In the $\eta - \eta'$ sector the predictions of the two models for the separate mass eigenvalues nicely coincide. It was emphasised that for a complete specification of the parameters of the model one needs a single extra information on the dependence of scalar spectra on the masses of the pseudoscalar nonet. This information is beyond ChPT, therefore in general only *ad hoc* assumptions can be made and tested through the consequences.

According to Gribov's confinement theory in addition to normal $\bar{q}q$ states also superbound states do exist, which contain quarks with *negative kinetic energy*. Some repulsive interaction between such quarks increases their energy above zero and makes these associations physically observable [5]. These states are of smaller size and higher energy than the normal scalar meson states, therefore it was suggested to associate them with f_0 in the isoscalar and with a_0 in the isovector channel. In the $SU(2) \times SU(2)$ model they form another $O(4)$ quartet in addition to σ, π. It is rather natural to assume that the masses of the superbound states are not sensitive to the variation of m_π.

In the three flavor case a similar "doubling" of the multiplet structure can be assumed, if the strange quark is light enough. In this paper we shall explore the boundary of the first order transition region around the chiral point in the three-flavor QCD [6] taking into account the extra requirement to have scalars in the spectra whose mass does not vary with m_π, m_K. The consequences of identifying the physically observable scalars with some mixture of the pure normal and pure superbound states will be investigated.

Numerical investigations of the chiral symmetry restoration were done in the framework of lattice QCD and systematically improved for the 3-flavor degenerate case $m_u = m_d = m_s \neq 0$. The initial estimate for the critical pseudoscalar meson mass, $m_{\rm crit}(\text{diag}) \approx 290$ MeV [7] was seen to be reduced to $60 - 70$ MeV [8] or may be to even further down [9] when finer lattices and improved lattice actions are used. Very recently de Forcrand and Philipsen reported an estimate $m_{crit} \approx (0.1 - 0.2)T_c$, which very conservatively means $m_{crit} \approx 15 - 30$MeV [10]. One should be conscious of the fact, that it is extremely difficult to reach continuum results in this mass regime.

Effective models (linear or non-linear sigma models, Nambu–Jona-Lasinio model) represent another, in a sense complementary, approach to the study of the phase structure, which one expects to work the better the lighter quark masses are used [11,12,13]. It is surprising that only moderate effort was invested to date to improve the pioneering studies of the $SU(3) \times SU(3)$ linear sigma model by Meyer-Ortmanns and Schaefer [11] which used a saddle point approximation valid in the limit of infinite number of flavors, and derived $m_{\rm crit}(\text{diag}) \lesssim 51$ MeV. An extension of their work to unequal pion and kaon masses was achieved by C. Schmidt [14]. He found $m_{\rm crit}(\text{diag}) = 47$ MeV and a phase boundary approaching the m_K-axis rather sharply. The phase boundary was calculated also by Lenaghan [15] using the Hartree-approximation to the effective potential derived in CJT-formalism. For the complete determination of the couplings of the three-flavor chiral meson model he fixed the $T = 0$ mass of the σ particle in addition to the experimental mass spectra of the pseudoscalar sector. The emerging phase boundary is rather sensitive to this mass. The estimate for $m_{\rm crit}(\text{diag})$ which one can extract from Fig. 3 of [15] for $m_\sigma = 900$ MeV is compatible with [11,14].

Our method of parametrisation and solution of $L\sigma M$ was described in detail in [4]. Therefore here we shall only review the set of the equations to be solved for the determination of the transition point. The changes arising from the implementation of the insensitivity of Gribov's superbound scalars to the variation of the fundamental masses will be emphasised. We conclude by giving the most characteristic features of the phase boundary.

2. $L\sigma M$ parametrisation consistent with ChPT

We outline first, how one obtains with ChPT the m_π, m_K-dependence of the masses and decay constants in the pseudoscalar sector. The dependence of the pion and kaon masses as well as of their decay constants on the quark masses were determined in [16] for the $SU(3) \times SU(3)$ nonlinear sigma model:

$$m_\pi^2 = 2A\left[1+\frac{1}{f^2}\left(\mu_\pi - \frac{1}{3}\mu_\eta + 16A(2L_8-L_5)+ \right.\right.$$
$$\left.\left. +16A(2+q)(2L_6-L_4)\right)\right], \qquad (1)$$
$$m_K^2 = A(1+q)\left[1+\frac{1}{f^2}\left(\frac{2}{3}\mu_\eta + 8A(1+q)(2L_8-L_5)\right.\right.$$
$$\left.\left. +16A(2+q)(2L_6-L_4)\right)\right], \qquad (2)$$
$$f_\pi = f\left[1+\frac{1}{f^2}\left(-2\mu_\pi - \mu_K + 8AL_5 + 8A(2+q)L_4\right)\right], \qquad (3)$$
$$f_K = f\left[1-\frac{1}{f^2}\left(\frac{3}{4}(\mu_\pi+\mu_\eta+2\mu_K) - 4A(1+q)L_5 - 8A(2+q)L_4\right)\right]. \qquad (4)$$

Here A, q are related to the quark masses, f is the coupling of the non-linear sigma model. μ_{PS} defines the so-called chiral logarithm for each pseudoscalar meson (PS) proportional to $\ln(m_{PS}/M_0)$ with $M_0 = 4\pi f$. The determination of the low energy chiral constants L_i is discussed in depth in [4].

One inverts the first two equations with $O(f^{-2})$ accuracy and finds the following m_π, m_K-dependence for the decay constants:

$$f_\pi = f\left[1-\frac{1}{f^2}(2\mu_\pi + \mu_K - 4m_\pi^2(L_4+L_5) - 8m_K^2 L_4)\right], \qquad (5)$$
$$f_K = f\left[1-\frac{1}{f^2}\left(\frac{3}{4}(\mu_\pi+\mu_\eta+2\mu_K) - 4m_\pi^2 L_4 - 4m_K^2(L_5+2L_4)\right)\right]. \qquad (6)$$

The extension to the $U(3)\times U(3)$ ChPT is somewhat more complicated. It was worked out in [17,18,19,20] and allows the determination of $m_\eta(m_\pi, m_K), m_{\eta'}(m_\pi, m_K)$. For the parametrisation of $L\sigma M$ one more independent relation can be obtained from the mixing $\eta-\eta'$ sector, for which in [4] we have chosen the trace of the 2×2 squared mass matrix, denoted by $M_\eta^2 \equiv m_\eta^2 + m_{\eta'}^2$:

$$\begin{aligned} M_\eta^2 &= 2m_K^2 - 3v_0^{(2)} + 2(2m_K^2+m_\pi^2)(3v_2^{(2)} - v_3^{(1)}) \\ &+ \frac{1}{f^2}\Big[8v_0^{(2)}(2m_K^2+m_\pi^2)(L_5+3L_4) + m_\pi^2(\mu_\eta - 3\mu_\pi) - 4m_K^2\mu_\eta \\ &+ \frac{16}{3}(6L_8 - 3L_5 + 8L_7)(m_\pi^2 - m_K^2)^2 \\ &+ \frac{32}{3}L_6(m_\pi^4 - 2m_K^4 + m_K^2 m_\pi^2) + \frac{16}{3}L_7(m_\pi^2+2m_K^2)^2\Big]. \end{aligned} \qquad (7)$$

Also the constants $v_i^{(j)}$ were determined in [4] from the $T = 0$ properties of the Goldstone particles.

Next, we turn to the problem of parameterising $L\sigma M$ to provide spectra the closest possible to ChPT. The Lagrangian of the $SU_L(3) \times SU_R(3)$ symmetric linear sigma model with explicit symmetry breaking terms is given [21] by

$$L(M) = \frac{1}{2}\mathrm{tr}(\partial_\mu M^\dagger \partial^\mu M + \mu_0^2 M^\dagger M) - f_1 \left(\mathrm{tr}(M^\dagger M)\right)^2 \\ - f_2 \mathrm{tr}(M^\dagger M)^2 - g\left(\det(M) + \det(M^\dagger)\right) + \epsilon_0\sigma_0 + \epsilon_8\sigma_8 \,, \qquad (8)$$

where M is a complex 3×3 matrix, defined by the σ_i scalar and π_i pseudoscalar fields $M := \frac{1}{\sqrt{2}}\sum_{i=0}^{8}(\sigma_i + i\pi_i)\lambda_i$, with $\lambda_i \,:\, i = 1\ldots 8$ the Gell-Mann matrices and $\lambda_0 := \sqrt{\frac{2}{3}}\mathbf{1}$. The last two terms of (8) break the symmetry explicitly, the possible isospin breaking term $\epsilon_3\sigma_3$ is not considered.

A detailed analysis of the symmetry breaking patterns which might occur in the system described by this Lagrangian can be found in [12]. The field expectation values $\langle\sigma_0\rangle, \langle\sigma_8\rangle$ both contain strange ($\equiv y$) and non-strange ($\equiv x$) components:

$$x = (\sqrt{2}\langle\sigma_0\rangle + \langle\sigma_8\rangle)/\sqrt{3}, \qquad y = (\langle\sigma_0\rangle - \sqrt{2}\langle\sigma_8\rangle)/\sqrt{3}. \qquad (9)$$

With help of the PCAC relations and the tree level mass formulae (see Table 1, where the $x - y$ basis is used instead of $0 - 8$) the following expressions can be derived for the couplings of $L\sigma M$:

$$\begin{aligned} x &= f_\pi, \qquad y = (2f_K - f_\pi)/\sqrt{2}\,, \qquad (10)\\ f_2 &= \frac{(6f_K - 3f_\pi)m_K^2 - (2f_K + f_\pi)m_\pi^2 - 2(f_K - f_\pi)M_\eta^2}{4(f_K - f_\pi)(8f_K^2 - 8f_K f_\pi + 3f_\pi^2)}\,,\\ g &= \frac{2f_K m_K^2 + 2(f_K - f_\pi)m_\pi^2 - (2f_K - f_\pi)M_\eta^2}{\sqrt{2}(8f_K^2 - 8f_K f_\pi + 3f_\pi^2)}\,,\\ M^2 &\equiv -\mu_0^2 + 4f_1(x^2 + y^2)\\ &= \frac{1}{2}M_\eta^2 + \frac{g}{\sqrt{2}}(2f_K - f_\pi) - 2f_2[(f_\pi - f_K)^2 + f_K^2]. \end{aligned}$$

The sources $\epsilon_x = (\sqrt{2}\epsilon_0 + \epsilon_8)/\sqrt{3}, \epsilon_y = (\epsilon_0 - \sqrt{2}\epsilon_8)/\sqrt{3}$, which explicitly break chiral symmetry are determined with help of the Gell-Mann–Oakes–Renner relations:

$$\epsilon_x = m_\pi^2 x, \qquad \epsilon_y = \frac{\sqrt{2}}{2}(m_K^2 - m_\pi^2)x + m_K^2 y. \qquad (11)$$

Table 1. Squared masses of the pseudoscalar boson nonet and their parity partners

m_π^2	$= -\mu_0^2 + 2(2f_1 + f_2)x^2 + 4f_1y^2 + 2gy$
$m_{a_0}^2$	$= -\mu_0^2 + 2(2f_1 + 3f_2)x^2 + 4f_1y^2 - 2gy$
m_K^2	$= -\mu_0^2 + 2(2f_1 + f_2)(x^2 + y^2) + 2f_2y^2 - \sqrt{2}x(2f_2y - g)$
m_κ^2	$= -\mu_0^2 + 2(2f_1 + f_2)(x^2 + y^2) + 2f_2y^2 + \sqrt{2}x(2f_2y - g)$
$m_{\eta xx}^2$	$= -\mu_0^2 + 2(2f_1 + f_2)x^2 + 4f_1y^2 - 2gy$
$m_{\eta yy}^2$	$= -\mu_0^2 + 4f_1x^2 + 4(f_1 + f_2)y^2$
$m_{\eta xy}^2$	$= -2gx$
$m_{\sigma xx}^2$	$= -\mu_0^2 + 6(2f_1 + f_2)x^2 + 4f_1y^2 + 2gy$
$m_{\sigma yy}^2$	$= -\mu_0^2 + 4f_1x^2 + 12(f_1 + f_2)y^2$
$m_{\sigma xy}^2$	$= 8f_1xy + 2gx$

Note: The expressions of the squared masses of parity partners having the same isospin and hypercharge appear in one block. Different isomultiplets are separated by double lines. In the lowest big block the matrix elements of the mixing in the $\eta-\eta'$ and $\sigma-f_0$ sectors are given in the $x-y$ base.

The requirement of the agreement of $L\sigma M$ with ChPT is fulfilled when the (m_π, m_K)-dependence of the couplings x, y, f_2, g, M^2 is determined with help of Eqs.(5,7). Before the procedure just described was first proposed in Ref. [4] $m_\pi - m_K$ mass tuning was taken into account only in (11). The quality of this not fully complete parametrisation (e.g. only the combination M^2 of f_1 and μ_0^2 is determined at this stage) can be assessed by comparing the separate the $m_\pi - m_K$ mass dependence of m_η and $m_{\eta'}$ obtained in the present parametrisation of $L\sigma M$ with the predictions of ChPT. In Fig. 1 the comparison is done for $m_\pi = 0$ and the agreement is fairly good up to $m_K \approx 800$MeV.

The combination M^2 of f_1 and μ_0^2 can be split up only by making use of the expression of the admixed scalars, therefore the use of one characteristics of the mixed scalar spectra is unavoidable [22].

For the separate determination of μ_0^2 and f_1 we have fixed first the mass of the f_0 meson. This meson mass is identified with the heavier eigenvalue calculated in the mixing scalar subspace. Since there is the possibility of mixing between the normal and the superbound multiplets we have chosen as a second possibility also fixing the trace $m_\sigma^2 + m_{f_0}^2 = m_{\sigma_{xx}}^2 + m_{\sigma_{yy}}^2$ (see Table 1) in the squared mass subspace of admixed scalars. In order to test the sensitivity of the results to the scalar masses (which are not very accurately known) we have performed the calculations for both alternatives with several numerical values.

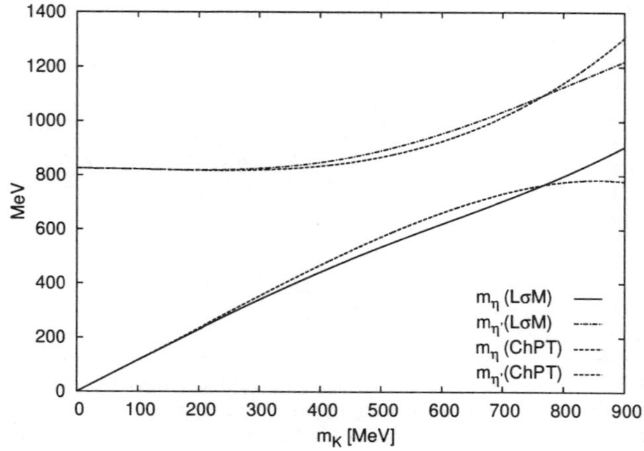

Figure 1. The tree–level kaon mass dependence of m_η and $m_{\eta'}$ for $m_\pi = 0$. The labels refer to the results of ChPT and the predictions of linear sigma model (LσM), respectively.

3. Quasiparticle thermodynamics and phase diagram

The renormalised set of the equations of state and of the selfconsistent pion propagator determine the temperature dependence of the vacuum expectation values x and y. The scheme of the perturbation theory applied here agrees with the Optimal Perturbation Theory of Chiku and Hatsuda [23], which was renormalised using the approach of [24].

The tree level mass of π involves now the thermal mass parameter:

$$m_\pi^2 = M^2(T) + 2(2f_1 + f_2)x^2 + 4f_1y^2 + 2gy, \tag{12}$$

and all other meson masses to be used in the tadpole integrals below agree with the formulas appearing in Table 1 with the replacement $-\mu_0^2 \to M^2(T)$. If all quantum corrections are condensed into $M^2(T)$, then the tree–level masses of other mesons are expressible through the mass of the pion. One might expect that the pion has the lowest mass and therefore for $M^2(T) > 0$ these squared masses are all positive, which is not the case when $-\mu_0^2 < 0$ is used in the propagators. We define a physical region of x and y where all tree-level mass squares are positive, and thus the one-loop contribution of the meson fluctuations to EoS is real. This region is most severely restricted by the mass of σ, which strongly decreases near the phase transition. We will restrict our attention to the solution of the EoS's in the physical region.

For the determination of the thermal mass we use the Schwinger–Dyson equation for the inverse pion propagator at zero external momentum. At one-loop it receives the contribution $\Pi(M(T), p = 0)$, which is the self-energy function of the pion at zero external momentum, plus the counterterm contribution $-\mu_0^2 - M^2(T)$. We apply the principle of minimal sensitivity (PMS) [23], that is we require that the pion mass be given by its tree-level expression:

$$\Pi(M(T), p = 0) - \mu_0^2 - M^2(T) = 0. \tag{13}$$

$\Pi(M(T), p)$ itself is a linear combination of the tadpole and bubble diagrams (the latter not included in the treatment of [12]), with coefficients derived with help of the 4-point and 3-point couplings among mass eigenvalue fields.

The self-energy can be represented as a linear combination of tadpole integrals, which gives when substituted into Eq. (13):

$$0 = -M^2(T) - \mu_0^2 + \sum_{i=\pi,\, K,\, \eta,\eta'}^{\alpha=\sigma,\, \pi} c^{\pi}_{\alpha_i} I(m_{\alpha_i}(T), T)\,. \tag{14}$$

Here $c^{\pi}_{\alpha_i}$ are the weights of the renormalised tadpole contributions evaluated with different mass eigenstate mesons $\alpha_i = \sigma_i, \pi_i$. The integrals over the corresponding propagators are evaluated with effective tree-level masses where $M^2(T)$ replaces $-\mu_0^2$. In this way (14) is actually a gap equation which determines the thermal mass parameter, $M^2(T)$. With help of Eq. (12) this equation can be also understood as a gap equation for the pion mass (the pion mass is present also in the expressions of $I(m_{\alpha_i}, T)$ through m_{α_i}!):

$$m_\pi^2 = -\mu_0^2 + 2(2f_1 + f_2)x^2 + 4f_1 y^2 + 2gy + \sum_{i=\pi,\, K,\, \eta,\eta'}^{\alpha=\sigma,\, \pi} c^{\pi}_{\alpha_i} I(m_{\alpha_i}(T), T)\,. \tag{15}$$

For the determination of the order parameters x, y we solved the two renormalised equations of state using the solution of the gap equation in the propagator masses:

$$\begin{aligned} -\epsilon_x - \mu_0^2 x + 2gxy + 4f_1 xy^2 + 2(2f_1 + f_2)x^3 + \\ + \textstyle\sum_{i=\pi,\, K,\, \eta,\eta'}^{\alpha=\sigma,\pi} J_i t^x_{\alpha_i} I(m_{\alpha_i}(T), T) = 0\,, \end{aligned} \tag{16}$$

$$\begin{aligned} -\epsilon_y - \mu_0^2 y + gx^2 + 4f_1 x^2 y + 4(f_1 + f_2)y^3 + \\ + \textstyle\sum_{i=\pi,\, K,\, \eta,\eta'}^{\alpha=\sigma,\pi} J_i t^y_{\alpha_i} I(m_{\alpha_i}(T), T) = 0\,. \end{aligned} \tag{17}$$

The quantities $t^x_{\alpha_i}$ and $t^y_{\alpha_i}$ give the corresponding weights. J_i is the isospin multiplicity factor: $J_\pi = 3$, $J_K = 4$, and $J_{\eta,\eta'} = 1$. The coefficients c and t were listed in [4]. One finds $c^\pi_{\alpha_i} = J_i t^x_{\alpha_i}/x$, which ensures that the solution for the mass of the pion obeys Goldstone's theorem.

The solution of Eqs. (15), (16), (17) for given m_π, m_K allows to establish the nature of the temperature driven transitions. First order transitions are signalled by multivaluedness in the temperature evolution of both the non-strange and strange condensates. For large values of the kaon mass, we claim that the phase transition is driven by the variation of the non-strange condensate, since each of the multiple solutions of the strange condensate are very close to each other, and all stay at high values. Subsequent decrease of the strange condensate at higher temperature displays only a crossover.

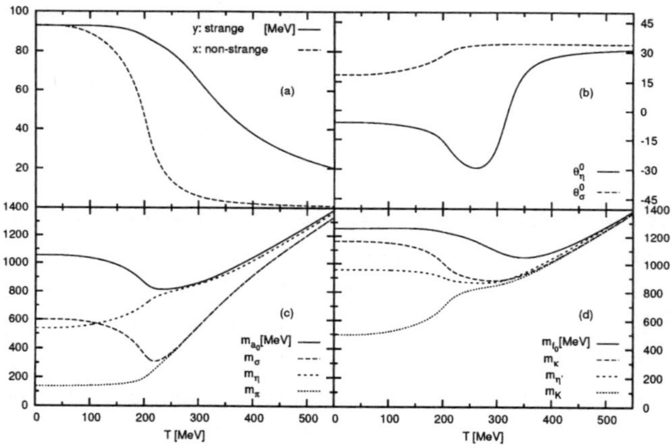

Figure 2. The temperature dependence in the physical point of: (a) the non-strange (x) and strange (y) condensates; (b) the pseudoscalar (θ_η) and scalar (θ_σ) mixing angles (in the (0-8) basis); (c) the mass of the chiral partners (π, σ) and (a_0, η); (d) the mass of f_0, κ, η', K mesons.

In the physical point the system exhibits clear crossover as can be seen in Fig. 2. Let us discuss next what happens near the $m_\pi = m_K$ diagonal. In Fig. 3 the gradual deformation of the boundary line is shown when four different (m_π, m_K)-independent conditions (listed on the figure) are imposed on the spectra of the admixed scalars. The boundary reaches the diagonal in the range $m_{crit} \approx 20$ MeV independently of the condition imposed. When moving away from the diagonal for $m_K > m_\pi$ the larger is the scalar mass scale the farther the boundary goes away from the $m_\pi = 0$

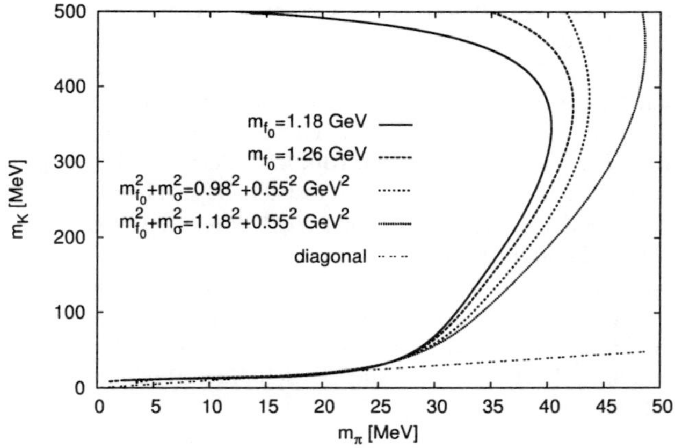

Figure 3. Variation of the boundary of first order transitions near and above the $m_\pi = m_K$ diagonal (note the rather different units on the two axes!).

axis. Therefore with this method one cannot make any definite statement on the location of the tricritical point on the m_K-axis defined as the point where the transition changes from a discontinuous nature into a continuous one.

In conclusion of this study we find that the assumption of the $m_\pi - m_K$ independence of the mass of the heavier isoscalar-scalar member of the nonet, suggested by Gribov's confinement picture leads to a rather unique conclusion, that the $m_\pi = m_K$ diagonal crosses the critical curve of the chiral phase transitions of the 3-flavor QCD at a rather low value: $15 \text{ MeV} \leq \text{m}_{\text{crit}} \leq 25 \text{ MeV}$.

Acknowledgement

We thank the organisers of the Gribov-75 conference for the opportunity to discover again the actual interest of the confinement theory of V.N. Gribov.

References

1. for a recent review see, P. Petreczky, *Nucl. Phys. B (Proc. Suppl.)* **140**, 78 (2005), Arxiv:hep-lat/0409139.
2. V.N. Gribov, *Eur. Phys. J.* **C10**, 91 (1999).
3. Y.L. Dokshitzer and D.E. Kharzeev, *Ann. Rev. Nucl. Part. Sci.* **54**, 487 (2004).
4. T. Herpay, A. Patkós, Zs. Szép and P. Szépfalusy, *Phys. Rev.* **D 71**, 125017 (2005).

5. F.E. Close, Yu.L. Dokshitzer, V.N. Gribov, V.A. Khoze and M.G. Ryskin, *Phys. Lett.* **B319**, 291 (1993).
6. R.D. Pisarski and F. Wilczek, *Phys. Rev.* **D29**, 334 (1984).
7. F. Karsch, E. Laermann and A. Peikert, *Phys. Lett.* **B520**, 41 (2001).
8. F. Karsch, C.R. Allton, S. Ejiri, S.J. Hands, E. Laermann and C. Schmidt, *Nucl. Phys. B (Proc. Suppl.)* **129&130**, 614 (2004), ArXiv:hep-lat/0309116.
9. C. Bernard, T. Burch, C. deTar, S. Gottlieb, E.B. Gregory, U.M. Heller, J. Osborn, R.L. Sugar and D. Toussaint, *Nucl. Phys. B (Proc. Suppl.)* **140**, 538 (2005), ArXiv:hep-lat/0409097.
10. O. Philipsen, *talk at the 29th Johns Hopkins Workshop in Theoretical Physics,* August 1-3, 2005, Budapest, to appear in the Proceedings.
11. H. Meyer-Ortmanns and B. J. Schaefer, *Phys. Rev.* **D53**, 6586 (1996).
12. J.T. Lenaghan, D.H. Rischke and J. Schaffner-Bielich, *Phys. Rev.* **D62**, 085008 (2000).
13. A. Barducci, R. Casalbuoni, G. Pettini and L. Ravagli, ArXiv:hep-ph/0508117
14. C. Schmidt, PhD Thesis, Bielefeld University, 2003;
result quoted in [8], available online at ftp://ftp.uni-bielefeld.de/pub/papers /physik/theory/e6/dissertationen/schmidt.ps.gz
15. J.T. Lenaghan, *Phys. Rev.* **D63**, 037901 (2001).
16. J. Gasser and H. Leutwyler, *Nucl. Phys.* **B250**, 465 (1985).
17. P. Herrera-Siklódy, J.I. Latorre, P. Pascual and J. Taron, *Phys. Lett.* **B419**, 326 (1998).
18. P. Herrera-Siklódy, J.I. Latorre, P. Pascual and J. Taron, *Nucl. Phys.* **B497**, 345 (1997).
19. B. Borasoy and S. Wetzel, *Phys. Rev.* **D63**, 074019 (2001).
20. N. Beisert and B. Borasoy, *Eur. Phys. J.* **A11**, 329 (2001).
21. L.-H. Chan and R.W. Haymaker, *Phys. Rev.* **D7**, 402 (1973).
22. N.A. Törnqvist, *Eur. Phys. J.* **C11**, 359 (1999).
23. S. Chiku and T. Hatsuda, *Phys. Rev.* **D58**, 076001 (1998).
24. A. Jakovác and Zs. Szép, *Phys. Rev.* **D71**, 105001 (2005).

ON GRIBOV'S BREMSSTRAHLUNG OF HADRONS AT HIGH ENERGIES

G.P. KORCHEMSKY

Laboratoire de Physique Théorique,* Université de Paris XI*
91405 Orsay Cédex, France

1. Introduction

Almost forty years ago, Vladimir Naumovich Gribov published a short, 4 pages long paper [1] entitled "Bremsstrahlung of Hadrons at High Energies". The paper appeared in pre-QCD era when the theory of strong interactions has not been formulated yet but it is rather remarkable that it reported on findings which had a tremendous impact on our understanding of QCD dynamics at high energy. In this note I shall try to summarize a recent progress.

The main finding of Ref. [1] can be summarized in the following quote: *"independently of what distances are important in the strong interactions, the cross section for bremsstrahlung can be calculated by the usual formulae for the radiation accompanying an interaction, over a comparatively broad range of energies and of angles of radiation"*. In application to the process of scattering of two energetic particles carrying the momenta p_1^μ and p_2^μ, the Gribov's bremsstrahlung theorem states that the amplitude $\mathcal{M}(p_1, p_2, k)$ of emission of a soft gluon with momentum k^μ satisfying

$$\frac{2(p_1 \cdot k)}{s} \ll 1\,, \qquad \frac{2(p_2 \cdot k)}{s} \ll 1\,, \qquad k_\perp^2 = \frac{4(p_1 k)(p_2 k)}{s} \ll p_{1,2}^2\,, \qquad (1)$$

(with $s = (p_1 + p_2)^2$) can be factorized into the product of elastic amplitude $\mathcal{M}_{\rm el}(p_1, p_2)$ and the eikonal current

$$\mathcal{M}(p_1, p_2, k) = g_s \left(\frac{ep_1}{(p_1 k)} - \frac{ep_2}{(p_2 k)} \right) \mathcal{M}_{\rm el}(p_1, p_2)\,. \qquad (2)$$

*Unité Mixte de Recherche du CNRS (UMR 8627).

The physical reason for this factorization is that the classical radiation from rapidly moving charged particles does not depend neither on their internal structure, nor on the details of the interaction.

It is straightforward to generalize (2) for multiple soft gluon emission. Since soft gluons carry the color charge, additional care has to be taken to ensure the conservation of the color charge during the emission. To this end, it proves convenient to switch from the momentum to the coordinate representation and consider the soft gluon emission as a space-time process. As a relevant physical example, let us consider a propagation of an energetic quark through a cloud of soft gluons. In the limit when its energy goes to infinity, the quark behaves as a point-like charged particle that moves along a straight line and interacts with soft gluons. Assuming that the quark undergoes a hard collision at the point x_0^μ its world-line C can be parameterized as

$$C: \qquad x^\mu(s) = s\, p_1^\mu + x_0^\mu \tag{3}$$

with $-\infty < s \leq 0$ being the proper time and $p_1^\mu = \partial_s x^\mu(s)$ being its four-dimensional momentum. Interacting with soft gluons described by gauge field $A_\mu^a(x)$, the quark acquires an eikonal phase that we shall denote as W_C. According to Gribov's bremsstrahlung theorem, this phase comes from interaction of soft gluons with the color current $j_\mu^a(x) = g_s t^a \int_C dx_\mu(s)\delta^{(4)}(x - x(s))$ created by quark

$$\begin{aligned} W_C &= \mathrm{T}\exp\left(ig_s \int d^4x\, A^{a,\mu}(x) j_\mu^a(x)\right) \\ &= \mathrm{T}\exp\left(ig_s p_1^\mu \int_{-\infty}^0 ds\, A_\mu^a(x_0 + sp_1)t^a\right). \end{aligned} \tag{4}$$

Here T stands both for the time ordering of soft gluon fields and for ordering of their color charges t^a along the world-line trajectory of quark, Eq. (3). The eikonal phase W_C is a unitary operator known in the literature as the Wilson line. Substituting the soft gluon field in (4) by a plane wave, $A_\mu^a(x) = e_\mu^a(k)\,\mathrm{e}^{ikx}$, one recovers "one-half" of the eikonal current entering (2)

$$W_C = 1 + g_s \frac{(p_1 \cdot e(k))}{(p_1 \cdot k)}\, \mathrm{e}^{ikx_0} + \mathcal{O}(g_s^2)\,, \tag{5}$$

with $e_\mu(k) = e_\mu^a(k) t^a$. To reproduce the eikonal current in (2), one has to take into account the eikonal phase induced by interaction of the outgoing quark with momentum p_2^μ with soft gluon radiation. Similar to (4), this phase is given by the path-ordered exponential (=Wilson line) evaluated along world-line of outgoing quark $x^\mu(s) = x_0^\mu + sp_2^\mu$ with $0 \le s < \infty$. Thus, all effects of soft gluon bremsstrahlung in QCD are factorized into eikonal phases given by Wilson line operators.

In what follows we shall consider two important QCD observables, in which soft gluon radiation plays a crucial role: the Isgur-Wise heavy-meson form factor, $\xi(\theta)$, and parton distributions in a hadron, $f(x)$, at the edge of the phase space, $x \to 1$. As we will demonstrate below, in both cases the conditions (1) are satisfied and the Gribov's bremsstrahlung theorem is at work: both observables are given by an expectation value of a Wilson line W_C with the integration contour C fixed by the kinematics of the process. A unique feature of the contour C is that it contains a few cusps at points in Minkowski space-time where the interaction with a large momentum has occured in the underlying process. In this way, Wilson lines with cusps, being fundamental objects in gauge theories, naturally appear from QCD bremsstrahlung.

2. Isgur-Wise form factor

The Isgur-Wise form factor $\xi(\theta)$ describes the electromagnetic transition of a heavy meson $|M(v)\rangle$ with mass m and momentum $p_\mu \equiv mv_\mu$, built from a heavy quark and a light component, to the same meson with momentum $p'_\mu \equiv mv'_\mu$ (with $v_\mu^2 = {v'_\mu}^2 = 1$) [2]

$$\langle M(v')|\overline{\Psi}(0)\gamma_\mu\Psi(0)|M(v)\rangle = \xi(\theta)(v+v')_\mu\,. \tag{6}$$

In the heavy-quark limit, $m \to \infty$, it depends only on the product of velocities $v'\cdot v = \cosh\theta$, or equivalently on the angle θ between v_μ and v'_μ in Minkowski space-time. A crucial observation is that in the heavy quark limit, $m \to \infty$, the meson $|M(v)\rangle$ can be considered as the "hydrogen atom" of QCD – the light component of the meson is propagating in the color field created by a heavy quark.

The heavy quark behaves as a classical charged particle with the velocity v_μ. It interacts with the light component of the meson through its eikonal current, $J_\mu^{a,\rm eik}(x) = g_s\int_{-\infty}^0 d\tau\, v_\mu t^a \delta^{(4)}(x - v\tau)$ with t^a being the quark color charge. This allows one to apply the eikonal approximation and replace the heavy quark field in (6) as

$$\begin{aligned}
&\Psi(x) \to \mathrm{e}^{-im(v\cdot x)}\, b_v\, W_{[x;-\infty]}\,,\\
&W_{[x;-\infty]} \equiv \mathrm{T}\exp\left(ig_s\int_{-\infty}^0 d\tau\, v\cdot A(x+v\tau)\right)\,,
\end{aligned} \tag{7}$$

where $W_{[x;-\infty]}$ is the eikonal phase (4) of a heavy quark in the fundamental representation of the $SU(N_c)$ and b_v amputates this quark inside the heavy meson. Applying similar transformation to the quark field in the final state meson $|M(v')\rangle$, one obtains the following expression for the form factor [3]

$$\xi(\theta) = \langle \widetilde{M}(v')|W_{[\infty;0]}W_{[0;-\infty]}|\widetilde{M}(v)\rangle \equiv \left\langle \mathrm{T}\exp\left(ig_s \int_\wedge dx_\mu A^\mu(x)\right)\right\rangle, \tag{8}$$

with the state $|\widetilde{M}(v)\rangle = b_v|M(v)\rangle$ denoting the light component of the meson with the amputated heavy quark. Here the net effect of nonperturbative interaction with the light component of the heavy meson is accumulated only via the Wilson line evaluated along infinite Λ–shaped contour consisting of two rays that run along the meson velocities v_μ and v'_μ. It is important to notice that the contour has a cusp at the point 0, in which the interaction with the external probe has occurred.

The Isgur-Wise form factor $\xi(\theta)$ is a nonperturbative observable in QCD. It depends on hadronic, long-distance scales as well as on the ultraviolet cut-off $\mu \sim m$, which sets up the maximal energy of soft gluons. Although $\xi(\theta)$ can not be calculated at present in QCD from the first principles, its dependence on μ can be found from the renormalization group equation [4,5]

$$\left(\mu\frac{\partial}{\partial\mu} + \beta(g_s)\frac{\partial}{\partial g_s} + \Gamma_{\mathrm{cusp}}(\theta;\alpha_s)\right)\xi(\theta) = 0\,, \tag{9}$$

where $\alpha_s = g_s^2/(4\pi)$ is the QCD coupling constant and $\Gamma_{\mathrm{cusp}}(\theta;\alpha_s)$ is the cusp anomalous dimension. To the lowest order in α_s

$$\Gamma_{\mathrm{cusp}}(\theta;\alpha_s) = \frac{\alpha_s C_F}{\pi}\left(\theta\coth\theta - 1\right) + \mathcal{O}(\alpha_s^2)\,, \tag{10}$$

where $C_F = (N_c^2-1)/(2N_c)$ is the Casimir operator of the $SU(N_c)$ group in the fundamental representation. The two-loop correction to (10) has been calculated in [5] and its dependence on θ is more involved.

Eq. (9) follows from renormalization properties of the Wilson line in the right-hand side of (8). It acquires the anomalous dimension due to the presence of a cusp on the integration contour. The cusp anomalous dimension $\Gamma_{\mathrm{cusp}}(\theta;\alpha_s)$ determines universal features of soft-gluon radiation and is known as the *QCD bremsstrahlung function*. As such, it is a positive definite function of the cusp angle (for real Minkowski angle θ) at arbitrary value of the coupling constant

$$\Gamma_{\mathrm{cusp}}(\theta;\alpha_s) \geq 0\,. \tag{11}$$

To see this we recall that at the cusp point the heavy quark suddenly changes its velocity from v_μ to v'_μ and, due to instantaneous acceleration, it starts to radiate soft (virtual and real) gluons with momentum $k < \mu$ and cut-off $\mu \sim m$. Denoting the eikonal phase of the heavy quark as $W \equiv W_{[\infty;0]}W_{[0;-\infty]}$ and using its unitarity, $W^\dagger W = 1$, one calculates the total probability for the heavy quark to undergo the scattering (the Bjorken sum rule) as

$$1 = \langle \widetilde{M}(v)|W^\dagger W|\widetilde{M}(v)\rangle = |\xi(\theta)|^2 + \sum_X \left|\langle \widetilde{M}_X(v')|W|\widetilde{M}(v)\rangle\right|^2 , \qquad (12)$$

where in the right-hand side we inserted the decomposition of the unity operator over the physical hadronic states and separated the contribution of the ground state meson, $|\widetilde{M}(v)\rangle$, from excited states $|\widetilde{M}_X(v)\rangle$. The Wilson line (8) defines the probability of the elastic transition, $|\xi|^2 \sim \exp(-w)$ with $w = 2\int^\mu (dk/k)\,\Gamma_{\rm cusp}(\theta;\alpha_s(k))$. For $\theta \neq 0$, depending on the sign of $\Gamma_{\rm cusp}(\theta;\alpha_s)$, it either vanishes or goes to infinity for $\mu \to \infty$. In order to preserve the unitarity condition $|\xi|^2 \leq 1$ that follows from (12), one has to require that $\xi \to 0$ for $\mu \to \infty$ leading to (11). At $\theta = 0$ the cusp vanishes, that is the heavy meson stays intact, the sum in (12) equals zero and $\xi(\theta = 0) = 1$. This implies that the cusp anomalous dimension vanishes for $\theta \to 0$.

3. Deeply inelastic scattering for $x \to 1$

Our second example is provided by deeply inelastic scattering of a hadron $H(p)$ with momentum p_μ off a virtual photon $\gamma^*(q)$ with momentum $-q_\mu^2 = Q^2 \gg p^2$ in the semi-exclusive limit $x_{\rm Bj} = Q^2/(2p\cdot q) \to 1$, i.e. when the invariant mass of the final state system becomes small $(q+p)^2 \ll Q^2$. In the scaling limit, $Q^2 \to \infty$, the cross section of the process is expressed in terms of the twist-two quark distribution function [6]

$$f(x) = \int_{-\infty}^{\infty} \frac{d\xi}{2\pi}\, \mathrm{e}^{-ix\xi} \langle H(p)|\overline{\Psi}(\xi n)\,\Gamma\, W_{[\xi n;0]}\Psi(0)|H(p)\rangle\,, \qquad (13)$$

describing the probability to find a quark inside the hadron $H(p)$ with the fraction x of its momentum p. The Wilson line $W_{[\xi n;0]}$ stretched in between the quark fields makes the bilocal operator gauge invariant. It goes along the light-like direction $n_\mu = (q_\mu + p_\mu x_{\rm Bj})/(p{\cdot}q)$, so that $n^2 = 0$ and $n{\cdot}p = 1$.

The matrix $\Gamma = \not{n}$ in (13) serves to select the quark states with opposite helicities. For our purposes, we will not specify Γ and treat it as a free parameter. The Mellin moments of the distribution function (13) are related to the matrix elements of local twist-two operators

$$\int_0^1 dx\, x^N f(x;\mu^2) = \langle H(p)|\overline{\Psi}(0)\Gamma\,(i\,n\cdot\mathcal{D})^N\,\Psi(0)|H(p)\rangle \equiv \langle\mathcal{O}_N^\Gamma(\mu^2)\rangle\,, \quad (14)$$

where $\mathcal{D}_\mu = \partial_\mu - ig_sA_\mu$ is a covariant derivative. The dependence of the distribution function on the ultraviolet cut-off μ is described by Dokshitzer-Gribov-Lipatov-Altarelli-Parisi equations. Their solution reads in terms of the anomalous dimensions

$$\langle\mathcal{O}_N^\Gamma(\mu^2)\rangle = (\mu/\mu_0)^{-\gamma_N^\Gamma(\alpha_s)}\,\langle\mathcal{O}_N^\Gamma(\mu_0^2)\rangle\,, \quad (15)$$

where we assumed for simplicity that the coupling constant does not run, $\beta = 0$.

The anomalous dimension of the twist-two operators, $\gamma_N^\Gamma(\alpha_s)$, depends on the choice of the matrix Γ. In particular, in case when Γ selects the same helicities of the quark fields in (14), $\Gamma = (1+\gamma_5)\not{n}\gamma_\perp$, the anomalous dimension is

$$\gamma_N(\alpha_s) = \frac{\alpha_s}{\pi}C_F\Big(2\psi(N+2) + 2\gamma_{\rm E} - \frac{3}{2}\Big) + \mathcal{O}(\alpha_s^2)\,, \quad (16)$$

where $\psi(x) = d\ln\Gamma(x)/dx$ is the Euler psi-function and $\gamma_{\rm E} = -\psi(1)$ is the Euler constant. For other choices of Γ, the anomalous dimensions have extra (rational in N) terms in addition to the ψ-function, see, e.g., [7]. As we will argue below, Eq. (16) has a hidden symmetry and leads to integrability of the evolution equations for three-quark (baryonic) composite operators.

As follows from (14), the asymptotics of the distribution function for $x \to 1$ is related to the contribution of twist-two operators of large Lorentz spins $N \sim 1/(1-x) \gg 1$. One finds from (16) that the anomalous dimension scales in this limit as

$$\gamma_N(\alpha_s) = \frac{\alpha_s}{\pi}C_F\Bigg\{\ln(N+2) + \gamma_{\rm E} - \frac{3}{4} - \frac{1}{2(N+2)} - \sum_{n=1}^{\infty}\frac{B_{2n}}{2n}(N+2)^{-2n}\Bigg\} + \ldots\,, \quad (17)$$

where B_n's are the Bernoulli numbers, $B_2 = 1/6$, $B_4 = -1/30$, $\ldots$, and the ellipsis stands for higher order terms in α_s. It turns out that the leading scaling behavior $\gamma_N^\Gamma(\alpha_s) \sim \ln N$ is a universal property of the anomalous dimensions of the twist-two operators (14) for arbitrary Γ. It holds to all orders in α_s and is intrinsically related to the cusp anomaly of the Wilson loops.

The reason for this is that analyzing deeply inelastic scattering for $x \to 1$ one encounters the same physical phenomenon as in the case of the Isgur-Wise form factor, i.e. the struck quark carries almost the whole momentum of the hadron and, therefore, it interacts with other partons by exchanging soft gluons. In these circumstances, in complete analogy to the previous case, Eq. (7), the quark field can be approximated by an eikonal phase evaluated along the classical path in the direction of its velocity $p_\mu = m v_\mu$,

$$f(x) = \int_{-\infty}^{\infty} \frac{d\xi}{2\pi} \mathrm{e}^{i(1-x)\,\xi} \langle \widetilde{H}(p) | W_\Pi(v\cdot n\,\xi\mu - i0) | \widetilde{H}(p) \rangle \,, \tag{18}$$

where $|\widetilde{H}(p)\rangle$ is the state of the target hadron with amputated energetic quark and the causal $-i0$ prescription ensures the correct spectral property, $f(x) = 0$ for $x > 1$. The п-shaped Wilson line in Eq. (18) consists out of two rays and one segment: a link from $-\infty$ to 0 along the velocity of the incoming quark, next along the light-cone direction n_μ to the point ξn_μ and, then, along $-v_\mu$ from 0 to ∞,

$$W_\Pi(v\cdot n\,\xi\mu) = W_{[-\infty;\xi]}\, W_{[\xi;0]}\, W_{[0;-\infty]} \,. \tag{19}$$

Substituting (18) into (14), one finds the following relation between the matrix elements of local composite operators at large spin N and the Wilson loop expectation value [8]

$$\langle \mathcal{O}_N^\Gamma(\mu^2) \rangle = \langle \widetilde{H}(p) | W_\Pi(-iN) | \widetilde{H}(p) \rangle \equiv \langle \mathrm{T} \exp \left(i g_s \int_\Pi dx_\mu A^\mu(x) \right) \rangle \,. \tag{20}$$

Here, the large Lorentz spin of the local operator defines the length of the light-cone segment:

$$v\cdot n\,\xi\,\mu \to -iN \,. \tag{21}$$

We would like to stress that Eq. (20) only holds for $N \gg 1$.

According to Eq. (20), the μ-dependence of the twist-two operators follows from the renormalization of the Wilson line (19). The latter has two cusps located at the points 0 and ξn_μ. In distinction with the previous case, one of the segments attached to the cusps lies on the light-cone, $n^2 = 0$, and the corresponding cusp angle is infinite, $\theta \sim \frac{1}{2} \ln[(v\cdot n)^2/n^2] \to \infty$. In this limit, the cusp anomalous dimension scales to all orders in α_s as [5]

$$\Gamma_{\rm cusp}(\theta;\alpha_s) = \theta\, \Gamma_{\rm cusp}(\alpha_s) + \mathcal{O}(\theta^0) \,. \tag{22}$$

Here $\Gamma_{\rm cusp}(\alpha_s)$ is a universal anomalous dimension independent on θ. At weak coupling, it has the following form in QCD in the $\overline{\rm MS}$ scheme

$$\Gamma_{\rm cusp}(\alpha_s) = \frac{\alpha_s}{\pi} C_F + \left(\frac{\alpha_s}{\pi}\right)^2 C_F \left\{ N_c \left(\frac{67}{36} - \frac{\pi^2}{12} \right) - n_f \frac{5}{18} \right\} + \mathcal{O}(\alpha_s^3) \,, \tag{23}$$

where n_f is the number of quark flavors.

The divergence of the anomalous dimension (22) for $\theta \to \infty$ indicates that the Wilson line with a light-like segment satisfies an evolution equation different from (9). The modified equation looks like [8]

$$\left(\mu\frac{\partial}{\partial\mu} + \beta(g_s)\frac{\partial}{\partial g_s} + 2\Gamma_{\rm cusp}(\alpha_s)\ln[i(v\cdot n)\,\xi\mu] + \Gamma(\alpha_s)\right)\langle W_{\Pi}(v\cdot n\,\xi\mu)\rangle = 0\,. \tag{24}$$

Here the factor of 2 stems from the presence of two cusps on the п-shaped line contour and $\Gamma(\alpha_s)$ is a process-dependent anomalous dimension. The explicit dependence of the anomalous dimension on the renormalization scale μ implies the absence of the multiplicative renormalizability of the light-like Wilson line.

Combining together Eqs. (24) and (20), we obtain the renormalization group equation for local composite operators $\langle\mathcal{O}_N^\Gamma(\mu^2)\rangle$ at large N. Matching its solution into (15), we find the asymptotic behavior of the anomalous dimensions of the twist-two quark operators for $N \to \infty$

$$\gamma_N^{(qq)}(\alpha_s) = 2\Gamma_{\rm cusp}(\alpha_s)\ln N + \mathcal{O}(N^0)\,. \tag{25}$$

Repeating a similar analysis for the twist-two gluon operators, one can show that their matrix elements satisfy (20) with the Wilson line defined in the adjoint representation of the $SU(N_c)$. Therefore, their anomalous dimension satisfies (25) upon replacing $C_F \to N_c$ leading to [8]

$$\gamma_N^{(gg)}(\alpha_s) = \frac{N_c}{C_F}\,\gamma_N^{(qq)}(\alpha_s) + \mathcal{O}(N^0)\,. \tag{26}$$

In general, the quark and gluon operators mix with each other. However, at large N the mixing occurs through the exchange of a soft quark with momentum $\sim 1/N$. Its contribution to the corresponding anomalous dimensions is suppressed by a power of $1/N$ leading to

$$\gamma_N^{(gq)}(\alpha_s) = \mathcal{O}(1/N)\,, \qquad \gamma_N^{(qg)}(\alpha_s) = \mathcal{O}(1/N)\,. \tag{27}$$

We would like to stress that the relations (25)–(27) are valid to all orders in α_s. Remarkably enough, they hold both in QCD and its supersymmetric extensions. In the latter case, the mixing matrix has a bigger size due to the presence of additional scalar fields. Nevertheless, this matrix remains diagonal at large N. Since the fields in supersymmetric Yang-Mills theories belong to the adjoint representation, the diagonal matrix elements are the same

$$\gamma_N^{(ab)} = 2\delta_{ab}\Gamma_{\rm cusp}(\alpha_s)\ln N + \mathcal{O}(N^0) \tag{28}$$

with $a, b = (q, g, s)$ and $\Gamma_{\rm cusp}(\alpha_s)$ defined in the adjoint representation of the gauge group. As we will show in the next Section, the fact that the anomalous dimension (16) turns out to be the Euler ψ-function and, as a consequence, has a universal scaling behaviour (28) leads to a hidden integrability of the evolution equations in QCD [9,10,11,12].

4. Integrability in high-energy QCD

Scale dependence of the structure functions of deep inelastic scattering and hadronic light-cone wave functions can be studied in QCD using the Operator Product Expansion. It can be reformulated as a problem of calculating the anomalous dimensions of the composite operators of a definite twist. The operators of the lowest twist have the following general form

$$\begin{aligned} \mathcal{O}^{(2)}_{N,k}(0) &= (n\mathcal{D})^k \Phi_1(0) (n\mathcal{D})^{N-k} \Phi_2(0), \\ \mathcal{O}^{(3)}_{N,\boldsymbol{k}}(0) &= (n\mathcal{D})^{k_1} \Phi_1(0) (n\mathcal{D})^{k_2} \Phi_2(0) (n\mathcal{D})^{N-k_1-k_2} \Phi_3(0), \end{aligned} \tag{29}$$

where $\boldsymbol{k} \equiv (k_1, k_2)$ denotes the set of integer indices k_i, n_μ is a light-cone vector such that $n_\mu^2 = 0$. Φ_k denotes elementary fields (quarks, antiquarks, gluons), and $\mathcal{D}_\mu = \partial_\mu - i g_s A_\mu$ is a covariant derivative. The operators of a definite twist mix under renormalization with each other. In order to find their scaling dependence one has to diagonalize the corresponding matrix of the anomalous dimension and construct linear combination of such operators, the so-called conformal operators [7,13]

$$\mathcal{O}^{(\rm conf)}_{N,q}(0) = \sum_{\boldsymbol{k}} C_{\boldsymbol{k},\boldsymbol{q}} \cdot \mathcal{O}_{N,\boldsymbol{k}}(0)\,. \tag{30}$$

A unique feature of these operators is that they have an autonomous RG evolution

$$\mu \frac{d}{d\mu} \mathcal{O}^{(\rm conf)}_{N,q}(0) = -\gamma_{N,q} \cdot \mathcal{O}^{(\rm conf)}_{N,q}(0)\,. \tag{31}$$

Here μ^2 is a UV cut-off and $\gamma_{N,q}$ is the corresponding anomalous dimension depending on some set of quantum numbers $\boldsymbol{q}$ to be specified below. It turns out that the problem of calculating the spectrum of the anomalous dimensions $\gamma_{N,q}$ to one-loop accuracy becomes equivalent to solving the Schrödinger equation for the $SL(2,\mathbb{R})$ Heisenberg spin magnet [9,10,11,12]. The number of sites in the magnet is equal to the number of fields entering into the operators under consideration.

To explain this correspondence it becomes convenient to introduce nonlocal light-cone (mesonic and baryonic) operators built from (two and three) quark fields of the same chirality:

$$F(z_1, z_2) = \sum_{i=1}^{N_c} (\bar{\Psi}_i^{\uparrow} \not{n})_\alpha (z_1 n)(\not{n} \Psi_i^{\uparrow})_\beta (z_2 n) \tag{32}$$

$$F(z_1, z_2, z_3) = \sum_{i,j,k=1}^{N_c} \epsilon_{ijk} (\not{n} \Psi_i^{\uparrow})_\alpha (z_1 n)(\not{n} \Psi_j^{\uparrow})_\beta (z_2 n)(\not{n} \Psi_k^{\uparrow})_\gamma (z_3 n)\,, \tag{33}$$

where $(\Psi_i^{\uparrow})_\alpha = (1+\gamma_5)(\Psi_i)_\alpha/2$ is a quark field of the ith color and αth flavour. Here n_μ is a light-like vector ($n_\mu^2 = 0$) defining certain direction on the light-cone and the scalar variables z_i serve as coordinates of the fields along this direction. The quaks fields in (32) are transformed under the gauge transformations. It is tacitly assumed that the gauge invariance of the nonlocal operators $F(z_i)$ is restored by including the Wilson lines between the fields in the appropriate (fundamental or adjoint) representation. The conformal operators appear in the OPE expansion of the nonlocal operators (32) for small $z_1 - z_2$ and $z_2 - z_3$.

The field operators entering the definition of $F(z_i)$ are located on the light-cone. This leads to the appearance of the additional light-cone singularities. They modify the renormalization properties of the nonlocal light-cone operators (32) and lead to nontrivial evolution equations which as we will show below are related to integrable chain models. We notice that there exists the following relation between the conformal three-particle operators (30) and the nonlocal operators (32)

$$\mathcal{O}_{N,q}^{(\mathrm{conf})}(0) = \Psi_{N,q}(\partial_{z_1}, \partial_{z_2}, \partial_{z_3}) F(z_1, z_2, z_3)\Big|_{z_i=0}\,, \tag{34}$$

where $\Psi_{N,q}(x_1, x_2, x_3)$ is a homogenous polynomial in x_i of degree N

$$\Psi_{N,q}(x_1, x_2, x_3) = \sum_{\boldsymbol{k}} C_{\boldsymbol{k},\boldsymbol{q}} \cdot x_1^{k_1} x_2^{k_2} x_3^{N-k_1-k_2} \tag{35}$$

with the expansion coefficients $C_{\boldsymbol{k},\boldsymbol{q}}$ defined in (30). Similar relations hold for the twist-2 operators. The problem of defining the conformal operators is reduced to finding the polynomial coefficient functions $\Psi_{N,q}(x_i)$ and the corresponding anomalous dimensions $\gamma_{N,q}$.

Using the renormalization properties of the nonlocal light-cone operators (32) one can show [9,10,11,12], that to the one-loop accuracy the QCD evolution equation for the conformal operators (32) can be rewritten in the form of a Schrödinger equation

$$\mathcal{H} \cdot \Psi_{N,q}(x_i) = \gamma_{N,q} \Psi_{N,q}(x_i)\,, \tag{36}$$

where the Hamiltonian $\mathcal{H}$ acts on the x_i−variables which are conjugated to the derivatives ∂_{z_i} and, therefore, have the meaning of light-cone projection $(n \cdot p_i)$ of the momenta p_i carried by particles described by the quark fields in (32). For mesonic operator $F(z_1, z_2)$, Eq. (32), the Hamiltonian $\mathcal{H}$ is given by [7]

$$\mathcal{H}^{(2)} = \frac{\alpha_s}{\pi} C_F \left[2H_{qq}(J_{12}) + \frac{1}{2} \right] , \tag{37}$$

$$H_{qq}(J_{12}) = \psi(J_{12}) - \psi(2) , \tag{38}$$

where $C_F = (N_c^2 - 1)/(2N_c)$ and the operator J_{12}, the so-called conformal spin, is defined as

$$J_{12}(J_{12} - 1) = - \left(\partial_{x_1} - \partial_{x_2}\right)^2 x_1 x_2 \, . \tag{39}$$

The eigenfunctions for the Hamiltonian (38) take the form

$$\Psi_N^{(2)}(x_1, x_2) = (x_1 + x_2)^N \mathrm{C}_N^{3/2} \left(\frac{x_1 - x_2}{x_1 + x_2} \right) \tag{40}$$

where $\mathrm{C}_N^{3/2}$ is the Gegenbauer polynomial. The corresponding eigenvalues define the anomalous dimensions of the twist-2 mesonic operators built from two quarks with the same helicity

$$\gamma_N^{(2)} = \frac{2\alpha_s}{\pi} C_F \left[\psi(N+2) - \psi(2) + \frac{1}{4} \right] = \frac{\alpha_s}{\pi} C_F \left[\sum_{k=1}^{N} \frac{2}{k+1} + \frac{1}{2} \right] , \tag{41}$$

which coincides with (16). At large N this expression has well-known asymptotic behaviour $\gamma_N^{(2)} \sim (\alpha_s C_F/\pi) 2 \ln N$.

For baryonic operator built from three quark fields of the same chirality the evolution kernel is given by [9,10]

$$\mathcal{H}^{(3)} = \frac{\alpha_s}{2\pi} \left\{ (1 + 1/N_c) \left[H_{qq}(J_{12}) + H_{qq}(J_{23}) + H_{qq}(J_{31}) \right] + 3C_F/2 \right\} \tag{42}$$

with H_{qq} given by (38) and J_{ik} being the two-particle conformal spins. It is conformal symmetry of QCD Lagrangian which dictates that the two-particle Hamiltonian in (38) and (42) is a function of the conformal two-particle spin, but it does not fix this function. According to (38), the latter is given in QCD by the Euler ψ-function and it leads to the following remarkable property of the evolution equations for anomalous dimensions of baryonic operators [9,10,11,12]. The Schrödinger equation (36) with the Hamiltonian defined in this way has a hidden integral of motion, $[\mathcal{H}^{(3)}, q] = 0$,

$$q = i \left(\partial_{x_1} - \partial_{x_2}\right) \left(\partial_{x_2} - \partial_{x_3}\right) \left(\partial_{x_3} - \partial_{x_1}\right) x_1 x_2 x_3 \tag{43}$$

and, as a consequence, the Schrödinger equation (36) is completely integrable. Moreover, one can identify (42) as the Hamiltonian of a quantum XXX Heisenberg magnet of $SL(2,\mathbb{R})$. This model is a generalization of the celebrated Heisenberg spin$-1/2$ magnet which has been studied in 1931 by Bethe as a model of one-dimensional metal and which has been solved by the method well-known nowadays as the Bethe Ansatz. In application to QCD, the Bethe Ansatz technique allows one to solve the evolution equations and the detailed results can be found in [14]. The obtained solutions have direct applications to phenomenology of the baryon wave functions, scale dependence of twist-3 spin structure functions *etc.*

Acknowledgements

I would like to thank Julia Nyiri, Yuri Dokshitzer and Peter Levai for their kind invitation to contribute to the Gribov Memorial Volume.

References

1. V.N. Gribov, *Sov. J. Nucl. Phys.* **5**, 280 (1967) [*Yad. Fiz.* **5**, 399 (1967)].
2. N. Isgur and M. Wise, *Phys. Lett.* **B237**, 527 (1990);
   M.A. Shifman and M.B. Voloshin, *Sov. J. Nucl. Phys.* **45**, 292 (1987).
3. S.V. Ivanov, G.P. Korchemsky and A.V. Radyushkin, *Sov. J. Nucl. Phys.* **44**, 145 (1986);
   G.P. Korchemsky and A.V. Radyushkin, *Phys. Lett.* **B279**, 359 (1992).
4. A.M. Polyakov, *Nucl. Phys.* **B164**, 171 (1980).
5. G.P. Korchemsky and A.V. Radyushkin, *Nucl. Phys.* **B283**, 342 (1987).
6. J.C. Collins and D.E. Soper, *Nucl. Phys.* **B194**, 445 (1982).
7. A.P. Bukhvostov, G.V. Frolov, L.N. Lipatov and E.A. Kuraev, *Nucl. Phys.* **B258**, 601 (1985).
8. G.P. Korchemsky, *Mod. Phys. Lett.* **A4**, 1257 (1989);
   G.P. Korchemsky and G. Marchesini, *Nucl. Phys.* **B406**, 225 (1993).
9. V.M. Braun *et al.*, *Phys. Rev. Lett.* **81**, 2020 (1998).
10. V.M. Braun, S.E. Derkachov, G.P. Korchemsky and A.N. Manashov, *Nucl. Phys.* **B553**, 355 (1999).
11. A.V. Belitsky, *Phys. Lett.* **B453**, 59 (1999); *Nucl. Phys.* **B558**, 259 (1999); *Nucl. Phys.* **B574**, 407 (2000).
12. S.E. Derkachov, G.P. Korchemsky and A.N. Manashov, *Nucl. Phys.* **B566**, 203 (2000).
13. S.J. Brodsky *et al.*, *Phys. Lett.* **B91**, 239 (1980); *Phys. Rev.* **D33**, 1881 (1986); Yu.M. Makeenko, *Sov. J. Nucl. Phys.* **33**, 440 (1981);
    Th. Ohrndorf, *Nucl. Phys.* **B198**, 26 (1982);
    V.M. Braun and I.E. Filyanov, *Z. Phys.* **C48**, 239 (1990).
14. V.M. Braun, G.P. Korchemsky and D. Müller, *Prog. Part. Nucl. Phys.* **51**, 311 (2003).

PROGRESS IN LATTICE STUDIES, HADRON SPECTRUM AND COLOUR CONFINEMENT

EXACT CHIRAL SYMMETRY IN LATTICE QCD

F. NIEDERMAYER*

Institute for Theoretical Physics, University of Bern, Sidlerstr. 5, CH-3012, Switzerland

We discuss the problem of how to preserve the chiral symmetry of the continuum theory by a lattice regularization of QCD – a goal which was thought to be impossible to achieve for a long time. We also discuss the amazing properties of the corresponding lattice Dirac operator.

1. Introduction

The lattice regularization was not the favorite approach to Quantum Chromodynamics of Volodya Gribov, and one can guess why. He had the rare ability to see the physics behind the mathematical expressions, to see the space-time development behind a Feynman diagram, he wanted to understand confinement in physical terms. The lattice regularization is not quite suited for this type of approach. One works with Euclidean time, and the mathematical expressions are usually much less transparent when the lattice regularization is used.

In this talk I shall discuss the problem of chiral symmetry for which the lattice approach did not seem to be adequate. Long time it was thought (supported by a no-go theorem) that any lattice regularization has to break chiral symmetry, and one would recover it only in the continuum limit $a \to 0$, where a is the lattice spacing. (In lattice field theory one usually works for convenience with "lattice units $a = 1$". In most cases I shall use this convention.)

For some years we know that the problem has an unexpected, elegant solution, preserving the mathematical beauty of the continuum approach, with the additional benefit of being a well defined, non-perturbatively regularized theory.

*On leave from Eötvös University, HAS Research Group, Budapest, Hungary

Since the written version of Peter Hasenfratz's talk published in these Proceedings discusses many details I shall try to avoid here the unnecessary overlap[a]. The reader is asked to look at that paper for the references. I would like to mention that I do not follow here the historical path of the subject – this is also described to some extent in the above paper.

In lattice QCD we have the gauge fields (gauge connections) $U_\mu(x)$ which are elements of color SU(3). Here x labels a lattice site and $\mu = 1,\ldots,4$ is the direction of the link $(x, x+\hat{\mu})$. Further we have the fermion fields – Grassmann variables – $\psi(x)$ and $\overline{\psi}(x)$ having Dirac and color indices. These variables are defined on a hypercubic lattice in Euclidean space-time, with a finite (but sufficiently small) lattice spacing a.

The lattice QCD action has the form

$$S(U, \overline{\psi}\psi) = S_g(U) + \sum_f \overline{\psi}_f D_{m_f}(U)\psi_f, , \tag{1}$$

where the sum over flavors is written out explicitly.

For the massless case $m = 0$ in the continuum formulation one has chiral symmetry due to the relation $\gamma_5 D_{\text{cont}} + D_{\text{cont}}\gamma_5 \equiv \gamma_5(\gamma_\mu \nabla_\mu) + (\gamma_\mu \nabla_\mu)\gamma_5 = 0$ where ∇_μ is the covariant derivative. In the lattice regularization one has to give up this relation: it is not consistent with some basic requirements as locality and absence of extra, unwanted particles ("doublers"), etc. A simple argument shows that there is no legal lattice Dirac operator which anticommutes with γ_5: in that case the flavor-singlet axial current would have no anomaly. The naive discretization – using just the lattice version of the covariant derivative – avoids the anomaly by introducing doublers (at the ends of the Brillouin zone). To cure this problem of extra particles, Wilson suggested to add a lattice Laplacean, giving a cut-off mass to the doublers. The price one has to pay is that chiral symmetry is explicitly broken by this extra term. The chiral symmetry (for the non-singlet axial currents) is recovered in the continuum limit. Although in principle there is nothing wrong with this procedure – one recovers the physics of the hadronic world by taking a sufficiently fine lattice – this is definitely not an elegant solution to the problem of chiral symmetry. One has to break the singlet chiral symmetry due to the presence of anomaly but one also breaks this way the non-singlet symmetry as well (and recovers this later only as $a \to 0$).

---

[a]Which is a doubling problem as well.

The special role of the non-singlet chiral symmetry that it is broken spontaneously (in the limit of massless quarks) and this breaking is responsible for the (quasi) Goldstone bosons which dominate the low energy physics of hadrons. The spontaneous breakdown of chiral symmetry is connected to the existence of topological charge – a topological invariant – of the gauge configuration, the integral of the topological charge density $q(x) \propto \epsilon_{\mu\nu\rho\sigma} F_{\mu\nu} F_{\rho\sigma}$. Due to the Atiyah-Singer theorem the (continuum) massless, chiral invariant Dirac operator D_{cont} has exact zero eigenmodes in a gauge background field with a non-zero topological charge. The number and the chirality of these modes depends on the topological charge of the configuration. For the gauge configuration contains well separated topological charges of opposite signs the two (would-be) zero eigenmodes of opposite chirality will split into two eigenmodes with tiny imaginary eigenvalues $\lambda = \pm i\alpha$ of zero chirality. A soup of $\pm$ topological charges ("instanton soup") is expected to yield a non-zero eigenvalue density of D_{cont} at $\lambda = 0$. This – according to the Banks-Casher relation – leads to a $q\bar{q}$ condensate, i.e. the spontaneous chiral symmetry breaking.

It seems that this scenario is not compatible with the lattice regularization. Already the notion of topological charge seems to be alien to the lattice formulation. In the continuum theory the gauge field sectors with different topological charges are separated – one cannot change the topological charge by continuously changing the gauge configuration. On the other hand, in the lattice formulation any configuration can be reached continuously from any other one. A way out is to use some interpolating procedure to extend the lattice field to the continuum space-time and define the topological charge on this. Obviously, the topological charge defined in this way will be a discontinuous function of the lattice field because the mapping will be necessarily discontinuous. Further, if one would like to maintain the Atiyah-Singer index theorem – which is essential for the spontaneous braking of the exact chiral symmetry – one should have a very special Dirac operator $D(U)$ which also should be a discontinuous function of the gauge field U.

2. The Ginsparg-Wilson relation

In spite of the Nielsen-Ninomiya no-go theorem and the problems mentioned above, such Dirac operators have been constructed. Interestingly the solution came from two quite different starting points: from the domain wall/overlap construction and from the fixed-point (FP) action of a renormalization group (RG) transformation. I discuss first briefly the FP approach.

For the gauge part of the QCD action one has the saddle point (i.e. the classical) equation

$$S_{\rm g}^{\rm FP}(V) = \min_{U} \left\{ S_{\rm g}^{\rm FP}(U) + T_{\rm g}(V,U) \right\} . \tag{2}$$

Here the gauge field V lives on the coarse lattice, U on the fine one, and $T_{\rm g}(V,U)$ is the blocking kernel. This is a recursive definition, which converges very rapidly since the minimizing field $U = \overline{U}(V)$ is much smoother than the original V. Actually, on $\overline{U}(V)$ one can define uniquely a topological charge. One can show that $S_{\rm g}^{\rm FP}(V)$ has exact instanton solutions with the same value of the action as the continuum theory, and in general it represents exactly the solutions of the classical equations of motion. The necessary discontinuity of the topological charge $Q_{\rm top}(V)$ discussed above is related to the discontinuity of the function $\overline{U}(V)$: at some V's there is a degenerate minimum of the rhs. of Eq. (2). This happens as the size of an instanton is decreased and the "instanton falls through the lattice".

In the classical approximation the fermionic part remains quadratic, and the corresponding propagator satisfies the recursion relation

$$D_{\rm FP}^{-1}(V) = \omega(U) D_{\rm FP}^{-1}(U) \omega^{\dagger}(U) + \frac{1}{\kappa} . \tag{3}$$

Here $U = \overline{U}(V)$ given by the pure gauge part, κ is the coefficient and $\omega(U)$ is the smearing function in the blocking kernel $\kappa(\overline{\chi} - \overline{\psi}\omega^{\dagger})(\chi - \omega\psi)$. (Note that the $1/\kappa$ in Eq. (3) is a contact term: $\delta_{xy} \cdot 1/\kappa$.)

It turned out that $D_{\rm FP}$ has remarkable features, in particular concerning its chiral properties. Consider the anticommutator

$$\gamma_5 D_{\rm FP}^{-1} + D_{\rm FP}^{-1} \gamma_5 \equiv 2R\gamma_5 . \tag{4}$$

For a general acceptable Dirac operator the anticommutator will be non-local, since D^{-1} is non-local (in contrast to D which has to be local.) However, one can show from Eq. (3) that for $D_{\rm FP}$ the operator R *is local.* (Actually, depending on the block transformation considered, it is a contact term $\propto \delta_{xy}$ or it lives on the hypercube.) As a consequence $D \equiv D_{\rm FP}$ satisfies the Ginsparg-Wilson (GW) relation

$$\gamma_5 D + D\gamma_5 = 2DR\gamma_5 D , \qquad R \text{ is local} . \tag{5}$$

Without much loss of generality one can consider the simplest case when $2R = 1$ (i.e. $2R_{xy} = \delta_{xy}$):

$$\gamma_5 D + D\gamma_5 = D\gamma_5 D . \tag{6}$$

This is a special form of the GW relation which simplifies the discussion of the chiral properties.

Independently, Neuberger arrived at a surprisingly simple form of the "overlap Dirac operator"

$$D = 1 - \mathbf{V}\,, \tag{7}$$

where the unitary operator $\mathbf{V}$ is given by

$$\mathbf{V} = A\left(A^\dagger A\right)^{-1/2}, \qquad \text{and} \quad A = 1 - D_{\mathrm{W}}\,. \tag{8}$$

Here D_{W} is the standard Wilson Dirac operator (or any acceptable one). Obviously, $\mathbf{V}$ is unitary and using the trivial relation $\gamma_5 D_{\mathrm{W}} \gamma_5 = D_{\mathrm{W}}^\dagger$ it is easy to show that the overlap Dirac operator D satisfies the GW relation (6). Which is far less obvious is the locality of D – this is a necessary feature of any acceptable Dirac operator. As was shown by Lüscher, under some reasonable smoothness conditions on the gauge field U the overlap operator $D(U)$ is indeed local. Note that due to the inverse square root $D(U)$ is a discontinuous function of the gauge field U. The discontinuity occurs when an eigenvalue of D_{W} passes through 1.

The spectrum of $D(U)$ is on the circle with unit radius, $\lambda = 1 - \mathrm{e}^{i\alpha}$ which touches the imaginary axis at $\lambda = 0$ and cuts the real axis at $\lambda = 2$. (For brevity we use here the lattice units "$a = 1$". In physical units one has in fact a circle $\lambda = (1 - \mathrm{e}^{i\alpha})/a$ of radius $1/a$ which grows as one approaches the continuum limit $a \to 0$. Only the part of this circle which is close to the imaginary axis is physical. The lattice regularization turns the spectrum of the continuum theory which is on the infinitely long infinite imaginary axis into a finite circle.)

The spectrum of D consists of three parts

- $\lambda = 0$, with $\gamma_5 u_\lambda = \pm u_\lambda$ (denote their numbers by n_+, n_-),
- $\lambda = 2$, with $\gamma_5 u_\lambda = \pm u_\lambda$ (with multiplicities n'_+, n'_-),
- $\lambda =$ complex, with $\gamma_5 u_\lambda = u_{\lambda^*}$ and $(u_\lambda, \gamma_5 u_\lambda) = 0$.

These simple properties have surprising consequences.

3. Topological charge and the index theorem

These properties allow us to define a topological charge and to prove the index theorem on the lattice. Define the topological charge density as

$$q(x) = \frac{1}{2}\mathrm{tr}(\gamma_5 D(x, x; U)) \tag{9}$$

where the trace is taken over color and Dirac indices. It is easy to show that the total topological charge is integer and equals to the index of D:

$$Q(U) = \sum_x q(x) = \frac{1}{2}\mathrm{Tr}\,(\gamma_5 D(U)) = -\frac{1}{2}\mathrm{Tr}\,(\gamma_5(2 - D(U))) =$$
$$-\frac{1}{2}\sum_\lambda (2-\lambda)(u_\lambda, \gamma_5 u_\lambda) = n_- - n_+ \equiv \mathrm{index}(D)\,. \quad (10)$$

Here we have used the properties of the eigenvectors of D listed above. Obviously, $Q(U)$ is a topological invariant since since the integer $n_- - n_+$ cannot change smoothly.

We arrived at a highly non-trivial result: we defined (using the Dirac operator) by Eq. (9) a local function $q(x;U)$ of the gauge field which sums up to an integer (equal to the index of the Dirac operator). It can be shown that as a consequence of the GW relation $q(x)$ reproduces the known topological charge density in the continuum limit:

$$q(x) = \frac{1}{32\pi^2}\epsilon_{\mu\nu\rho\sigma}\mathrm{tr}\,(F_{\mu\nu}(x)F_{\rho\sigma}(x)) + O(a^2)\,. \quad (11)$$

The construction defines a topological charge of the gauge configuration. Obviously, this definition will be different for different solutions of the GW relation. However, for sufficiently extended topological objects they give the same value for Q – only the point where "instantons fall through the lattice" will be different – but instantons of size of the lattice spacing a will be suppressed as one approaches the continuum limit. As discussed, any realization of the GW relation defines a topological charge. In addition, for the FP realization the topological charge defined by D_{FP} agrees with the one defined by the pure gauge theory.

4. Exact chiral symmetry on the lattice

At this point we have a peculiar situation: all signs of chiral symmetry are present – we have a lattice theory which reproduces the features of the continuum chiral symmetric situation (e.g. the index theorem) but the chiral symmetry, $\gamma_5 D + D\gamma_5 = 0$ is not satisfied.

Lüscher made the important observation that the lattice action with a Dirac operator satisfying the GW relation has an exact symmetry which can be considered as a generalization of the chiral symmetry of the continuum theory. Perform an infinitesimal change of variables $\psi \to \psi + i\epsilon\delta\psi$ and

$\overline{\psi} \to \overline{\psi} + i\epsilon\delta\overline{\psi}$ with a global flavor singlet transformation

$$\delta\psi = \gamma_5 \left(1 - \frac{1}{2}aD\right)\psi\,, \qquad (12)$$
$$\delta\overline{\psi} = \overline{\psi}\left(1 - \frac{1}{2}aD\right)\gamma_5\,.$$

From the GW relation Eq. (6) it follows that the action is invariant

$$\delta(\overline{\psi}D\psi) = 0\,. \qquad (13)$$

The corresponding flavor non-singlet transformation is given by

$$\delta\psi = T\gamma_5 \left(1 - \frac{1}{2}aD\right)\psi\,, \qquad (14)$$
$$\delta\overline{\psi} = \overline{\psi}\left(1 - \frac{1}{2}aD\right)\gamma_5 T\,,$$

where T is a generator in flavor space, is also a symmetry of the action. (In these relations we write out explicitly the lattice spacing a in order to see the formal continuum behavior $a \to 0$.)

Now it seems that we have more symmetry than possible – the flavor singlet transformation should be anomalous. Where is the anomaly hidden? As pointed out by Lüscher, the fermionic integration measure is *not* invariant under the singlet transformation in Eq. (12)

$$\begin{aligned}\delta[d\overline{\psi}d\psi] &= \mathrm{Tr}(a\gamma_5 D)[d\overline{\psi}d\psi] \\ &= 2N_\mathrm{f}(n_- - n_+)[d\overline{\psi}d\psi]\,.\end{aligned} \qquad (15)$$

That is, the measure breaks the flavor singlet chiral symmetry in a topologically non-trivial gauge field where $\mathrm{index}(D) = n_- - n_+ \neq 0$. The corresponding (global) Ward identity for the flavor singlet transformation for an arbitrary operator O is

$$\langle \delta O \rangle_\mathrm{F} = 2N_\mathrm{f}\nu \langle O \rangle_\mathrm{F}\,, \qquad (16)$$

where $\nu = Q(U)$ is the topological charge, N_f is the number of flavors and $\langle \ldots \rangle_\mathrm{F}$ denotes the integral over the fermionic fields in a fixed gauge background.

Note that the non-singlet symmetry is not anomalous, $\langle \delta O \rangle_\mathrm{F} = 0$ since Eq. (15) for this case contains $\mathrm{Tr}(a\gamma_5 DT) \propto \mathrm{tr}(T) = 0$. Hence we have here a situation similar to the continuum case: only the flavor singlet symmetry is broken explicitly by the anomaly, and this breaking is due to the fermionic measure. We would like to stress, however, that this is now on the level of a non-perturbatively defined regularized theory. No extra regularization beyond the lattice regularization was used here as opposed to the continuum case where Fujikawa introduced for this purpose the heat kernel regularization. Of course, the continuum limit remains a delicate question, but it is reassuring that one can maintain the exact chiral symmetry already at finite lattice spacing.

5. The chiral decomposition

In the continuum formulation chiral symmetry allows one to decompose the fermion field into left and right chiral components, transforming independently. Besides yielding a convenient formalism for vector theories like QCD, this decomposition is a necessary first step towards the chiral gauge theories. Due to the exact chiral symmetry of a GW Dirac operator this decomposition can be done in the lattice regularization as well. One can define the operator

$$\hat{\gamma}_5 = \gamma_5(1 - aD) = \gamma_5 \mathbf{V} \,. \tag{17}$$

Due to the relation $\gamma_5 \mathbf{V} \gamma_5 = \mathbf{V}^\dagger$ we have

$$\gamma_5 D = -D\hat{\gamma}_5 \,, \quad \text{and} \quad \hat{\gamma}_5^2 = 1 \,. \tag{18}$$

Introducing the projectors

$$P_\pm = \frac{1}{2}(1 \pm \gamma_5) \,, \quad \hat{P}_\pm = \frac{1}{2}(1 \pm \hat{\gamma}_5) \,, \tag{19}$$

the Dirac operator can be decomposed as

$$D = P_+ D \hat{P}_- + P_- D \hat{P}_+ \,. \tag{20}$$

As a next step we introduce the chiral components of the fermionic field:

$$\psi_L = \hat{P}_- \psi \,, \quad \overline{\psi}_L = \overline{\psi} P_+ \,, \tag{21}$$

$$\psi_R = \hat{P}_+ \psi \,, \quad \overline{\psi}_R = \overline{\psi} P_- \,. \tag{22}$$

In terms of these components the fermionic part of the action assumes the form

$$\overline{\psi} D \psi = \overline{\psi}_L D \psi_L + \overline{\psi}_R D \psi_R \,, \tag{23}$$

in close analogy to the continuum case. Note that Eqs. (21,22) are not symmetric in ψ and $\overline{\psi}$. This asymmetric definition plays an essential role when one tries to define chiral gauge theories – these questions, however, will not be discussed here.

Finally, we mention that now we are in a position to introduce the θ-parameter into the QCD lattice action in an exact manner, even at nonzero lattice spacing. Define the lattice Lagrangean with a complex mass m and the θ-parameter:

$$L = L_{\rm g} + \overline{\psi} D \psi + m \overline{\psi}_R \psi_L + m^* \overline{\psi}_L \psi_R - i\theta q(x)\,, \qquad (24)$$

where $L_{\rm g}$ is the gauge action and $q(x)$ is defined through D by Eq. (9). By changing the variables

$$\psi_L \to \mathrm{e}^{-i\alpha} \psi_L\,, \quad \overline{\psi}_L \to \mathrm{e}^{i\alpha} \overline{\psi}_L\,. \qquad (25)$$

one can restore the invariance of the action by $m \to \mathrm{e}^{i\alpha} m$ and the non-invariance of the measure can be compensated by $\theta \to \theta - N_{\rm f}\alpha$, exactly as in the continuum theory.

6. Summary

We have shown that a Dirac operator satisfying the GW relation provides a mathematically elegant solution to the problem of chiral symmetry in the framework of the lattice regularization. Note that the lattice regularization is not only a numerical tool to simulate the strong interaction on a computer – it is the only non-perturbative regularization based on first principles. Therefore the fact that exact chiral symmetry can be maintained by the lattice regularization is an important theoretical achievement. It is also useful for practical purposes – it avoids a few annoying properties of the Wilson action: the additive mass renormalization of the fermion masses with the corresponding fine tuning, the mixing of operators of different chiral representations and other artifacts connected with the explicit breaking of the chiral symmetry. On the other hand, the Dirac operators satisfying (even approximately) the GW relation are much more complicated that the Wilson action. Due to new numerical methods and the increasing computer power, however, Monte Carlo simulations using this type of Dirac operators became now possible.

Acknowledgments

I thank the organizers for the inspiring workshop.

References

1. P. Hasenfratz, *CP, T and CPT in non-perturbative chiral gauge theories*, *This proceedings*, and references therein.

CONFINEMENT AND FINE TUNING

V.I. ZAKHAROV

MPI, Föhringer Ring 6, 80805, München, Germany
E-mail:xxz.mppmu.mpg.de

We briefly review and comment on the lattice data which indicate that confining vacuum fields are fine tuned, exhibiting non-trivial dependences both on the infrared and ultraviolet scales.

1. Introduction

This mini-review of lattice data on confining field configurations and their implications is intended as a contribution to a volume dedicated to the memory of Vladimir Naumovich Gribov. The problem of confinement occupied his mind for many years and the solutions proposed are outlined in Ref. [1] [a]. We cannot match the deepness of the Gribov's analysis. Instead, we emphasize lattice data on the confining fields [b]. The data suggest that the confining field configurations in the vacuum state of lattice Yang-Mills theory are fine tuned, combining dependences on the infrared and ultraviolet scales. The notion of fine tuning is borrowed actually from the Higgs physics and, generically, refers to the case when the mass of a state is much smaller than its inverse radius. We are driven to introduce this notion into physics of QCD [5] not by inspiration but rather by the data. It is even more amusing that V.N. Gribov did argue on pure theoretical grounds that in QCD the infrared and ultraviolet regimes are strongly correlated to produce a confining solution [c].

[a]For a recent review see Ref. [2].
[b]The data are mostly due to the ITEP lattice group, see, in particular, Ref. [3].
[c]This point, as a building block of the Gribov's vision of the confinement theory, has been recently emphasized by A.I. Vainshtein [4]. During my last conversation on physics with V.N. Gribov – while walking through the streets of Munich – I did ask him, in very vague terms, about 'fine tuning'. He did approve the idea immediately. I was so much surprised that postponed detailed discussion till later times. There happened to be no later times and I have not learned what Vladimir Naumovich had actually in mind. I still believe that similarity of views is in fact only generic, not extending to details.

It goes without saying that much more efforts is needed to appreciate confinement in terms of fine-tuned fields. In these notes we present only rather scattered observations on the lattice phenomenology of the confining fields.

2. Case study: compact U(1)

In Abelian case, confinement is well understood in terms of fine tuning, both theoretically [6] and in the lattice simulations [7].

The case study is the compact $U(1)$ gauge theory. The action of the model in the continuum limit is the same as for free electromagnetic field:

$$S = \frac{1}{4e^2}\int d^4x F_{\mu\nu}^2 \quad . \tag{1}$$

There exists a classical solution which corresponds to the Dirac monopole The corresponding radial magnetic field is proportional to $\mathbf{B} \sim \mathbf{r}/e \cdot r^3$. 'Compactness' of the $U(1)$ is just invisibility of the Dirac string provided that the Dirac quantization condition is satisfied. If the monopole has a trajectory of length L, then the corresponding action is equal to:

$$S_{mon} = L \cdot M(a) = (\text{const})\frac{L}{ae^2} \tag{2}$$

where $M(a)$ is the monopole radiative mass which is proportional to integral from the magnetic field squared and divergent in the ultraviolet. To make the radiative mass of the monopole well defined one has to introduce a cut off. Within the lattice formulation, the cut off is provided by finiteness of the lattice spacing a.

If one keeps the length of the trajectory finite the suppression due to the action (2) is infinitely strong in the limit $a \to 0$ and the monopoles seem irrelevant. However, one has to take also into account the effect of the entropy as well. Namely, the probability to observe a trajectory of length L is proportional to:

$$W(L) \sim \exp(-S_{mon}) \cdot N_L \sim \exp\Big(\frac{L}{a}\big(-\frac{\text{const}}{e^2} + \ln 7\big)\Big) , \tag{3}$$

where N_L is the number of different trajectories of length L. This number depends in fact on details of the regularization in the ultraviolet and the factor $\ln 7$ is valid (approximately) for a cubic lattice. Indeed, the monopole trajectory can be continued in seven various directions at each step and be accumulating the same total length L [d].

---

[d] The backward motion is not counted since if a link is covered in the both directions, it is not counted as a part of trajectory at all.

It might worth emphasizing that the 'lattice' so far has been introduced for purely theoretical reasons, to count the number of trajectories of length L. The whole theoretical framework is called polymer approach to field theory and can be developed much further, see e.g. Ref. [8].

It just happens so that this formulation is very convenient for comparison with the lattice data since the monopole trajectories are defined on the lattice as violations of the Bianchi identities:

$$\partial_\mu \tilde{F}_{\mu\nu} = j_\nu^{mon} \quad , \tag{4}$$

where j_ν^{mon} is the monopole current. Thus, monopoles are defined as trajectories [e]. Violation of the Bianchi identities assumes singular fields. However, the whole procedure is well defined in the lattice regularization [9].

Coming back to Eq (3) , we see that there exists a critical value of the coupling e^2_{crit} when the effects of the suppression due to the action of the enhancement due to the entropy cancel each other:

$$\frac{\text{const}}{e^2_{crit}} \approx \ln 7 \, . \tag{5}$$

Then any length of the trajectory is allowed, including the infinite one.

The central point is that only fine tuned theories with

$$|e^2 - e^2_{crit}| \equiv \epsilon \ll 1 \tag{6}$$

are interesting. Indeed, the natural scale for the lengths of the trajectories is of order $L_{typ} \sim a/\epsilon$ and it can be maintained independent of a with $a \to 0$ only by tuning ϵ.

3. Vacuum defects

Now we proceed to a much more subtle non-Abelian case. At first sight, a direct generalization of the Abelian case would read:

$$D_\mu G_{\mu\nu} = 0 \, , \qquad D_\mu \tilde{G}_{\mu\nu} \equiv j_\nu^{mon} \, , \tag{7}$$

where D_μ is the covariant derivative, $G_{\mu\nu}$ is non-Abelian field strength tensor and we suppressed the color indices. However, this set of equations looks self-contradictory. Indeed, the monopole current now has also a color index. Thus, we introduce in fact a new colored object and the equations of motion should have also been changed.

---

[e]A non-trivial point is that the magnetic current is defined on the quantum level, for each field configuration. The classical equations of motion, $\partial_\mu F_{\mu\nu} = 0$ are true, however, only on average.

Let us make a step back and first assume only that the equations of motion are true. Then one can inverse them and express potential in terms of the field strength tensor [10]:

$$A_\mu = \frac{1}{g} G^{-1}_{\mu\nu}(\partial_\lambda G_{\lambda\nu}) \quad , \tag{8}$$

where $G^{-1}_{\mu\lambda}$ is the matrix inverse to the matrix of the field strength tensor. As far as the potential (8) is well defined the Bianchi identities would be satisfied as well. Indeed, singular potential are needed to violate the Bianchi identities.

Thus, it is natural to define defects in terms of singularities of the matrix G^{-1} [11,12]. Consider first three-dimensional case and gauge group SU(2). Then, in general case there exist three independent magnetic-field vectors $\mathbf{B}^a$ where a is the color index, a=1,2,3. Introduce determinant constructed on these fields,

$$\Delta \equiv \epsilon_{abc}\epsilon_{ikl} B^a_i B^b_k B^c_l \quad . \tag{9}$$

Singularities of the matrix G^{-1} are associated with zeros of the determinant. The condition

$$\Delta = 0 \tag{10}$$

defines then a vacuum 'defect'. Note that the definition of the defect is perfectly gauge and rotational invariant.

Condition (10) constitutes a single constraint and, generically, defect is a 2d dimensional surface in 3d space [f]. Alternatively, the defects are lines on the dual lattice. In case of 3d non-Abelian theory the defects have not been studied on the lattice yet. However, one can argue that Abelian compact Higgs model with charge $Q = 2$ matter field imitates the non-Abelian case. Within this Higgs model the defects are indeed 1d lines on the dual lattice [14].

Coming back to the four dimensional case, we notice first that the group of rotations in four dimensions is a direct product of two independent rotation groups in three dimensions, $O(4) = O(3) \times O(3)$. It is well known that the chiral gluon fields, $(G^\pm)^a_i$ are irreducible representations of these groups. Respectively, there are two independent determinants, $\Delta^\pm$ similar to (9). Therefore, one can introduce two types of 3d defects:

$$\Delta^+ = 0 \; or \; \Delta^- = 0 \; . \tag{11}$$

[f] Note that this way of counting the dimensionality of the defects goes back to Ref. [13].

These 3d defects are distinguished by chirality of the gluon fields. Namely, if $\Delta^+ = 0$ then the three independent color vectors which can be defined on the corresponding 3d subspace are fields with negative chirality, $(G^-)^a_i$, $a = 1, 2, 3;\ i = 1, 2, 3$.

Note that the vanishing of one of the determinants $\Delta^\pm$ does not induce, generally speaking, any singularity of the matrix G^{-1}. Thus, chirality of the gluon fields is the only signal for these defects. In an alternative language, these 3d subspaces are expected to possess positive, respectively, negative topological charge. On the lattice, the distribution of the topological charge can be probed by studying properties of eigenfunctions of the Dirac operator constructed on the vacuum gluonic fields, see e.g. Ref. [15]. Very recently, there appeared data supporting the hypothesis that the topological charge is concentrated on lower-dimensional defects, see, in particular, Ref. [16] and the last paper in Ref. [3]. However, at this moment it is too early to judge whether these, newly emerging defects are associated with 3d subspaces of vanishing of Δ^+ (Δ^-).

In case of 4d singularity of the matrix G^{-1} arises if both determinants are equal to zero:

$$\Delta^\pm = 0\,. \tag{12}$$

These defects are two dimensional subspaces with following properties:

* The fields are singular, generally speaking;

** The non-Abelian fields are aligned with the surface. Indeed there are two independent field-strength tensor on the surface defined by Eq. (12) and locally the surface is spanned on these vectors.

Defects with such properties are indeed observed in lattice simulations, see, in particular the first paper in Ref. [3]. The surfaces are so the called central vortices and were discovered first in the process of search for confining field configurations, for review see Ref. [17].

As the next step in classification of the defects, one can introduce zeros of second order:

$$\Delta^+ \sim 0^2\,, \Delta^- \sim 0^1\,,\ or\ \Delta^- \sim 0^2,\ \Delta^+ \sim 0^1\,. \tag{13}$$

These three conditions single out 1d defects, or lines. There is only one independent field-strength vector associated with these defects. It is natural to identify these defects with the lattice monopoles which were discovered in the course of the confinement studies, for review see, e.g. Ref. [18].

To reiterate, the confining field configurations are visualized as infinitely thin two-dimensional sheets carrying singular magnetic fields. The surfaces are populated by particles, or monopoles which are represented by trajectories lying on the surfaces.

4. Fine tuning

It might worth emphasizing that the classification scheme above is based on classical-field considerations. In particular, we assumed that the equations of motion are valid. Quantum mechanically, this is true only on average. Similarly, the properties which follow from the classification scheme are expected to be true only on average. For a given vacuum field configuration the monopoles and vortices are defined in a specific lattice language. In particular, to identify monopoles one projects original non-Abelian fields into the closest Abelian field configuration [g].

Both monopoles and vortices percolate through the vacuum. The total length, respectively area of these infinite clusters is in physical units, for references see [17,18]:

$$L^{mon}_{perc} \approx 30\ (\mathrm{fm})^{-3}\ V_{total}\,, \tag{14}$$

$$A_{vort} \approx 24\ (\mathrm{fm})^{-2}\ V_{total}\,, \tag{15}$$

where V_{total} is the total volume of the lattice.

On the other hand, the non-Abelian action associated with the defects is singular in the continuum limit $a \to 0$. For the vortices one finds:

$$S_{vort} \approx 0.54 \cdot \frac{A_{vort}}{a^2}\,, \tag{16}$$

see first paper in Ref. [3]. Moreover, for the monopoles one finds [20]:

$$S_{mon} \approx \ln 7 \cdot \frac{L}{a}\,, \tag{17}$$

where we quote the numerical coefficient in the form convenient for comparison with (3). Note that in the $U(1)$ case one has to tune the coupling e^2 to ensure mass of the monopole independent on the lattice spacing. In the non-Abelian case there is no free parameter to tune. Indeed, the coupling runs and cannot be tuned. The data (17) indicate, however, that there exists self-tuning ensured by dynamics of the non-Abelian fields.

We pause here to reiterate that the ultraviolet divergence (17) does not imply that there is no continuum limit for the monopoles. The mass of a particle corresponds, in the language of statistical physics, to free energy,

[g] Note that the monopoles implied by the classification scheme of the preceding section are indeed Abelian since they are characterized by a single non-vanishing field-strength vector. However, unlike pure Abelian monopoles, the monopoles defined through (13) are not spherically symmetrical. And, indeed, the field of the lattice monopoles defined through Abelian projection is not spherically symmetrical in terms of non-Abelian fields [19].

which is, in turn, difference between energy and entropy. In case of fine tuning, the free energy does have a smooth continuum limit. On the other hand, both energy and entropy are ultraviolet divergent.

The next question is then, what is the significance of the divergence (17) if only free energy has direct physical meaning. The answer is that measuring action (17) is relevant to measuring size of the excitation. Indeed, particle data tables contain not only data on masses of particles but on their sizes as well. These two parameters are independent, generally speaking. In particular, there are strong bound on possible finite size of quarks. Note that the size is not directly related to any matrix element. Rather, one relies on evaluating theoretically a cross section (mostly, inclusive).

Similarly, observing ultraviolet divergence (17) signals zero-size fluctuations. The size is smaller or of order of the lattice spacing a.

Sensitivity to the ultraviolet physics makes the vacuum defects subject to stringent theoretical constraints, for detailed discussion see [5,21]. Indeed, the observation (17) looks exactly the same as the condition (5) for fine tuning. However Eq (5) refers explicitly to an elementary particle. While in case of QCD no other particles but gluons are allowed to exist at short distances. To make the constraint precise, one should consider ultraviolet divergences in local matrix elements. In particular, if one introduces a magnetically charged field ϕ_M, then for elementary particle one would have

$$\langle 0| \; |\phi_M|^2 \; |0\rangle \;\sim\; a^{-2} \quad . \tag{18}$$

Such divergence is not allowed for our 'monopoles' since there are no point-like particles but gluons at short distances. What is allowed is

$$\langle 0| \; |\phi_M|^2 \; |0\rangle \;\sim\; a^{-2} \;\sim\; \Lambda^2_{QCD} \quad . \tag{19}$$

Eq. (19) is the simplest example of constraints due to asymptotic freedom.

Eq. (19), in turn, can be converted into a constraint on the total length of the monopole trajectories:

$$L_{total} \;\leq\; (\text{const}) \cdot \Lambda^2_{QCD} a^{-1} \cdot V_{total} \quad . \tag{20}$$

Note that it is not sufficient to compare this constraint with the data on the percolating monopole cluster (14). Indeed, apart from the infinite percolating cluster there exist finite, or short cluster. Moreover, for an elementary particle it is short cluster that dominate the total length,

$$L_{total} \;\sim\; a^{-3} V_{total} \quad . \tag{21}$$

It is most remarkable that the constraint (20) is satisfied by the data on the lattice monopoles:

$$L_{total} \approx 6.2\,(fm)^{-2}\cdot a^{-1}V_{total} \quad , \tag{22}$$

see second paper in Ref. [3]. From the geometrical point of view Eq. (22) just imply that the 'monopoles' live on a two-dimensional surface, as is implies by the classification scheme of the preceding section, see Eq. (12).

To summarize, the confining field configurations are infinitely thin crumpled 2d surfaces which preserve the total area in physical units if the lattice spacing is changed. The latter observation implies cancellation between action and entropy. Generically, similar cancellation is postulated within field theory in the polymer representation, where all the factors are explicit.

5. Vortices and $\langle (A^a_\mu)^2\rangle_{min}$

Amusingly enough, the data (22) allow to find the vacuum expectation value of the operator of a magnetically charged scalar field living on the vortices:

$$\langle 0|\ |\phi_M|^2\ |0\rangle \approx (0.2\ \mathrm{GeV})^2 \quad . \tag{23}$$

Moreover, definition of the 'monopoles' – within the classification scheme we are pursuing– is perfectly gauge invariant. Thus, one can wonder whether (23) can be related to matrix elements of the gluon fields. The point is that the first explicitly gauge-invariant has dimension four, $(G^a_{\mu\nu})^2$.

A natural guess is that (23) is related to the minimal value of the gauge field squared, $\langle (A^a_\mu)^2\rangle_{min}$, introduced in Ref. [22]. The minimization implies a kind of non-locality. In particular, in case of pure electromagnetic field there is an explicit relation for $\langle (A^a_\mu)^2\rangle_{min}$ in terms of the field-strength tensor [h]:

$$\langle\ A^2_\mu\ \rangle = (\mathrm{const})\int \frac{d^4x d^4x'}{(x-x')^2}F_{\mu\nu}(x)F_{\mu\nu}(x') \quad . \tag{24}$$

Eq. (24) is written for classical fields. In the quantum case, the condensate $\langle (A^a_\mu)^2\rangle_{min}$ suffers from ultraviolet divergences.

It is quite obvious, on the other hand, that the ultraviolet divergent piece, $\langle (A^a_\mu)^2\rangle_{min} \sim a^{-2}$ is not related to the physics of confinement since it is saturated by zero-point fluctuations. Moreover, it was speculated [5,12] that only that part of $\langle (A^a_\mu)^2\rangle_{min}$, which cannot be related to the field

[h] For a direct generalization to the non-Abelian case see Ref. [23].

strength tensor is specific for the non-Abelian case and related to the confinement. From considerations of sect. 3 it follows that the potential associated with the two-dimensional defects cannot be related to the field strength tensor. Indeed, the surface is defined in terms of singularity of the matrix G^{-1} and these singularities preclude the potential from being related to the field strength tensor, see (8). Moreover, the potential itself is expected to be singular in the continuum limit, just because of the singularity of the G^{-1}. Note that the situation here is similar to the case of, say, Dirac monopole in the Abelian case. Indeed, in the latter case we also have a singular potential and violation of Bianchi identities.

On the lattice, the singularities are regularized by finiteness of the lattice spacing and the $\langle (A^a_\mu)^2\rangle_{min}$ associated with the 2d defects can be estimated as

$$\left(\langle (A^a_\mu)^2\rangle_{min}\right)_{vort} \sim a^{-2}\cdot(a\cdot\Lambda_{QCD})^2 \sim \Lambda^2_{QCD}, \qquad (25)$$

where the factor $(a\cdot\Lambda_{QCD})^2$ stands for the probability of a given link to belong to the vortex and is a direct consequence of the observation (14).

Thus, we see that the value of $\langle (A^a_\mu)^2\rangle_{min}$ which *cannot* be related to the field strength tensor is indeed of order Λ^2_{QCD}, as one would expect. Moreover, recent measurements [24] of the variation of $\langle (A^a_\mu)^2\rangle_{min}$ within the confining string (which extends between external heavy quarks) do indicate strongly that this variation is of order

$$\delta\left(\langle (A^a_\mu)^2\rangle_{min}\right) \sim \Lambda^2_{QCD} \ .$$

There is argument [24] that the quantity $\langle (A^a_\mu)^2\rangle_{min}$ takes over the role of the Higgs field usually introduced within the monopole mechanism of confinement (for review see [18]). Further link between vortices, $\langle (A^a_\mu)^2\rangle_{min}$ and confinement is provided by results of measurements of the density of the vortices inside the confining tube [25]. Namely, it was observed that the vortices are 'scared off' by the electric field of external quarks so that the density of vortices inside the tube is strongly suppressed. Now, we can add that the vortices carry away with themselves $\langle (A^a_\mu)^2\rangle_{min}$, as is observed in Ref. [24].

Thus, we see that consideration of the two-dimensional vacuum defects and (see section 3) provides links between singular fields, condensate of dimension two, and the monopole mechanism of confinement.

6. Vortices and confinement

Phenomenological evidence in favor of the vortex mechanism of confinement can be found, e.g., in Ref. [17]. However, confinement phenomena at large distances are not sensitive to the ultraviolet scale. In particular, all the evidence in favor of the vortex mechanism of confinement discussed in [17] would remain unchanged if measurements are made at a fixed value of the lattice spacing, $a \sim \Lambda_{QCD}^{-1}$. In this sense, the ultraviolet facet of the vortices discussed in these notes cannot be derived directly from the confinement. Moreover, one can wonder how singular fields, see (14) can be relevant to confinement. To confront these doubts we consider here stochastic model of confinement, for a review see [26]. We will see that the model is actually simplified and appears more consistent for singular fields. Also, we will see that a string tension of order $\sigma \sim \Lambda^2_{QVD}$ comes out absolutely natural from considering two-dimensional defects of section 3.

For our purposes, we can understand confinement as area law for the Wilson loop, $\langle W \rangle$. Moreover, to introduce the stochastic model it is convenient to start with an Abelian case. Then one uses the following chain of identities and approximations to evaluate it:

$$
\begin{aligned}
W &= \langle \exp\big(ig \oint A_\mu dx_\mu\big)\rangle \\
&\approx \exp\big(-g^2/2 \oint dx_\mu \oint dx'_\nu \langle A_\mu, A_\nu\rangle\big) = \\
&= \exp\big(\int d\sigma_{\mu\nu} d\sigma'_{\alpha\beta}\langle G_{\mu\nu}, G_{\alpha\beta}\rangle\big) \ , \qquad (26)
\end{aligned}
$$

where $d\sigma$ are elements of area and we used the Stokes theorem. Moreover, for the correlator of the field strength tensors one substitutes

$$\langle G(x), G(x')\rangle = \langle G^2\rangle_{conf} \exp - |x - x'|/T \ , \qquad (27)$$

where we suppressed for simplicity the Lorentz indices, and introduced notation $\langle G^2\rangle_{conf}$ for the vacuum expectation value of the confining component of the vacuum fields. Moreover, assuming that the surface in Eq. (26) is in fact the minimal area one obtains the area law for the Wilson loop, with the string tension

$$\sigma \sim \langle G^2\rangle_{conf} \cdot T^2 \ . \qquad (28)$$

In the realistic, non-Abelian case one assumes similar estimates to hold with

$$\langle G^2\rangle_{conf} \sim \Lambda^4_{QCD}, \ T \sim \Lambda^{-1}_{QCD} \ , \qquad (29)$$

for justification and applications see, e.g., review in Ref. [26].

From algebraic point of view, derivation of (29) is much more complicated because the fields do not commute any longer and one uses the P-exponent and the corresponding non-Abelian version of the Stokes theorem, for details and references see [26]. It is not difficult, however, to figure out which bilinear correlator enters equations like (26). Indeed, one begins with the Wilson loop which is gauge invariant and, therefore, it can be expressed in terms of gauge invariant correlators alone. If we keep only bilinear in $G^a_{\mu\nu}$ terms then this correlator is nothing else but

$$D(x,x') \;=\; \langle G^a(x)\;\Phi_{a,b}\;G^b(x')\rangle \quad , \tag{30}$$

where we suppressed the Lorentz indices, a, b are color indices and $\Phi_{a,b}$ is the Dirac string, or parallel color transport which ensures the gauge invariance. The central point now is that $\Phi_{a,b}$ has infinite self energy. Indeed, it is like Wilson line itself. In case of Wilson line the self energy is subtracted as far as one is interested in the heavy-quark potential. In our case of using the stochastic approach we should conclude, however, that

$$D(x,x') \;\sim\; \exp(-(\text{const})|x-x'|/a) \quad , \tag{31}$$

where a is the lattice spacing used as an ultraviolet regulator for the self energy just mentioned.

The stochastic model may still be valid, with account of (31), provided that

$$\langle G^2\rangle_{conf} \;\sim\; \Lambda^2_{QCD}\cdot a^{-2} \quad . \tag{32}$$

In other words the confining fields are to be singular in the ultraviolet. For normalization, let us mention that in the limit $a \to 0$ the vacuum expectation value $\langle (G^a_{\mu\nu})^2\rangle$ is dominated by perturbative fluctuations:

$$\langle G^2\rangle_{p.th.} \;\sim a^{-4} \quad , \tag{33}$$

while the instanton contribution is of order

$$\langle G^2\rangle_{inst} \;\sim\; \Lambda^4_{QCD} \quad , \tag{34}$$

where by instantons we mean any quasiclassical or soft field configuration. Although the estimate (31) might look absolutely unrealistic, it does agree with the lattice data on 2d defects quoted above, see (14), (16). Moreover, combining all the numerical factors, we find out that the confining string tension evaluated in the stochastic model approximation is about one half of the actual value:

$$\sigma_{stoch} \;\approx\; \frac{1}{2}\sigma_{tot} \quad . \tag{35}$$

Since the stochastic model invokes assumptions which can hardly be exact, (in particular, choosing the minimal area) the agreement of the model with the data is quite good.

Most recently, the stochastic component of the vacuum fields has been studied by measuring the so called inverse participation ratio (IPR). Namely, one solves numerically equations

$$D^2\phi_n^a \;=\; \lambda_n^2\phi_n^a \tag{36}$$

where D^2 is the covariant Laplacian constructed on the vacuum fields, generated numerically, λ_n^2 and ϕ_n^a are eigenvalues and eigenfunctions of a colored scalar particle, interacting with the vacuum gluonic fields. The eigenfunctions are normalized by the condition:

$$\int d^4x\rho_n(x) \;\equiv\; 1\,, \tag{37}$$

$$\rho_n(x) \;=\; (\text{const})|\phi_n^a(x)|^2\,. \tag{38}$$

The IPR is then defined as

$$(IPR)_n \;\equiv\; V_{total}\int d^4x\rho_n^2(x) \quad . \tag{39}$$

where V_{total} is the total volume of the lattice. Note that for a plane wave the $IPR \sim 1$. For a particle localized on the volume l_{loc}^4, on the other hand:

$$IPR \;\sim\; \frac{V_{total}}{l_{loc}^4} \quad . \tag{40}$$

The test scalar particle is fictitious and the values of $\lambda_n, (IPR)_n$ are gauge-invariant functions of the vacuum gluonic fields.

The mechanism of localization in our case is strictly speaking not known. However, the whole terminology and ideas are borrowed from the well known example of the Anderson localization. In that case, the localization is due to random fields.

In case of scalar test particles (36) the lowest eigenfunctions turn to be localized as well [27]. This is in accord with the stochastic model of confinement. Moreover, if the stochastic fields are singular one can expect that the IPR values depend non-trivially on the lattice spacing as well. And, indeed, the IPR values do depend on the lattice spacing.

However, the measurements revealed unexpected features of the localization as well. Namely, it turns out that the localization length depends crucially on the color spin of the test particles [27]. Roughly speaking:

$$l_{loc}(T=\frac{1}{2}) \sim \Lambda_{QCD}^{-1},\;\; l_{loc}(T=1) \sim \sqrt{a\Lambda_{QCD}^{-1}},\;\; l_{loc}(T=\frac{3}{2}) \sim a\,. \tag{41}$$

Straightforward interpretation would be that there exist various components of the stochastic fields in the vacuum, both singular and non-singular. However, it is too early to make final conclusions on the nature of the stochastic fields in the vacuum. There are no doubts, however, that the picture with soft fields alone cannot be true.

7. Conclusions

Lattice measurements have revealed highly non-trivial dependences on the ultraviolet cut off of physically interesting quantities. The observations lead to a concept of fine tuning of the confining fields. The evidence in favor of the fine tuning is pure phenomenological. It is natural to think about the fine tuning as of a kind of symmetry. Knowing only the standard formulation of Yang-Mills theories, it is very difficult to appreciate this symmetry and explain huge cancellation between ultraviolet divergent actions and entropies associated with the vacuum defects. If one is to speculate about a hidden symmetry which could explain these cancellations, the dual, or string formulation [28] comes to mind. Much more work is needed however to make this relation to strings explicit.

References

1. V.N. Gribov, *"The theory of quark confinement"* in "Erice 1996, Effective theories and fundamental interactions" p 30;
   V.N. Gribov, *Eur. Phys. J.* **C10** (1999) 91 [arXiv:hep-ph/9902279];
   V.N. Gribov, *"Orsay lectures on confinement"* [arXiv:hep-ph/9905285].
2. Yu.L. Dokshitzer, D.E. Kharzeev, *Ann. Rev. Nucl. Part. Sci.* **54**, 487 (2004).
3. F.V. Gubarev, A.V. Kovalenko, M.I. Polikarpov, S.N. Syritsyn and V.I. Zakharov, *Phys. Lett.* **B574**, 136 (2003) [arXiv:hep-lat/0212003];
   V.G. Bornyakov, P.Yu. Boyko, M.I. Polikarpov and V.I. Zakharov, *Nucl. Phys.* **B672**, 222 (2003) [arXiv:hep-lat/0305021];
   A.V. Kovalenko, M.I. Polikarpov, S.N. Syritsyn and V.I. Zakharov, *Phys. Rev.* **D71**, 054511 (2005) [arXiv:hep-lat/0402017];
   M.I. Polikarpov, S.N. Syritsyn and V.I. Zakharov, *JETP Lett.* **81**, 143 (2005) [arXiv:hep-lat/0402018];
   A.V. Kovalenko, M.I. Polikarpov, S.N. Syritsyn, V.I. Zakharov *Phys. Lett.* **B613**, 52 (2005) [arXiv:hep-lat/0408014];
   M.I. Polikarpov, F.V. Gubarev, S.M. Morozov and V.I. Zakharov, arXiv:hep-lat/0510098.
4. A.I. Vainshtein, arXiv:hep-ph/0106200.

5. V.I. Zakharov, *in "Moscow 2003, I. Ya. Pomeranchuk and physics at the turn of the century"*, p.177. arXiv:hep-ph/0306262;
M.N. Chernodub and V.I. Zakharov, *Nucl. Phys.* **B669**, 233 (2003);
V.I. Zakharov, *in "Wako 2003, Color confinement and hadrons in quantum chromodynamics"* p.112. arXiv:hep-ph/0309301.
6. A.M. Polyakov, *Phys. Lett.*, **B59**, 82 (1975);
M.E. Peskin, *Annals Phys.* **113**, 122 (1978).
7. J.D. Stack and R.J. Wensley, *Nucl. Phys.* **B371**, 597 (1992).
8. G. Parisi, *"Statistical Field Theory"*, Addison-Wesley, (1987) Chapter 16.
9. T.A. DeGrand and D. Toussaint, *Phys. Rev.* **D22**, 2478 (1980).
10. M.B. Halpern, *Phys. Rev.* **D16**, 1798 (1977); *ibid.* **D19**, 517 (1979);
O. Ganor and J. Sonnenschein, *Int. J. Mod. Phys.* **A11**, 5701 (1996).
11. V.I. Zakharov, *Phys. Atom. Nucl.* **68**, 573 (2005).
12. F.V. Gubarev and S.M. Morozov, *Phys. Rev.* **D71**, 114514 (2005).
13. G. 't Hooft, *Nucl. Phys.* **B190**, 455 (1981).
14. M.N. Chernodub, R. Feldmann, E.M. Ilgenfritz and A. Schiller, *Phys. Lett.* **B605**, 161 (2005).
15. Ch. Gattringer, *Phys. Rev. Lett.* **88**, 221601 (2002).
16. C. Aubin, arXiv:hep-lat/0410024;
C. Bernard *et al.*, arXiv:hep-lat/0510025;
Y. Koma *et al.*, arXiv:hep-lat/0509164.
17. J. Greensite, *Prog. Part. Nucl. Phys.* **51**, 1 (2003).
18. M.N. Chernodub, F.V. Gubarev, M.I. Polikarpov and A.I. Veselov, *Progr. Phys. Suppl.* **131**, 309 (1998);
A. Di Giacomo, *Progr. Phys. Suppl.* **131**, 161 (1998);
T. Suzuki, *Progr. Phys. Suppl.* **131**, 633 (1998).
19. J. Ambjorn. J. Giedt, J. Greensite, *JHEP* **0002**, 033 (2000).
20. V.G. Bornyakov, *et al. Phys. Lett.* **B537**, 291 (2002).
21. V.I. Zakharov, arXiv:hep-ph/0309178.
22. F.V. Gubarev, L. Stodolsky and V.I. Zakharov, *Phys. Rev. Lett.* **86**, 2220 (2001); F.V. Gubarev and V.I. Zakharov, *Phys. Lett.* **B501**, 28 (2001).
23. M.A.L. Capri *et al.*, arXiv:hep-th/0510240.
24. M.N. Chernodub *et al.*, arXiv:hep-lat/0508004;
M.N. Chernodub *et al.*, arXiv:hep-lat/0509089.
25. V.G. Bornyakov, A.V. Kovalenko, M.I. Polikarpov and D.A. Sigaev, *Nucl. Phys. Proc. Suppl.* **119B**, 739 (2003).
26. A. Di Giacomo, H.G. Dosch, V.I. Shevchenko and Yu.A. Simonov, *Phys. Rep.* **372**, 319 (2002).
27. J. Greensite *et al.*, *Phys. Rev.* **D71**, 114507 (2005); arXiv:hep-lat/0509070; arXiv:hep-lat/0510097.
28. J.M. Maldacena, *Adv. Theor. Math.* **2**, 231 (1998);
A.M. Polyakov, *Int. J. Mod. Phys.* **A14**, 645 (1999).

THE EFFECTIVE BOSONIC STRING ACTION IN QUANTUM CHROMODYNAMICS

JULIUS KUTI
Department of Physics, University of California, San Diego
La Jolla, California 92093, United States

Bosonic string formation in gauge theories is reviewed with particular attention to the confining flux in lattice QCD and its string theory description. Recent results on the Casimir energy of the ground state and the string excitation spectrum are analyzed in the Dirichlet string limit of large separation between static sources. The closed string-soliton (torelon) with electric flux winding around a compact dimension is also discussed.

Remembering Volodya Gribov

When Julia Nyiri kindly invited me to the Gribov-75 Memorial Workshop, I was hoping to participate and enjoy the opportunity in paying tribute to Volodya with my modest contribution. Unfortunately, I was not able to attend the meeting but Julia asked me to send this short note as my token participation. I have known Volodya for more than two decades before his untimely death. I had the good fortune in my life to interact and work with some outstanding scientists and it was a great honor to count Volodya among them. We never worked together but we had continued discussions about physics, intermittently, for many years. There are many great personal memories, joyful and sad, but I would like to mention briefly only two encounters with Volodya which were about basic physics and left a very deep impression on me for the rest of my life.

In 1977 the meeting of the European Physical Society was held in Budapest. It was a big event, and the highlight of the conference was a special evening lecture by Dirac to honor him and his science. Dirac chose to talk in the overflown room about his growing dissatisfaction with mathematical inadequacies in quantum field theory. It was a field Dirac helped to develop from its early infancy into one of the most remarkable success stories of twentieth century physics. Dirac was gradually building his case against the theory, and Volodya was sitting in the audience with visibly growing unrest and dissatisfaction. He did not go to the blackboard to argue but I saw his face. After the lecture we talked because I was curious about his views and objections. I think I vaguely suspected what he would say. This was a time when critical phenomena were not common knowledge or acquired tools among most practitioners of quantum field theory, but Volodya being a renaissance man, interested in everything deep and challenging, knew better. He understood perhaps better than anyone else in that room that Dirac's view was a disconnection: if there were something seriously wrong with quantum field theory, universality and critical phenomena would not work in condensed matter systems. There was a deep connection between field theory and critical phenomena, not to be ignored.

Many years later, Julia and Volodya visited us in San Diego, California. It was a joyful occasion with many fond memories. Volodya gave a talk at UCSD on his work on quark confinement and we spent a good part of the week in discussions at the blackboard. His style was captured brilliantly in Andor Frenkel's talk in this Proceedings, and I witnessed a mini version of it in San Diego. Volodya had some problems with the way he wanted to describe quark propagation in the vacuum and tried to work it out during his visit. He used the blackboard as a painter would use his canvas and it was an experience to observe. I remain hopeful that someone will continue his unfinished work. My short note here is another way of looking at the problem of quark confinement, pursuing the connection between QCD and string theory. It is hard for me to believe that Volodya would approve but it fills me with great sadness that I cannot even try. We miss him every day, but we also keep him in our hearts and memories.

1. String Formation in Lattice Gauge Theories

This report will include the discussion of recent lattice work on the Casimir energy and the excitation spectrum of the bosonic Dirichlet string, and the closed string with unit winding number (string-soliton). I will use an effective string theory description for the interpretation of lattice simulations which should provide the appropriate framework for future investigations in the field. A more comprehensive survey can be found in my review talk at Lattice 2005 which also includes further references [1].

To illustrate the origin of string formation in field theory, I will describe some results from 3D Z(2) lattice gauge theory which will also help in interpreting string spectra in lattice QCD. Dual transformation maps the 3D Z(2) gauge model into the 3D Ising model [2] with a nontrivial fixed point which separates the confined and deconfined phases of the $Z(2)$ gauge field. The universality class of three-dimensional ϕ^4 field theory describes the fluctuations of the confining electric flux of the closed string-soliton (torelon) with unit winding number in a compact dimension, and the Dirichlet string with static color sources as well [1].

A classical soliton solution ϕ_s of the ϕ^4 field equation in the broken (confining) phase represents the ground state of a string with electric flux connecting static sources [1]. Similarly, string-soliton solution can be obtained for the winding string around a compact dimension. Although the soliton solutions are classical in the Ising variables, they represent complicated Z(2) gauge field distributions which would be difficult to understand without the dual representation. The fluctuation matrix $\mathcal{M} = -\nabla^2 + \mathcal{U}''(\phi_s)$ around the soliton solution ϕ_s is derived from the second derivative of the field potential $\mathcal{U}$. From the diagonalization of the matrix $\mathcal{M}$ we find the energy eigenvalues and the wavefunctions of the excitations around the string-soliton of unit winding,

$$E^2_{m,n} = \lambda^2_m + 4\sin^2(q_n/2), \quad \psi_n(x,y) = f_m(y)\exp(iq_n x),$$
$$q_n = \frac{2\pi n}{L}, \quad n = 1,2,3,..., \tag{1}$$

where the λ^2_m eigenvalues and the corresponding $f_m(y)$ eigenfunctions are obtained from the semi-classical quantization of the one-dimensional soliton. There is a tower of states for every value of m, with quantized momenta q_n, $n = 1,2,3,...$, along the compact x-direction.

The zero mode, $m = 0$, $\lambda^2_0 = 0$, $f_0(y)$, is responsible for the restoration of broken translational symmetry of the soliton in the y-direction, hence the terminology of massless Goldstone modes of left and right movers. A new collective field variable can be introduced which will have the geometric interpretation of transverse displacements $\xi(\sigma,\tau)$ on the string world sheet. To get the effective string theory action, we integrate out the massive fields and the resulting non-local action of the collective string variable $\xi(\sigma,\tau)$ can be approximated by the derivative expansion,

$$S_{z\text{-}string} = \frac{1}{2\pi\alpha'}\int d\tau \int d\sigma\Big\{1 - \frac{1}{2}(\partial\xi)^2 - \frac{1}{8}((\partial\xi)^2)^2 + \mathcal{O}(((\partial\xi)^2)^3)\Big\}$$
$$+ S_{FP}(\xi)\ , \tag{2}$$

where $1/2\pi\alpha'$ is the string tension. Contributions from the Fadeev-Popov determinant $S_{FP}(\xi)$ will only effect derivatives starting in sixth order.

The first three terms of the effective action in Eq. 2 match the first three terms of the $(\partial\xi)^2$ expansion of the 3D classical Nambu-Goto string action given by $S_{NG} = \sqrt{1+(\partial\xi)^2}$ in our variables. In higher orders, the expansion of $S_{z\text{-}string}$ deviates from S_{NG}. For example, some sixth order derivative terms in $S_{\text{z-string}}$ are expected to match the extrinsic geometric curvature term introduce by Polyakov [3] and related to the negative stiffness of the Z(2) string.

Lüscher and Weisz (LW) developed a string effective action which is model independent and applicable in any dimension D, with boundary operators included to describe end distortions of the Dirichlet string [4]. The LW effective action has the form

$$\begin{aligned} S_{LW} &= \sigma RT + \mu T \\ &+ \frac{1}{2\pi\alpha'} \int_0^T d\tau \int_0^R d\sigma \frac{1}{2} \partial_a\xi\partial_a\xi \\ &+ \frac{1}{4} b \int_0^T d\tau \Big\{ (\partial_1\xi\partial_1\xi)_{\sigma=0} + (\partial_1\xi\partial_1\xi)_{\sigma=R} \Big\} \\ &+ \frac{1}{4} c_2 \int_0^T d\tau \int_0^R d\sigma \frac{1}{2} (\partial_a\xi\partial_a\xi)(\partial_b\xi\partial_b\xi) \\ &+ \frac{1}{4} c_3 \int_0^T d\tau \int_0^R d\sigma \frac{1}{2} (\partial_a\xi\partial_b\xi)(\partial_a\xi\partial_b\xi) + ..., \end{aligned} \tag{3}$$

to fourth order in the derivatives (the string displacement vector ξ has D-2 components and the self-energy mass term μT is irrelevant for our discussion). A duality argument was invoked [4] to set the boundary coupling to $b = 0$, in addition to a duality relation between c_2 and c_3. In D=4 dimensions, this leaves in fourth order one independent interaction term with arbitrary c_2 coupling. In the analysis below, I will set the free coupling to $c_2 = \frac{1}{2\sigma}$, as expected in the effective string approch of Polchinski and Strominger [6] (PS) which is based on Poincare invariance and reparameterization symmetry [1]. In D=3 dimensions, the two derivative interactions in fourth order are kinematically not independent, both c_2 and c_3 are fixed by symmetries, and the effective string action S_{LW} in Eq. 3 becomes identical to $S_{z\text{-}string}$ of Eq. 2 up to and including fourth order derivatives. The effective string actions S_{LS} and $S_{z\text{-}string}$ are very similar in their scope of use.

In the discussion below, I will use the PS effective string approach which in effect is almost identical to LW predictions, except the $c_2 = \frac{1}{2\sigma}$ constraint and the full restoration of Poincare invariance for the closed string with winding in the compact dimension. The asymptotic form of the string-soliton spectrum with winding number w along the compact dimension was worked out in the PS effective string approach [6,7,8],

$$E^2_{n,N+\widetilde{N}} = \sigma^2 R^2 w^2 - \frac{\pi}{3}\sigma(D-2) + 4\pi\sigma(N+\widetilde{N}) + \frac{4\pi^2 n^2}{R^2} + \vec{p}_T^{\,2} \ , \tag{4}$$

where $N+\widetilde{N}$ describes the sum of right and left movers along the compactified dimension of length R. The momentum square $\vec{p}_{n\|}^{\,2} = (2\pi n/R)^2$, $n = 0, 1, 2, ...$, along the compactified direction is quantized, and for any integer winding number w there exists a relation $N - \widetilde{N} = nw$ between the discrete compact momentum and the difference in the number of right and left movers. The transverse momentum $\vec{p}_T$ of the winding string is a D-2 dimensional vector, with $\vec{p}_T = 0$ in the simulations of $w = 1$ torelon spectra which I will discuss later. The spectrum of the Dirichlet string is given by

$$E_N = \sigma R \sqrt{1 - \frac{\pi}{12\sigma R^2}(D-2) + \frac{2\pi N}{\sigma R^2}} \ , \tag{5}$$

which agrees with the old result of Arvis [5], but there will be correction terms to Eq. 5 starting at order $O(R^{-5})$ in the effective string approach. Similar remarks apply to the torelon spectrum as well.

2. Casimir Energy

The effective string prediction for the ground state energy of two static sources at separation R is given asymptotically by

$$E_0(R) = \sigma R - \frac{\pi(D-2)}{24R} - \frac{\pi^2(D-2)^2}{1152\sigma R^3} \ , \tag{6}$$

where the model dependent correction $\mathcal{O}(R^{-5})$ and a cutoff dependent constant, not contributing to $E_0'(R)$, are not included. There is some evidence, without rigorous proof, that the three terms of $E_0(R)$ are universal for bosonic strings [1]. Corrections to $E_0(R)$, starting at $\mathcal{O}(R^{-5})$, are not predictable without new terms and parameters in the effective string action. New physics, like string stiffness, will be coded in the higher order terms. Deviations are certainly expected from the Nambu-Goto form which has no new free parameters but becomes inconsistent on the $\mathcal{O}(R^{-5})$ scale in non-critical physical dimensions [5].

The effective Casimir term, defined by $C_{eff}(R) = -\frac{1}{2}R^3 E_0''(R)$, has the asymptotic behavior $C_{eff}(R) = \frac{\pi(D-2)}{24}(1 + \frac{\pi(D-2)}{8\sigma R^2} + ...)$ which isolates the Casimir energy with leading R^{-2} correction. The definitive lattice simulation of $C_{eff}(R)$ came from an efficient method to measure the ground state energy with high precision [9] as shown in Fig. 1. The results had the

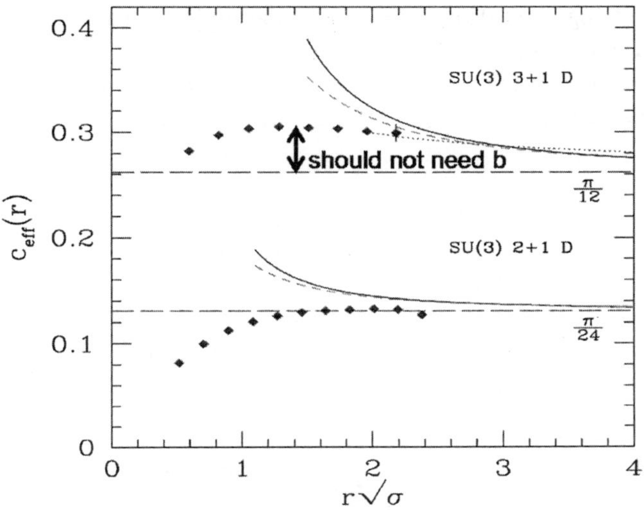

Figure 1. The solid lines are the NG predictions from the full square root expression of Eq. 5. The dashed lines, merging with the NG curves, designate $E_0(R)$ of Eq. 6. The dotted line includes boundary operator corrections [9]. The black diamond data points were obtained with high precision and the dotted line includes boundary operator corrections [9].

popular interpretation of bosonic string formation on the scale $R \lesssim 1$ fm. In D=3 dimensions the agreement seemed to be a perfect match to $\pi/24$, but boundary operators had to be added from Eq. 3 to explain deviations from $\pi/12$ when D=4 [9]. In fact, the interpretation should have been the reverse. The universal $O(R^{-2})$ asymptotic behavior, represented by the dashed line in Fig. 1, is a measurable correction to $\pi/12$ and apparently reached by the data for D=4. Therefore the D=4 result does not require any boundary terms which were later ruled out by LW duality regardless [4]. It is in D=3 dimensions that the lattice results remain noticeably below the universal dashed line.

3. Dirichlet String Spectrum

3D Z(2) and 3D SU(2) gauge models. The simulation of the Dirichlet string spectrum was implemented in the dual Ising representation of the 3D Z(2) gauge model. In the presence of fixed Polyakov loops of two sources, the plaquette-plaquette correlators of the Z(2) gauge field are measured to capture string excitations. With this method, several excitations were determined with great precision [10]. Dirichlet string theory for the lowest four string excitations is tested in Fig. 2 with reach to $R = 7$ fm separation. The first excitation energy gap of the Dirichlet string state in the 3D SU(2)

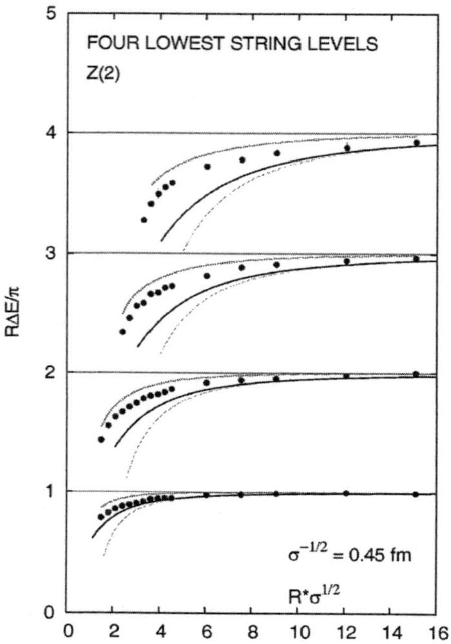

Figure 2. $R{\cdot}\Delta E/\pi$ is plotted for the first four Dirichlet string excitations in the 3D Z(2) gauge model at coupling $\beta = 0.23$ in the dual Ising representation. Solid black lines show the full NG prediction which will break down at small R values. Dashed lines represent universal predictions, accurate on the scale $\mathcal{O}(R^{-3})$ of the effective string action, as explained earlier. On the scale $O(R^{-5})$ string predictions become model dependent. Dotted lines show the energy eigenvalues as calculated numerically from the fluctuation operator $\mathcal{M}$ of section 1 in soliton quantization. Their deviations from the integer lines are good indicators of end distortion effects [1,10].

gauge model is shown in Fig. 3 [11].

Conclusions from the 3D Z(2) and 3D SU(2) Dirichlet string spectrum: (1) With similar trends for different gauge groups in D=3 dimensions, the expected asymptotic string limit has been reached for large values of R, (2)

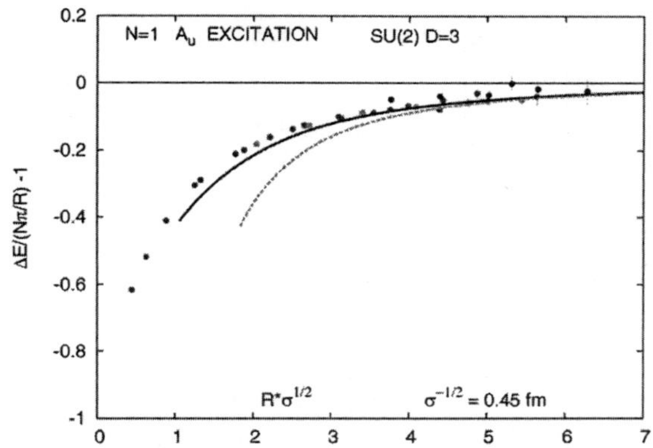

Figure 3. The N=1 Dirichlet string excitation is displayed in the 3D SU(2) model with reach to R=3 fm. The definitions of the solid black line and the dashed line were given in Fig. 2. The data points of different shades correspond to different lattice spacings, indicating scaling behavior.

the energy spectrum of the fluctuation operator $\mathcal{M}$ of section 1 indicates the presence of noticeable end effects in the spectrum, (3) without control on end distortions in the effective string action, it is difficult to disentangle boundary effects from asymptotic $O(R^{-3})$ behavior in the energy gaps which makes it problematic how to probe higher order new physics with Dirichlet boundary conditions.

4D SU(3) gauge model. In the 4D SU(3) Yang-Mills gauge model, which is the gluon sector of QCD, three exact quantum numbers determine the classification scheme of the gluon excitation spectrum in the presence of a static $q\bar{q}$ pair [12]. Restricted to the $R = 0.2 - 2$ fm range of selected simulations, the energy gap ΔE above the ground state is compared to string predictions for the first excited state in Fig. 4. The quantity $R \cdot \Delta E/N\pi - 1$ is plotted to show percentage deviations from the asymptotic string levels. In sharp contrast, for $R \ll 1$ fm it was shown that the multipole expansion of the gluon field can be applied with success in the short-distance operator product expansion of gluon excitations around static color sources [12]. The observed short distance level ordering is very different from the string spectrum at short distances. For 0.5 fm $< R <$ 2 fm, a dramatic crossover of the energy levels toward a string-like spectrum is observed as R increases. For example, the N=3 Σ_u^- state breaks rapidly away from its N=1 Π_u short-distance $O(3)$ degeneracy partner to approach the ordering and degeneracies expected from bosonic string theory [1,12].

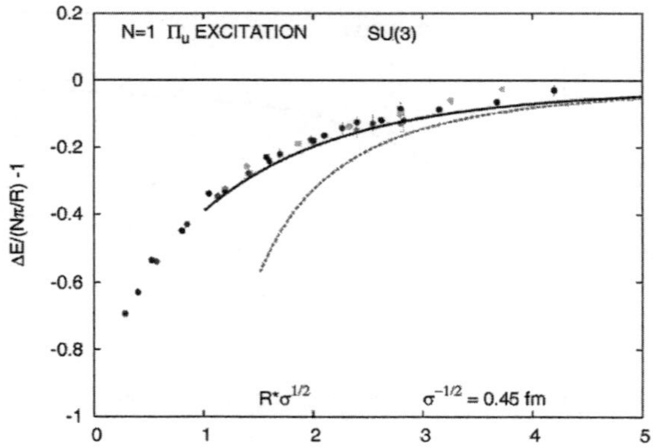

Figure 4. The lowest Dirichlet string excitation is displayed in the 4D SU(3) gauge model with reach to R=2 fm. Solid black line shows the full NG prediction which will break down at small R values. The data points of different shades correspond to different gauge couplings, demonstrating scaling behavior in the spectrum. The dashed line represents universal prediction on the $\mathcal{O}(R^{-3})$ scale of the effective string action, as it was discussed earlier.

4. Torelon string spectrum

3D Z(2) gauge model. The simulation of the torelon spectrum is a straightforward application of the procedure outlined earlier. The seam creating the torelon runs across the periodic x-direction of the lattice creating one unit of winding with $w = 1$. This seam can be moved by gauge transformations and the torelon is free to move in the y-direction, with zero transverse momentum projected. The excitation spectrum of the torelon should be compared with the predictions of effective string theory as given in Eq. 4. Fig. 5 shows two torelon states with the same principal quantum number, $N + \tilde{N} = 2$, but different $\vec{p}_n$ momenta. A single right mover with $N = 2, \tilde{N} = 0$ and running with two units of momentum in the compactified direction has positive $\mathcal{O}(R^{-3})$ energy correction to the asymptotic string level and split from the state of the composite state of back to back right and left movers with zero total momentum in the compact dimension. The dramatic fine structure, predicted in Eq. 4, is in agreement with lattice simulations [1].

4D SU(3) gauge model. The main author of the 4D SU(3) torelon excitation spectrum is Maresca with details of her work reviewed earlier [8,13]. The method is similar to the lattice technology which was used to obtain the Dirichlet string spectrum. Given the difficulty of these calculations, the

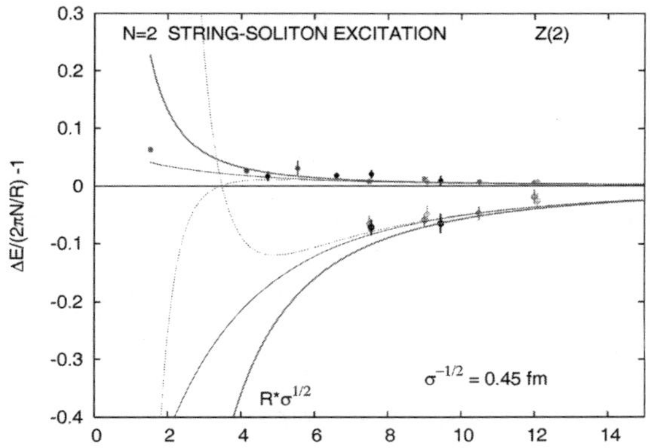

Figure 5. The spectra correspond to two different torelon excitations with $w = 1, N + \widetilde{N} = 2$. Above the null line the spectrum of the $n = 2, N = 2, \widetilde{N} = 0$ state is depicted with $p_n = 4\pi/R$ momentum in the compact dimension. Below the null line, the spectrum of the state $N = 1, \widetilde{N} = 1, p_n = 0$ is shown. There are two pairs of solid lines. The thin lines correspond to the spectrum of Eq. 4 which coincides with the exact NG prediction. The thick lines show the universal $\mathcal{O}(R^{-3})$ predictions of the effective string action after the square root of the right hand side in Eq. 4 is expanded in inverse powers of R. The dotted lines are $\mathcal{O}(R^{-5})$ NG terms which were left in the plot only to indicate the divergence of the expansion as the tachion singularity is approached.

results are in reasonably good agreement with string theory expectations, but there are some distortions. Although the evidence for the torelon string spectrum in the 4D SU(3) gauge model is not as strong as the 3D Z(2) simulations, the full complex lattice technology is in place for extended and more comprehensive simulations which hopefully will be reported in the near future.

Acknowledgements

I would like to thank Julia Nyiri and Peter Levai for making the workshop a success and for helping me to prepare my contribution. This work was supported by the DOE, Grant No. DE-FG03-97ER40546.

References

1. J. Kuti, *Lattice QCD and String Theory*, Proceedings of Lattice 2005, PoS(LAT2005)001 [hep-lat/0511023].
2. M. Caselle, R. Fiore, F. Gliozzi, M. Hasenbusch, P. Provero, *String effects in the Wilson loop: a high precision numerical test*, *Nucl. Phys.* **B486**, 245 (1997).
3. A. M. Polyakov, *Fine structure of strings*, *Nucl. Phys.* **B268**, 406 (1986).

4. M. Lüscher and P. Weisz, *String excitation energies in SU(N) gauge theories beyond the free-string approximation*, *JHEP* **0407**, 014 (2004).
5. J.F. Arvis, *The exact $q\bar{q}$ potential in Nambu string theory*, *Phys. Lett.* **B127**, 106 (1983).
6. J. Polchinski and A. Strominger, *Effective string theory*, *Phys. Rev. Lett.* **67**, 1681 (1991).
7. J.M. Drummond, *Universal subleading spectrum of effective string theory*, [hep-th/0411017]; J. Kuti, *unpublished.*
8. F. Maresca, *Comparing the excitations of the periodic flux tube with effective string models*, Ph.D. Thesis, Trinity College, Dublin, 2004.
9. M. Lüscher and P. Weisz, *Quark confinement and the bosonic string*, *JHEP* **0207**, 049 (2002).
10. K.J. Juge, J. Kuti, C.J. Morningstar, *QCD string formation and the Casimir energy*, Proceedings of *Color confinement and hadrons in quantum chromodynamics*, Wako, Japan (2004) 233 [hep-lat/0401032].
11. K.J. Juge, J. Kuti, C.J. Morningstar, *Excitations of the static quark anti-quark system in several gauge theories*, Proceedings of *Color confinement and hadrons in quantum chromodynamics*, Wako, Japan (2004) 221 [hep-lat/0312019].
12. K.J. Juge, J. Kuti, C.J. Morningstar, *The fine structure of the QCD string spectrum*, *Phys. Rev. Lett.* **90**, 161601 (2003).
13. K.J. Juge, J. Kuti, F. Maresca, C.J. Morningstar, M. Peardon, *Excitations of torelon*, *Nucl. Phys. Proc. Suppl.* **129**, 703 (2004).

THE LOW-MASS σ-MESON: IS IT AN EYEWITNESS OF CONFINEMENT?

V.V. ANISOVICH AND V.A. NIKONOV
Petersburg Nuclear Physics Institute,
Gatchina, 188300, Russia
E-mail: Anisovic@thd.pnpi.spb.ru

In the framework of the dispersion relation N/D-method we restore the low–energy $\pi\pi$ ($IJ^{PC} = 00^{++}$)-wave amplitude sewing it with the previously obtained K-matrix solution for the region 450–1900 MeV. The restored N/D-amplitude has a pole on the second sheet of the complex-s plane, near the $\pi\pi$ threshold. We discuss the hypothesis that this low–mass pole, or the low-mass σ-meson, corresponds to the dynamical state related to the confinement forces, that is the eyewitness of confinement.

1. Introduction

In the region of strong interactions, the fundamental objects of QCD, quarks and gluons, may substantially change their characteristics as compared to those at small distances. Phenomenological investigations, as well as theoretical estimates, show that in the strong interaction region we deal with constituent quarks and effective massive gluons (for more detail see, for example [1], and references therein). In the paper [2], V.N. Gribov drew the attention of the reader to the fact that strong interactions, especially interactions caused by the confinement forces, may form particles which are different from the standard (non-exotic) mesons and baryons. First of all, it concerned the sector of scalar mesons. In the present paper, we discuss the hypothesis of whether the low-mass σ meson, if it exists, may point to the existence of the amplitude singularities related to the confinement.

The experimental data on meson spectra accumulated by Crystal Barrel Collaboration [3], GAMS [4] and BNL [5] groups provided a good basis for the study of the ($IJ^{PC} = 00^{++}$)-wave, and the combined K-matrix analysis of the reactions $\pi\pi \to \pi\pi$, $K\bar{K}$, $\eta\eta$, $\eta\eta'$, $\pi\pi\pi\pi$ was performed over the mass range 450–1900 MeV [6]. Then the K-matrix analysis has been expanded to the waves $\frac{1}{2}0^{+}$ [7] and 10^{+} [8], thus making it possible to establish the $q\bar{q}$

systematics of scalars for $1^3P_0q\bar{q}$ and $2^3P_0q\bar{q}$ multiplets.

The K-matrix amplitude analysis is a good instrument to perform the $q\bar{q}$ nonet classification of mesons in terms of "bare states", namely, the states without inclusion of the decay channels (detailed discussion can be found in Ref. [9]). The decay processes cause a strong mixing of scalar states, for in the transitions $(q\bar{q})_1 \to$ *real mesons* $\to (q\bar{q})_2$ the orthogonality of the coordinate wave functions does not work. The other important effect caused by the decay is the accumulation of widths of neighbouring resonances by one of them: as a result, a broad scalar/isoscalar state appears at the mass region 1200-1600 MeV.

According to Refs. [7,8], the lightest scalar multiplet $1^3P_0q\bar{q}$ of bare states is located in the region 950–1200 MeV. After the mixing originated from the decay processes, the bare states of the $1^3P_0q\bar{q}$ nonet are transformed into a set of resonances: $f_0(980)$, $a_0(980)$, $f_0(1300)$, $K_0(1000)$ (or $K_0(1400)$) which are the descendants of the lightest pure $q\bar{q}$ states. The scalar/isoscalar resonances $f_0(1500)$ and $f_0(1750)$ are descendants of the bare $2^3P_0q\bar{q}$ states. At the same region there is a broad resonance, with a mass about 1200–1600 MeV, which is the descendant of the lightest scalar glueball: the gluonium and scalar/isoscalar $q\bar{q}$ states are strongly mixed (according to the rules of $1/N$ expansion [10], the transition *gluonium* $\to q\bar{q}$ is not suppressed), and the $q\bar{q}$ component in the broad state reaches 50%, while the scalars $f_0(1300)$ and $f_0(1500)$ have considerable admixtures of gluonium components. In $f_0(980)$ and $f_0(1750)$, the $s\bar{s}$ component is predominant.

An important result of Refs. [6,8] is that the K-matrix 00^{++}-amplitude has no pole singularities in the 500–800 MeV region. The $\pi\pi$-scattering phase δ_0^0 increases smoothly in this energy region reaching 90° at 800–900 MeV. A straightforward explanation of such a behaviour of δ_0^0 might be the presence of a broad resonance, with mass about 600–900 MeV and width $\Gamma \sim 500$ MeV [11,12,13,14]. However, according to the K-matrix solution [6,8], the 00^{++}-amplitude does not contain pole sigularities on the second sheet of the complex-$M_{\pi\pi}$ plane inside the interval $450 \le Re\ M_{\pi\pi} \le 900$ MeV: the K-matrix amplitude has a low-mass pole only, which is located on the second sheet either near the $\pi\pi$ threshold or even below it. In Refs. [6,8], the presence of this pole was not emphasized, for the left-hand cut, which is important for a reconstruction of the analytic structure of the low-energy partial amplitude, was taken into account in an indirect way only. A proper way for the description of the low-mass amplitude must be the use of the dispersion relation representation.

In this paper the dispersion relation $\pi\pi$ scattering amplitude is reconstructed in the region of small $M_{\pi\pi}$ being sewed with the K-matrix solution of Refs. [6,8] found for $M_{\pi\pi} \sim 450 - 1950$ MeV. Namely, on the basis of the data for δ_0^0, we construct the N/D amplitude below 900 MeV sewing it with the K-matrix amplitude with an objective to make a continuation to the region $s = M_{\pi\pi}^2 \sim 0$. By this sewing we strictly follow the results obtained for the K-matrix amplitude in the region 450-900 MeV, that is, the region where we can be confident in the results of the K-matrix representation. Let us remind that the K-matrix representation allows us to restore correctly the analytic structure of the amplitude in the region $s > 0$ (threshold and pole singularities) but not the left-hand singularities at $s \leq 0$ (singularities related to forces). Therefore, being cautious, we cannot be quite confident in the K-matrix results below the $\pi\pi$ threshold.

In Section 2 the dispersion relation amplitude is found using the method of an approximation of the left-hand cut suggested in [15]. The constructed N/D-amplitude provides a good description of δ_0^0 from threshold to 900 MeV, thus including the region $\delta_0^0 \sim 90°$. This amplitude does not have a pole in the region 500–900 MeV; instead, the pole is located near the $\pi\pi$ threshold.

We suggest that the low-mass pole in the scalar/isoscalar wave is related to a fundamental phenomenon at large distances (in hadronic scale). In Section 3 we argue that the low-mass pole corresponds to a white composite particle which is inherent to subprocesses responsible for the colour confinement forces.

2. Dispersion relation solution for the $\pi\pi$-scattering amplitude below 900 MeV

The pion-pion scattering partial amplitude being a function of the invariant energy squared, $s = M_{\pi\pi}^2$, can be represented as a ratio $N(s)/D(s)$, where $N(s)$ has a left-hand cut, which is due to the "forces" (the interactions owing to t- and u-channel exchanges), while the function $D(s)$ is determined by the rescatterings in the s-channel. $D(s)$ is given by the dispersion integral along the right-hand cut in the complex-s plane:

$$A(s) = \frac{N(s)}{D(s)}, \quad D(s) = 1 - \int\limits_{4\mu_\pi^2}^{\infty} \frac{ds'}{\pi} \frac{\rho(s')N(s')}{s' - s + i0}. \tag{1}$$

Here $\rho(s)$ is the invariant $\pi\pi$ phase space, $\rho(s) = (16\pi)^{-1}\sqrt{(s - 4\mu_\pi^2)/s}$. It is supposed in (1) that $D(s) \to 1$ with $s \to \infty$ and CDD-poles are absent (a detailed presention of the N/D-method can be found in [16]).

The N-function can be written as an integral along the left-hand cut as follows:

$$N(s) = \int\limits_{-\infty}^{s_L} \frac{ds'}{\pi} \frac{L(s')}{s'-s}\,, \tag{2}$$

where the value s_L marks the beginning of the left-hand cut. For example, for the one-meson exchange diagram $g^2/(m^2-t)$, the left-hand cut starts at $s_L = 4\mu_\pi^2 - m^2$, and the N-function in this point has a logarithmic singularity; for the two-pion exchange, $s_L = 0$.

Below we work with the amplitude $a(s)$, which is defined as follows:

$$a(s) = \frac{N(s)}{8\pi\sqrt{s}\left(1 - P\int\limits_{4\mu_\pi^2}^{\infty} \frac{ds'}{\pi}\frac{\rho(s')N(s')}{s'-s}\right)}\,. \tag{3}$$

The amplitude $a(s)$ is related to the scattering phase shift: $a(s)\sqrt{s/4-\mu_\pi^2} = \tan\delta_0^0$. In Eq. (3) the threshold singularity is singled out explicitly, so the function $a(s)$ contains only a left-hand cut and poles corresponding to zeros of the denominator in the right-hand side (3): $1 = P\int\limits_{4\mu_\pi^2}^{\infty}(ds'/\pi)\cdot\rho(s')N(s')/(s'-s)$. The pole of $a(s)$ at $s > 4\mu_\pi^2$ corresponds to the phase shift value $\delta_0^0 = 90°$. The phase of the $\pi\pi$ scattering reaches the value $\delta_0^0 = 90°$ at $\sqrt{s} = M_{90} \simeq 850$ MeV. Because of that, the amplitude $a(s)$ may be represented in the form:

$$a(s) = \int\limits_{-\infty}^{s_L} \frac{ds'}{\pi}\frac{\alpha(s')}{s'-s} + \frac{C}{s-M_{90}^2} + D. \tag{4}$$

For the reconstruction of the low-mass amplitude, the parameters D, C, M_{90} and $\alpha(s)$ have been determined by fitting to the experimental data. In the fit we have used the method, which has been approved in the analysis of the low-energy nucleon-nucleon amplitudes [15]. Namely, the integral in the right-hand side of (4) has been replaced by the sum

$$\int\limits_{-\infty}^{s_L} \frac{ds'}{\pi}\frac{\alpha(s')}{s'-s} \to \sum_n \frac{\alpha_n}{s_n - s} \tag{5}$$

with $-\infty < s_n \leq s_L$.

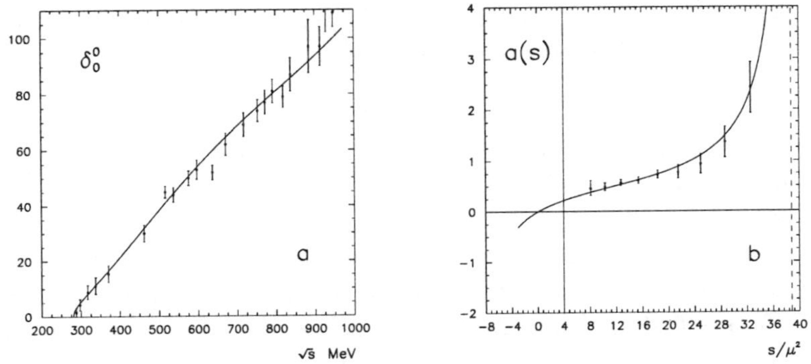

Figure 1. a) Fit to the data on δ_0^0 by using the N/D-amplitude. b) Amplitude $a(s)$ in the N/D–solution (solid curve) and the K-matrix approach [7,8] (points with error bars).

The description of data within the N/D-solution, which uses six terms in the sum (5), is demonstrated on Fig. 1a. Parameters of the solution are as follows:

$s_n\ \mu_\pi^{-2}$	-9.56	-10.16	-10.76	-32	-36	-40
$\alpha_n\ \mu_\pi^{-1}$	2.21	2.21	2.21	0.246	0.246	0.246

$$M_{90} = 6.228\ \mu_\pi, \qquad C = -13.64\ \mu_\pi, \qquad D = 0.316\ \mu_\pi^{-1} \tag{6}$$

The scattering length in this solution is equal to $a_0^0 = 0.22\ \mu_\pi^{-1}$, the Adler zero is at $s = 0.12\ \mu_\pi^2$. The N/D-amplitude is sewed with the K-matrix amplitude of Refs. [6,8], and figure 1b demonstrates the level of the coincidence of the amplitudes $a(s)$ for both solutions (the values of $a(s)$ which correspond to the K-matrix amplitude are shown with error bars determined in [6,8]).

The dispersion relation solution has a correct analytic structure at the region $|s| < 1$ GeV^2. The amplitude has no poles on the first sheet of the complex-s plane; the left-hand cut of the N-function after the replacement given by Eq. (5) is transformed into a set of poles on the negative piece of the real s-axis: six poles of the amplitude (at $s/\mu_\pi^2 = -5.2,\ -9.6,\ -10.4,\ -31.6,\ -36.0,\ -40.0$) represent the left-hand singularity of $N(s)$. On the second sheet (under the $\pi\pi$-cut) the amplitude has two poles: at $s \simeq (4 - i14)\mu_\pi^2$ and $s \simeq (70 - i34)\mu_\pi^2$ (see Fig. 2). The second pole, at $s = (70 - i34)\mu_\pi^2$, is located beyond the region under consideration, $|s| < 1$ GeV^2 (nevertheless, let us stress that the K-matrix amplitude [6,8] has a set of poles just in the region of the second pole of the N/D-amplitude). The pole near threshold, at

$$s \simeq (4 - i14)\mu_\pi^2 \,, \tag{7}$$

is what we discuss. The N/D-amplitude has no poles at $Re\sqrt{s} \sim 600 - 900$ MeV despite the phase shift δ_0^0 reaching $90°$ here.

The data for δ_0^0 do not fix the N/D-amplitude rigidly. The position of the low-mass pole can be easily varied in the region $Re\, s \sim (0-4)\mu_\pi^2$, and there are simultaneous variations of the scattering length in the interval $a_0^0 \sim (0.21 - 0.28)\mu_\mu^{-1}$ and Adler zero at $s \sim (0-1)\mu_\pi^2$. The discussed solution (6) is chosen because of recent data for the scattering length $a_0^0 \sim (0.228 \pm 0.015)\mu_\mu^{-1}$ [17] that coincide with the value found in (6).

Let us stress that the way of reconstruction of the dispersion relation amplitude used here differs from the mainstream attempts of determination of the N/D-amplitude. In the classic N/D procedure, that is the bootstrap one, the pion-pion amplitude is to be determined by analyticity, unitarity and crossing symmetry. That means an unique determination of the left-hand cut by the crossing channels. However, the bootstrap procedure is not realized till now; the problems which face the bootstrap program nowadays are discussed in Ref. [18] and references therein. Nevertheless, one can try to saturate the left-hand cut by known resonances in the crossing channels. Usually it is supposed that the dominant contribution into the left-hand cut comes from the ρ-meson exchange supplemented by some $f_2(1275)$ and σ exchanges. Within this scheme, the low-energy amplitude is restored being corrected by available experimental data. In the scheme used here,

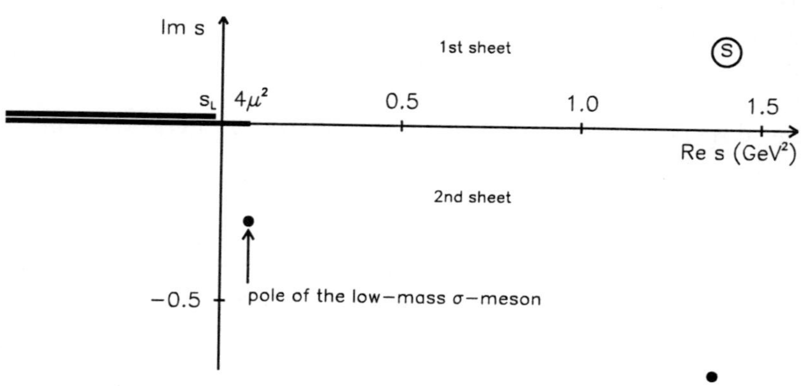

Figure 2. Complex-s plane and singularities of the N/D-amplitude.

the amplitude in the physical region at 450-1950 MeV is supposed to be known (the result of the K-matrix analysis) - then a continuation of the amplitude is made from the region 450-900 MeV to the region of smaller masses; the continuation is corrected by the data. As a result, we restore the pole near the threshold (the low-mass σ-meson) and the left-hand cut (although with a less accuracy, actually, on a qualitative level).

In the approaches which take into account the left-hand cut as a contribution of some known meson exchanges, the following low-mass pole positions were obtained:
(i) dispersion relation approach, $s \simeq (0.2 - i22.5)\mu_\pi^2$ [19],
(ii) meson exchange models, $s \simeq (3.0 - i17.8)\mu_\pi^2$ [20], $s \simeq (0.5 - i13.2)\mu_\pi^2$ [21], $s \simeq (2.9 - i11.8)\mu_\pi^2$ [22],
(iii) linear σ-model, $s \simeq (2.0 - i15.5)\mu_\pi^2$ [23].

In Refs. [24,25], the pole positions were found in the region of the higher masses, at $Re\ s \sim (7 - 10)\ \mu_\pi^2$.

3. Low-mass pole as the eyewitness of confinement

It was suggested in [26,27] that the existence of the light σ-meson may be due to a singular behaviour of forces between quark and antiquark at large distances (in quark models they are conventionally called "confinement forces"). The scalar confinement potential, which is needed for the description of the spectrum of the $q\bar{q}$-states in the region 1000–2000 MeV, behaves at large hadronic distances as $V^{(c)}_{confinement}(r) \sim \alpha r$, where $\alpha \simeq 0.19\,\text{GeV}^2$. In the momentum representation, such a growth of the potential is associated with singular behaviour at small q:

$$V^{(c)}_{confinement}(q) \ \sim \ \frac{1}{q^4}\ . \tag{8}$$

In the colour space, the main contribution comes from the component $c = 8$, i.e. the confinement forces should be the octet ones. The question that is crucial for the structure of σ-meson is as follows: is there a component with the colour singlet $V^{(1)}_{confinement}(q)$ in the singular potential (8)?

If the singular component with $c = 1$ exists, then it must reveal itself in hadronic channels as well, that is, in the $\pi\pi$-channel. In hadronic channels this singularity should not be exactly the same as in the colour octet ones, because the hadronic unitarization of the amplitude (which is absent in the channel with $c = 8$) should modify somehow the low-energy amplitude. One may believe that, as a result of the unitarization in the channel $c = 1$, i.e. due to the account for hadronic rescattering, the singularity of $V^{(1)}_{confinement}(q)$ may appear in the $\pi\pi$-amplitude on the second sheet being split into several poles. The modeling of the scalar confinement potential, taken into account for the decay of unstable levels, confirms the pole splitting. One may believe that this singularity is what we call "the light σ-meson".

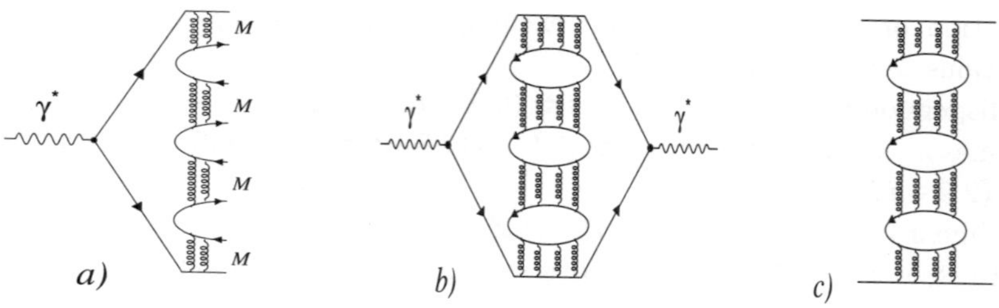

Figure 3. a) Quark–gluonic comb produced by breaking a string by quarks flowing out in the process $e^+e^- \to \gamma^* \to q\bar{q} \to mesons$. b) Convolution of the quark–gluonic combs. c) Example of diagrams describing interaction forces in the $q\bar{q}$ systems.

Therefore, the main question consists in the following: the $V^{(1)}_{confinement}(q^2)$, does it have the same singular behaviour as $V^{(8)}_{confinement}(q^2)$? The observed linearity of the (n, M^2)-trajectories, up to the large-mass region, $M \sim 2000 - 2500$ MeV [1], favours the idea of the universality in the behaviour of potentials $V^{(1)}_{confinement}$ and $V^{(8)}_{confinement}$ at large r, or small q. To see that (for example, in the process $\gamma^* \to q\bar{q}$, Fig. 3a) let us discuss the colour neutralisation mechanism of outgoing quarks as a break of the gluonic string by newly born $q\bar{q}$-pairs. At large distances, that corresponds to the formation of states with large masses, several new $q\bar{q}$-pairs should be formed. It is natural to suggest that a convolution of the quark–gluon combs governs the interaction forces of quarks at large distances, see Fig. 3b. The mechanism of the formation of new $q\bar{q}$-pairs to neutralise colour charges does not have a selected colour component. In this case, all colour components $3 \otimes \bar{3} = 1 + 8$ behave similarly, that is, at small q^2 the singlet and octet components of the potential are uniformly singular, $V^{(1)}_{confinement}(q^2) \sim V^{(8)}_{confinement}(q^2) \sim 1/q^4$. This is seen in Fig. 3a: the quark–gluon ladder ensures the t-channel flow of colour charge $c = 3$, so the quark–antiquark interaction block, being the convolution of ladder diagrams $3 \otimes \bar{3} = 1 + 8$ contains two equivalent, singlet and octet, components. This points to a similarity of $V^{(1)}_{confinement}$ and $V^{(8)}_{confinement}$.

4. Conclusion

We have analysed the structure of the low-mass $\pi\pi$-amplitude in the region $M_{\pi\pi} \le 900$ MeV using the dispersion relation N/D-method, which provides us with a possibility to take the left-hand singularities into consideration. The dispersion relation N/D-amplitude is sewed with that given by the K-matrix analysis performed at $M_{\pi\pi} \sim 450 - 1950$ MeV [6]. Obtained in this way the N/D-amplitude has a pole on the second sheet of the complex-s plane near the $\pi\pi$ threshold. This pole corresponds to the existence of the low–energy bound state.

On the basis of the conventional confinement model, we argue that the confinement interaction can produce low-mass pole singularities both for the colour–octet and colour–singlet states. Then the colour–singlet scalar state, which reveals itself in the low-mass $\pi\pi$-amplitude, may be named the eyewitness of confinement.

Acknowledgement

We thank A.V. Anisovich, Ya.I. Azimov, D.V. Bugg, Yu.L. Dokshitzer, H.R. Petry, A.V. Sarantsev and V.V. Verestchagin for fruitful discussions.

References

1. V.V. Anisovich, J. Nyiri, M.N. Kobrinsky and Yu.M. Shabelski, *"Quark Model and High Energy Collisions"*, 2nd edition, World Scientific, Singapore, 2004.
2. V.N. Gribov, Preprint LU-TP 91-7, Lund University, 1991.
3. V.V. Anisovich *et al.*, *Phys. Lett.* **B323**, 233 (1994); C. Amsler *et al.*, *Phys. Lett.* **B333**, 277 (1994); **B355**, 425 (1995).
4. D. Alde *et al.*, *Phys. Lett.* **B241**, 600 (1990); *Zeit. Phys.* **C66**, 375 (1995); Yu.D. Prokoshkin *et al.*, *Physics-Doklady* **342**, 473 (1995); F. Binon *et al.*, *Nuovo Cim.* **A78**, 313 (1983); **A80**, 363 (1984).
5. S.J. Lindenbaum and R.S. Longacre, *Phys. Lett.* **B274**, 492 (1992); A. Etkin *et al.*, *Phys. Rev.* **D25**, 1786 (1982).
6. V.V. Anisovich, Yu.D. Prokoshkin and A.V. Sarantsev, *Phys. Lett.* **B389**, 388 (1996).
7. A.V. Anisovich and A.V. Sarantsev, *Phys. Lett.* **B413**, 137 (1997).
8. V.V. Anisovich, A.A. Kondashov, Yu.D. Prokoshkin, S.A. Sadovsky and A.V. Sarantsev, *Yad. Fiz.* **60**, 1489 (2000) [*Phys. of Atom. Nucl.* **60**, 1410 (2000)]; V.V. Anisovich and A.V. Sarantsev, *Eur. Phys. J.* **A16**, 229 (2003).
9. A.V. Anisovich, V.V. Anisovich and A.V. Sarantsev, *Zeit. Phys.* **A359**, 173 (1997).

10. G. 't Hooft, *Nucl. Phys.* **B72**, 161 (1974);
G. Veneziano, *Nucl. Phys.* **B117**, 519 (1976).
11. S.D. Protopopescu *et al.*, *Phys. Rev.* **D7**, 1279 (1973).
12. P. Estabrooks, *Phys. Rev.* **D19**, 2678 (1979).
13. K.L. Au, D. Morgan and M.R. Pennington, *Phys. Rev.* **D35**, 1633 (1987).
14. S. Ishida *et al.*, *Prog. Theor. Phys.* **98**, 1005 (1997).
15. V.V. Anisovich, M.N. Kobrinsky, D.I. Melikhov and A.V. Sarantsev, *Nucl. Phys.* **A544**, 747 (1992).
16. G.F. Chew, *The analytic S-matrix*, NY, Benjamin, 1966.
17. S. Pislak *et al.*, *Phys. Rev. Lett.* **87**, 221801 (2001).
18. A.V. Verestchagin and V.V. Verestchagin, *Phys. Rev.* **D58**, 016002 (1999).
19. J.L. Basdevant, C.D. Frogatt and J.L. Petersen, *Phys. Lett.* **B41**, 178 (1972).
20. J.L. Basdevant and J. Zinn-Justin, *Phys. Rev.* **D3**, 1865 (1971);
D. Iagolnitzer, J. Justin and J.B. Zuber, *Nucl. Phys.* **B60**, 233 (1973).
21. B.S. Zou and D.V. Bugg, *Phys. Rev.* **D48**, R3942 (1994); **D50**, 591 (1994).
22. G. Janssen, B.C. Pearce, K. Holinde and J. Speth, *Phys. Rev.* **D52**, 2690 (1995).
23. N.N. Achasov and G.N. Shestakov, *Phys. Rev.* **D49**, 5779 (1994).
24. V.E. Markushin and M.P. Locher, *"Structure of the light scalar mesons"*, Talk given at the Workshop on Hadron Spectroscopy, Frascati, 1999, preprint PSI-PR-99-15.
25. N.A. Törnquist and M. Roos, *Phys. Rev. Lett.* **76**, 1575 (1996).
26. V.V. Anisovich and V.A. Nikonov, *Eur. Phys. J.* **A8**, 401 (2000); arXiv:hep-ph/0008163.
27. V.V. Anisovich, *Usp. Fiz. Nauk* **174**, 49 (2004) [*Physics-Uspekhi* **47**, 45 (2004)].

CP, T AND CPT IN NON-PERTURBATIVE CHIRAL GAUGE THEORIES

P. HASENFRATZ

Institute for Theoretical Physics
University of Bern
Sidlerstrasse 5, CH-3012 Bern, Switzerland

As a non-perturbative and gauge invariant regularization the lattice became a powerful tool for a deeper understanding of vector and chiral gauge theories. Exact chiral symmetry on the lattice is a prerequisite of these developments. I will discuss briefly the history of chiral symmetry on the lattice which was long and troubled and had an amazingly nice upshot. Then an unexpected feature of lattice regularized chiral gauge theories will be discussed: although CPT remains intact, CP and T are violated at any finite resolution. It is not yet understood whether and under what conditions this symmetry breaking disappears in the continuum limit and whether a lattice formulation exists where these symmetries are explicit from the start.

1. Introduction

Gribov, Migdal and Polyakov were among the first who undestood the deep relation between quantum field theories and critical statistical systems. This relation lies at the heart of the lattice regularization of QCD which opened the way towards non-perturbative problems in this theory.

In spite of his historical role in these developments Gribov was somewhat sceptical concerning lattice QCD. Not only that the Euclidean space formulation of lattice QCD was inappropriate for his approach on confinement, but he had also serious doubts whether an analytic continuation from Minkowski to Euclidean space is possible at all. Indeed, no rigorous proof exists beyond perturbation theory.

Although quantum field theories are very successful to describe the fundamental laws of Nature as we know them today, not much is rigorous there, unfortunately. One should keep this in mind in general and in particular in the context of chiral gauge theories.

2. Gauge fields and fermions on the lattice

A hyper-cubic lattice is a natural regularization when path integrals are used to formulate a quantum field theory in $d = 4$ Euclidean space. The first and decisive steps have been done by Wilson [1]. For summaries with details on different aspects of theoretical an numerical issues, see [2,3,4,5].

We start with a vector (as opposed to chiral) gauge theory. The fermion field ψ lives on the lattice points indexed by an integer vector $x \in Z_4$ while the gauge fields are associated with the links $(x, \hat{\mu}), \mu = 1, \ldots, 4$ of the lattice. Derivatives are replaced by finite difference operators using, for example the nearest neighbor forward and backward difference operators[a]

$$\partial^{\mu}_{x,y} = \delta_{x+\hat{\mu},y} - \delta_{x,y}, \qquad \partial^{\mu *}_{x,y} = \delta_{x,y} - \delta_{x-\hat{\mu},y} \,. \tag{1}$$

A simple guess for the action of a massless free fermion on the lattice might read as

$$\sum_{x,y} \overline{\psi}_x \frac{1}{2} \gamma^{\mu} \left(\partial^{\mu}_{x,y} + \partial^{\mu *}_{x,y} \right) \psi_y \,. \tag{2}$$

Switching on the gauge interaction a parallel transporter is included in the finite difference operators to assure gauge invariance

$$\nabla^{\mu}_{x,y} = U(x, \mu)\, \delta_{x+\hat{\mu},y} - \delta_{x,y} \,, \qquad \nabla^{\mu *}_{x,y} = \delta_{x,y} - U(x - \hat{\mu}, \mu)^{\dagger}\, \delta_{x-\hat{\mu},y} \,, \tag{3}$$

where $U(x, \mu)$ is an element of the gauge group (for QCD, as an example, $SU(3)$). The group element $U(x, \mu)$ transforms under a gauge transformation as

$$\tilde{U}(x, \mu) = V(x) U(x, \mu) V^{\dagger}(x + \hat{\mu}) \,. \tag{4}$$

Eq. (4) implies that the trace of the product of gauge matrices along a closed path on the lattice is a gauge invariant quantity. Locality requires that the extension of the gauge action is $O(a)$. Taking the smallest loop, the plaquette, we get a discretized, local and gauge invariant action

$$S = -\frac{2}{g^2} \sum_{p} \mathrm{ReTr}\, U_p + \sum_{x,y} \overline{\psi}_x\, D_{x,y}\, \psi_y \,, \tag{5}$$

where the sum runs over the plaquettes, U_p is the product of four directed link matrices around the plaquette p and the Dirac operator has the form

$$D = \frac{1}{2} \gamma^{\mu} \left(\nabla^{\mu} + \nabla^{\mu *} \right) \,. \tag{6}$$

[a]Here and in most of the following expressions we suppress the lattice unit a. All the fields and parameters are made dimensionless by appropriate a powers: in the mass m a factor of a, in the fermion field ψ a factor of $a^{3/2}$, etc, is absorbed.

Taking the formal continuum limit $a \to 0$ of eq. (5) we obtain the standard continuum form of an action for a massless quark in interaction with gauge fields. Obviously, there are infinitely many local and gauge invariant lattice actions which have this property. The continuum predictions are expected not to depend on this freedom, however (universality).

In order to define the quantum theory we have to fix the integration measure of the path integral. Here we take the natural choice of the group invariant measure for every link variable $U(x,\mu)$ and integrate over the Grassmann variables $\overline{\psi}(x)$ and $\psi(x)$ according to the standard rules of Grassmann integration. This measure is also gauge invariant.

The action in eq. (5) has a problem, however. The Dirac operator in eq. (6) describes actually 16 massless fermion species - a problem called doubling. This is the first small surprise along a tortured way around chiral symmetry.

3. Doublers, chiral symmetry and a no-go theorem

The free fermion Dirac operator D in Fourier space has the form

$$D(p) = \sum_{\mu} \gamma^{\mu} \sin(p^{\mu}) \tag{7}$$

which has 16 zeros in the first Brillouin zone at $(0,0,0,0),\ldots,(\pi,\pi,\pi,\pi)$. It can be shown [6] that these species couple to the axial current with alternating sign and their total contribution to the $U(1)$ axial anomaly cancels [7,8]. This observation resolves a paradox: the theory is fully regularized, D anticommutes with a standardly defined γ_5, so we have a U(1) chiral symmetry without anomaly. Without the doublers this would be a contradiction.

The question remained whether there exists a clever lattice discretization such that the doublers are removed but (at least the non-singlet part of) chiral symmetry survives on the lattice. This question was answered by a no-go theorem [9] (see also ref. [6]). The no-go theorem states that for free fermions the following four conditions can not hold simultaneously:

1. $D(x)$ is local;
2. the Fourier transformed D behaves for $p \ll 1$ as $i\gamma^{\mu}p^{\mu} + O(p^2)$;
3. there are no doublers;
4. $\gamma_5 D + D\gamma_5 = 0$.

If we do not want to violate flavor symmetry for massless fermions then $D(x)$ is diagonal in the flavor indices and the statement is valid for each flavor separately. Locality is the most important property of a quantum field theory, so we can not give up this condition. The 2nd point just declares that we have a Dirac particle. If we do not want doublers (since there are no such copies in nature), there remains only the possibility to violate condition 4, i.e. to give up chiral symmetry on the lattice.

Accepting chiral symmetry breaking terms in the action there are different possibilities to follow. Wilson [10] suggested to add a dimension 5 operator to the action in eq. (5) which modifies the Dirac operator as

$$D_{\mathrm{W}} = \frac{1}{2}\left[\gamma^\mu(\nabla^\mu + \nabla^{\mu *}) - \nabla^\mu \nabla^{\mu *}\right] . \tag{8}$$

The new term (Wilson term) gives large $\propto$ cutoff masses to the doublers, but - at least classically - leaves the $p = 0$ pole unchanged. The Wilson term has an effect on the renormalization of bare parameters (in particular, for the bare fermion mass $m_q = 0$ is not protected), creates mixings between operators in different chiral multiplets and it also influences the way and speed the continuum is approached. The final physical predictions of the continuum theory are, however, independent of this term.

4. The Ginsparg-Wilson relation

The no-go theorem seemed to close all the ways towards a chiral symmetric lattice regularization. As the authors formulated [9]: 'The important consequence of our work is to *discourage any attempt to construct chiral invariant lattice models for QCD*'. No-go theorems, however, are frequently circumvented in an unexpected way. This is what happened with chiral symmetry on the lattice.

Soon after the no-go theorem was presented Ginsparg and Wilson [11] suggested a condition for the Dirac operator which, as the authors formulated, implies a 'remnant' chiral symmetry on the lattice. More than 15 years later this condition, called the Ginsparg-Wilson relation, became very important in the theoretical developments concerning global and local chiral symmetry on the lattice and beyond.

Ginsparg and Wilson treated free fermions, used some basic notions of Wilson's renormalization group theory and then, with a bold move, suggested that the result should be valid in general [11]: if the Dirac operator satisfies the relation

$$S_B = \sum_{x_B, y_B} \overline{\chi}_{x_B}\, D_{x_B, y_B}\, \chi_{y_B} \tag{9}$$

and the Dirac operator is constrained by the relation

$$D\gamma^5 + \gamma^5 D = D\gamma^5 2RD\,, \tag{10}$$

where R is a local operator (in the simplest case R = 1), then the predictions for physical quantities obey the constraints of chiral symmetry on the lattice even before the continuum limit is taken.

Assume there exists a local, gauge covariant solution of this equation. The first trivial observation is that the inverse of D satisfies

$$D^{-1}\gamma^5 + \gamma^5 D^{-1} = \gamma^5 2R\,. \tag{11}$$

The inverse of D, the quark propagator, from which the physical correlators are constructed is a non-local quantity. The equation above shows, however, that its anticommutator with γ^5 is local, i.e. has an extension of $O(a)$ only. The violation of chiral symmetry is a contact term. It has been shown in ref. [11] also that a Dirac operator satisfying eq. (10) reproduces the $U(1)$ chiral anomaly.

The following simple consideration strongly suggests that we are indeed on a good track. Consider an arbitrary correlator where the operators are separated by physical distances. Write down the corresponding Ward identity using the standard method in path integral formulation: introduce new Grassmann integration variables which are related to the original ones by an infinitesimal γ^5 chiral rotation. The action is not invariant and, according to eq. (10), gives an insertion $\propto \overline{\chi} D\gamma^5 2RD\chi$ to the correlation function. The fermion field χ ($\overline{\chi}$) will be paired with one of the antifermion (fermion) fields of the operators. The two propagators which bind the breaking term to the operators are canceled by the two D operators in the breaking term. What remains is $\gamma^5 2R_{x,y}$, where x and y are the positions of the operators. Since R is local and the operators are separated at physical distances, the contribution from the breaking term is zero.

Eq. (10) is a highly non-trivial equation. The chance of finding a local solution in the interactive case was rather dim. The paper and its results were hardly noticed and got completely forgotten[b].

It was noticed in 1997 [12] that the fixed point Dirac operator in QCD, which is defined by classical saddle point equations obtained through renormalization group considerations [13,14], satisfies the Ginsparg-Wilson (GW) relation. Soon it was found [15] that the domain wall fermions [16] and the related overlap fermions [17] satisfy these relations also. Actually, the domain wall and overlap fermions were already known to imply chiral symmetry. The GW relation, however, created a unifying powerful prescription.

---

[b]According to SLAC Spires the paper was not cited at all over twelve years between 1986 and 1997 September when the first solution of the GW equation in the presence of gauge fields was identified.

5. Index theorem on the lattice

It is a simple exercise to show that the solutions of the Ginsparg-Wilson relation satisfy the index theorem on the lattice [18]: the zero modes of D are chiral and the associated index is a topological invariant which represents the topological charge on the lattice.

For notational simplicity consider the case $2R = 1$ in eq. (10) and assume the hermiticity property in Euclidean space $D^\dagger = \gamma^5 D \gamma^5$ which is satisfied by all the Dirac operators. It is then easy to show that the spectrum $\{\lambda\}$ of D lies on a unit circle around the point $z = 1$ in the complex plane with the following properties. The real modes $\lambda = 0$ and $\lambda = 2$ are chiral (i.e. the corresponding eigenvectors are also eigenvectors of γ^5 with ± 1 eigenvalues). If ϕ_λ is an eigenvector with a complex eigenvalue λ then $\phi_{\lambda^*} = \gamma^5 \phi_\lambda$ and they are orthogonal: $(\phi_\lambda, \gamma^5 \phi_\lambda) = 0$

Define the density

$$q(x) = \frac{1}{2}\mathrm{tr}(\gamma^5 D(x,x))\,, \tag{12}$$

where the trace is taken in color and Dirac space. Consider now

$$\sum_x q(x) = \frac{1}{2}\mathrm{Tr}(\gamma^5 D) = -\frac{1}{2}\mathrm{Tr}\left[\gamma^5(2-D)\right] = $$
$$= -\frac{1}{2}\sum_\lambda (2-\lambda)(\phi_\lambda, \gamma^5 \phi_\lambda)\,, \tag{13}$$

where Tr denotes trace in all the indices and in the first step we subtracted zero: $\mathrm{Tr}\gamma^5 = 0$. Using the orthogonality properties discussed above only the $\lambda = 0$ modes contribute giving

$$\sum_x q(x) = n_- - n_+\,, \tag{14}$$

where n_- (n_+) is the number of left (right) handed zero modes of the Dirac operator. The index of D is, therefore, a sum over x of the pseudoscalar density $q(x)$. It can be shown [19] that on smooth configurations $q(x)$ is the continuum topological charge density

$$q(x) = \frac{1}{32\pi^2}\epsilon_{\mu\nu\rho\sigma}\mathrm{tr}^c(F_{\mu\nu}F_{\rho\sigma}) + O(a^2)\,, \tag{15}$$

where the trace is taken in color space. In the case of the fixed point Dirac operator one can show in addition that $\sum_x q(x)$ is the fixed-point topological charge Q_{FP} which takes integer values on any gauge configuration, i.e. the topological charge from the gauge and from the fermion sector is always consistent on the lattice if the FP action is used.

6. Exact chiral symmetry transformation

As I discussed before, one can raise arguments that the Ginsparg-Wilson relation implies the physical consequences of chiral symmetry. One can derive Ward identities also and demonstrate that they lead to the same physical predictions as those with exact chiral symmetry [20]. This approach is, however, cumbersome. In 1998 Lüscher has found an exact symmetry transformation which could be identified as the chiral symmetry transformation on the lattice [21]. This was a theoretical breakthrough. It gave an elegant technical tool to derive Ward identities, study anomalies and to handle the problem of chiral gauge theories. Chiral symmetry is realized on the lattice in an unusual, but beautiful way.

Consider a vector gauge theory with N_f flavors and take $2R = 1$ in eq. (10) for simplicity. If the Dirac operator D satisfies the Ginsparg-Wilson relation then the infinitesimal non-singlet transformation

$$\psi' = \psi + i\epsilon T\gamma^5 \left(1 - \frac{1}{2}D\right)\psi\,, \qquad \overline{\psi}' = \overline{\psi} + i\epsilon\overline{\psi}\left(1 - \frac{1}{2}D\right)\gamma^5 T \tag{16}$$

and the singlet transformation

$$\psi' = \psi + i\epsilon\gamma^5 \left(1 - \frac{1}{2}D\right)\psi\,, \qquad \overline{\psi}' = \overline{\psi} + i\epsilon\overline{\psi}\left(1 - \frac{1}{2}D\right)\gamma^5 \tag{17}$$

leave the fermion action $\overline{\psi}D\psi$ invariant. In eq. (16) T is an $SU(N_f)$ generator. It is trivial to demonstrate this statement. Less trivial is the way this symmetry transformation avoids the Nielsen-Ninomiya no-go theorem: the chiral transformation is not a simple γ^5 rotation as the no-go theorem assumed. The transformation depends on D and so it depends on the gauge field over which the fermions propagate. In the formal continuum limit the transformation goes over to the standard γ^5 rotation[c].

The action is invariant under the singlet transformation also, but the fermion integration measure is not [21]. A non-trivial Jacobi determinant is generated depending on the gauge field

$$\prod_x d\overline{\psi}'_x d\psi'_x = \left(1 + i\epsilon\mathrm{Tr}(\gamma^5 D)\right)\prod_x d\overline{\psi}_x d\psi_x\,. \tag{18}$$

For a non-singlet transformation the Jacobi determinant is 1 due to $\mathrm{tr}(T) = 0$. According to the index theorem on the lattice, a factor of $1 + i\epsilon 2N_f Q_{\mathrm{top}}$ is produced by the measure for a singlet transformation,

[c] In eqs. (16) and (17) D is multiplied by the lattice unit a as it is obvious from dimensional analysis.

where $Q_{\rm top}$ is the topological charge of the gauge configuration as defined by the index of the Dirac operator. Consider the transformation in eq. (17) as a change of variable in the fermionic path integral. The expectation value $\langle \mathcal{O} \rangle$ of an arbitrary operator remains unchanged which leads to the correct anomalous Ward identity on the lattice

$$\langle \delta\mathcal{O} \rangle + 2N_f \langle Q_{\rm top} \mathcal{O} \rangle = 0\,, \tag{19}$$

where $\delta\mathcal{O} = \left[\mathcal{O}(\overline{\psi}', \psi') - \mathcal{O}(\overline{\psi}, \psi)\right]/i\epsilon$ is the variation of the operator $\mathcal{O}$ under the transformation.

7. Left and right handed fermions, the mass term and the θ parameter

In the formal continuum the massless fermion action of a vector theory falls into left handed and right handed parts

$$\overline{\psi} D \psi = \overline{\psi}_L D \psi_L + \overline{\psi}_R D \psi_R\,. \tag{20}$$

This is also possible on the lattice although in this case the projectors on the ψ field depend on the gauge configuration [22,23,2]:

$$\psi_L = \hat{P}_L \psi\,, \qquad \psi_R = \hat{P}_R \psi\,, \qquad \overline{\psi}_L = \overline{\psi} P_R\,, \qquad \overline{\psi}_R = \overline{\psi} P_L\,, \tag{21}$$

where

$$\hat{P}_{R/L} = \frac{1}{2}(1 \pm \hat{\gamma}_5)\,, \qquad P_{R/L} = \frac{1}{2}(1 \pm \gamma_5)\,. \tag{22}$$

Here $\hat{\gamma}_5$ depends on the gauge field through the Dirac operator

$$\hat{\gamma}_5 = \gamma_5(1 - D) \tag{23}$$

and satisfies $(\hat{\gamma}_5)^2 = 1$ due to the Ginsparg-Wilson relation (I consider $2R = 1$ for simplicity). Using these definitions one obtains the decomposition in eq. (20) on the lattice.

The relations above are striking in many ways. The decomposition $\psi(x) = \psi_L(x) + \psi_R(x)$ depends on the gauge field in the neighborhood of x. Even for free fermions, the projectors depend on the Dirac operator. Beyond that, there is an asymmetry between the fermions and antifermions: the projectors on $\overline{\psi}$ are standard, they are identical to those in the continuum. This asymmetry is deeply related to the fermion number anomaly in chiral gauge theories [24].

The scalar and pseudo-scalar densities are introduced as usual

$$S = \overline{\psi}_L \psi_R + \overline{\psi}_R \psi_L = \overline{\psi} \left(1 - \frac{1}{2} D \right) \psi \,, \tag{24}$$

$$P = \overline{\psi}_L \psi_R - \overline{\psi}_R \psi_L = \overline{\psi} \gamma_5 \left(1 - \frac{1}{2} D \right) \psi \,. \tag{25}$$

These densities behave correctly under chiral transformations. A mass term in the action of a vector theory like QCD can be introduced as

$$S = S_g(U) + \int dx \left[\overline{\psi} D \psi + \overline{\psi}_L m \psi_R + \overline{\psi}_R m^\dagger \psi_L - i\theta q(x) \right] \,, \tag{26}$$

where a CP breaking θ-term is added also and a sum over flavors is assumed. For a real mass m the massive Dirac operator can be written as $(1 - m/2)(D + m)$. The mass shifts the eigenvalue circle by m and modifies the radius in such a way that the point $\lambda = 2$, where the ultraviolet eigenvalues are concentrated remains fixed. Starting from eq. (26) the known Ward identities can be derived, but this time in a controlled environment. The Ward identities are valid at any finite value of the cutoff (and so also in the continuum limit) as far as the Dirac operator is a local solution of the Ginsparg-Wilson relation. Conserved vector, axial vector and chiral currents can be defined [25,26]. It became possible to clarify theoretically opaque quantities like the topological susceptibility [27,28] and to make progress on the related Witten-Veneziano relation [29]. The Ginsparg-Wilson relation has also been used to construct actions with improved chiral and scaling properties, like the chirally improved (CI) action [30].

8. Chiral gauge theories

Standard perturbatively defined regularizations have problems with chiral gauge theories. Of course, in principle one can give up gauge symmetry and introduce all the counter terms needed to get gauge invariant renormalized predictions in perturbation theory. This might not be the best way to proceed, not even practically. In addition, our ultimate goal is to understand these theories (like the standard model) beyond perturbation theory.

The decomposition eq. (20) is the first step towards a gauge invariant formulation of chiral gauge theories on the lattice. The projectors in eq. (21) are gauge field dependent, so the dividing surface between the left and right handed fermions in the space of fermion degrees of freedom is moving as the gauge configuration is changing. Even the number of fermions with a certain chirality, say left handed, is changing. Since the number of left handed antifermions is constant (their projectors are gauge field independent), the total fermion number is changing as we are flying over the gauge configuration space. Indeed, the difference between the fermion and antifermion degrees of freedom is

$$\mathrm{Tr}\hat{P}_L - \mathrm{Tr}P_R = \frac{1}{2}\mathrm{Tr}(\gamma_5 D)\,, \tag{27}$$

which, according to the index theorem discussed before, takes different values in the different topological sectors. On one hand, eq. (27) shows that fermion number violation occurs naturally in this formulation and this a welcome feature. On the other hand, having different degrees of freedom in the different topological sectors might imply difficulties to connect these sectors with each other in establishing the theory.

In the last few years important results were obtained in this field. Lüscher has demonstrated that a U(1) chiral gauge theory with fermions in anomaly free representation can be fully defined on the lattice so that gauge symmetry is exactly preserved [31]. The same could be established for a general compact group in every order of perturbation theory [32]. Beyond the $U(1)$ case it is not known whether and under which conditions these theories are free of non-perturbative obstructions (anomalies). Such obstructions might exist as has been shown in an $SU(2)$ chiral theory with a single left handed fermion in the fundamental representation [33]. This was also demonstrated on the lattice [34]. A different approach which is pursued on the lattice since a long time is to give up explicit gauge symmetry. I refer here to a recent paper [35] and to the references therein.

9. CP, T and CPT in non-perturbatively formulated chiral gauge theories

In spite of significant recent progress on the non-perturbative formulation of chiral gauge theories there remained several unsolved problems. One of them is the puzzle that the left- and right-handed projectors, and so the left- and right-handed actions, break CP- and T-symmetry on the lattice, while CPT remains intact.

As we discussed above the Ginsparg-Wilson relation implies exact chiral symmetry on the lattice where, unlike in the continuum, the chiral symmetry transformation and also the projectors to left and right handed fermions depend on the Dirac operator and so on the gauge field.

Actually, the form of the projectors is not unique. One might introduce a continuous parameter s and for each value of s the fermion action is decomposed correctly

$$\overline{\psi} D \psi = \overline{\psi}_L^{(s)} D \psi_L^{(s)} + \overline{\psi}_R^{(s)} D \psi_R^{(s)}. \tag{28}$$

where

$$\psi_{L/R}^{(s)} = P_{\mp}^{(s)} \psi, \qquad \overline{\psi}_{L/R}^{(s)} = \overline{\psi} \bar{P}_{\pm}^{(s)}. \tag{29}$$

and the projectors are defined as

$$\begin{aligned} P_{\mp}^{(s)} &= \frac{1}{2}\left(\mathbb{1} \mp \Gamma_5^{(s)}\right), & \bar{P}_{\mp} &= \frac{1}{2}\left(\mathbb{1} \mp \overline{\Gamma}_5^{(s)}\right), \\ \Gamma_5^{(s)} &= \left(\mathcal{N}^{(s)}\right)^{-1} \gamma^5 \left(\mathbb{1} - sD\right), & \overline{\Gamma}_5^{(s)} &= \left(\mathcal{N}^{(s)}\right)^{-1} \left(\mathbb{1} - (1-s)D\right) \gamma^5, \\ \mathcal{N}^{(s)} &= \sqrt{\mathbb{1} - s(1-s) D D^{\dagger}}. \end{aligned} \tag{30}$$

Beyond being gauge field dependent, the projectors in eq.(30) are asymmetric with respect to ψ and $\overline{\psi}$. One might interpret this asymmetry as a signal for a possible fermion number anomaly in chiral gauge theories [24]. Indeed, as we discussed above, using the index theorem [18], it is easy to connect the fermion number violation to the topological charge of the gauge field configurations. On the other hand, the decomposition in eq.(28) violates CP-symmetry also. Under the CP transformation

$$\psi(n) \to W \overline{\psi}^T(\mathcal{P}n), \qquad \overline{\psi}(n) \to -\psi^T(\mathcal{P}n) W^{-1},$$

$$U_\mu(n) \to U_\mu^{(CP)}(n) \doteq \begin{cases} [U_\mu(\mathcal{P}n - \mu)]^T & (\mu = 1,2,3) \\ U_\mu^*(\mathcal{P}n) & (\mu = 4) \end{cases} \tag{31}$$

and with the CP-symmetry condition for the Dirac operator

$$W^{-1} D(n, n'; U^{(CP)})^T W \doteq D(\mathcal{P}n', \mathcal{P}n; U) \tag{32}$$

the left-handed action in eq.(28) is not invariant [36]

$$\sum_{n,n'} \overline{\psi}_L^{(s)}(n) D(n,n';U) \psi_L^{(s)}(n') \overset{CP}{\to} \sum_{n,n'} \overline{\psi}_L^{(1-s)}(n) D(n,n';U) \psi_L^{(1-s)}(n'). \tag{33}$$

At $s = \frac{1}{2}$ the CP-symmetry is restored, but just at this value of s the projectors become non-local: the $\lambda = 2$ point, where the normalization operator $\mathcal{N}^{(\frac{1}{2})}$ becomes zero, is an accumulation point of the ultraviolet modes. It can be shown that time reflection symmetry T has the same problem: it is violated at any value of s except at $s = \frac{1}{2}$, where, however locality is violated [37]. On the other hand, CPT is conserved [37].

The significance of this CP-symmetry violation on the physical properties of the theory is an interesting, but not yet completely understood issue [38]. I make here a short and trivial remark only on this difficult problem. Due to a technical constraint in the treatment of chiral gauge theories on the lattice the theory falls into topological sectors where the relative weight factors remain undetermined. A theory defined on the Gaussian fixed-point has a finite number of free parameters only and so these complex overall factors weighting the partition function in the different topological sectors can not be simply *chosen* to satisfy the requirement of symmetry. Actually, the topological charge $Q = 0$ sector should give informations on the $Q \neq 0$ sectors due to the clustering property of separated topological objects.

Acknowledgements

I thank the organizers for the invitation and many of the participants for the discussions at this meeting. This work was supported by the Schweizerischer Nationalfonds.

References

1. K.G. Wilson, *Phys. Rev.* **D10**, 2445 (1974).
2. F. Niedermayer, *Nucl. Phys. Proc. Suppl.* **73**, 105 (1999) [arXiv:hep-lat/9810026].
3. M. Lüscher, *Nucl. Phys. Proc. Suppl.* **83**, 34 (2000) [arXiv:hep-lat/9909150]; L. Giusti, *Nucl. Phys. Proc. Suppl.* **119B**, 149 (2003) [arXiv:hep-lat/0211009].
4. M. Creutz, *Rev. Mod. Phys.* **73**, 119 (2001) [arXiv:hep-lat/0007032]; S. Chandrasekharan and U.-J. Wiese, arXiv:hep-lat/0405024.
5. T. DeGrand, *Int. J. Mod. Phys.* **A19**, 1337 (2004) [arXiv:hep-lat/0312241].
6. L.H. Karsten and J. Smit, *Nucl. Phys.* **B192**, 100 (1981).
7. S.L. Adler, *Phys. Rev.* **177**, 2426 (1969).
8. J.S. Bell and R. Jackiw, *Nuovo Cim.* **60A**, 47 (1969).
9. N.B. Nielsen and M. Ninomiya, *Nucl. Phys.* **B185**, 20 (1981).
10. K.G. Wilson, in *New Phenomena in Subnuclear Physics*, ed. A. Zichichi, (Plenum Press, New York) Part A, 69 (1977).
11. P.H. Ginsparg and K.G. Wilson, *Phys. Rev.* **D25**, 2549 (1982).
12. P. Hasenfratz, *Nucl. Phys. Proc. Suppl.* **63A-C**, 53 (1998) [arXiv:hep-lat/9709110].
13. P. Hasenfratz and F. Niedermayer, *Nucl. Phys.* **B414**, 785 (1994).
14. T. DeGrand, A. Hasenfratz, P. Hasenfratz and F. Niedermayer, *Nucl. Phys.* **B454**, 587 (1995) [arXiv:hep-lat/9506030]; W. Bietenholz and U.-J. Wiese, *Nucl. Phys.* **B464**, 319 (1996) [arXiv:hep-lat/9510025]; T. DeGrand, A. Hasenfratz, D. Zhu, *Nucl. Phys.* **B475**, 321 (1996) [arXiv:hep-lat/9603015].

15. H. Neuberger, *Phys. Lett.* **B427**, 353 (1998) [arXiv:hep-lat/9801031].
16. D.B. Kaplan, *Phys. Lett.* **B288**, 342 (1992) [arXiv:hep-lat/9206013].
17. R. Narayanan and H. Neuberger, *Phys. Lett.* **B302**, 62 (1993) [arXiv:hep-lat/9212019]; *Phys. Rev. Lett.* **71**, 3251 (1993) [arXiv:hep-lat/9308011]; *Nucl. Phys.* **B412**, 574 (1994) [arXiv:hep-lat/9307006]; *Nucl. Phys.* **B443**, 305 (1995) [arXiv:hep-lat/9411108].
18. P. Hasenfratz, V. Laliena and F. Niedermayer, *Phys. Lett.* **B427**, 125 (1998) [arXiv:hep-lat/9801021].
19. K. Fujikawa, *Nucl. Phys.* **B546**, 480 (1999) [arXiv:hep-lat/9811235]; H. Suzuki, *Prog. Theor. Phys.* **102**, 141 (1999) [arXiv:hep-lat/9812019]; K. Fujikawa, *Int. J. Math. Phys.* **B16**, 1931 (2002) [arXiv:hep-lat/0205024]; D. H. Adams, *Annals Phys.* **296**, 131 (2002) [arXiv:hep-lat/9812003].
20. P. Hasenfratz, *Nucl. Phys.* **B525**, 401 (1998) [arXiv:hep-lat/9802007].
21. M. Lüscher, *Phys. Lett.* **B428**, 342 (1998) [arXiv:hep-lat/9802011].
22. R. Narayanan, *Phys. Rev.* **D58**, 097501 (1998) [arXiv:hep-lat/9802018].
23. M. Lüscher, *Nucl. Phys.* **B538**, 515 (1999) [arXiv:hep-lat/9808021].
24. G. 't Hooft, *Phys. Rev.* **D14**, 3432 (1976).
25. Y. Kikukawa and A. Yamada, *Nucl. Phys.* **B547**, 413 (1999) [arXiv:hep-lat/9810024].
26. P. Hasenfratz, S. Hauswirth, K. Holland, T. Jörg and F. Niedermayer, *Nucl. Phys.* **B643**, 280 (2002) [arXiv:hep-lat/0205010].
27. L. Giusti, G. C. Rossi and M. Testa, *Phys. Lett.* **B587**, 157 (2004) [arXiv:hep-lat/0402027].
28. M. Lüscher, arXiv:hep-lat/0404034.
29. L. Giusti, G.C. Rossi, M. Testa and G. Veneziano, *Nucl. Phys.* **B628**, 234 (2002) [arXiv:hep-lat/0108009].
30. C. Gattringer, *Phys. Rev.* **D63**, 114501 (2001) [arXiv:hep-lat/0003005]; C. Gattringer, I. Hip and C.B. Lang, *Nucl. Phys.* **B597**, 451 (2001).
31. M. Lüscher, *Nucl. Phys.* **B549**, 295 (1999) [arXiv:hep-lat/9811032]; *Nucl. Phys.* **B568**, 162 (2000) [arXiv:hep-lat/9904009].
32. H. Suzuki, *Progr. Theor. Phys.* **101**, 1147 (1999) [arXiv:hep-lat/9901012]; *Nucl. Phys.* **B585**, 471 (2000) [arXiv:hep-lat/0002009]; M. Lüscher, *JHEP* **06**, 028 (2000) [arXiv:hep-lat/0006014].
33. E. Witten, *Phys. Lett.* **B117**, 324 (1982).
34. O. Bär and I. Campos, *Nucl. Phys. Proc. Suppl.* **83**, 594 (2000) [arXiv:hep-lat/0001025].
35. M. Golterman and Y. Shamir, arXiv:hep-lat/0404011.
36. P. Hasenfratz, *Nucl. Phys. Proc. Suppl.* **106**, 159 (2002) [arXiv:hep-lat/0111023].
37. M. Bissegger and P. Hasenfratz *Phys. Lett.* **B613**, 57 (2005) [arXiv:hep-lat/0501010].
38. K. Fujikawa, M. Ishibashi and H. Suzuki, *JHEP* **0204**, 046 (2002) [arXiv:hep-lat/0203016]; *Nucl. Phys. Proc. Suppl.* **119**, 781 (2003) [arXiv:hep-lat/0203016];
K. Fujikawa and H. Suzuki, *Phys. Rev.* **D67**, 034506 (2003) [arXiv:hep-lat/0210013].

GRIBOV'S PICTURE OF CONFINEMENT AND CHIRAL SYMMETRY BREAKING

C. EWERZ

Dipartimento di Fisica, Università di Milano and INFN, Sezione di Milano
Via Celoria 16, I-20133 Milano, Italy

ECT, Strada delle Tabarelle 286*
I-38050 Villazzano (Trento), Italy
E-mail: ewerz@ect.it

Gribov's scenario of confinement caused by supercritical charges is described and the present status of Gribov's equation for the Green function of light quarks is discussed.

1. Introduction

One of the most challenging problems in quantum field theory is to understand the confinement mechanism in QCD. V.N. Gribov attacked this problem in a longlasting effort in which he developed both a physical picture and a quantitative approach to describing the confinement mechanism. His main concern was not just to find a confinement mechanism but rather *the* confinement mechanism relevant to the real world in which we live. It was a key observation for him[1] that the existence of very light quarks can drastically change the vacuum structure of QCD. Therefore a solution of the confinement problem appeared impossible to him without understanding the effects of light quarks. From this point of view the question of confinement in pure gluodynamics is interesting but probably irrelevant for our real world.

From the early 90's onwards the problem of light quarks therefore became the main focus of Gribov's research [2]. In this context he addressed several key questions. One of them is the way in which the Pauli principle can provide a binding mechanism even for almost massless quarks due to the occurrence of supercritical charges. Another one is the relation between the infrared and ultraviolet regimes of the theory which appear to be linked much more closely than one might naively expect. He also addressed the question whether and how the perturbative expansion has to be reconsidered in a theory in which the fundamental fields of the Lagrangian do not exist as asymptotic states. This is in fact a very deep problem as one recognizes by recalling that the usual treatment of propagators with the Feynman $i\epsilon$-prescription requires that the perturbative vacuum is stable – which is not true in QCD where quarks and gluons cannot propagate freely.

In 1997 V.N. Gribov was convinced that he had finally found answers to these questions and that he understood the physics of light quarks sufficiently well. He then collected his results in two papers [3,4] which were supposed to conclude his study of confinement in QCD. Unfortunately, the second paper had to remain unfinished when he passed away.

In the present contribution we describe the idea of supercritical charges in QCD, give a review of the Gribov equation for the propagator of light quarks, and summarize the present status of that equation. For a more detailed exposition of the motivations and physical ideas behind Gribov's approach we refer the reader to the recent review [5]. All of Gribov's original papers on confinement (as well as many of his papers and lectures on high energy scattering) are collected in the book [6].

2. Light Quarks and Supercritical Charges

The phenomenon of supercritical charges is well known in QED. Consider the energy levels that an electron would have in the static external field of a heavy nucleus A_Z with charge Z. When Z is increased the energy levels are lowered. At $Z > 137$ the lowest (1s) level sinks below $-m_e c^2$ and hence dives into the Dirac sea [a]. In this situation an electron from the Dirac sea fills this level and forms a strongly bound state with the nucleus while a positron is emitted. This is also reflected in the quantum mechanical energy of the electron as it results from the Dirac equation which becomes complex at $Z > 137$, signaling an instability that corresponds to the so-called 'falling onto the center'. In summary, the heavy nucleus decays into a supercritically bound state A_{Z-1} with a smaller charge $Z-1$ and a positron,

$$A_Z \to A_{Z-1} + e^+ \,. \tag{1}$$

Starting from a highly charged nucleus that process continues for higher energy levels until the resulting bound state is subcritically charged.

[a]The critical value $Z_c = 137$ holds for a pointlike nucleus, in the more realistic situation of an extended nucleus it would be somewhat higher.

Gribov's idea is that a similar mechanism might apply to the color charges of QCD if light quarks exist with a Compton wave length that is larger than the typical color charge radius or confinement radius. He proposed the intriguing possibility that in QCD already the color charge of a *single* quark could be supercritical, in contrast to QED where a very large charge is required. Note that in QED the relevant parameter for the phenomenon of supercritical charges is in fact $\alpha_{\rm em} Z$ and that also a large electromagnetic coupling $\alpha_{\rm em}$ could trigger the instability. In QCD the role of supercritical charge is in fact played by the strong coupling constant α_s, which becomes large at low momentum scales. As we will see in section 4, supercritical behavior is expected if the strong coupling constant exceeds a critical value α_c given by [2]

$$\frac{\alpha_c}{\pi} = C_F^{-1}\left(1 - \sqrt{\frac{2}{3}}\right) = 0.137\,, \tag{2}$$

with the Casimir operator $C_F = (N_c^2 - 1)/(2N_c) = 4/3$. Note that this critical value is remarkably small.

Let us now briefly discuss how supercritical color charges change the vacuum structure of light quarks. (A more detailed account of the confinement mechanism caused by light quarks can be found in Refs. [7,8].) The emerging picture is again most conveniently discussed in the language of the Dirac sea. For subcritical coupling we have the usual situation with empty positive energy quark states and a filled Dirac sea of negative energy states. If the coupling is larger than the critical value α_c additional states appear. Supercritical bound states can form consisting of a quark and an antiquark which both have positive kinetic energy. But due to the supercritical binding, the bound state has a *negative* total energy. In order to avoid having infinitely many mesonic negative energy states the corresponding quark and antiquark states have to be filled in the vacuum. In particular, the vacuum then contains occupied quark states of *positive kinetic energy*, in addition to the negative energy states in the Dirac sea. In such a vacuum state the Pauli exclusion principle prevents single quarks from propagating freely since the states required for that are already occupied. In that sense the Pauli principle can bind even massless quarks which usually tend to move at the speed of light. A single quark put into the QCD vacuum would undergo a decay into a supercritically bound meson and another quark, $q \to M + q$. Since the resulting quark is again supercritically charged that process would continue and eventually shield the color charge completely, such that a single quark would only exist as a resonance but not as an asymptotic state.

Obviously, this confinement mechanism based on the Pauli principle works only for quarks, that is for fermions in general, but does not apply to gluons. It remains to be investigated how gluons are affected by the light quarks in such a picture. It is conceivable that the confinement of gluons is, at least to some extent, caused by their coupling to confining light quarks, and would then be an indirect or second-order effect.

From a technical point of view the situation in QCD is considerably more complicated than the example of a heavy nucleus in QED. While a heavy quark can be treated in a similar way as the heavy nucleus, the supercritical charge of a light quark can clearly not be considered as an external field. In view of this problem Gribov invented a new approach to the Dyson-Schwinger equation for light quarks that we discuss in the next section.

Interestingly, the dispersive approach [9] to power corrections to event shape variables in e^+e^- collisions and in deep inelastic scattering suggests that the integrated strong coupling constant in the infrared region (assuming of course that such a concept exists) satisfies

$$\alpha_0 = \frac{1}{2\,\text{GeV}} \int_0^{2\,\text{GeV}} \alpha_s(k)\, dk \simeq 0.5\,, \tag{3}$$

see Ref. [10] for a recent review. From (3) one can deduce that the strong coupling becomes in fact larger than the critical value (2) at least in some range of momenta, but remains comparatively small in most if not all of the small-momentum region. This observation raises some hopes that – unexpectedly – some aspects of low-momentum phenomena like for example hadronization might be accessible at least qualitatively in a semi-perturbative way.

3. Gribov's Equation for the Green Function of Light Quarks in Feynman Gauge

Let us now turn to the Gribov equation (or Gribov-Dyson-Schwinger equation) for the Green function of light quarks in Feynman gauge [3]. For Gribov this equation was mainly a technical step on the way to a quantitative support of the physical picture discussed in the previous section. But it can also be viewed as a new approach to the quark's Dyson-Schwinger equation which is very interesting already by itself, independently of the supercritical confinement scenario.

The Green functions of QCD satisfy a tower of coupled Dyson-Schwinger equations. Since it is virtually impossible to solve this system of integral equations as a whole one usually has to resort to simplifications. Most current approaches to the Dyson-Schwinger equations are based on truncation schemes (based for example on certain topological classes of diagrams), but their accuracy is often very difficult to control. Gribov discovered that it is possible to attack the Dyson-Schwinger equation for the quark's Green function in Feynman gauge based on an approximation rather than a mere truncation scheme. The guiding principle in this approximation scheme is to collect the most singular contributions from the infrared momentum region, that is those contributions that are expected to cause chiral symmetry breaking and confinement.

In Feynman gauge the gluon propagator has the general form

$$D_{\mu\nu}(k) = -\frac{g_{\mu\nu}}{k^2}\,\alpha_s(k^2)\,, \tag{4}$$

which can be interpreted as a nonperturbative definition of the strong coupling constant α_s at low momentum scales. We now make the (rather mild) assumption that $\alpha_s(k^2)$ is not divergent as $k^2 \to 0$ and does not vary too strongly. For spacelike momenta a typical example for such a coupling is the model

$$\alpha_s(k^2) = \frac{4\pi}{\left(11 - \frac{2}{3}n_f\right)\,\ln(-k^2/\Lambda^2_{\rm QCD} + a)}\,, \tag{5}$$

which for $a = 6$ satisfies (3) and agrees with the perturbative one-loop running at large momenta. It has been shown [8] that the qualitative results discussed below are independent of the particular model used for α_s at low momentum scales.

One now starts with the Dyson-Schwinger equation for the quark's inverse Green function $G^{-1}(q)$ and applies the d'Alembert operator $\partial^2 = \partial^\mu\partial_\mu$ with

$$\partial_\mu = \frac{\partial}{\partial q^\mu}\,. \tag{6}$$

Let us illustrate this procedure for the one-loop self-energy diagram

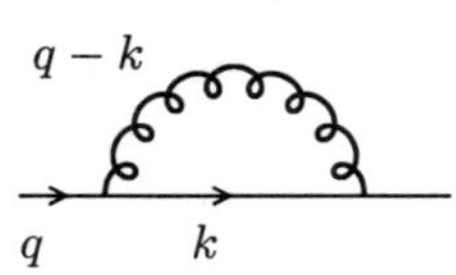

which is given by

$$\Sigma = -i\,C_F\frac{\alpha_s}{\pi}\int\frac{d^4k}{4\pi^2}\,\gamma^\mu G(k)\gamma_\mu\,\frac{1}{(q-k)^2}\,. \tag{7}$$

Firstly, differentiating twice with respect to the external momentum q_μ makes this integral convergent. Secondly, since $1/q^2$ is the Green function of the four-dimensional d'Alembert operator ∂^2,

$$\partial^2 \frac{1}{(q-q')^2 + i\epsilon} = -4\pi^2 i\, \delta^{(4)}(q-q')\,, \tag{8}$$

we can eliminate the integral in (7) and obtain to first order in α_s

$$\partial^2 G^{-1}(q) = C_F \frac{\alpha_s}{\pi} \gamma^\mu G(k) \gamma_\mu\,. \tag{9}$$

One can now systematically proceed to higher order diagrams. In each diagram the most singular contribution from the infrared is obtained when both derivatives are applied to the same gluon line, since only those terms give rise to a delta function according to (8). All other terms are clearly less singular and still involve integrals over loop momenta.

Including higher orders one builds up two full quark-gluon vertices Γ_μ which due to the delta functions have to be taken at vanishing gluon momentum. Note that as a consequence of the differentiations we obtain two full vertices in spite of the fact that the original Dyson-Schwinger equation involves one bare and one full vertex. The two zero-momentum vertex functions $\Gamma_\mu(q,q,0)$ can be related to the inverse Green function via the Ward identity

$$\Gamma_\mu(q,q,0) = \partial_\mu G^{-1}(q)\,, \tag{10}$$

and the corrections to the Ward identity due to the nonabelian structure of QCD can be argued to be subleading in the Dyson-Schwinger equation in the approximation scheme used here. We hence obtain the Gribov equation for the light quark's Green function in Feynman gauge,

$$\partial^2 G^{-1} = g\,(\partial^\mu G^{-1})\, G\,(\partial_\mu G^{-1}) + \cdots\,, \tag{11}$$

where

$$g = C_F \frac{\alpha_s(q)}{\pi}\,. \tag{12}$$

and the ellipsis in (11) stands for less IR-singular terms of order g^2 which can – at least in principle – be systematically computed. Note that via (12) we have inserted the running strong coupling constant $\alpha_s(q)$ in the Gribov equation. Following the derivation outlined above one would first obtain the coupling at zero momentum due to the delta function (8), but the difference between the two is subleading in the sense of our approximation.

With the running coupling the Gribov equation has the remarkable property that it can reproduce the usual renormalization group behavior for the quark propagator in the perturbative region of large spacelike momenta. Therefore the equation collects the most important terms both in the infrared as well as in the ultraviolet, and hence has a good chance of giving a valid description of the quark propagator for all momenta.

The derivation of the equation strongly relies on the mathematical identity (8) and hence on the Feynman gauge. Most other approaches to the Dyson-Schwinger equations use Landau gauge such that a direct comparison with those approaches is rather difficult. If one succeeds in deriving a genuinely gauge-independent phenomenon like chiral symmetry breaking or confinement in one gauge, however, it will clearly hold in any gauge.

We finally point out that the same equation (11) applies also for electrons in QED after replacing C_F by 1 and α_s by $\alpha_{\rm em}$. However, the behavior of the coupling constants and the boundary conditions are very different in the two theories.

4. Chiral Symmetry Breaking

On general grounds the quark's Green function has the form

$$G^{-1}(q^2) = Z^{-1}(\not{q} - M)\,, \tag{13}$$

where $M(q^2)$ is the dynamical mass function of the quark and $Z(q^2)$ stands for the wave function renormalization.

In QCD the infrared and ultraviolet limits of Green functions are very closely related [3]. In order to see this let us consider the limit of vanishing coupling $\alpha_s \to 0$, corresponding to large spacelike momenta, $-q^2 \to \infty$, for which we obtain the free equation $\partial^2 G^{-1} = 0$. One readily verifies that this equation has the solution

$$G^{-1}(q) = Z^{-1}\left(\not{q} - m - \frac{\nu_1^3}{q^2} - \frac{\nu_2^4}{q^4}\,\not{q}\right)\,. \tag{14}$$

Note that in addition to the usual two constants, namely the wave function renormalization Z and the bare mass m, we find two further integration constants ν_1^3 and ν_2^4 since the Gribov equation is a second order differential equation. According to their mass dimension one expects that they can be related (up to factors) to the quark and gluon condensates. The additional terms involving these two constants are singular at low q^2. In QED one can simply drop these terms since the behavior of the Green function is perturbative at low momentum scales and the perturbative degrees of freedom in the Lagrangian coincide with the observable fields. In QCD the situation is drastically different since there are no freely propagating quarks at low momenta. Gribov's differential equation (11) hence relates the asymptotic solution (14) at high momenta to a complicated and qualitatively different structure at low momenta. In order to obtain the full and correct solution at low momenta one must therefore in general keep all four terms of (14) even in the regime of asymptotic freedom. The potential consequences of this result for applications of perturbative QCD still need to be explored.

We now turn to the behavior of the dynamical mass function of the quark in the whole region of spacelike momenta in order to see how chiral symmetry breaking occurs.[3] One finds from Gribov's equation (11) that the dynamical mass function can have oscillations and change its sign in an interval of momenta if the coupling constant exceeds its critical value (2) in that interval [2,3]. That critical value can be obtained analytically from (11). The full mass function can be calculated numerically for spacelike momenta [8]. Choosing the model (5) with $a = 6$ for the running coupling one obtains[b] the mass function $M(q^2)$ shown in Figure 1 for three different initial values at $q^2 = 0$. In the figure q denotes the spacelike momentum,

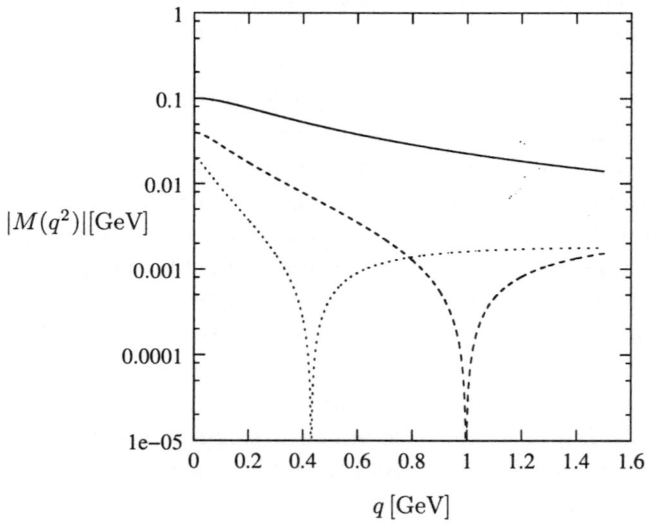

Figure 1. The mass function for three different values of the renormalized mass

[b]This figure has in fact been obtained using the modified equation (18) that includes pion corrections instead of the original equation (11). For spacelike momenta the mass functions resulting from the two equations have qualitatively the same behavior and their difference is rather small, see section 5 below.

$q = \sqrt{-q^2}$. Depending on the initial value the mass function can stay positive (upper curve) or can have oscillations resulting in the spikes in the logarithmic plot in the two lower lines in Figure 1.

We now define the renormalized mass m_R as the value of the dynamical mass function at vanishing quark momentum,

$$m_R = \lim_{q \to 0} M(q^2)\,, \tag{15}$$

and further define a perturbative mass m_P as the value of the mass function at a large (perturbative) momentum scale λ,

$$m_P = M(\lambda^2)\,. \tag{16}$$

The precise value chosen for the scale λ is not relevant for the present discussion as long as the coupling is no longer supercritical at that scale. In practice λ can therefore be chosen as small as 2 GeV, for example.

Recall that by solving the Gribov equation (11) one obtains a certain value for m_P for each particular choice of m_R. We can hence consider the dependence of the renormalized mass m_R on the perturbative mass m_P (or vice versa). If the coupling constant is subcritical in the whole range of momenta (for example if a in (5) is chosen sufficiently large) this dependence is one-to-one. If on the other hand the coupling is supercritical, the mass function can oscillate and different m_R can correspond to the same m_P at the scale λ. In this situation the dependence of m_R on m_P is no longer one-to-one and instead becomes multivalued, as is shown in Figure 2. It can be easily seen that even for vanishing perturbative quark mass m_P a large renormalized mass m_R is generated. Hence chiral symmetry breaking occurs dynamically if the coupling is supercritical in the infrared.

One can then proceed and compute the solutions to the Gribov equation (11) in the whole complex q^2-plane in order to study their analytic structure. In general this requires further assumptions about the behavior of α_s and hence of the gluon propagator $D_{\mu\nu}$, see (4), for complex q^2. If one assumes that $D_{\mu\nu}$ does not have dramatic singularities off the real q^2-axis one finds that even for supercritical coupling the analytic structure of the quark's Green function does not correspond to a confined quark.[12] Instead, the Green function exhibits a pole and a cut on the positive real q^2-axis corresponding to the free propagation of the quark.

5. Pion Effects on the Quark's Green Function

In the process of chiral symmetry breaking pions are generated as Goldstone bosons and appear in the physical spectrum as massless states. Their Bethe-

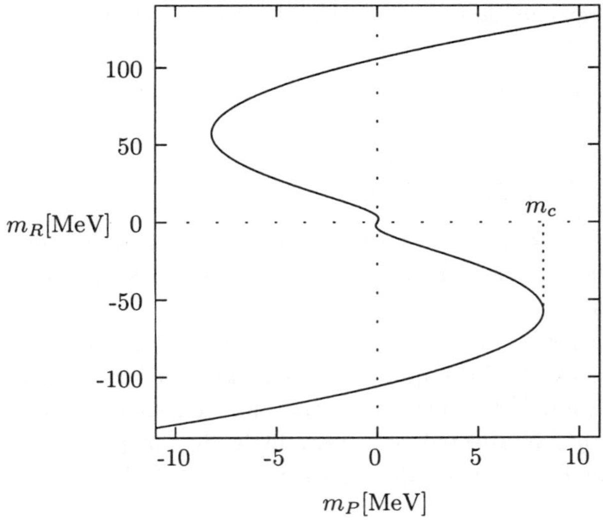

Figure 2. The dependence of the renormalized mass m_R on the perturbative mass m_P reflecting chiral symmetry breaking

Salpeter amplitude ϕ_π is described by an equation [3] that is derived in the same approximation as the equation (11) for the quark's Green function described above. For massless pions the solution of that equation can be found analytically in terms of the quark's Green function,

$$\phi_\pi \sim \{\gamma_5, G^{-1}\}\,. \tag{17}$$

Once pions appear as massless states in the physical spectrum due to chiral symmetry breaking they might have a considerable effect on the quark's Green function. This backreaction is not properly taken into account in the original equation (11). In order to obtain an improved equation for the Green function of light quarks one should therefore add their contribution separately [4]. Incidentally, the pion propagator is proportional to $1/k^2$ such that one can treat pion loops on the quark using the same mathematical identity (8) as for the gluon propagator in Feynman gauge, hence again extracting the most IR-singular contributions. The coupling of the pion to the quark can be related to the pion decay constant f_π via a Goldberger-Treiman relation, and taking into account the proper isospin factor for two light quark flavors one obtains the improved Gribov equation

$$\partial^2 G^{-1} = g\,(\partial^\mu G^{-1})\,G\,(\partial_\mu G^{-1}) - \frac{3}{16\pi^2 f_\pi^2}\{i\gamma_5, G^{-1}\}\,G\,\{i\gamma_5, G^{-1}\}\,. \tag{18}$$

The pion decay constant f_π can further be shown to fulfill in a selfconsistent way the equation

$$f_\pi^2 = \frac{1}{8}\int \frac{d^4q}{(2\pi)^4 i}\,\mathrm{Tr}\left[\{i\gamma_5, G^{-1}\}\,G\,\{i\gamma_5, G^{-1}\}\,G\,(\partial_\mu G^{-1}G)^2\right] + \frac{1}{64\pi^2 f_\pi^2}\int \frac{d^4q}{(2\pi)^4 i}\,\mathrm{Tr}\left[\left(\{i\gamma_5, G^{-1}\}\,G\right)^4\right]. \qquad (19)$$

Note that also the modified Gribov equation (18) is a differential equation rather than an integral equation and that it again involves only the quark's Green function.

Choosing for f_π in (18) its phenomenological value one can repeat the numerical analysis of the quark's Green function based on the improved Gribov equation [11,12]. One finds that for spacelike momenta the mass function $M(q^2)$ is affected only very little by the pion correction, as is illustrated in Figure 3 for the choice $m_R = 0.1\,\mathrm{GeV}$. As a consequence

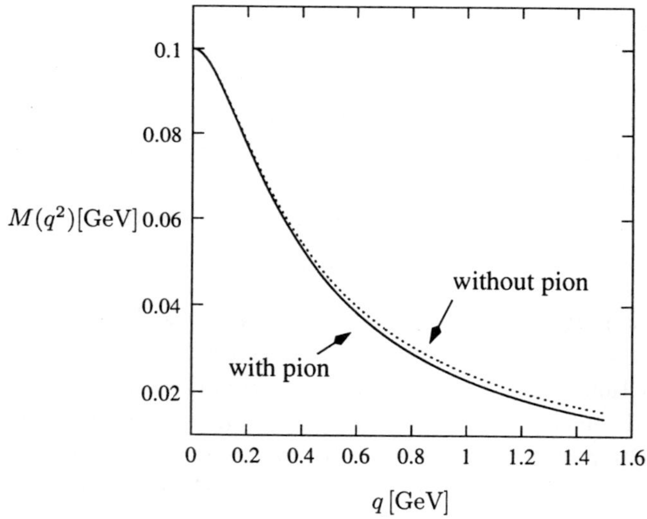

Figure 3. Change in the mass function due to pion corrections

the general behavior of the Green function for spacelike momenta as it was found already with the original equation (11) remains the same when pion corrections are taken into account. In particular, the oscillations of the mass functions for supercritical coupling again lead to chiral symmetry breaking in exactly the same way as discussed above.

Preliminary numerical results indicate that the situation is quite different when one moves away from spacelike momenta into the complex q^2-plane. Here the pion corrections become important and appear to induce considerable changes in the analytic structure of the Green function of light quarks. Gribov argued that the modified equation (18) will lead to an analytic structure of the Green function corresponding to confined quarks and hence to the general confinement of color charges in QCD [4]. A full numerical study of this conjecture remains to be carried out.

6. Summary and Outlook

Gribov has developed an attractive physical picture of QCD in which chiral symmetry breaking and confinement are caused by the existence of light quarks which can trigger the mechanism of supercritical color charges. He has derived an equation for the Green function of light quarks that collects the most important contributions both in the infrared and in the ultraviolet momentum region. The equation describes chiral symmetry breaking and gives rise to pions as Goldstone bosons. There are some indications that the backreaction of the pions on the quark leads to confinement.

The next step in a systematic study of Gribov's approach should be a full investigation of the equation (18) which includes pion corrections, in particular a study of the resulting analytic structure of the quark's Green function. A more difficult but extremely important future step should be to develop equations that describe the gluon sector of the theory in a similar approximation. Another crucial goal is to find observables that are sensitive to the particular confinement mechanism due to supercritical charges and to devise phenomenological tests of Gribov's scenario.

In the course of his study of confinement Gribov has raised a number of questions concerning the general treatment of gauge theories in which the elementary fields of Lagrangian are not the ones that can propagate freely. In Gribov's picture a satisfactory solution of the confinement problem can only be obtained if these questions are addressed at the same time. That makes it a very challenging task to fully understand his ideas and to develop his program further in the future, but it is worth the effort.

Acknowledgments

I am very grateful that I had the opportunity to learn so much about the subject of this contribution and about quantum field theory in general from Vladimir Gribov in countless discussions and lectures when he was the supervisor of my diploma thesis. It was a great experience to see his unconventional way of thinking and his passion for physics from a close distance and to benefit from his deep understanding and insight.

I would like to thank Julia Nyiri, Yuri Dokshitzer and Péter Lévai for organizing this workshop and for inviting me to contribute to it. This work was supported by a Feodor Lynen fellowship of the Alexander von Humboldt Foundation.

References

1. V.N. Gribov, *Phys. Scripta* **T15**, 164 (1987).
2. V.N. Gribov, LU-TP-91-7, *Presented at Perturbative QCD Workshop, Lund, Sweden, May 1991*; also contained in Ref. [6].
3. V.N. Gribov, *Eur. Phys. J.* **C10**, 71 (1999) [arXiv:hep-ph/9807224].
4. V.N. Gribov, *Eur. Phys. J.* **C10** (1999) 91 [arXiv:hep-ph/9902279].
5. Yu.L. Dokshitzer and D.E. Kharzeev, *Ann. Rev. Nucl. Part. Sci.* **54**, 487 (2004) 487 [arXiv:hep-ph/0404216].
6. V.N. Gribov, *'Gauge Theories and Quark Confinement'*, Phasis Publishing House, Moscow, 2001.
7. V.N. Gribov, *Orsay lectures on confinement (I - III)*,
   LPT Orsay 92-60, hep-ph/9403218;
   LPT Orsay 94-20, hep-ph/9404332;
   LPT Orsay 99-37, hep-ph/9905285.
8. C. Ewerz, *Eur. Phys. J.* C **C13** (2000) 503 [arXiv:hep-ph/0001038].
9. Yu.L. Dokshitzer, G. Marchesini and B.R. Webber, *Nucl. Phys.* **B469**, 93 (1996) [arXiv:hep-ph/9512336].
10. Yu.L. Dokshitzer, *Acta Phys. Polon.* **B36**, 361 (2005) [arXiv:hep-ph/0411227].
11. C. Ewerz, work in progress
12. C. Ewerz, in Proceedings of the *6th International Conference on Quark Confinement and the Hadron Spectrum*, Villasimius, Italy, 2004, *AIP Conf. Proc.* **756**, 336 (2005) [arXiv:hep-ph/0411374].

MASS GAP AND GLUON CONFINEMENT IN THE YANG-MILLS THEORY

V. GOGOHIA

HAS CRIP RMKI, Department for Theoretical Physics,
P.O.B. 49, Budapest 114, H-1525, Hungary
E-mail: gogohia@rmki.kfki.hu

The general iteration solution (i.e., when the relevant skeleton loop integrals, contributing into the gluon self-energy, have to be iterated, which means that no any truncations/approximations have been made) of the Schwinger-Dyson equation of motion for the full gluon propagator unavoidably becomes the exact sum of the two terms. The first term is the Laurent expansion in the inverse powers of the gluon momentum squared, starting necessarily from the simplest one $1/(q^2)^2$. Each power-type severe (i.e., more singular than $1/q^2$) IR singularity is accompanied by the corresponding powers of the mass squared parameter (the mass gap) multiplied by the corresponding residues. The mass gap is responsible for the large scale structure of the true QCD vacuum. The standard second term is always as much singular as $1/q^2$ and otherwise remaining undetermined. The inevitable existence of the first term makes just the principal difference between the nonlinear dynamics of non-Abelian QCD and the linear one of Abelian QED. Moreover, the infrared renormalization program of the theory leads to the gluon confinement criterion in a gauge-invariant way.

1. Introduction

If QCD itself is a confining theory, then a characteristic scale is very likely to exist. It should be directly responsible for the large scale structure of the true QCD vacuum in the same way as Λ_{QCD} is responsible for the nontrivial perturbative (PT) dynamics there (asymptotic freedom (AF) [1,2]). The main purpose of this talk is just to show how the above-mentioned characteristic scale (the mass gap in what follows, for simplicity) responsible for the nonperturbative (NP) dynamics in the infrared (IR) region may explicitly appear in QCD. This especially becomes imperative after Jaffe and Witten have formulated their famous theorem [3].

The propagation of gluons is one of the main dynamical effects in the true QCD vacuum. The gluon Green's function is (Euclidean signature here and everywhere below)

$$D_{\mu\nu}(q) = i\left\{T_{\mu\nu}(q)d(q^2,\xi) + \xi L_{\mu\nu}(q)\right\}\frac{1}{q^2} , \tag{1}$$

where ξ is the gauge fixing parameter ($\xi = 0$ - Landau gauge and $\xi = 1$ - Feynman gauge) and $T_{\mu\nu}(q) = \delta_{\mu\nu} - q_\mu q_\nu/q^2 = \delta_{\mu\nu} - L_{\mu\nu}(q)$. Evidently, $T_{\mu\nu}(q)$ is the transverse ("physical") component of the full gluon propagator, while $L_{\mu\nu}(q)$ is its longitudinal (unphysical) one. The free gluon propagator is obtained by setting simply the full gluon form factor $d(q^2,\xi) = 1$ in Eq. (1.1), i.e., $D^0_{\mu\nu}(q) = i\left\{T_{\mu\nu}(q) + \xi L_{\mu\nu}(q)\right\}(1/q^2)$. The main tool of our investigation is the Schwinger-Dyson (SD) equation of motion [1,4,5] (see below) for the full gluon propagator (1) since its solutions reflect the quantum-dynamical structure of the true QCD ground state.

2. Gluon SD equation

The general structure of the SD equation for the full gluon propagator [1,5] in the pure Yang-Mills (YM) theory can be written down symbolically as follows (for our purposes it is more convenient to consider the SD equation for the full gluon propagator and not for its inverse):

$$D(q) = D^0(q) + D^0(q)T_g[D](q)D(q) . \tag{2}$$

Here and in some places below, we omit the dependence on the Dirac indices, for simplicity. A pure gluon contribution $T_g[D](q)$ into the gluon self-energy is a sum of the four pure gluon skeleton loops, and consequently they contain explicitly the full gluon propagators. Precisely this makes the gluon SD equation highly nonlinear (NL), and this is one of the reasons why it cannot be solved exactly. Let us present now the NL pure gluon part, which was symbolically denoted as $T_g[D](q)$ in the gluon SD equation (2). It is

$$T_g[D](q) = \frac{1}{2}T_t + \frac{1}{2}T_1(q) + \frac{1}{2}T_2(q) + \frac{1}{6}T_2'(q) , \tag{3}$$

where the so-called constant tadpole term is

$$T_t = g^2 \int (id^4q_1/(2\pi)^4)T_4^0 D(q_1) . \tag{4}$$

Three other skeleton loop integrals describe all the independent interactions between massless gluons, namely $T_1(q)$ contains only three gluon vertices, $T_2(q)$ contains three- and four-gluon vertices as well, while $T_2'(q)$ contains only four-gluon vertices. Their explicit expressions are not important here. However, they can be found in Refs. [5,6,7].

The general iteration solution (i.e., when the above-mentioned skeleton loop integrals are to be iterated) of the gluon SD equation (2) looks like

$$\begin{aligned} D(q) &= D^{(0)}(q) + D^{(1)}(q) + D^{(2)}(q) + D^{(3)}(q) + ... = D^0 \\ &+ D^0(q)T_g[D^0 + D^{(1)} + D^{(2)} + D^{(3)}(q) + ...](q) \\ &\times [D^0(q) + D^{(1)}(q) + D^{(2)}(q) + D^{(3)}(q) + ...] \, . \end{aligned} \tag{5}$$

Evidently, $D^{(0)}(q) = D^0(q)$. It is nothing but the skeleton loops expansion. If one knows how to sum up this expansion, so he knows the full gluon propagator and vice-versa, i.e., any solution to the gluon SD equation should be compatible with this expansion.

3. Regularization

Due to AF [1,2] all the skeleton loop integrals as well as those which will appear in the formal iteration solution (5) are divergent, so the general problem of their regularization arises. For our future purpose it is convenient to regularize the above-mentioned skeleton loop integrals by subtracting as usual their values at a safe space-like point $q^2 = \mu^2$ (Euclidean signature is already chosen). Thus, one obtains

$$\begin{aligned} T_1^R(q) &= T_1(q) - T_1(q^2 = \mu^2), \\ T_2^R(q) &= T_2(q) - T_2(q^2 = \mu^2), \\ T_2'^R(q) &= T_2'(q) - T_2'(q^2 = \mu^2) \, , \end{aligned} \tag{6}$$

where all the divergent constants $T_1(q^2 = \mu^2)$, $T_2(q^2 = \mu^2)$ and $T_2'(q^2 = \mu^2)$ have the dimensions of the mass squared. At the same time, the introduction of the ultraviolet (UV) cutoff Λ^2 in these integrals as well as in the constant tadpole term T_t, which has also the dimensions of the mass squared, is assumed. The formal subtraction procedure for this term leads to the conclusion that in general its regularization can be discarded since formally $T_t^R = T_t - T_t = 0$. Let us remind that in the dimensional regularization method [1,2] $T_t(D^0) = 0$, indeed, but not in our case when integrals are cut off by hand at upper limit.

The subtractions (6) mean that the decomposition of the regularized quantities into the independent tensor structures can be written down as follows:

$$T_1^R(q) \equiv T_{(1)\mu\nu}^R(q) = \delta_{\mu\nu} q^2 T_1^{(1)}(q^2) + q_\mu q_\nu T_1^{(2)}(q^2),$$
$$T_2^R(q) \equiv T_{(2)\mu\nu}^R(q) = \delta_{\mu\nu} q^2 T_2^{(1)}(q^2) + q_\mu q_\nu T_2^{(2)}(q^2),$$
$$T_2^{\prime R}(q) \equiv T_{(2)\mu\nu}^{\prime R}(q) = \delta_{\mu\nu} q^2 T_2^{\prime(1)}(q^2) + q_\mu q_\nu T_2^{\prime(2)}(q^2) \ . \quad (7)$$

where all the invariant functions $T_1^{(n)}(q^2), T_2^{(n)}(q^2), T_2^{\prime(n)}(q^2)$ at $n = 1, 2$ are dimensionless ones. In the region of small q^2 they are to be represented in the form of the corresponding Taylor expansions and remaining arbitrary otherwise. Due to the definition $q_\mu q_\nu = q^2 L_{\mu\nu}$, instead of the independent structures $\delta_{\mu\nu}$ and $q_\mu q_\nu$ in Eqs. (7) and below, one can use $T_{\mu\nu}$ and $L_{\mu\nu}$ as the independent structures with their own invariant functions. From these relations it follows that the regularized skeleton loop contributions are of the order q^2 always, i.e., $T_1^R(q) = O(q^2), T_2^R(q) = O(q^2)$ and $T_2^{\prime R}(q) = O(q^2)$.

The NL pure gluon part (3) thus can be exactly decomposed as the sum of the two terms as follows:

$$T_g[D](q) = T_g[D] + T_g^R[D](q), \quad (8)$$

where $T_g[D] = (1/2)T_t(D) + (1/2)T_1(D^2) + (1/2)T_2(D^4) + (1/6)T_2'(D^3)$, and $T_g^R[D](q) = (1/2)T_1^R(q) + (1/2)T_2^R(q) + (1/6)T_2^{\prime R}(q) = O(q^2; D)$. In the first relation we introduce the following notations: $T_t = T_t(D), T_1(q^2 = \mu^2) = T_1(D^2), T_2(q^2 = \mu^2) = T_2(D^4)$ and $T_2'(q^2 = \mu^2) = T_2'(D^3)$, showing their dependence on the corresponding number of the gluon propagators only, for simplicity. In the second relation the last equality shows explicitly that this term is of the order q^2, depending again on D in general. Thus the gluon SD equation (2) becomes divided into the two principally different terms, namely

$$D(q) = D^0(q) + D^0(q) T_g[D] D(q) + D^0(q) O(q^2; D) D(q) \ . \quad (9)$$

Let us note that the both blocks $T_g[D]$ and $O(q^2; D)$ which should be iterated with respect to D depend themselves on D. This means that there is no way to sum up, in general, these iteration series into the corresponding geometrical progression. The latter possibility appears only when the blocks to be iterated do not depend on D like in quantum electrodynamics (QED). The difference between these two cases lies, of course, in the NL dynamics of QCD in comparison with the linear one in QED.

4. Iteration

The formal iteration solution of the gluon SD equation (9) now looks like

$$\begin{aligned} D(q) = D^{(0)}(q) + D^{(1)}(q) + D^{(2)}(q) + D^{(3)}(q) + ... = D^0(q) + \\ +D^0(q)T_g[D^0 + D^{(1)} + D^{(2)} + D^{(3)}(q) + ...] \\ \times[D^0(q) + D^{(1)}(q) + D^{(2)}(q) + D^{(3)}(q) + ...] \\ +D^0(q)O_g(q^2; D^0 + D^{(1)} + D^{(2)} + D^{(3)}(q) + ...) \\ \times[D^0(q) + D^{(1)}(q) + D^{(2)}(q) + D^{(3)}(q) + ...] , \end{aligned} \tag{10}$$

where

$$\begin{aligned} D^{(0)}(q) &= D^0(q), \\ D^{(1)}(q) &= D^0(q)F_1[D^0]D^0(q) + D^0(q)O(q^2; D^0)D^0(q), \\ D^{(2)}(q) &= D^0(q)F_2[D^0, D^{(1)}][D^0(q) + D^{(1)}(q)] \\ &+ D^0(q)O(q^2; D^0 + D^{(1)})[D^0(q) + D^{(1)}(q)], \\ D^{(3)}(q) &= D^0(q)F_3[D^0, D^{(1)}, D^{(2)}][D^0(q) + D^{(1)}(q) + D^{(2)}(q)] \\ &+ D^0(q)O(q^2; D^0 + D^{(1)} + D^{(2)})[D^0(q) + D^{(1)}(q) + D^{(2)}(q)] \end{aligned} \tag{11}$$

and so on. Let us consider now in more detail the divergent constants $F_1[D^0], F_2[D^0, D^{(1)}], F_3[D^0, D^{(1)}, D^{(2)}], ...$ Evidently, we have introduced the following notations: $F_1[D^0] = T_g[D^0] = (1/2)T_t(D^0) + (1/2)T_1((D^0)^2) + (1/2)T_2((D^0)^4) + (1/6)T_2'((D^0)^3)$, $F_2[D^0, D^{(1)}] = T_g[D^0 + D^{(1)}] = (1/2)T_t(D^0 + D^{(1)}) + (1/2)T_1((D^0 + D^{(1)})^2) + (1/2)T_2((D^0 + D^{(1)})^4) + (1/6)T_2'((D^0 + D^{(1)})^3)$, $F_3[D^0, D^{(1)}, D^{(1)}] = T_g[D^0 + D^{(1)} + D^{(2)}] = (1/2)T_t(D^0 + D^{(1)} + D^{(2)}) + (1/2)T_1((D^0 + D^{(1)} + D^{(2)})^2) + (1/2)T_2((D^0 + D^{(1)} + D^{(2)})^4) + (1/6)T_2'((D^0 + D^{(1)} + D^{(2)})^3)$, and so on. As underlined above, each of them has the dimensions of the mass squared, so on general ground one can represent them as follows:

$$\begin{aligned} F_1 \equiv F_1[D^0] &= \Delta^2 C_1(\lambda, \nu, \xi, g^2), \\ F_2 \equiv F_2[D^0, D^{(1)}] &= \Delta^2 C_2(\lambda, \nu, \xi, g^2), \\ F_3 \equiv F_3[D^0, D^{(1)}, D^{(2)}] &= \Delta^2 C_3(\lambda, \nu, \xi, g^2) , \end{aligned} \tag{12}$$

and so on. In these relations Δ^2 is the above-mentioned mass gap. The dimensionless constants $C_1, C_2, C_3, ...$ depend in general on the dimensionless UV cutoff λ introduced as follows: $\Lambda^2 = \lambda\Delta^2$. They depend on the dimensionless renormalization point ν introduced as follows: $\mu^2 = \nu\Delta^2$. Evidently, via the corresponding subscripts these constants depend on which iteration for the gluon propagator D is explicitly done, which in its turn gives the dependence on the gauge fixing parameter ξ. They also depend on the dimensionless coupling constant squared g^2 (see the corresponding skeleton loop integrals in Refs. [5,6,7]). The parameters λ and ν may in principle depend on ξ and g^2 (and hence vise-versa) since they have been itroduced by hand, however, this is not important at this stage. Evidently, above we use the short-hand notation $D^0 \equiv D^0(q)$.

It is convenient, however, to introduce the short-hand notations as follows: $D^0(q) \equiv D_0(q) \equiv D_0$, $O_1(q^2) \equiv O(q^2; D^0)$, $O_2(q^2) \equiv O(q^2; D^0 + D^{(1)})$, $O_3(q^2) \equiv O(q^2; D^0 + D^{(1)} + D^{(2)})$, and so on. Then the formal iteration solution (10) becomes

$$
\begin{aligned}
D(q) &= D^{(0)}(q) + D^{(1)}(q) + D^{(2)}(q) + D^{(3)}(q) + ... = D_0 \\
&+ [D_0F_1D_0 + D_0O_1(q^2)D_0] \\
&+ [D_0F_2D_0 + D_0F_2D_0F_1D_0 + D_0F_2D_0O_1(q^2)D_0 \\
&+ D_0O_2(q^2)D_0 + D_0O_2(q^2)D_0F_1D_0 + D_0O_2(q^2)D_0O_1(q^2)D_0] \\
&+ [D_0F_3D_0 + D_0F_3D_0F_2D_0O_1(q^2)D_0 + D_0O_3(q^2)D_0F_2D_0O_1(q^2)D_0 \\
&+ D_0O_3(q^2)D_0O_2(q^2)D_0O_1(q^2)D_0 + ...] + ... \,, \qquad (13)
\end{aligned}
$$

where for $D^{(3)}(q)$ explicitly shown a few terms only, for simplicity. Since $D_0 \equiv D_0(q) \sim (q^2)^{-1}$, there are three formally different types of terms, namely

$$
\begin{aligned}
D(q) &= D_0^2(F_1 + F_2 + F_3 + ...) + D_0^3(F_1F_2 + F_3F_1 + F_2F_3 + ...) \\
&+ D_0^4(F_1F_2F_3 + ...) + [D_0F_2D_0O_1(q^2)D_0 + D_0F_3D_0F_2D_0O_1(q^2)D_0 \\
&+ D_0O_3(q^2)D_0F_2D_0O_1(q^2)D_0 + ...] + O(D_0) \,. \qquad (14)
\end{aligned}
$$

The first type of terms are the terms which are as much singular as $D_0^{2+k} \sim (q^2)^{-2-k}$, $k = 0, 1, 2, 3, ...$ The third type of terms is as much singular as the free gluon propagator, i.e., they are of the order $D_0 \sim (q^2)^{-1}$ since all functions $O_n(q^2), n = 1, 2, 3, ...$ are of the order q^2 always. The second type of terms are the so-called mixed up terms since they may contain the above-mentioned both types of terms, indeed. In order to show this explicitly, let us recall that we can equivalently represent all functions $O_n(q^2), n = 1, 2, 3, ...$ in the form of the corresponding Taylor expansions in integer powers of small q^2. It is convenient to present such kind of expansions in terms of $D_0^{-1} \sim q^2$ as follows:

$$O_n(q^2) = D_0^{-1} \sum_{m=0}^{k-1} (D_0\Delta^2)^{-m} O_n^{(m)}(1) + O_n(D_0^{-k-1}), \quad n = 1, 2, 3, \dots . \tag{15}$$

Here $O_n(D_0^{-k-1})$ denotes the terms of the corresponding order in powers of D_0^{-1} and otherwise remaining arbitrary. They always have the dimensions of D_0^{-1}, so that $O_n(D_0^{-k-1}) \sim D_0^{-1}(D_0\Delta^2)^{-k}$. For the first mixed up term $D_0F_2D_0O_1(q^2)D_0$, one gets

$$\begin{aligned} D_0F_2D_0O_1(q^2)D_0 &= D_0^{2+1}F_2[D_0^{-1}O_1^{(0)}(1) + O_1(D_0^{-2})] \\ &\sim D_0^2F_2O_1^{(0)}(1) + D_0F_2\Delta^{-2} \\ &= (D^0)^2F_2O_1^{(0)}(1) + O(D_0) \ , \end{aligned} \tag{16}$$

since for this term $k = 1$ in the exponent $3 = 2 + k$. The dimensionless combination $F_2\Delta^{-2}$ is included into the term $O(D_0)$. Thus, one concludes that this mixed up term becomes the sum of the two different terms. The first one should be included into the terms of the order D_0^2 shown as the first term in the iteration solution (14), while the second term should be combined with the terms which always are of the order D^0.

For the mixed up term $D_0F_3D_0F_2D_0O_1(q^2)D_0$, we get

$$\begin{aligned} &D_0 \ F_3D_0F_2D_0O_1(q^2)D_0 = \\ &= \ D_0^{2+2}F_2F_3[D_0^{-1}O_1^{(0)}(1) + D_0^{-2}\Delta^{-2}O_1^{(1)}(1) + O_1(D_0^{-3})] \\ &\sim \ D_0^3F_2F_3O_1^{(0)}(1) + D_0^2F_2F_3\Delta^{-2}O_1^{(1)}(1) + D_0F_2F_3\Delta^{-4} \\ &= \ D_0^3F_2F_3O_1^{(0)}(1) + D_0^2F_2F_3\Delta^{-2}O_1^{(1)}(1) + O(D_0) \ , \end{aligned} \tag{17}$$

since for this term $k = 2$ in the exponent $4 = 2 + k$. The combination $F_2F_3\Delta^{-2}$ has the dimensions of the mass squared (see Eqs. (12)). Again the dimensionless combination $F_2F_3\Delta^{-4}$ is included into the term $O(D_0)$. The first term from this sum is to be shifted into the second term of Eq. (14). The second term from this sum is to be shifted into the first term of Eq. (14) and the last one is to be shifted into the last term of the general iteration solution (14).

For how to formulate the above-described "shifting" procedure for the terms which contain the product of the two and more $O(q^2)$ functions, see the extended version of this report given in Ref. [7]. So there is a possibility to treat all other mixed up term in the same way. Let us also note that all the combinations of different masses squared (like the above-mentioned $F_2\Delta^{-2}, F_2F_3\Delta^{-2}$, etc.) can be reduced to the corresponding powers of the mass gap Δ^2 and dimensionless coefficients $C_1, C_2, C_3, ...$. They are to be multiplied by the different dimensionless constants coming from the "shifting" procedure and depending in general on the same quantities as in Eqs. (12). It is worth underlining that the exact and unique separation between the two kind of terms ($\sim D_0^{2+k}(q)$, $k = 0, 1, 2, 3, ...$ and $\sim D_0(q)$) is achieving by keeping the necessary number of terms from the corresponding Taylor expansions (15). This "shifting" method in its general form has been formulated and applied in our previous publications [5,8].

Omitting all this tedious algebra and summing up all terms, one gets that the full gluon propagator becomes the exact sum of the two different terms, namely

$$D(q) = D_0^2(q)\Delta^2 \sum_{k=0}^{\infty} [D^0(q)\Delta^2]^k a_k(\lambda, \nu, \xi, g^2) + O(D_0(q)) \ , \tag{18}$$

where the q^2-independent factors $a_k(\lambda, \nu, \xi, g^2)$ themselves are the sum of an infinitely number of terms, i.e., $a_k(\lambda, \nu, \xi, g^2) = \sum_{m=0}^{\infty} a_{k,m}(\lambda, \nu, \xi, g^2)$. The last sum over m indicates that all iterations with respect to D contribute into each k. Let us emphasize that in getting this exact result no any approximations/truncations have been made since all the relevant skeleton loop contributions into the gluon self-energy have been iterated. Also all the different mass squared parameters in the expansion (14) have to be expressed in terms of the mass gap via the relations (12). It is worth emphasizing here that the terms $O(D_0(q))$ is the sum of the terms which are of the order $D_0(q)$ from the very beginning (the third type of terms in the expansion (14)) and the terms of the same order having appeared due to the above-described "shifting" procedure.

5. Exact structure of the full gluon propagator

Restoring the tensor structure, omitting the tedious algebra and again taking into account that $D_0 \sim (q^2)^{-1}$, the general iteration solution of the gluon SD equation (18) for the full gluon propagator can be algebraically (i.e., exactly) decomposed as the sum of the two principally different terms, namely

$$
\begin{aligned}
D_{\mu\nu}(q) &= D^{INP}_{\mu\nu}(q,\Delta^2) + D^{PT}_{\mu\nu}(q) \\
&= iT_{\mu\nu}(q)\frac{\Delta^2}{(q^2)^2}\sum_{k=0}^{\infty}(\Delta^2/q^2)^k\phi_k(\lambda,\nu,\xi,g^2) + \\
&\quad + i\Big[T_{\mu\nu}(q)d^{PT}(q^2;\xi) + \xi L_{\mu\nu}(q)\Big]\frac{1}{q^2} \ , \qquad (19)
\end{aligned}
$$

where $d^{PT}(q^2;\xi) = \sum_{m=0}^{\infty} a_m(q^2;\xi)$ and

$$
\phi_k(\lambda,\nu,\xi,g^2) = \sum_{m=0}^{\infty}\phi_{k,m}(\lambda,\nu,\xi,g^2) \ . \qquad (20)
$$

In Eq. (19) the superscript "INP" stands for the intrinsically NP part of the full gluon propagator which depends only on the transverse ("physical") degrees of freedom of gauge bosons. Evidently, we distinguish between the two terms in Eq. (19) by the character of the corresponding IR singularities. The power-type severe (or equivalently NP) IR singularity is defined as more singular than the power-type IR singularity of the free gluon propagator, i.e., more singular than $(q^2)^{-1}$. The PT IR singularity is defined as much singular as $(q^2)^{-1}$ always. That is why the longitudinal component of the full gluon propagator is included into its PT part. It remains undetermined. Anyway, we are not responsible for this part and it will not be discussed in what follows. Let us recall again that this term is the sum of the "old" terms and the "new" ones coming from the "shifting" procedure. Also the INP part vanishes when the mass gap goes to zero, while the PT counterpart survives. Thus, one can conclude that the separation between these two terms is unique and exact.

The INP part is nothing else but the corresponding Laurent expansion in integer powers of severe IR singularities, accompanied by the corresponding powers of the mass gap. The important feature of the general iteration solution is that the skeleton loop integrals are to be iterated. This means that there are no any assumptions and approximations made since all the relevant contributions have been taken into account. In addition, in the INP term the sum over m simply shows that an infinite number of iterations (all iterations) invokes each severe IR singularity (and hence the mass gap) in the full gluon propagator. In Refs. [5,6,8] we came to the same structure (19) but in a rather different way, though arguments of Refs. [5,6] have not been so convincing we should admit.

The decomposition (19) of the full gluon propagator into the two terms is exact. Due to the NL dynamics of the QCD ground state, we were able to establish exactly the functional structure (i.e., the dependence on q^2 and hence on the mass gap squared) of the first term $D^{INP}(q)$. Let us emphasize that its Laurent expansion form is necessary since it allows one to put each power-type NP IR singularity (which are independent distributions) under firm mathematical control [5,6,7,8] (see below as well). The both terms are valid in the whole energy/momentum range, i.e., they are not asymptotics. At the same time, we achieved the exact separation between the two terms responsible for the NP (dominating in the IR) and PT (dominating in the UV) dynamics in the true QCD vacuum.

The unavoidable presence of the first term in Eq. (19) makes the principal difference between non-Abelian QCD and Abelian QED, where such kind of term in the full photon propagator is certainly absent (in the former theory there is direct coupling between massless gluons, while in the latter one there is no direct coupling between massless photons). Precisely this term violets the cluster properties of the Wightman functions [9], and thus validates the Strocchi theorem [10], which allows for such IR singular behavior of the full gluon propagator. For some interesting aspects of the comparison between the NL dynamics of the QCD and the linear dynamics of the QED vacua see Ref. [11].

Though the coefficients of the Laurent expansion (the so-called residues at poles $\phi_k(\lambda, \nu, \xi, g^2)$) may explicitly depend on the gauge fixing parameter ξ, the zero momentum modes enhancement (ZMME) effect itself (represented in the INP part) does not depend on it. Since we do not specify explicitly the value of the gauge fixing parameter ξ, the ZMME effect takes place at its any value. In this sense this effect is gauge invariant. This is very similar to AF. It is well known that the exponent which determines the logarithmic deviation of the full gluon propagator from the free one in the UV region ($q^2 \gg \Lambda^2_{QCD}$) explicitly depends on the gauge fixing parameter. At the same time, AF itself does not depend on it, i.e., it takes place at any ξ.

Also due to the arbitrariness of the above-mentioned residues there is no way to sum up these Laurent series into the function with regular behavior at small q^2. It can be done apparently for the special choice of a gauge (for example, Landau gauge), summing up a special class of the corresponding diagrams contributing to the residues, etc. In other words, in order to get smooth gluon propagator the different kinds of the approximation/truncation have to be made [12]. Within the general iteration solution, there is no smooth gluon propagator. It is inevitably severely IR singular at origin.

The QCD Lagrangian does not contain a mass gap. However, we discovered that the mass scale parameter responsible for the NP dynamics in the IR region should exist in the true QCD ground state. At the level of the gluon SD equation it is hidden in the skeleton loop contributions into the gluon self-energy. It explicitly shows up (and hence the corresponding severe IR singularities) when the gluon momentum goes to zero. At the fundamental quark-gluon (i.e., Lagrangian) level the dynamical source of a mass gap (and hence of severe IR singularities) is the self-interaction of massless gluons, i.e., the NL dynamics of QCD. The triple gluon vertex vanishes when all the gluon momenta involved go to zero ($T_3(0,0) = 0$), while its four-gluon counterpart survives ($T_4(0,0,0) \neq 0$). Then one may think that the latter one plays more important role than the former one in the IR structure of the gluon SD equation and thus in the arising of the mass gap mainly from the quartic gluon potential (Feynman [3,13] has also arrived at the same conclusion but on a different basis). To show this on general ground is a rather complicated task since explicitly we can calculate only simplest contributions to the residues $\phi_k(\lambda, \nu, \xi, g^2)$. It is worth emphasizing, however, that such kind of the concrete calculations is neither important nor necessary, only the general dependence of the residues $\phi_k(\lambda, \nu, \xi, g^2)$ on their arguments is all that matters within our approach.

6. IR renormalization and gluon confinement

The power-type NP IR singularities represent a rather broad and important class of functions with algebraic singularities. They regularization should be done within the dimensional regularization (DR) method [14] correctly implemented into the theory of distributions [15]. The crucial observation is that the regularization of these singularities does not depend on their powers [5,6,7,8,15], namely

$$(q^2)^{-2-k} = \frac{1}{\epsilon}\Big[a(k)[\delta^4(q)]^{(k)} + O(\epsilon)\Big], \quad \epsilon \to 0^+ \,, \tag{21}$$

where $a(k)$ is a finite constant depending only on k and $[\delta^4(q)]^{(k)}$ represents the kth derivative of the δ-function. Here ϵ is the IR regularization parameter, introduced within the DR method [5,14], and which should go to zero at the end of the computations. So, it follows that each NP IR singularity scales as $1/\epsilon$ as ϵ goes to zero. This regularization expansion takes place only in four-dimensional QCD with Euclidean signature. In other dimensions and signature it is more complicated [5,15].

In the presence of such severe IR singularities all the quantities should depend in principle on ϵ. Thus the general IR renormalization program is needed in order to express all the quantities in terms of their IR renormalized versions. For this purpose it is convenient to rewrite the INP part of the full gluon propagator as follows:

$$D^{INP}(q,\Delta^2) = \sum_{k=0}^{\infty}(\Delta^2)^{k+1}(q^2)^{-2-k}\phi_k(\lambda,\nu,\xi,g^2) , \tag{22}$$

and here and below we again suppressed the tensor indices (i.e., the transverse tensor $iT_{\mu\nu}(q)$), for simplicity. Let us introduce further the following relations: $\Delta^2 = X(\epsilon)\bar{\Delta}^2$, $\phi_k(\lambda,\nu,\xi,g^2) = Z_k(\epsilon)\bar{\phi}_k(\lambda,\nu,\xi,g^2)$, where and below all quantities with bar are IR renormalized, i.e., they exist as ϵ goes to zero, by definition, while $X(\epsilon)$ and $Z_k(\epsilon)$ are the corresponding IR multiplicative renormalization (IRMR) constants. The dependence of both, the IR regularized (Δ^2 and $\phi_k(\lambda,\nu,\xi,g^2)$) and the IR renormalized ($\bar{\Delta}^2$ and $\bar{\phi}_k(\lambda,\nu,\xi,g^2)$) on ϵ is assumed but not explicitly shown, for simplicity. Substituting further these relations into the Laurent expansion (21), in terms of the IR renormalized quantities, it then becomes

$$D^{INP}(q,\bar{\Delta}^2) = \sum_{k=0}^{\infty}(\bar{\Delta}^2)^{k+1}(q^2)^{-2-k}\bar{\phi}_k(\lambda,\nu,\xi,g^2)X^{k+1}(\epsilon)Z_k(\epsilon), \quad \epsilon \to 0^+ . \tag{23}$$

Due to the distribution nature of the NP IR singularities, which appear in the full gluon propagator, the two different cases should be distinguished.

I. If there is an explicit integration over the gluon momentum, then from the dimensional regularization (21) and Eq. (23), it finally follows

$$D^{INP}(q,\bar{\Delta}^2) = \sum_{k=0}^{\infty}(\bar{\Delta}^2)^{k+1}a(k)[\delta^4(q)]^{(k)}\bar{\phi}_k(\lambda,\nu,\xi,g^2)\bar{B}_k(\epsilon), \quad \epsilon \to 0^+ , \tag{24}$$

provided the INP part will not depend on ϵ at all as it goes to zero. For this we should put

$$X^{k+1}(\epsilon)Z_k(\epsilon) = \epsilon\bar{B}_k(\epsilon), \quad k = 0,1,2,3..., \quad \epsilon \to 0^+ , \tag{25}$$

then the cancellation with respect to ϵ will be guaranteed term by term (each NP IR singularity is completely independent distribution) in the Laurent skeleton loop expansion (22), that is dimensionally regularized and IR renormalized in Eq. (24).

II. If there is no explicit integration over the gluon momentum, then the functions $(q^2)^{-2-k}$ in the Laurent skeleton loops expansion (23) cannot be treated as the distributions, i.e., there is no scaling as $1/\epsilon$. The INP part of the full gluon propagator, expressed in the IR renormalized terms, in this case disappears as ϵ, namely

$$D^{INP}(q,\bar{\Delta}^2) = \epsilon \sum_{k=0}^{\infty} (\bar{\Delta}^2)^{k+1} (q^2)^{-2-k} \bar{\phi}_k(\lambda,\nu,\xi,g^2) \bar{B}_k(\epsilon) \sim \epsilon, \quad \epsilon \to 0^+ . \tag{26}$$

This means that any amplitude for any number of soft-gluon emissions (no integration over their momenta) will vanish in the IR limit in our picture. In other words, there are no transverse gluons in the IR, i.e., at large distances (small momenta, $q^2 \to 0$) there is no possibility to observe physical gluons experimentally as free particles. Let us emphasize that the PT part is to be totally neglected in comparison with the INP one in the $q^2 \to 0$ limit, i.e., the full gluon propagator D is reduced to its INP part in this limit before taking the $\epsilon \to 0^+$ limit. So color gluons can never be isolated. This behavior can be treated as the gluon confinement criterion. Evidently, it does not depend explicitly on the gauge choice in the full gluon propagator, i.e., it is gauge-invariant. It is also general one, since even going beyond the gluon sector nothing can invalidate it. For the first time it has been derived in Ref. [5] (see Refs. [6,7,8] as well).

7. Conclusions

The physical meaning of our mass gap is the scale directly responsible for the NP dynamics in the true QCD ground state just as Λ^2_{QCD} is responsible for the nontrivial PT dynamics there. In this way the former determines the deviation of the full gluon propagator from the free one in the IR, while the latter makes this in the UV. The full gluon propagator (1) multiplied by q^2 is power-type enhanced in the IR and logarithmical weakened in the UV. The coupling constant squared (which *a priori* is not small) itself plays no role in the presence of the mass gap, so its dynamical origin goes beyond the PT. This is also a direct evidence for the "dimensional transmutation" [1,2], which occurs whenever a massless theory acquires a mass scale parameter dynamically. This is especially important, since there is none in the QCD Lagrangian. Let us emphasize that the mass gap survives after summing up an infinite number of the relevant contributions (skeleton loop diagrams) and performing the IR renormalization program. Moreover, it provides gluon confinement (26), that is, there must be no transverse gluons at large distances. At the same time, severe IR singular structure of the full gluon propagator (19) and the fact that all the orders of the coupling constant squared contribute into the mass gap (skeleton loops expansion) explain why the interaction in our picture is strong but short-ranged (24) (the δ function and its derivatives). It is interesting to note that Gribov [16] by differentiating twice the quark SD equation, arrived, in fact, to the same δ-type potential as well. However, the principal feature of our approach–the mass gap–was missing in this procedure.

Our mass gap appears in the NL part of the Yang-Mills sector of full QCD, however, its relation to the mass gap introduced by Jaffe and Witten in Ref. [3] is still to be understood, though there is a great similarity between them. It is worth also emphasizing that our mass gap and the Jaffe-Witten (JW) mass gap [3] cannot be interpreted as the gluon mass, i.e., they always remain massless within our approach. If quantum YM with compact simple gauge group $G = SU(3)$ exists on $\mathbf{R^4}$, then it exhibits a mass gap in the sense discussed by Jaffe and Witten, indeed. Moreover, it confines gluons as well. Color confinement of gluons is the IR renormalization gauge-invariant effect within our approach. Just the fundamental self-interaction of massless gluons (i.e., the NL dynamics of QCD) makes the full gluon propagator so singular in the IR and thus necessarily exhibits a mass gap.

Concluding, a few remarks are in order. The "shifting" procedure described above makes it possible to fix exactly the INP part of the full gluon propagator at the expense of its PT one which remains undetermined. However, just the INP part is responsible for confinement of gluons within our approach. So that's we do not know the PT part of the full gluon propagator is not important for our purpose, namely to establish exactly and uniquely the NL dynamical mechanism of color confinement. We will finish this investigation in Ref. [17] which will soon appear. In particular, we will complete the IRMR program of the YM theory within our approach. We will also introduce the UV and the IR renormalized mass gap. The latter will be achieved by fixing the dependence of $X(\epsilon)$ on its argument.

Concluding finally, let us recall once more that this talk in its extended version is also present in Ref. [7].

Acknowledgements

Support from HAS-JINR Scientific Collaboration Fund is to be acknowledged.

References

1. W. Marciano and H. Pagels, *Phys. Rep.* **C 36**, 137 (1978).
2. M.E. Peskin and D.V. Schroeder, *An Introduction to Quantum Field Theory*, AW, Advanced Book Program, 1995.
3. A. Jaffe and E. Witten, *Yang-Mills Existence and Mass Gap*, http://www.claymath.org/prize-problems/, http://www.arthurjaffe.com .
4. P. Maris and G.D. Roberts, arXiv:nucl-th/0301049.
5. V. Gogohia, arXiv:hep-ph/0311061.
6. V. Gogohia, *Phys. Lett.* **B618**, 103 (2005).
7. V. Gogohia, arXiv:hep-ph/0511156, v3.
8. V. Gogohia, *Phys. Lett.* **B584**, 225 (2004).
9. R. Streater and A. Wightman, *Spin and Statistics and all That*, W.A. Benjamin, New York, 1964.
10. F. Strocchi, *Phys. Lett.* **B62**, 60 (1976).
11. G.V. Efimov, *Russ. Jour. Elem. Part. and Nucl. Phys.* **35**, 1116 (2004).
12. D.V. Shirkov, arXiv:hep-ph/0208082.
13. R. Feynman, *Nucl. Phys.* **B188**, 479 (1981).
14. G. 't Hooft and M. Veltman, *Nucl. Phys.* **B44**, 189 (1972).
15. I.M. Gel'fand and G.E. Shilov, *Generalized Functions*, Academic Press, New York, 1968, Vol. I.
16. V.N. Gribov, *Gauge Theories and Quark Confinement*, PHASIS, Moscow, 2002.
17. V. Gogohia, arXiv:hep-ph/0502206, v3.

SPINODAL INSTABILITY AND CONFINEMENT

JANOS POLONYI

Institute for Theoretical Physics, Louis Pasteur University, Strasbourg, France

It is pointed out that inhomogeneous condensates or spinodal instabilities suppress the propagation of elementary excitations due to the absorptive zero mode dynamics. This mechanism is shown to be present in the scalar ϕ^4 model and in Quantum Gravity. It is conjectured that the plane wave states of color charges have vanishing scattering amplitude owing to the color condensate in the vacuum.

1. Introduction

The mechanism of quark confinement continues to be one of the stumbling block in particle physics and one tends to shelve it by the help of more or less ad hoc models motivated by QCD. Instead of a detailed construction of "The Mechanism" which obviously involves some highly technical and non-perturbative issues of QCD we suggest in this talk a rather well known phenomenon, spinodal instability, as a natural ingredient of confinement.

There are actually two mechanisms for quark confinement. One is established by means of studying Wilson-loops in quenched lattice QCD without dynamical quarks and is operating with a linearly rising potential between static fundamental color charges[1]. One may call this the hard mechanism because it is based on large energy scales to suppress localized quarks. Another mechanism, proposed by V. Gribov[2], is conjectured in QCD with dynamical, light quarks and is reminiscent of the supercritical vacuum in QED. The separation of a quark from a hadron triggers an increase in the running coupling constant which in turn generates screening by the antiquark of a $q-\bar{q}$ vacuum-polarization. We may call this the soft mechanism because the screening is achieved by vacuum fluctuations of the energy of a light meson. The soft mechanism is the real one, observed in hadrons and based on the non-perturbative phenomena in the Dirac sea caused by the increase of the running coupling constant at large distances, the key phenomenon of the hard mechanism. An inherent difficulty of both mechanisms is the explanation of gluon confinement. We argue below that the colored condensate of the QCD vacuum should drive a spinodal instability which automatically confines color charges, both gluons and quarks.

We shall discuss this mechanism in the framework of two theories. One is the the ϕ^4 scalar model in the symmetry broken phase. We argue that when a homogeneous external source is coupled to the field $\phi(x)$ and its value is chosen in such a manner that $|\langle\phi(x)\rangle|$ is decreased, then the elementary excitations become confined. The second example is Quantum Gravity where confinement of gravitons is expected. QCD is touched upon briefly at the end only.

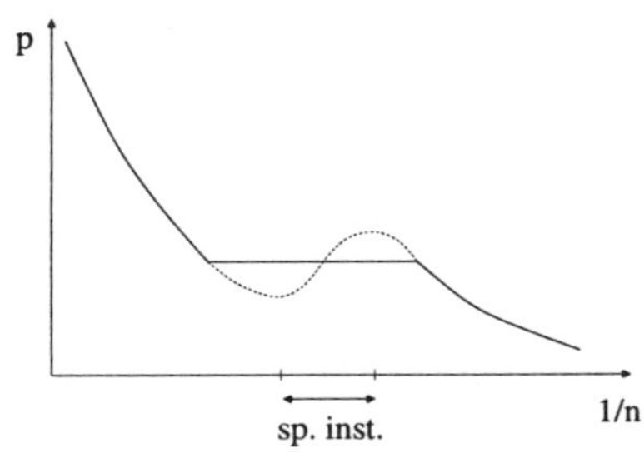

Figure 1. A qualitative sketch of the pressure p as the function of the density n for the van der Waals equation of state. The density interpolates between the values corresponding to the two phases while the pressure remains unchanged on the horizontal line, the Maxwell-cut.

The basic idea is very simple. Let us start with the spinodal instability shown in Fig. 1 which is usually observed in first order phase transitions.The spinodal unstable region is characterized by the apparent violation of the sum rule which assures that the pressure is a non-decreasing function of the density. The Maxwell-cut restores stability and renders the pressure independent of the density in the mixed phase where domains made up by the two stable phases are formed. The sound velocity,

$$c = \sqrt{-\frac{1}{n^2}\frac{\partial p}{\partial 1/n}}, \tag{1}$$

is clearly vanishing in the mixed phase. The vanishing is due to the domain walls which can be displaced without energy absorption from the density waves. The density fluctuations are therefore non-propagating and the sound waves become "confined".

2. Scalar model

One can gain more insight into the dynamics of the instabilities of Fig. 1 by considering a scalar model where a homogeneous external source, j, coupled linearly to the field $\phi(x)$ is introduced in order to dial a desired value of the condensate. The partition function of the Euclidean model is

$$Z = \int D[\phi] e^{-\int dx[\frac{1}{2}(\partial\phi(x))^2 + U(\phi(x))] + j\int dx\phi(x)} \tag{2}$$

where the potential is an even function, $U(-\phi) = U(\phi)$, and is double-well shaped, as shown in Fig. 2. We shall determine the vacuum which is supposed to be homogeneous for any value of the source j in the tree-level approximation. The vacuum of the theory with $j = 0$ is dominated by a single homogeneous configuration $\langle\phi(x)\rangle = \phi_{\rm vac} > 0$ satisfying the conditions $U'(\phi_{\rm vac}) = 0$ and $U''(\phi_{\rm vac}) > 0$.

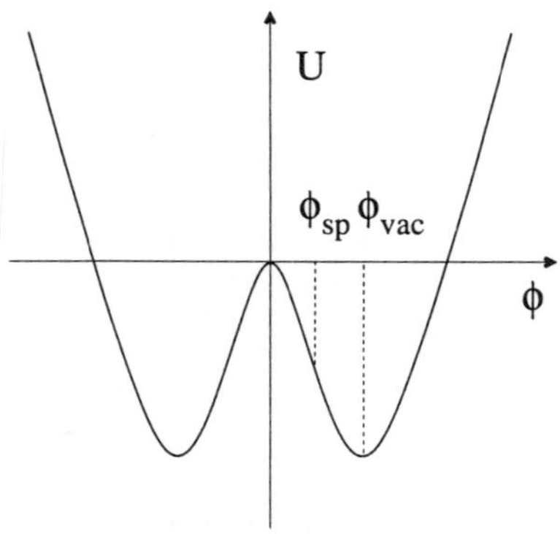

Figure 2. The double-well potential of the scalar model.

The gradual decrease of the external source towards more negative values decreases the saddle-point $\phi(x) = \phi_j > 0$ which is defined by the equation $U'(\phi_j) = j$, as shown in Fig. 3. The homogeneous saddle-point vacuum remains stable, $U''(\phi_j) > 0$, as long as

$$\phi_{\rm sp} \le \phi_j \le \phi_{\rm vac}, \tag{3}$$

where the lower bound is defined by the equation $U''(\phi_{\rm sp}) = 0$. Notice that the vacuum constructed in such a manner is stable against infinitesimal fluctuations but decays into the true vacuum, given by the absolute minimum of $U(\phi) + j\phi$ when fluctuations with sufficiently large amplitude,

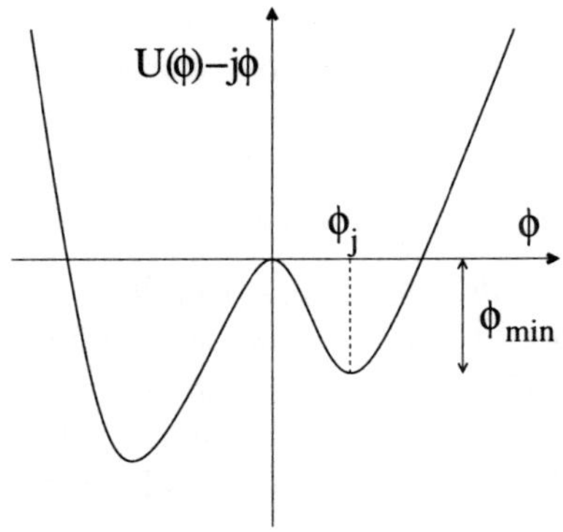

Figure 3. The vacuum is dominated by a single homogeneous saddle point in the nucleation phase, given by Eq. (3).

$\Delta\phi > \phi_{\min}$ are formed. This regime of the model is called nucleation phase because the large amplitude fluctuations which destabilize the false vacuum are energetically stable above a certain size. All modes are massive in this phase.

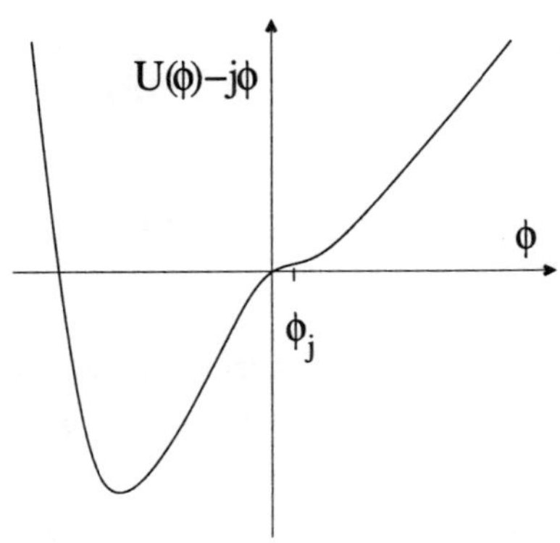

Figure 4. The potential energy in the spinodal phase.

As the external source is pushed further towards the negative direction the saddle-point reaches the lower bound in Eq. (3) and we enter into the spinodal instability region, cf. Fig. 4, characterized by instabilities against infinitesimal fluctuations. The saddle-points are inhomogeneous and support domains where $\phi(x)$ assumes opposite sign. The breakdown of the external, space-time symmetries induces zero modes, the location and the direction of the domain walls. We shall argue in the next Section that these soft modes make the "sound waves" non-propagating, ie. remove the plane waves from the asymptotical scattering states of the theory.

3. Instability induced renormalization

The inhomogeneous saddle-points appearing in the spinodal phase require a more powerful method than the mean-field approximation. It is a time honored strategy to deal with modes one-by-one in a sequential manner, as in the renormalization group method[3], instead of facing all of them in the same time. Such a scheme proves to be useful for the saddle-points, too[4].

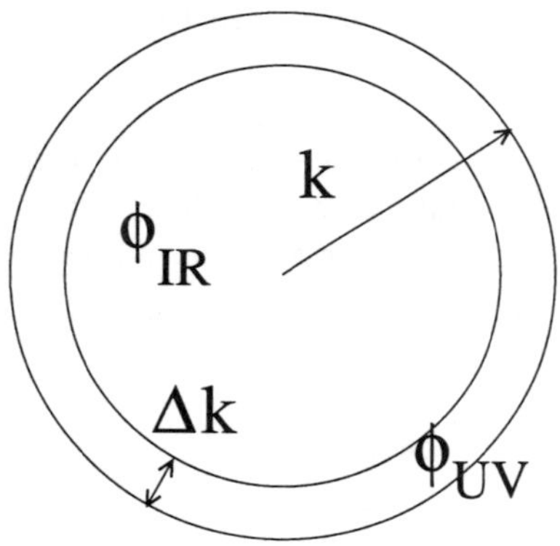

Figure 5. The split of the field variable into an infrared and ultraviolet component is defined in the Fourier space.

Let us consider an infinitesimal decrease of the UV cutoff, $k \to k - \Delta k$, in momentum space which induces the change $S_k[\phi] \to S_{k-\Delta k}[\phi]$ in the bare action of the scalar model,

$$e^{-\frac{1}{\hbar}S_{k-\Delta k}[\phi_{\rm IR}]} = \int D[\phi_{\rm UV}]e^{-\frac{1}{\hbar}S_k[\phi_{\rm IR}+\phi_{\rm UV}]}, \tag{4}$$

where the split $\phi(x) = \phi_{\rm IR}(x) + \phi_{\rm UV}(x)$ of the field variable is defined by requiring that the Fourier transform of $\phi_{\rm UV}(x)$ is non-vanishing within the shell $k - \Delta k < p < k$ only, as shown in Fig. 5. The tree-level contribution to the evolution equation (4) is

$$S_{k-\Delta k}[\phi_{\rm IR}] = S_k[\phi_{\rm IR} + \phi_{\rm UV}[\phi_{\rm IR}]]. \tag{5}$$

A non-trivial saddle-point, $\phi_{\rm UV}[\phi_{\rm IR}] \neq 0$, induces an evolution in the bare action, $S_{k-\Delta k}[\phi_{\rm IR}] \neq S_k[\phi_{\rm IR}]$.

Two remarks are in order now about the saddle-points. The first is that the non-vanishing saddle-point is the hallmark of spontaneous symmetry breaking, the appearance of a condensate in the theory. In fact, the inverse propagator, $\delta^2 S_k[\phi]/\delta\phi\delta\phi$, approaches its renormalized form as $k \to 0$. Spontaneous symmetry breaking is driven by the negative values of the renormalized propagator for small momenta computed in the symmetrical vacuum, $\phi(x) = 0$. Hence $\phi(x) = 0$ ceases to be the absolute minimum of the bare action at sufficiently small values of the cutoff k in the symmetry broken phase and a non-trivial saddle point is found for Eq. (4). The other remark is about the circumstance that these non-trivial saddle-points appear in a scalar model in any dimension even if the whole partition function supports no solitons or instantons. The saddle-points mentioned again and again in this Section correspond to a constrained functional integral where the blocked variables are held fixed. It is this constrain which induces the block variable-dependent saddle-points and the partition function as a complicated integral does not reveal their existence.

Let us find the tree-level solution of the evolution equation in the local potential approximation where the bare action is assumed to be of the form

$$S_k[\phi] = \int dx \left[\frac{1}{2}(\partial\phi(x))^2 + U_k(\phi(x))\right]. \tag{6}$$

This ansatz allows us to seek the solution only for $\phi_{\rm IR}(x) = \Phi$ because

$$U_{k-\Delta k}(\Phi) = \min_{\phi_{\rm UV}} \int dx \left[\frac{1}{2}(\partial\phi_{\rm UV})^2 + U_k(\Phi + \phi_{\rm UV}(x))\right]. \tag{7}$$

As a further simplification, we search for the minimum among the plane wave configurations only,

$$\phi_{\rm UV}(x) = \rho_k \cos(kn_k \cdot x + \theta_k). \tag{8}$$

Note that the unit vector n_k and the phase θ_k are zero modes arising from the breakdown of the rotational and translational symmetry. Eq. (7) now reads as

$$U_{k-\Delta k}(\Phi) = \min_{\rho_k} \left(\frac{1}{4}k^2\rho_k^2 + \frac{1}{\pi}\int_0^\pi dy U_k(\Phi + \rho_k \cos y)\right). \tag{9}$$

In order to obtain the renormalization group flow we impose the initial condition

$$U_k(\phi) = -\frac{m^2}{2}\phi^2 + \frac{g}{4!}\phi^4, \tag{10}$$

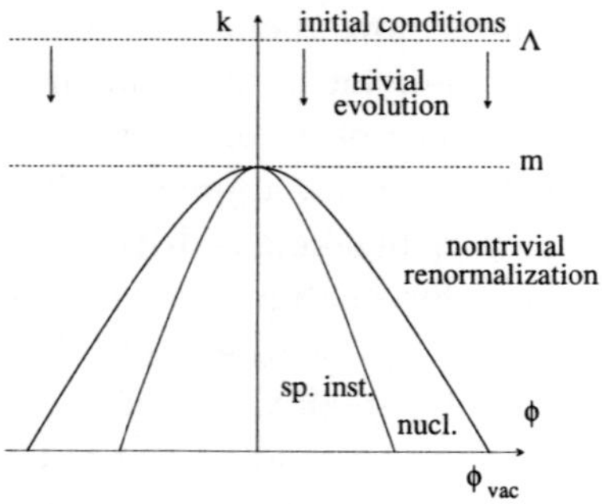

Figure 6. Tree-level evolution of the potential $U(\phi)$. The initial condition is set at $k = \Lambda \gg \sqrt{-m^2}$ but the tree-level evolution starts only for $k^2 = -m^2$. The saddle-point is non-vanishing within the area bounded by the outer curve. The naive reasoning would suggest the possibility of separating nucleation and spinodal phases in the instable regime. This proved to be wrong by the numerical solution, the spinodal region spreads over the whole unstable domain.

with $m^2 < 0$ at $k = \Lambda \gg \sqrt{-m^2}$ for Eq. (9). There is no evolution for $k \approx \Lambda$ because the kinetic energy dominates the inverse propagator

$$G(p) = \frac{1}{p^2 - m^2 + \frac{g}{2}\Phi^2} \tag{11}$$

for $p^2 \gg -m^2$, cf. Fig. 6. The inhomogeneous saddle-point becomes non-trivial and the potential evolves for $k^2 < -m^2$.

The following observations can be made after inspecting the numerical solution of Eq. (9):

(1) Eq. (9) goes over a differential equation in the limit $\Delta k \to 0$ and $S_k[\phi]$ is continuous[5] in k.

(2) The instability is always followed by the appearance of inhomogeneous saddle-points and their zero modes. The zero modes are soft and can be excited by arbitrarily small energies, indicating spinodal phase separation in the whole instability region. The nucleation phase expected naively inbetween the stable and the spinodal region disappears.

(3) The bare action $S_k[\phi]$ is always flat for modes with $p = k$, in the variable ϕ_{UV}, due to the cancellation between the kinetic energy and the potential energy which is found to be

$$U_k(\Phi) = -\frac{1}{2}k^2\Phi^2 \tag{12}$$

within the instability region. The resulting flat action, the lack of restoring force to any equilibrium position, becomes the usual Maxwell-cut as $k \to 0$.

(4) The vacuum, $\phi_{\rm vac}$, is at the edge of the unstable region.

(5) These tree-level results, Eq. (12) in particularly, hold independently of the choice of the potential at the cut-off.

The unexpected feature is the appearance of zero modes, n_k and θ_k in Eq. (8), within the unstable region in the one-component model with the discrete internal symmetry $\phi(x) \to -\phi(x)$. These are Goldstone modes arising from the breakdown of translational and rotational symmetry by the inhomogeneous saddle-points. The plane-wave is a poor-man's approach to the domain wall structure and its phase and direction control the location and the direction of the walls. The integration over the zero modes appearing independently at each value of k restores the external symmetries and renders the vacuum homogeneous.

It is point 3 which is relevant from the point of view of confinement. What we have obtained here is a layer-by-layer scan of the bare action for this theory. In fact, the dynamics of a mode with wave vector p is not contained explicitly or appears in a rather approximative manner only in $S_k[\phi]$ for $|p| > k$ or $|p| < k$. This dynamics is best described by the bare action with $k = |p|$. Let us suppose that the external homogeneous source stabilizes the vacuum at $|\langle\phi(x)\rangle| < \phi_{\rm vac}$ and the model is in the spinodal phase. The cancellation between the kinetic and the potential energy should therefore be kept as a feature characterizing all unstable modes and no plane waves exist with small wave numbers. One may say in the more formal language of the reduction formulae that there are no mass-shell singularities in the Green functions for these modes and the scattering amplitude is vanishing, ie. the low energy elementary quanta of the theory are non-propagating, confined.

What happens with the high energy modes belonging to the stable regime of Fig. (6)? These modes are supposed to propagate on the background field of the inhomogeneous condensate and the integration over the zero modes appears as a quenched average. One expects that this is a strongly disordered system where the Euclidean analogy of the Anderson localization prevents these states from being extended. Having no detailed study it is sufficient to say that the restored external symmetries impose the cancellation of the scattering amplitude at any energy once it is found vanishing at low energy. In other words, all particle modes are confined as long as there is a finite low energy range with inhomogeneous condensate.

4. Savvidy vacuum in Quantum Gravity

There is another model where confinement of the elementary excitations are expected in the semiclassical approximation, namely gravity[6]. Let us consider a point-like body surrounded by its horizon and suppose that the position of the body undergoes small amplitude oscillations. The acceleration sends gravitational waves into the space-time within the horizon. Though the non-propagating features of the solution of the Einstein equations may indicate the motion of the body outside of the horizon, the propagating gravitons are supposed to remain confined by the horizon.

It is not difficult to recognize the trace of spinodal instability in quantum gravity. Let us consider the partition function[7]

$$Z = \int D[g] e^{-S_E[g]}, \tag{13}$$

for the Euclidean signature metric tensor, where the Einstein action is given by

$$S_E[g] = -\kappa_B^2 \int dx \sqrt{g} R. \tag{14}$$

This theory poses serious problems both in the UV and the IR domains.

The difficulty in the UV region is that the theory is non-renormalizable. We shall keep the cutoff Λ at a large but finite value and consider (14) as an effective action valid for distances $\ell > \ell_{\min} \approx 1/\Lambda$ only. The regularization is carried out by restricting the functional integral for geometries into the vicinity of de Sitter spaces, characterized by their curvature R, and by suppressing modes whose eigenvalue with respect to the operator $-D^2$ is superior to Λ^2.

The IR problems come from the fact the the action (14) is unbounded form below in the conform modes, Ω, defined as $\tilde{g}_{\mu\nu} = g_{\mu\nu}\Omega^2$ because

$$S_E[\tilde{g}] = -\kappa^2 \int dx \sqrt{g} \left(\Omega^2 R + 6\Omega_{;\mu}\Omega_{;\nu} g^{\mu\nu} \right). \tag{15}$$

Instead of modifying the Wick-rotation[8], the quantization rules [9] or the functional integral measure[10] which lead to further problems with general covariance it is more illuminating to recognize the similarity to the instability of the Yang-Mills vacuum[11]. In fact, the interactions among the elementary excitations are attractive in both cases and populate modes macroscopically. These modes are inhomogeneous in order not to break gauge and external symmetries. The apparent difference, that the Yang-Mills action is bounded from below but the gravitational one is not, may not be important because the stabilization of the vacuum can be achieved by quantum fluctuations. An obvious example is the Hydrogen atom where the kinetic energy created by localization balances the attractive, unbounded Coulomb potential.

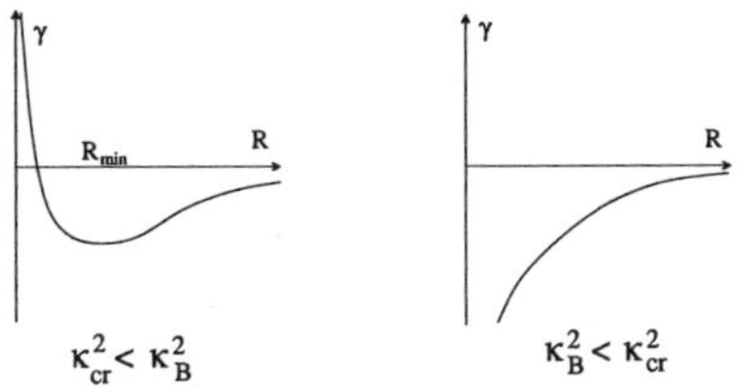

Figure 7. The effective potential as the function of the de Sitter curvature in the low and the high cutoff phases.

The one-loop contributions of the stable modes to the effective potential $\gamma_{\rm eff}(R) = -\ln Z(R)$ where $Z(R)$ is the perturbative partition function on the de Sitter space of curvature R give

$$\gamma(R) = \begin{cases} -\frac{v\kappa_B^2}{R} + c_1\Lambda^4 \ln\frac{c_2\kappa_B^2}{\Lambda^2}\left(\frac{1}{R^2} - \frac{1}{c_3^2\Lambda^4}\right) & R \le c_3\Lambda^2 \\ -\frac{v\kappa_B^2}{R} & R > c_3\Lambda^2 \end{cases}, \tag{16}$$

with $c_1 \approx 7.201$, $c_2 \approx 2.989$ and $c_3 \approx 0.665$, cf. Figs. 7. This result indicates a quantum phase transition at $\kappa_{cr}^2 = \frac{\Lambda^2}{c_2}$ and the curvature

$$R_{min} = \begin{cases} \frac{2c_1\Lambda^4}{v\kappa_B^2} \ln\frac{c_2\kappa_B^2}{\Lambda^2} & \kappa_{cr}^2 \ll \kappa_B^2 \\ 0 & \kappa_B^2 \ll \kappa_{cr}^2 \end{cases} \tag{17}$$

at the minimum of the effective potential. What is left to see is whether the conformal modes are stabilized at these vacua. This phase diagram agrees qualitatively with the one found in lattice simulations[12] suggesting a radiative corrections induced stabilization of the vacua of Eqs. (17). Notice that whatever stabilization mechanism prevails the vacuum contains inhomogeneous condensate in both phases, made of the inhomogeneous conformal modes, and one expects that the appearing zero modes make the plane waves, gravitons, non-propagating.

5. Conclusions

It has been argued that inhomogeneous condensate, in particular spinodal instability, leads to the confinement of elementary excitations by the absorptive dynamics of the zero modes. Two models have been mentioned, a scalar ϕ^4 model and Euclidean quantum gravity. The mechanism is established on a qualitative level only, the more satisfactory construction obviously requires a systematical and through study of the tree-level contributions to the renormalization group equations.

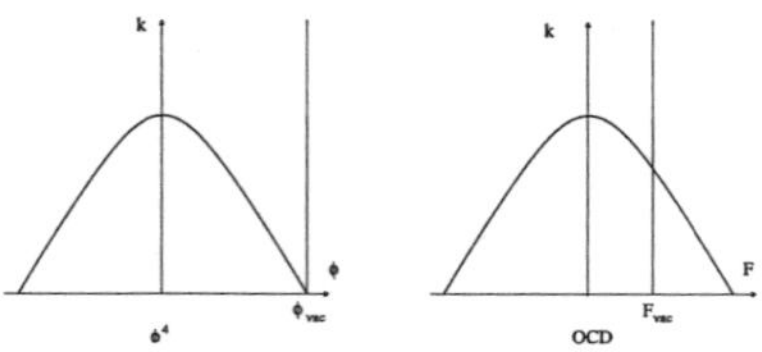

Figure 8. The tree-level evolution in the ϕ^4 model (left) and in QCD (right). The vertical axes show the cutoff, the horizontal ones correspond to the expectation value of the scalar field and some components of the field strength tensor. The inhomogeneous saddle-points appear within the regions bounded by the solid line.

Is this mechanism effective in the QCD vacuum? Let us compare the known tree-level renormalization flow of the scalar model with the expected flow in QCD, as shown in Figs. 8. The vacuum of the ϕ^4 model, shown in the first figure, is at the edge of the spinodal phase. The effects of the eventual zero modes within the unstable region are suppressed and no confinement occurs in the absence of the external source. The color condensate disappears at short distances or in the presence of strong external chromomagnetic field in QCD. The region with condensate is shown qualitatively in the second figure. What is important is that the true vacuum is within the unstable region. Hence one expects that the zero mode dynamics of the inhomogeneous condensate renders the effective action flat and removes colored plane wave states from the asymptotical sector of QCD.

References

1. K.G. Wilson, *Phys. Rev.* **D10**, 2445 (1974); M. Creutz, *Phys. Rev.* **D21**, 2308 (1980).
2. V.N. Gribov *Eur. Phys. J.* **C10**, 91 (1999).
3. K.G. Wilson and J. Kogut, *Phys. Rep.* **C12**, 77 (1974); K.G. Wilson, *Rev. Mod. Phys.* **47**, 773 (1975).
4. J. Alexandre, V. Branchina and J. Polonyi, *Phys. Lett.* **B445**, 351 (1999).
5. R.B. Israel, in *Random Fields* (Esztergom, 1979), J. Fritz, J.L. Lebowitz and D. Szász eds. (North-Holland, Amsterdam, 1981); A.C.D. van Enter, R. Fernández and A. Sokal, *J. Stat. Phys.* **72**, 879 (1993).
6. A. Salam and J. Strathdee, *Phys. Rev.* **D18**, 4596 (1978); E. Recami and P. Castorina, *Lett. Nuov. Cim.* **15**, 347 (1976).

7. J. Polonyi and E. Regős, *Casimir effect: running Newton constant or cosmological term*, submitted to Class. Quant. Grav.
8. S. Hawking, in *General Relativiy: An Einstein Centenary Survey*, eds. S. Hawking and W. Israel (Cambridge Univ. Press, Cambridge, 1979).
9. J. Greensite, *Nucl. Phys.* **B361**, 729 (1991); *Phys. Lett.* **B291**, 405 (1992); *Nucl. Phys.* **B390**, 439 (1993).
10. P. Mazur and E. Mottola, *Nucl. Phys.* **B341**, 187 (1990); E. Mottola, *J. Math. Phys.* **36**, 2470 (1995).
11. I.A. Batalin, S.G. Matinyan and G.K. Savvidy, *Sov. J. Nucl. Phys.* **26**, 214 (1977); G.K. Savvidy, *Phys. Lett.* **B71**, 133 (1977).
12. B.A. Berg, *Phys. Rev. Lett.* **55**, 904 (1985); *Phys. Lett.* **B176**, 39 (1986); H.W. Hamber, *Phys. Rev.* **D45**, 507 (1992); *Nucl. Phys.* **B400**, 347 (1993); E. Bittner, W. Janke and H. Markum, arXiv:hep-lat/0311031.

GENERAL FIELD THEORY, GRAVITY AND MACRO-WORLD

SUPERMAGNETS AND SIGMA MODELS

A.M. POLYAKOV

Joseph Henry Laboratories, Princeton University
Princeton, New Jersey 08544

We discuss new methods for non-compact sigma models with and without RR fluxes. The methods include reduction to one dimensional supermagnets, super-coset constructions and supertwistors. This work is a first step towards the solution of these models, which are important in several areas of physics. I dedicate it to the memory of Volodya Gribov.

1. Introduction

Nonlinear sigma models describe low energy interaction of Goldstone particles and are important in many parts of modern physics. The models in two dimensions are especially interesting since the infrared effects in this case are highly nontrivial. Such models are relevant in string theory, quantum Hall effect, spin chains, disordered systems etc.

It has been shown long ago that in general the models with the non-abelian symmetry groups acquire a non-perturbative mass gap [1]. This result contradicted to the well known fact that in the Heisenberg antiferromagnet (which has O(3) symmetry) spin waves are gapless. The puzzle was resolved seven years later [2,3] when it was argued that the Heisenberg model with the half integer spin must be described by the sigma model with the fine tuned topological term, which makes it gapless. In the case of integer spin, the gap found in [1] is expected to appear (and does so experimentally). At the same time, even for the integer spins, there are special forms of the interaction leading to a completely integrable gapless antiferromagnets. So, the Goldstone particles can remain massless when they want to.

In this paper I will begin developing new methods for analysis and interpretation of the massless sigma models. My refreshed interest in the old toy is related to the gauge/strings duality, the key elements of which are the supersymmetric sigma models. Existing methods are painfully inadequate for getting physical information in this case as well as in the other areas. The purpose of this paper is to outline new ways and hopes for their solution.

Let us begin with a brief summary of the Haldane-Affleck approach. One considers a Heisenberg antiferromagnet. Its hamiltonian has the form

$$\mathcal{H} = \sum S_x S_{x+1} , \tag{1}$$

where S are spin one half matrices. They can be expressed in terms of free fermions, $a_{x\alpha}, \alpha = 1, 2$ as

$$S_x = a_x^+ \sigma a_x , \tag{2}$$

$$a_x^+ a_x = 1 , \tag{3}$$

where σ are the Pauli matrices and the constraint (3) ensures that there is exactly one fermion at each site (half-filled band). One then can give a plausible argument that in the continuous limit the fermions become relativistic, left and right moving Dirac particles with isotopic spin, $\psi_{L\alpha}(x)$ and $\psi_{R\alpha}(x)$. This doubling of the fermions occurs since the continuous limit enforces the continuity of the even and odd sites separately. The Lagrangian for these fields has the standard Dirac form

$$L = \psi^+(\hat{\partial} + \hat{A})\psi . \tag{4}$$

Here the $U(1)$ gauge field A is a Lagrange multiplier ensuring the constraint (3). The spin field S_x has the following behavior in the continuos limit (reflecting its antiferromagnetic nature)

$$S_x = (-1)^x n(x) + l(x) , \tag{5}$$

where $l \ll n$ are continuous functions and in terms of the Dirac fermions

$$n(x) = \psi_L^+ \sigma \psi_R + c.c \ . \tag{6}$$

Without A field Dirac fermions generate the $SU(2)$ currents $J_{L,R} = \psi_{L,R}^+ \sigma \psi_{L,R}$ which are holomorphic and antiholomorphic and the $U(1)$ current. The latter is killed by the A integration. The resulting system is equivalent to the WZNW action (non-abelian bosonization [4]). For higher spins one considers several flavors of fermions and gauge out the flavor currents.

In order to relate these facts to the $O(3)$ sigma model one needs another representation of the spin operators, coming from the geometrical (Kirillov) quantization. It is based on the formula

$$\int Dne^{-W(n)} n_{a_1} ... n_{a_k} = Tr(S_{a_1} ... S_{a_k}) \ . \tag{7}$$

On the right side of this formula we have a trace of the spin matrices in a representation with the spin S, while on the left side we have an integral over the unit sphere, which is an orbit of the rotation group. The action has the form

$$W(n) = ik \int A(n)dt \tag{8}$$

with $A(n)$ being a vector potential of the magnetic monopole and $k = 2S$. The key feature of this action which ensures (7) is its variation under the group transformation

$$\delta n = \omega \times n \tag{9}$$

$$\delta W = \int (n \frac{d\omega}{dt})dt \ . \tag{10}$$

With these formulae the geometric quantization follows from the standard Ward identities, as we will discuss later in the general case. We can now replace the quantum spins by the classical orbit. Due to the antiferromagnetic order, we expect in the continuum limit

$$S_x \equiv (-1)^x N_x \Rightarrow (-1)^x n(x) + l(x) \ , \tag{11}$$

where $n(x) \gg l(x)$ are continuous functions. The action for our magnet in this variables takes the form

$$F = \int dt \{ \sum_x (-1)^x W(N_x) + \sum_x (N_x - N_{x+1})^2 + const \} \ . \tag{12}$$

In the continuous limit we should keep only such terms that the sign factor $(-1)^x$ disappears, otherwise the oscillations would kill the sum. That means that only the terms linear in l(x) must appear in the first term of (12). Using the variation formula (10) we get

$$F \simeq ik \int dt \int dx l(x) [n \times \frac{dn}{dt}] + \int dt dx \{ (\nabla n + 2l)^2 + l^2 \} \ . \tag{13}$$

Integrating out the field $l(x)$ we obtain the result of [2,3] — the n-field sigma model with the theta term at $\vartheta = 2\pi S$.

The reason why we need an antiferromagnet to describe a relativistic theory is the following. Spin changes sign under the time reversal. Hence for a ferromagnet the presence of the spontaneous magnetization implies breaking of the CPT-symmetry and the spectrum of the Goldstone particles is non-relativistic, $\omega = k^2$. In antiferromagnets we can combine CPT-reflection with the shift x to $x + 1$. This symmetry $\omega \Rightarrow -\omega$ is not broken and implies the relativistic spectrum $\omega^2 = k^2$ in the long wave limit. The sigma model is simply the most general effective lagrangian describing relativistic, $O(3)$ invariant spin waves.

2. General groups

Let us now generalize these considerations for arbitrary groups and supergroups. The first step is well known — it is Kirillov's theory. This theory allows to replace traces in a certain representation by integration over a particular orbit. To set the stage I rederive it by using rather pedestrian methods. The action has the form

$$W = \int dt Tr(K\Omega^{-1}\frac{d\Omega}{dt}) , \tag{14}$$

where everything is in the adjoint representation. The matrix K depends on the representation which we eventually want to describe. If this representation has the highest weight $\mu = (\mu_1 ... \mu_r)$ (r is the rank of the group) and if we have Cartan's generators H_k, then $K = (\mu_k H_k)$. If we consider the variation of Ω by right multiplication, $\delta\Omega = \varepsilon\Omega$, where $\varepsilon = \varepsilon^A \tau^A$ with τ^A the set of adjoint generators, it follows from (14) that

$$\delta W = \int dt T^A \frac{d\varepsilon^A}{dt} , \tag{15}$$

where $T^A = Tr(K\Omega^{-1}\tau^A\Omega)$. These quantities define an orbit of the group, since they are invariant (as well as the action) under the following gauge transformation

$$\Omega \Rightarrow \Omega h \tag{16}$$

$$hKh^{-1} = K . \tag{17}$$

The last equation explicitly shows dependence of the stability group on the highest weight. Indeed, if we look at the infinitesimal transformations then a generator defined by a root α belongs to the gauge group if $(\mu\alpha) = 0$. When μ is a fundamental weight there are non-trivial solutions of this equation, some of which will be discussed below. For a generic weight the stability group is generated by the Cartan subalgebra.

In order to derive the Ward identities we notice that under the above right multiplication we have a transformation law for T^A

$$\delta T^A = f^{ABC}\varepsilon^B T^C \tag{18}$$

(here f^{ABC} are the structure constants). Combining (15) and (18) we get the equations of motion

$$\frac{d}{dt}\langle T^A(t) T^{B_1}(t_1) ... T^{B_n}(t_n)\rangle$$
$$= \sum_k \delta(t - t_k) f^{AB_kC}\langle T^{B_1}(t_1) ... T^C(t_k) ... T^{B_n}(t_n)\rangle . \tag{19}$$

This is precisely the equation satisfied by the trace of the time-ordered product of matrices of a given representation, and hence the characters are given by the functional integrals with the action W. In other words, the matrices of representation can be replaced by the classical quantities T^A defining the orbit. So far we described the well known results in a slightly unusual way. Now we pass to something new.

3. Collectivization

We use this infamous word to describe formation of spin waves out of fluctuations of individual spins.

Consider a magnet with the Hamiltonian (1), except this time S_x^A are the matrices in a given representation of a given group. By the same logic as above we can write the action in the form

$$F = i\sum_x (-1)^x W[\Omega_x] + (T_x^A - T_{x+1}^A)^2 \quad . \tag{20}$$

The factor $(-1)^x$ tells us that the representations we use for the even/odd sites must be conjugate to each other.

Once again we assume that $\Omega_x \simeq (1 + (-1)^x l(x))\Omega(x)$ and that the adjoint matrix $l(x) \ll 1$.It is convenient to work with the matrices in the adjoint representation, defined as $T(x) = T^A \tau^A$. Using the formulae for variations, given above, we get (again keeping only the term in which $(-1)^x$ cancels)

$$F = \int dt dx Tr\{(l\partial_t T) + (\partial_x T + [l,T])^2 + [l,T]^2\} \quad . \tag{21}$$

Excluding l we get the equation

$$[T, \partial_x T] + [T,[T,l]] = 2\partial_t T \quad . \tag{22}$$

Since according to our definition $T = \Omega K \Omega^{-1}$, its derivative satisfies

$$\partial_\alpha T = [B_\alpha, T] \tag{23}$$

$$B_\alpha = \partial_\alpha \Omega \Omega^{-1} \quad . \tag{24}$$

If we set $l = B_1 + \omega$, we get

$$[T,[T,\omega]] = 2[B_0, T] = 2\partial_t T \quad . \tag{25}$$

The theta term Q, which contains both x and t derivatives has the form

$$Q \sim \int dt dx Tr(T[B_0, B_1]) = \int dt dx Tr(K[A_0, A_1]) \tag{26}$$

$$A_\alpha = \Omega^{-1}\partial_\alpha \Omega \quad . \tag{27}$$

The equaton for ω can be solved by "rotating" it back to the form

$$[K,[K,\widetilde{\omega}] = -2[K,A_0] \tag{28}$$

$$\widetilde{\omega} = \Omega^{-1}\omega\Omega \ . \tag{29}$$

From this we conclude that

$$[K,\widetilde{\omega}] = -2A_0 + h \ , \tag{30}$$

where h is a connection in the stability group, $[K,h]=0$. Excluding h from the action and performing some trivial rescalings, we arrive at the sigma model action, describing the above antiferromagnet

$$S = \int dtdx\{Tr[K,A_\alpha]^2 + i\pi\varepsilon_{\alpha\beta}Tr(K[A_\alpha,A_\beta])\} \ . \tag{31}$$

It must be noticed that in general this action is deformed by renormalization, since nothing prevents functions of T to appear under the trace (since T has classical dimension zero and is invariant under the gauge group). Also, they may and should be present in the original hamiltonian. However, from the definition of T it follows that it satisfies polynomial relations, defining the orbit of the form $TrT^k = C_k(\mu)$, where $C_k(\mu)$ is related to the value of the Casimir operators in the given representation. That reduces or eliminates the number of possible structures. A general invariant Lagrangian should have the form

$$L = Tr\{a(T)(\partial_\alpha T)^2 + \varepsilon_{\alpha\beta}b(T)\partial_\alpha T\partial_\beta T\} \ . \tag{32}$$

The second, parity violating term, is not in general a topological invariant. However, in some cases it is. For example, for the $O(3)$ model, the constraint is $T^2 = const$ and thus $b(T) \sim T$; it is easy to check that we get the standard theta term this way. However, for a general constraint the second term in (32) is parity violating but not topological. The parity violating terms (which we will call $\varepsilon-$terms) can be easily written in terms of the left-invariant connections (in a way invariant under the gauge group H)

$$L_\varepsilon = \varepsilon_{\alpha\beta}Tr(b(K)[K,A_\alpha][K,A_\beta]) \ . \tag{33}$$

In many instances this structure degenerates into a topological term, but in some important cases it doesn't.

Let us give a few less trivial examples of the sigma models related to antiferromagnets. Take first a vector representation of $SO(D)$, $D=2n$, which is described by the fundamental weight corresponding to the first node of the Dynkin diagram. If we denote the generators by $M_{\mu\nu}$, for this representation $K = M_{12}$. The generators commuting with K are M_{12} itself, as well as those of remaining $SO(D-2)$. Hence $T \in \frac{SO(D)}{SO(D-2)\times SO(2)}$ which is the Grassman manifold. It is conveniently described by the field of the antisymmetric tensor $t_{\mu\nu}$, satisfying conditions

$$t_{\mu\nu}^2 = 1; \qquad t \wedge t = 0 . \tag{34}$$

The corresponding sigma model with the theta term has the Lagrangian

$$L = \frac{1}{2\gamma}(\partial_\alpha t_{\mu\nu})^2 + i\vartheta \varepsilon_{\alpha\beta} t_{\mu\nu} \partial_\alpha t_{\mu\lambda} \partial_\beta t_{\lambda\nu} . \tag{35}$$

The claim is that this model describes low energy properties of the quantum $O(D)$ antiferromagnet in the vector representation. We will see below that, while being asymptotically free, it has a conformal fixed point described by free fermions.

Another interesting case is the spinor representation of $O(2n)$. For this representation $K = \frac{1}{2}(M_{12} + M_{34} + ...M_{2n-1,2n})$. The stability group is $U(n)$ and we are dealing with the coset $\frac{SO(2n)}{U(n)}$. This is the orbit of pure spinors, which are the Weyl spinors λ with the constraints, analogous to (34), $\overline{\lambda}\gamma_{\mu_1...\mu_p}\lambda = 0$ for $p < n$. Again, the conformal point is described by the Lagrangian for pure spinors and their conjugates and is closely related to the description of the Berkovits string.

Finally let us discuss supergroups, beginning with the $OSp(1|2)$. Its generators are obtained by adding to $SL(2)$ two spinorial supercharges Q_α. The matrix Ω can be conveniently written as $\Omega = e^{(\vartheta Q)}g$,where g is the matrix of $SL(2)$ and $\vartheta - s$ are two Grassman variables. It is quite easy to find the action $W(n, \vartheta)$. For that we notice that

$$\Omega^{-1}d\Omega = g^{-1}dg + g^{-1}e^{-(\vartheta Q)}de^{(\vartheta Q)}g \tag{36}$$

$$e^{-(\vartheta Q)}de^{(\vartheta Q)} = (Qd\vartheta) + \frac{1}{2}\tau^a\overline{\vartheta}\gamma^a d\vartheta . \tag{37}$$

Using these formulae we get

$$W(n, \vartheta) = \int dt\{A(n) + \frac{1}{2}\overline{\vartheta}(\gamma^a n^a)\frac{d\vartheta}{dt}\} . \tag{38}$$

This action performs geometrical quantization of $OSp(1|2)$, provided that we identify the generators as follow

$$S^a \Rightarrow (1 - \frac{1}{2}(\overline{\vartheta}\vartheta))n^a \equiv N^a \tag{39}$$

$$Q^\alpha \Rightarrow \vartheta^\alpha \tag{40}$$

$$(N^a)^2 + \overline{\vartheta}\vartheta = 1 . \tag{41}$$

4. Non-compact supermagnet

Let us describe the magnet and the sigma model of the last example. The resulting theory is a closed cousin of the AdS/CFT sigma models. There are two complications to be overcome. The first is non-compactness of the orbit and the second is the presence of the grassman dimensions. Let us discuss the first problem first. Non-compact magnets have been considered before in the important work [5], but we need a somewhat different perspective.

We are looking for a magnet, the hamiltonian of which acts in the Hilbert (positive norm) space. This is possible if the spin operator at each site acts in a unitary representation of $SL(2) = Sp(2)$. Such representations are necessarily infinite dimensional. If we perform the geometric quantization with the phase factor $\exp \pm i\kappa \int A(n)dt$ with real κ the unitarity is to be expected and we have to identify the representation. Parametrizing the unit vector n as $n = (\cosh\theta, \sinh\theta\cos\varphi, \sinh\theta\sin\varphi)$, we find that the phase factor is equal to $\exp \pm i\kappa \int \cosh\theta \frac{d\varphi}{dt} dt$ (in the Poincare coordinates it is even simpler, $\exp \pm i\kappa \int \frac{dx}{dt}\frac{dt}{y}$). This corresponds to the unitary representations with the lowest (highest) weights, described by the upper (lower) sheet hyperboloids. If κ is an integer, this is a discrete series of representation. Otherwise, this is a representation of the covering group of $SL(2)$. It is useful for our purposes to construct these representations explicitly as the Verma modulus. Namely, consider the Virasoro generators $(L_{\pm 1}, L_0)$ which form the $SL(2)$ subalgebra. The representation in question is defined by the lowest weight vector $|\kappa\rangle$ satisfying $L_1|\kappa\rangle = 0$ and $L_0|\kappa\rangle = \kappa|\kappa\rangle$. All other states in the representation are simply $|\kappa + m\rangle = (L_{-1})^m|\kappa\rangle$. This representation can be obtain from the finite dimensional $SU(2)$ representations by means of analytic continuation, setting the spin $S = -\kappa$. All these facts are easy to check by solving the Schrödinger equation for a particle on a hyperboloid with the phase factor and concentrating on the lowest energy level. The wave function in this case is proportional to the Jacobi function $P_{m\kappa}(\cosh\theta)$, and the limit of geometric quantization, when only the phase factor is left in the lagrangian, corresponds to taking $l = k$, $m \geq \kappa$.

It is well known that the spin chains of arbitrary integer spin can be made completely integrable by the special choice of the hamiltonian [6,7], which has the form

$$\mathcal{H} = \sum_x f(S_x \cdot \widetilde{S}_{x+1}) \ , \tag{42}$$

where the function f is expressed in terms of the derivative of the gamma function. In our case it should require a special consideration. The reason is that as a result of the factor $(-1)^x$ in (20), representations at the even and odd points are conjugate (denoted by tilde) to each other. The conjugate representation is generated from the highest weight state $|\widetilde{\kappa}\rangle$ satisfying $L_{-1}|\widetilde{\kappa}\rangle = 0$ and $L_0|\widetilde{\kappa}\rangle = -\kappa|\widetilde{\kappa}\rangle$. Integrability is not affected by the above analytic continuation to negative spins. Indeed the Lax operator has the standard form

$$\mathcal{L} = \lambda + i\vec{\sigma}\vec{S} \ , \tag{43}$$

where σ are the Pauli matrices and S are the spin operators in any representation. The fact that at the even points we have the lowest weight representation and at the odd ones the highest weight, doesn't prevent us from constructing integrals of motion generated by the $Tr\prod\mathcal{L}_x(\lambda)$. But constructing the hamiltonian, the Bethe ansatz and finding the physical vacuum, the excitation spectrum and the central charge remains an open problem. The technical difficulty is that the product of representations in (42) has neither highest nor lowest weight. Because of that we will discuss an alternative approach to Bethe ansatz below.

There is no guarantee that this system has a relativistic limit, but if it does, this magnet describes (in the infrared limit) conformal points of the non-compact n-field. These conformal points (if they exist) must be unusual. In the compact case of $SU(2)$ we know that such points are described by the WZNW action and contain holomorphic and antiholomorphic $SU(2) \times SU(2)$ currents. This can't be the case in the above model since the non-compact version of current algebra contains negative norm states, while the lattice model is unitary. Moreover, in the perturbation theory this hyperbolic n-field is obtained from the standard spherical n-field by changing the sign of the coupling constant. That reverses the asymptotic freedom to the "zero charge" (IR fixed point at zero coupling). That means that the non-trivial IR limit can arise only as a result of fine tuning of higher relevant operators.

Let us generalize these considerations for the $OSp(1|2)$ group. Again it is convenient to consider it as a subgroup of the super-Virasoro algebra. The generators are simply $(L_{\pm 1}, L_0, G_{\pm\frac{1}{2}})$. The lowest weight unitary representations are obtained by the action of the raising operator $G_{-\frac{1}{2}}$ on the state $|\kappa\rangle$, annihilated by $G_{\frac{1}{2}}$. The conjugate, highest weight, representations are defined as before. If we rename these generators by (S^a, q^α), with α being a spinor index and q a Majorana spinor, we can consider an integrable supermagnet in the form

$$\mathcal{H} = \sum_x f(S_x \widetilde{S}_{x+1} + \overline{q}_x \widetilde{\ } . q_{x+1}) \ . \qquad (44)$$

Once again the R -matrix for the $OSp(1|2)$ system has been found long ago [8], but the spectrum of the above (non-compact) antiferromagnet is not known. Let us find the corresponding sigma model. As clear from the above discussion, it contains a hyperbolic unit vector n and its superpartner ϑ. It is convenient to write it first in terms of connections. The corresponding coset space is $OSp(1|2)/O(2)$. The connections of $OSp(1|2)$ can be written down as $\{B^a_\mu, A_\mu, \psi^\alpha_\mu\}$ where $a = 1,2$ is the "vertical" index, A_μ is the $SO(2)$ connection and ψ^α_μ are the two spinor connections. They satisfy the standard Maurer-Cartan equations. The lagrangian must have $SO(2)$ gauge symmetry. The most general form of it is

$$\mathcal{L} = \frac{1}{2\alpha}\{(B^a_\mu)^2 + c_1 \overline{\psi}_\mu \psi_\mu + c_2 \varepsilon_{\mu\nu} \overline{\psi}_\mu \gamma_3 \psi_\nu\} \ , \qquad (45)$$

where α is a coupling constant; we don't include here the standard theta term because in the non-compact case it is trivial, but we have instead the epsilon term, which is not a total divergence. These models appeared in [9,10,11] in connection with gauge/strings duality. They describe the AdS_2 space with the RR fluxes. As was shown in [11], they are a part of the much more interesting cosets, $\frac{OSp(2|4)}{SO(4)\times SO(2)} \to AdS_4$, $\frac{SU(4|2)}{SO(5)\times SU(2)} \to AdS_5 \times S^1$, and $\frac{PSU(4|4)}{SO(5)\times SO(5)} \to AdS_5 \times S^5$. In all these cases the lagrangians are basically the same, except the the type of spinors are different, the matrix γ_3 is replaced by γ_5 etc. There exists a supermagnet representation for these models. In order to have conformal symmetries it is necessary to adjust the couplings c_1 and c_2. These conformal points are completely integrable, at least classically. For the action (45) the Lax representation is as following [11]

$$L_+ = \partial_+ + A_+ T_3 + \lambda^{-1} B^a_+ T_a + (\lambda^{-\frac{1}{2}} \psi^\alpha_{L+} + \lambda^{\frac{1}{2}} \psi^\alpha_{R+}) Q_\alpha \qquad (46)$$

$$L_- = \partial_- + A_- T_3 + \lambda B^a_- T_a + (\lambda^{-\frac{1}{2}} \psi^\alpha_{L-} + \lambda^{\frac{1}{2}} \psi^\alpha_{R-}) Q_\alpha \qquad (47)$$

where (T_3, T_a, Q_α) are the generators of $OSp(1|2)$, and $\psi_{L,R} = \frac{1\pm\gamma_3}{2}\psi$. As it was checked in [11] the equations of motion following from $[L_+, L_-] = 0$ correspond to the lagrangian (45) with $c_1 = 0$ and $c_2 = 1$. Amazingly, at the same point the $\kappa-$symmetry appears and the perturbative beta function is zero. We can add to this an assumption that at the same point the sigma model becomes equivalent to the supermagnet (44). The Lax representation for the last of the above cosets has been found in [12]. An interesting different class of cosets was found in [13]. It deals with the different manifolds, but more importantly with the cases $c_1 = c_2 = 1$. Perhaps the c_1 term, together with the added pure spinor lagrangian, can be looked at as fixing of the κ gauge symmetry, with the pure spinors playing the role of the corresponding ghosts.

5. Null vectors and linear sigma models

The above approach used the discretized theory. It is possible to avoid it. Let us again consider the n -field first. It is equivalent to the WZNW model at level one. On the other hand, the same conformal point must appear in the linear sigma model of the vector field $\overrightarrow{\phi}$, satisfying the equation of motion [12]

$$\partial^2 \overrightarrow{\phi} = \lambda \phi^2 \overrightarrow{\phi} , \tag{48}$$

where λ is a coupling constant and the physical mass of the ϕ field is tuned to zero. The reason for this "linearization" is that the constraint $n^2 = 1$ is relaxed by the infrared fluctuations. Whether this relaxation is relevant, depends on the infrared dynamics.

Let us "solve" this equation using the OPE of the WZNW model (similar approach was used in [14] for the Ising and other minimal models). It is based on the fact, understood at the very beginning of conformal field theory, that the equations of motion must be read from the right to the left, replacing the products of fields by their OPE.

Let g be an $SU(2)$ matrix entering the action and try to identify $\overrightarrow{\phi} \sim Tr(\overrightarrow{\sigma} g)$. First of all, in WZNW model at $k = 1$ we have the fusion rule [15]

$$g \times g \times g = [g] , \tag{49}$$

where brackets mean all possible operators obtained by the action of the Kac-Moody currents J^a_{-n} on the field g. So, the first conclusion is that the RHS of (48) belongs to the current block

$$\phi^2 \phi^a = c_1 \phi^a + c_2^{abcd} J^b_{-1} \overline{J}^c_{-1} \phi^d + c_3 L_{-1} \overline{L}_{-1} \phi^a + ... \tag{50}$$

where we dropped the higher order terms.

The third term in this equation is just what we need for the (48). The first term is the physical mass of the ϕ-field and must be eliminated by hands (in the language of critical phenomena that corresponds to setting the temperature to be critical). So, let us concentrate on the second term.

We will show that at $k = 1$, due to the structure of the null vectors, it reduces to the third term. Consider the states of the form

$$|\psi\rangle = J^a_{-1}|\alpha\rangle \tag{51}$$

where $|\alpha\rangle$ is a state with spin 1/2 and the projection of the spin equal to $\pm\frac{1}{2}$. The state $|\psi\rangle$ can be decomposed into spin $\frac{3}{2}$ and spin $\frac{1}{2}$ components. It is well known that the Kac-Moody algebras at the level k have the null vector of the form

$$(J^+_{-1})^{k-2j+1}|j,j\rangle = 0 \tag{52}$$

where the current acts on the highest weight state with the spin j. By choosing $j = \frac{1}{2}$ and $k = 1$, we conclude that the spin $\frac{3}{2}$ part of the state $|\psi\rangle$ is actually zero. As for the remaining spin $\frac{1}{2}$ part, we have the Knizhnik-Zamolodchikov relation

$$\sigma^a J^a_{-1}|\alpha\rangle = (k+2)L_{-1}|\alpha\rangle\ . \tag{53}$$

Hence the contribution of the second term in (50) is the same as of the third term. Replacing L_{-1} by ∂ we derive the original equation of motion!

The group theory of the above is as following (this is the pattern to be generalized below). With respect to left and right multiplication the matrix g transforms as $(\frac{1}{2}, \frac{1}{2}) \sim (0 \oplus 1)$ where we decomposed with respect to the diagonal subgroup. As we act on g with $J\overline{J}$ we get an object $(\frac{1}{2}\otimes 1, \frac{1}{2}\otimes 1) \sim (\frac{1}{2}\oplus\frac{3}{2}, \frac{1}{2}\oplus\frac{3}{2})$. The key observation is that the $\frac{3}{2}$ part is a null vector, while the $\frac{1}{2}$ part is a Virasoro descendant.

There is a parity odd relevant operator in the linear sigma model, $O = \varepsilon_{\mu\nu}\overrightarrow{\phi}[\partial_\mu\overrightarrow{\phi} \times \partial_\nu\overrightarrow{\phi}]$. It becomes a topological density in the limit when and if $\overrightarrow{\phi}^2 \to const.$ Because of the fusion rules [15], we conclude that in terms of the WZNW fields we have $O \Rightarrow Tr(g)$. This is consistent with the fact that the WZNW action is invariant under the simultaneous change of orientation, $\varepsilon \Rightarrow -\varepsilon$ and $g \Rightarrow -g$. According to the formula for the anomalous dimensions, $\Delta = \frac{j(j+1)}{k+2}$, we see that the dimension of the theta-term as well as the n-field itself is equal to $\frac{1}{4}$ (which is well known in the Haldane-Affleck approach). From the linear sigma model point of view, this critical point requires fine-tuning of two parameters, the mass term (temperature) and the theta term.

What happens at higher values of k and for other groups? Let us give some non-trivial examples. Consider first $k = 2$ and the matrix g in the vector representation, $g \sim (1,1) \sim (0 \oplus 1 \oplus 2)$. The current descendant transforms as

$$J\bar{J}g \sim (0 \oplus 1 \oplus 2, 0 \oplus 1 \oplus 2) \sim (0 \oplus 1, 0 \oplus 1) \ , \tag{54}$$

since the formula (52) tells us that the spin two part is a null vector. The formula (53) implies that the spin one object is a Virasoro descendent. There are, however no constraints on the spin zero part. Therefore we can draw the following conclusions. First, if we want to describe a linear sigma model with the vector field $\overrightarrow{\phi}$ by WZNW action with $k = 2$, we obtain the equation of motion with the non-vanishing Kac-Moody descendent coming from $(1,0) + (0,1)$. Generally this doesn't describe the conformal point of the $\overrightarrow{n}$ field. But, if we fine tune the parameters to eliminate this term, we do get a conformal sigma model. This conclusion is consistent with the well known fact that while generally spin 1 antiferromagnet has a mass gap, there is a special, completely integrable multicritical point, at which it is conformal (see [16] for a review).

This is not all, however. Consider a linear sigma model with the field of spin 2, described by a traceless symmetric tensor ϕ_{ij}. We see from the (54) that in this case spin zero descendents do not appear. Hence we arrive at the conclusion that spin 2 order parameter leads to a conformal theory without extra fine tuning. The equations of motion in this case are

$$\partial^2 \phi_{ij} = \lambda [\phi^3]_{ij} \ , \tag{55}$$

where the brackets mean the projection to spin 2 representation. We expect this theory to be described by the free fermions.

It is easy to generalize these considerations for other order parameters and symmetry groups. It might seem puzzling that we find conformal regimes in cases, like CP^N models, in which the large N expansion tells us that they don't exist. The resolution of the paradox lies in the fact that to reach the above fixed points one has to add to the action operators containing higher powers the fields and their derivatives. Such operators (neglected and intractable in the large N expansion) might seem hopelessly irrelevant. However, an old computation of their dimensions in [17] indicates that the one loop correction tends to compensate the growth of the naive dimension. For example, in the $O(3)$ case the operators in question have the form

$$O_{sp} = (\partial_z n \partial_{\bar{z}} n)^s (\partial_z n)^{2p} (\partial_{\bar{z}} n)^{2p} \tag{56}$$

with the dimension (in the one loop approximation; α_0 is a coupling constant)

$$\delta = s + 2p + \alpha_0 [2s(s-1) - 16p] \ . \tag{57}$$

I believe that some of these operators become relevant at large p (as already was conjectured in [17,18]) and drive the theories to higher fixed points. They are identified with the finite number of relevant operators of WZNW at level k and must be fine-tuned at the conformal point.

These methods are easy to generalize to the case when the field ϕ is defined via gauged WZNW theory. For example, the coset model

$$M = \frac{SU(2)_k \times SU(2)_1}{SU(2)_{k+1}} \tag{58}$$

gives rise to the scalar field $\phi = Tr(\, g_1^{-1} g_2)$ (where the indices 1 and 2 refer to the factors in the numerator of (58)) satisfying the equation

$$\partial^2 \phi = (\phi)^{2k+1} \tag{59}$$

in complete agreement with Zamolodchikov's results [14].

The last example in this section is the supergroup $OSp(1|2)$. This example is important for analyses of the gauge/string dualities. The algebra $OSp(1|2)$ is obtained by adding to the three generators of $SU(2)$, $J^{\pm}$, J^3, two spinor anticommuting generators, $q^{\pm}$. The finite dimensional representations are the pairs of symmetric spinors of ranks $j = 2s$ and $j-1$ which transforms as

$$q_\alpha \Phi_{\alpha_1 \ldots \alpha_j} = \sum \varepsilon_{\alpha\alpha_k} \Psi_{\alpha_1 \ldots \widehat{\alpha}_k \ldots \alpha_j} \tag{60}$$

$$q_\alpha \Psi_{\alpha_1 \ldots \alpha_{j-1}} = \Phi_{\alpha\alpha_1 \ldots \alpha_{j-1}} \,. \tag{61}$$

The dimension of these representations is $2s+1+2(s-\frac{1}{2})+1 = 4s+1$. The fundamental representation has dimension three and consists of a scalar and a two components spinor. The matrix g which enters the WZNW action is 3×3. If we denote these representation as $[s]$, they decompose into the $SU(2)$ representations, denoted by the round brackets, as $[s] = (s)+(s-\frac{1}{2})$.

We can identify the fields of the linear sigma model by the relations, directly generalizing the $SU(2)$ case

$$\phi_a = str(\sigma_a g) \tag{62}$$

$$\vartheta_\alpha = str(\sigma_\alpha g) \tag{63}$$

where the $\sigma - s$ are the 3×3 generalization of the Pauli matrices. The object we get transforms by the vector representation [1] which has $3+2$ dimensions. As before, we have to use the fusion rules to identify the linear sigma model. Once again, we need to know the null vectors for $OSp\{1|2\}$. All of them are known from the comprehensive work [19]. For our purposes we need only the simplest ones which are easy to get directly. The algebra has the form

$$[J_n^+, J_m^-] = 2J_{n+m}^3 + nk\delta_{n+m,0} \tag{64}$$

$$\{q_n^+, q_m^-\} = 2J_{n+m}^3 + 2nk\delta_{n+m,0} \tag{65}$$

$$\{q_n^\pm, q_m^\pm\} = \pm 2J_{n+m}^\pm \tag{66}$$

$$[J_n^\pm, q_m^\mp] = -q_{n+m}^\pm \tag{67}$$

while all other commutators are obvious. The null states at the first level are as following

$$|f_1\rangle = J_{-1}^+|s\rangle \tag{68}$$

$$|f_2\rangle = (q_{-1}^+ + J_{-1}^+ q_0^-)|s\rangle \tag{69}$$

where the highest weight state $|s\rangle$ satisfies the conditions $J_0^+|s\rangle = q_0^+|s\rangle = 0, J_0^3|s\rangle = s|s\rangle$. The above states are null, provided that $s = \frac{k}{2}$. Other null states at level 1 arise at $k = s - 1, -2s - 1, -s - \frac{3}{2}$. It is easy to write explicit expressions for all of them. What we need, however, is a more invariant meaning of these vectors. Namely, the multiplet of currents, (J, q) transforms according to representation of $OSp(1|2)$ [1]. When we act with the currents on the operators in representation $[s]$, we have to decompose the product into irreducible components. This is given by the formula

$$[1] \times [s] = [s + 1] + [s + \frac{1}{2}] + [s] + [s - \frac{1}{2}] + [s - 1] \tag{70}$$

(the only difference with the rotation group is that the half-integer spins appear in the middle). It is easy to see now that at $s = \frac{k}{2}$ the whole representation $[s + 1]$ is formed by the null states (and thus drops out). At $s = k + 1$ the term $[s + \frac{1}{2}]$ is dumped. This pattern continues, removing $[s - \frac{1}{2}]$ at $s = -k - \frac{3}{2}$ and $[s - 1]$ at $s = -\frac{k+1}{2}$.

The null vectors are important here for two reasons. First, they define the fusion rules. For example, a simple use of the Ward identities show that at $k = 1$ the primary field φ of the representation $[\frac{1}{2}]$ satisfies the fusion

$$\varphi \times \varphi = I + \varphi \tag{71}$$

where I is a unit operator; the difference with the $SU(2)$ fusions is in the second term. More over, $\varphi \times \varphi \times \varphi = I + 2\varphi$, and as a result the OSp models are very different from their $O(3)$ counterparts. A naive idea to describe the theory as a linear sigma model with the interaction $V \sim (\vec{\phi}^2 + \bar{\vartheta}\vartheta)^2$ fails because the equations of motion will contain descendants of the unit operator (proportional to the product of the two currents). This is just as well, since a model like that would have larger symmetry $OSp\ (3|2)$ (reflecting the absence of the spin-orbit interaction) which we certainly don't need. In order to get a conformal theory we have to combine the φ^4 and φ^3 terms in such a way as to cancel the unit operator in the equations of motion. The general structure of these cubic terms is

$$L_3 = c_1(\overline{\vartheta}\gamma^a\partial_\mu\vartheta)\partial_\mu\phi^a + c_2\varepsilon_{\mu\nu}(\overline{\vartheta}\gamma^a\partial_\mu\vartheta)\partial_\nu\phi^a \ . \tag{72}$$

These terms create the spin-orbit interaction and are analogous to the terms in the Green-Schwartz action. We should notice that this model necessarily contains the negative norms. Coupling to gravity and imposing κ-symmetry should eliminate them but the concrete mechanisms of this are not clear at present. Perhaps once again this linear sigma model is equivalent to the (45) but it is not proved.

6. Using connections

Let us describe yet another approach to the same class of problems. Let us look again at the $O(3)$ invariant n-field, describing it this time in terms of the $SU(2)$ connections, which we denote as $(B^a_\mu,\ A_\mu)$ where $a = 1, 2$ and the direction 3 is selected for the gauge group $U(1)$. These connections satisfy the Maurer-Cartan equations (the zero field strength condition)

$$\nabla_+ B_- - \nabla_- B_+ = 0 \tag{73}$$

$$\partial_- A_+ - \partial_+ A_- = \epsilon^{ab} B^a_- B^b_+ \tag{74}$$

where $\nabla_+ B^a = \partial_+ B^a + \epsilon^{ab} A_+ B^b$ is the $O(2)$ covariant derivative. The invariant lagrangian has the form

$$L = \frac{1}{2\alpha} B^a_+ B^a_- + \theta\epsilon^{ab} B^a_+ B^b_- \tag{75}$$

and the equations of motion are

$$\nabla_+ B_- + \nabla_- B_+ = 0 \ . \tag{76}$$

Let us compare them with the equations for the WZNW action for $SU(2)$, written in terms of the same three connections (A, B). Of course the first pair of the equations (the zero field strength) is precisely the same as (73,74). But the equations of motion this time are

$$\partial_- A_+ = \partial_- B_+ = 0 \tag{77}$$

meaning that we have three holomorphic currents. Unlike (76) there is no gauge symmetry in these equations.

To make a comparison of the two models let us chose a gauge $A_- = 0$ for the n-field. The equations of motion take the form

$$\partial_- B_+ = 0 \tag{78}$$

$$\partial_- A_+ = \epsilon^{ab} B_+^a B_-^b \tag{79}$$

and are, generally speaking, different from (77).

However, if we chose $k = 1$ in WZNW, something special happens. Let us remember that in this theory there are two sets of currents. The first one (which we have denoted A_+, B_+) are defined as $J_+ = g^{-1}\partial_+ g$ and $J_- = g\partial_- g^{-1}$ are holomorphic and have normal dimensions. Let us call them "direct currents" (DC). The second set consists of the "alternative currents" (AC), which are $K_+ = g\partial_+ g^{-1}$ and $K_- = g^{-1}\partial_- g$. For a general WZNW theory the alternative currents do not conserve and acquire the anomalous dimension $\Delta = 1 + \frac{b}{k+b}$, where b is the dual Coxeter number (adjoint Casimir). Notice that $K_+ = -g^{-1}J_+ g$. That implies that AC are the current descendants of the adjoint operator $g^{-1} \otimes g$.

The crucial point for us is that in the equation (79) the A_+ and B_+ are the components of the direct current while the B_- is an AC. As we already discussed in the last section, at $k = 1$ the adjoint operator decouples from the OPE. Hence, at this value of k the B_- field can be set to zero and the WZNW equations become identical to the n-field equations! Thus, as expected, the n-field theory has a conformal fixed point equivalent to the WZNW fixed point at $k = 1$. Now, let us take $k > 1$. This time the B_- terms don't disappear by themselves. They must be forced out by adding the higher derivative operators (56) and fine-tuning their couplings. This explains why the integrable antiferromagnets of higher spin S are described by the $k = 2S$ theory. It is straightforward to generalize this method for the general case G/H. It requires, however a careful analysis of the null vectors.

More conformal points arise when we use gauged WZNW to solve the coset theory, "cosets for cosets". Let us describe the general idea, leaving the details for the future. As is well known, gauging of WZNW refers to the diagonal group, $g \sim h^{-1}gh$, while the cosets we are interested in are always $g \sim gh$. Surprisingly there is a connection between the two. In general we should compare the diagonal coset G/F with the right coset G/H, but here we restrict ourselves with the simplest case of $SU(2)/U(1)$. We will also write the gauged action in a somewhat non-standard form

$$S = W_G[B_+, A_+] - W_H[A_+] \tag{80}$$

where W-s are the WZNW actions for the groups G and H as functionals of the left-invariant connections. Notice that in this definition both A and B are defined for the group G (and this A is substituted to the action for the group H). By the use of the standard variation formulae [20] it is easy to obtain the equations of motion

$$\partial_- A_+ - \partial_+ A_- = [B_+ B_-] \tag{81}$$

$$\nabla_- B_+ - \nabla_+ B_- = 0 \tag{82}$$

$$\widetilde{\nabla}_- B_+ = 0 \tag{83}$$

where $\nabla = \partial + A$ and $\widetilde{\nabla}=\partial + \widetilde{A}$ with the definition $\partial_+ \widetilde{A}_- - \nabla_- A_+ = 0$. These equations have obvious H gauge invariance. In the gauge $A_+ = 0$ they coincide with the standard gauged WZNW in the gauge $a_+ = 0$ (where a is a gauge field). Indeed in the latter case we have an action

$$S = W[g] + \int a_- A_+ \,. \tag{84}$$

Integration over a enforces $A_+ = 0$ condition and the extra term leads to the above equations with $\widetilde{A}_- = a_-$.

Once again by exploiting various null vectors at various k one can "solve" the local G/H theory by the gauged WZNW. We shall leave it for the future work.

7. Supertwistors and zero curvature

In the ordinary sigma models one of the ways of finding the Bethe ansatz is to reformulate the model in terms of the multi-flavored fermions [20]. It uses the fact that as the number of flavors go to infinity, the fermions in the external gauge fields impose the condition that the field strength is zero. For the lagrangians quadratic in the gauge fields they can be integrated out, giving the four-fermion interaction. For such models the Bethe ansatz is known and all one had to do to describe the sigma model was to take the limit of infinite flavors in the Bethe equations. In this section we will outline the first part of this procedure — reduction to the four-particle interaction.

Let us discuss again the case of $\frac{OSp(1|2)}{0(2)}$ (other cases are similar). The simplest representation which we shall use is a supertwistor $\Lambda_\pm = (\lambda_\pm^\alpha, \phi_\pm)$ which consists of the bosonic spinor λ (with two Majorana components) and fermionic scalar ϕ, which are spinors on the world sheet. The free lagrangian has the form

$$L = \overline{\lambda}\widehat{\partial}\lambda + \phi\widehat{\partial}\phi = \overline{\Lambda}\widehat{\partial}\Lambda \tag{85}$$

where $\widehat{\partial}$ is a two dimensional Dirac operator and as usual $\overline{\lambda}^{\alpha} = \epsilon^{\alpha\beta}\lambda^{\beta}$. Notice that the inverted statistics of these fields is a must — otherwise the lagrangian is a total derivative. There are five currents of $OSp(1|2)$ — three of $SL(2)$, $\overline{\lambda}\tau^{i}\lambda$ and two spinor currents $\lambda^{\alpha}\phi$. They couple to the five gauge fields $(A, B^{a}, \psi^{\alpha})$. The covariant derivatives have the form

$$\nabla\lambda = \partial\lambda + B^{a}\tau^{a}\lambda + A\tau^{3}\lambda + \psi\phi \tag{86}$$

$$\nabla\phi = \partial\phi + \overline{\psi}\lambda\ . \tag{87}$$

When we consider an infinite number of copies of these twistors and integrate them out, we get, as was shown in [20] a delta function of the field strength. Hence, if we add the lagrangian (85) to (45), in the infinite flavour limit we obtain the desired sigma model. On the other hand we can integrate out the connections first and get the model with quartic interaction of supertwistors which has the form

$$L_{int} \sim (\phi_{+}\phi_{-})(\overline{\lambda}_{+}\tau^{3}\lambda_{-}) + (\overline{\lambda}_{+}\tau^{a}\lambda_{-})^{2}\ . \tag{88}$$

This interaction is presumably completely integrable and should produce the R matrix for supertwistors satisfying the Yang-Baxter equations. This matrix should be of XXZ (trigonometric) type and perhaps coincide with the already known ones [8]. If so, finding the Bethe equations should be relatively straightforward. There is a slight complication because of the $O(2)$ gauge symmetry. In the bosonic case this difficulty can be overcome by adding the term A^2 to the action (thus getting the XXZ model) and then sending its coefficient to zero. The same should work in the present case.

8. Conclusions

The present work is a first step towards finding CFT-s describing sigma models. I think that it will require a work of many people to complete this task. I also believe that if we want to understand fundamental physics, there is no way avoiding it. The nature of space-time at large curvature is determined by gauge/strings correspondence and it, to the large extent, boils down to the sigma models, similar to the ones we discussed. It is possible that when better understood these ideas will influence our picture of the early universe.

Other important physical problems involving sigma models are the quantum Hall effect and the theory of 2d turbulence (important for meteorology). In both cases we have to deal with a non-compact CFT. In the case of turbulence there are general considerations [21] and recent interesting numerical data [22], supporting this view.

On the other hand it should be said that the problem is very difficult. It will require considerable amplification of the already esoteric technics of Bethe ansatz and/or some intricate analysis of the CFT algebras. Also, it is possible that some important fixed points are of the different nature than the ones described in this paper. Still I am hopeful that these difficulties will be resolved.

I would like to thank my friends, B. Altshuler, N. Berkovits, I. Klebanov, N. Nekrasov, L. Takhtajan, P. Wiegmann, and A. Zamolodchikov for very helpful discussions.

This work was partially supported by the NSF grant 0243680. Any opinions, findings and conclusions or recommendations expressed in this material are those of the authors and do not necessarly reflect the views of the National Science Foundation.

References

1. A.M. Polyakov, *Phys. Lett.* **B59**, 85 (1975).
2. F. Haldane, *Phys. Rev. Lett.* **50**, 1153 (1983).
3. I. Affleck, *Journ. of Phys. of Cond. Matt.* **1**, 3047 (1989).
4. E. Witten, *Nucl. Phys.* **B223**, 4222 (1983).
5. L. Faddeev and G. Korchemsky, *Phys. Lett.* **B342**, 311 (1995); G. Korchemsky, *Nucl. Phys.* **B550**, 397 (1999).
6. L. Takhtajan, *Phys. Lett.* **A87**, 479 (1982).
7. H. Babujan, *Nucl. Phys.* **B215**, 317 (1983).
8. P. Kulish, *J. Sov. Math.* **35**, 1111 (1986).
9. A. Polyakov, *Int. Journ. Mod Phys.* **A14**, 645 (1999).
10. H. Verlinde, PUPT/2113, hep-th/0403024.
11. A. Polyakov, *Mod. Phys. Lett.* **19**, 22 (2004).
12. A. Polyakov, *ZhETF* (*Sov. Phys. J.*) **55**, 1026 (1968).
13. N. Berkovits *et al.*, *Nucl. Phys.* **B567**, 61 (2000).
14. A. Zamolodchikov, *Sov. J. Nucl. Phys.* **44**, 529 (1986).
15. V. Knizhnik and A. Zamolodchikov, *Nucl. Phys.* **B247**, 83 (1984).
16. L. Faddeev and N. Reshetikhin, *Ann. Phys.* **167**, 363 (1986).
17. F. Wegner, *Z. Phys.* **B78**, 33 (1990).
18. V. Kravtsov, I. Lerner and V. Yudson, *Phys. Lett.* **A134**, 245 (1989).
19. V. Kac and Y. Wakimoto, *Proc. Nat. Acad. USA* **85**, 4956 (1986).
20. A. Polyakov and P. Wiegmann, *Phys. Lett.* **B131**, 121 (1983).
21. A. Polyakov, *Nucl. Phys.* **B396**, 367 (1993).
22. D. Bernard *et al.*, *Nature* (in print).

HIGH-ENERGY UNITARITY AND THE STANDARD MODEL*

ALAN. R. WHITE
Argonne National Laboratory,
9700 South Cass, Il 60439, USA
E-mail: arw@hep.anl.gov

High-energy unitarity is argued to select a special version of QCD as the strong interaction. Electroweak symmetry breaking has to be due to a new sextet quark sector — that will produce large cross-section effects at the LHC. The sextet sector embeds, uniquely, in a massless $SU(5)$ theory that potentially generates the states and interactions of the Standard Model within a bound-state S-Matrix. Infra-red chirality transitions of the massless Dirac sea play an essential dynamical role.

1. Introduction

Volodya Gribov devoted much of his life to studying the implications of unitarity for a high-energy S-Matrix. My own work is based heavily on Volodya's ideas and formalisms and in this article, written in honor of what would have been his 75th birthday, I will outline results suggesting that high-energy unitarity could be a determining constraint on a particle S-Matrix. I will first argue that a special version of QCD (QCD_S), with experimentally attractive features, is selected as the strong interaction [1]. A new color sextet quark sector is present that has just the right properties to produce electroweak symmetry breaking - with large cross-section effects predicted for the LHC. A unique embedding [2] of the sextet sector in a left-handed $SU(5)$ theory (GUT_S) removes the sextet electroweak anomaly and also, amazingly, includes "almost exactly" the lepton and triplet quark sectors of the Standard Model, with QCD_S included in it's entirety. Because some of the $SU(2)\otimes U(1)$ quantum numbers are not quite right, GUT_S can not be a conventional unified theory. Rather, it may provide an underlying *massless* field theory which dynamically generates, *within the bound-state*

*This work is supported by the U.S. Department of Energy under Contract No. W-31-109-ENG-38.

S-Matrix, both the triplet and sextet hadronic sectors and, also, the leptonic sector of the Standard Model. If this proves* to be the case then (it will be arguable that) unitarity uniquely determines the Standard Model !

My results imply that the infra-red chiral anomaly effects of a massless Dirac sea can have a deep significance in a bound-state S-Matrix. At high-energy, zero momentum chirality transitions (within reggeon vertices) produce states and amplitudes with dramatically different properties from those implied, at first sight, by the underlying field theory. In the S-Matrix of massless QCD_S, *the chirality transitions produce confinement and chiral symmetry breaking* - with a minimal hadronic (triplet and sextet) spectrum [1], and Critical Pomeron high-energy behavior [3]. Although much remains to be understood about the S-Matrix of GUT_S, it is already clear that, *just because of the Dirac sea interactions, the Standard Model S-Matrix could indeed emerge.* The $SU(5)$ symmetry is confined and so, very importantly, proton decay is not directly implied. In addition, because the chirality transitions occur only for vector coupled fermions, *the real* $SU(3) \otimes U(1)$ *part of the theory dominates the S-Matrix.* A resulting "wee gluon anomaly interaction" gives [1] a mass for SU(3) singlet left-handed vector bosons and the full interaction structure of the Standard Model appears. There is a related bound-state mass spectrum in which, because of the fermion representation structure, there are no (unwanted) symmetries.

I will emphasize where Gribov's work underlies the arguments that I present, beginning with the formulation of reggeon unitarity [4,5] and ending with the importance of the dynamical role [6] of the Dirac sea. The Critical Pomeron solution [3] of reggeon unitarity and the reggeization of non-abelian gauge theories (that has been established, in remarkable depth, by Lev Lipatov and collaborators [7]) are also crucial.

2. Reggeon Unitarity

The multi-regge behavior of scattering amplitudes is determined by partial-wave amplitudes analytically continued in complex angular momentum. The singularity structure in each "J-plane" is controlled by t-channel unitarity. Elastic unitarity generates regge poles which, in turn, generate regge cuts via multiparticle unitarity. The reggeon unitarity equations then determine the threshold discontinuity due to any combination of regge poles, in any partial-wave amplitude. If $\alpha(t)$ is a regge pole trajectory, the

*Most importantly, sextet electroweak symmetry breaking must be seen [1] at the LHC.

M-reggeon cut trajectory is $J = \alpha_M(t) = M\,\alpha(t/M^2) - M + 1$.

When the reggeon unitarity equations were first derived [4], they were a spectacular intellectual jump and a profound reformulation and generalization of existing low-order diagrammatic calculations. There were, however, many uncertainties in the derivation that were resolved only after the development [5] of multiparticle symptotic dispersion relations as a fundamental basis for multi-regge theory. The generality of reggeon unitarity makes it extremely powerful, particularly when applied to the problem of constructing the multi-regge region QCD S-Matrix. First we discuss the abstract Reggeon Field Theory (RFT) solution of these equations.

3. The Critical Pomeron

The analogy with conventional unitarity makes it straightforward to formulate an effective field theory solution of reggeon unitarity. In fact, Gribov developed [8] a direct (underlying) diagrammatic formulation of RFT only because [5] of the uncertainties that, for a long time, surrounded the derivation of reggeon unitarity. We consider an, even-signature, pomeron regge pole with trajectory $j = \alpha(t)$. If $\alpha(0) = 1$ then, also, $\alpha_M(0) = 1$ and all the multipomeron singularities accumulate at one point. In this case, a simultaneous solution of all the discontinuity formulae is essential and all RFT diagrams have to be summed. Remarkably, a renormalization group formalism can be applied and a fixed-point solution shown to exist.

Under a renormalization group transformation, which rescales both $(1 - J)$ and $k_\perp$, only the triple pomeron coupling r survives as a relevant coupling. A fixed-point exists at $\alpha(0) = 1$ and $r^2 = \frac{4\pi^2}{3}\epsilon + O(\epsilon^2)$ (where $\epsilon = 4 - D$ and D is the dimension of $k_\perp$). $\epsilon = 2$ gives an interacting pomeron theory with the "universality" property of a critical phenomenon. As a result, the asymptotic behavior can be precisely calculated without knowledge of the underlying "bare" parameters. Scaling laws can be derived for a wide variety of cross-sections and it can be shown that all known s- channel unitarity constraints on a theory of rising cross-sections are satisfied [9]. It can also be shown that, when t is positive, all the multipomeron cuts separate and reggeon unitarity is satisfied. Therefore, the Critical Pomeron provides a complete, unique, solution of high-energy unitarity which, in a sense, represents the ultimate achievement of the pre-QCD program to construct a unique S-Matrix from unitarity. *(It has been my goal, for a long time, to determinine how and when the Critical Pomeron occurs in a field theory.)*

4. The Supercritical Pomeron

The supercritical phase [10] (obtained when, formally, $\alpha(0) > 1$) plays a crucial role in linking the Critical Pomeron to an underlying field theory. Using a stationary point of the effective lagrangian introduces a "pomeron condensate" that generates new classes of RFT diagrams. Figure 1 contains a simple example of a new diagram. Reggeon unitarity determines

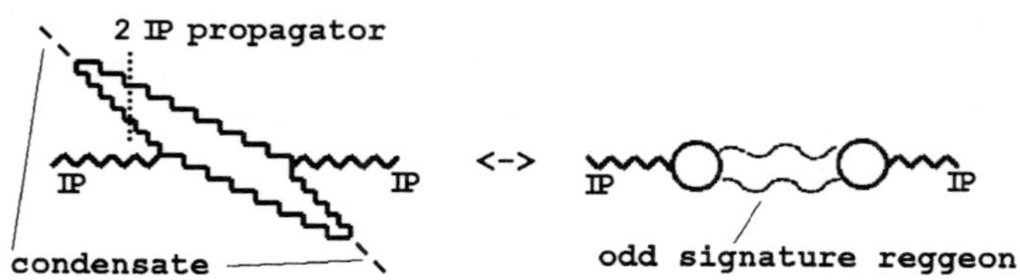

Figure 1. A new RFT diagram generated by the pomeron condensate.

that the $k_\perp$ poles produced by the two-pomeron propagators coupled to the condensate should be interpreted as particle poles lying on an odd-signature trajectory degenerate with that of the pomeron. The odd-signature reggeon couples pairwise to the pomeron and reggeon states involving many vector particle poles similarly appear in higher-order diagrams. In the supercritical phase, therefore, divergences in rapidity (produced by $\alpha(0) > 1$) are converted to vector particle divergences in $k_\perp$ that are associated with the "deconfinement of a vector particle" on the pomeron trajectory.

5. The Critical Pomeron and QCD$_S$.

If the Critical Pomeron is not present in a theory then, since perturbative cross-sections rise, it is unlikely that large momentum perturbation theory can match smoothly with unitary forward amplitudes. In fact, before even discussing dynamical details, we can give some general arguments that specifically link the Critical Pomeron to a special version of QCD (QCD$_S$).

Breaking $SU(3)$ color to $SU(2)$ gives "color superconducting QCD" (CSQCD) in which a massive, reggeized, vector particle (a "massive gluon") is deconfined, just as is expected in the supercritical pomeron phase†. However, at large momentum, a smooth transition from CSQCD to QCD is possible only if an asymptotically free scalar "Higgs" field can be employed.

†That the gauge group has to be $SU(3)$ can be seen [11], without the detailed construction of the next Section, by using cut RFT to determine the center of the group.

This is a strong requirement that is only possible [1,12] if the asymptotic freedom constraint on QCD is "saturated" - immediately implying that there must be a further quark sector beyond that present in the conventional version of the Standard Model. Also, in the infra-red transverse momentum region, the transition from CSQCD to QCD can give the Critical Pomeron only if the infra-red divergences of perturbative reggeized gluon diagrams can generate non-perturbative pomeron diagrams describing a confining theory. As we will see, an infra-red fixed-point, that again exists only when asymptotic freedom is saturated, is crucial - in addition to chiral anomalies.

The saturation constraint (unrealistically) requires sixteen color triplet quarks or, alternatively, two color sextets and six triplets (giving QCD_S). The resulting "sextet pions" provide, exactly, a "Higgs sector" producing electroweak mass generation [1,13]. Thus, requiring unitarity for the strong interaction leads directly to electroweak symmetry breaking and implies that there is a major sector of the strong interaction that has yet to be seen experimentally.

6. $CSQCD_S$ States and Amplitudes

The derivation of supercritical pomeron diagrams required [10] the study of multi-regge amplitudes in which the "pomeron condensate" could be understood as due to a "wee parton" component of scattering hadrons. Correspondingly, to derive supercritical states and amplitudes within $CSQCD_S$, we have to consider a multi-regge limit [14] in which the amplitude for the scattering of regge pole hadrons via pomeron exchange can emerge in it's entirety, as illustrated schematically in Figure 2.

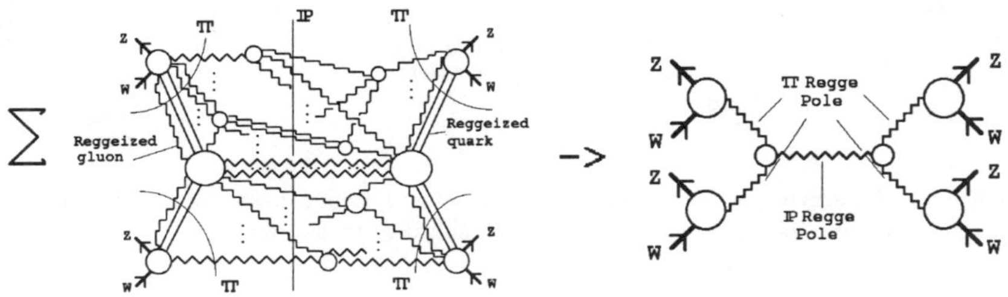

Figure 2. The transition from perturbative reggeon diagrams to reggeized pions scattering via pomeron exchange.

Fortunately, the generality of reggeon unitarity implies that the complicated reggeon diagrams involved can be constructed similarly to the well-known elastic scattering diagrams, the only difference being that vertices coupling distinct reggeon channels can contain anomalies not present in internal reggeon interactions.

The main infra-red divergence of massless gluon reggeon diagrams is due to reggeization. This divergence exponentiates to zero all amplitudes with non-zero $SU(2)$ color in any reggeon channel. We consider diagrams, of the form shown in Figure 3, in which each reggeon state has two components,

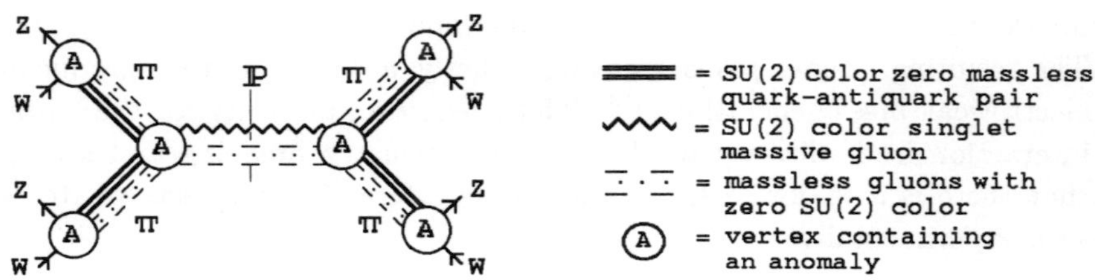

Figure 3. Diagrams with anomaly vertices that will produce the pion amplitude.

both of which carry zero $SU(2)$ color. One component contains massless gluons while the other is either a quark-antiquark pair or a massive gluon reggeon. We choose external left-handed vector boson vertices because, as illustrated in Figure 4, they directly contain [15] a triangle anomaly. Con-

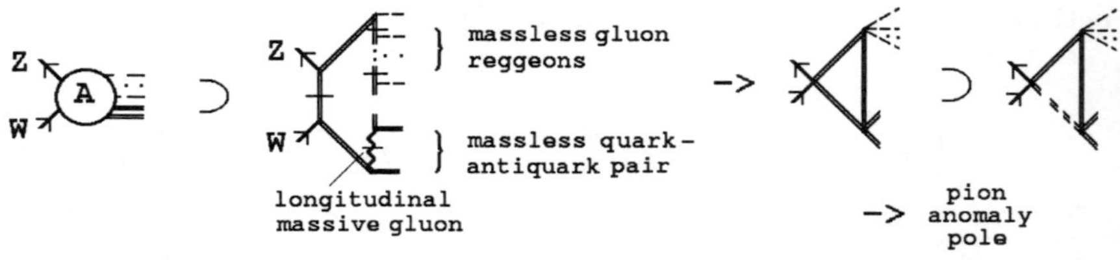

Figure 4. A reggeon vertex triangle diagram. The hatched lines are on-shell and the broken quark line indicates a zero momentum chirality transition.

sequently, an "anomaly pole" appears, as shown, when the gluon reggeons carry zero transverse momentum. The internal vertices in Figure 3 contain U(1) anomaly diagrams of the form shown in Figure 5.

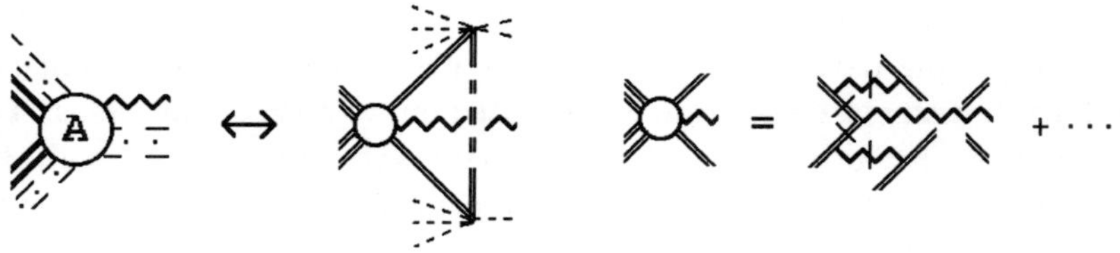

Figure 5. The reduction to a triangle diagram that involves the U(1) anomaly. The broken quark line again indicates the chirality transition giving the anomaly pole.

The initial effect of the anomalies is a large transverse momentum (non-unitary) power enhancement [15] of the high energy behavior. The enhancement can be removed with a transverse momentum cut-off but, because gauge invariance Ward identities are violated, a new infra-red divergence is produced. An overall (logarithmic) divergence is generated when all massless gluon transverse momenta are scaled uniformly to zero. (The $U(1)$ anomaly pole contributes as a transverse momentum conserving δ-function.) However, when the massless gluon component in any reggeon state carries normal color parity (= signature), interactions with the second component in the state again exponentiate the divergence. For anomalous ($\neq$ signature) color parity gluon components, these interactions are absent.

The infra-red fixed point, due to saturation, implies that color zero massless gluon reggeon interactions scale[‡] as all transverse momenta are scaled to zero. The overall "anomalous wee gluon" divergence is then preserved when interactions amongst the massless gluons are included. The residue of the divergence gives a physical CSQCD_S amplitude, in which the transverse momentum cut-off can be removed and all the massless gluons in Figure 3 contribute only as a "reggeon condensate". The condensate provides (crucial) zero transverse momentum, anomalous wee gluon, components in both the "pion" and the "pomeron".

$SU(2)$ anomalous gluons must have odd signature and so the CSQCD_S pomeron is an even signature regge pole which, because of the reggeon condensate, is exchange degenerate with a reggeized massive gluon, just as in supercritical RFT. Also because of the condensate, "pion" Goldstone boson anomaly poles[§], from the external vertices, appear in the physical amplitudes. By removing the transverse momentum cut-off *after the extraction of anomaly infra-red divergences*, we replace ultra-violet chirality

[‡] This is related to the conformal invariance of Green's functions at the fixed-point.
[§] Via the chirality transition, the condensate is absorbed by a zero momentum anti-quark (or quark) that becomes unphysical, as in Gribov's confinement picture [6].

violation (producing bad high-energy behavior) by infra-red chirality violation producing particle poles. This is how a confining, chiral symmetry breaking, bound-state spectrum is generated out of reggeon diagrams. The spectrum consists of triplet and sextet quark/antiquark "pions" and, due to the equivalence of conjugate $SU(2)$ representations, tripet and sextet quark/quark (and anti-quark/anti-quark) "nucleons".

7. Restoration of $SU(3)$ color

There is every indication [1] that the reggeon diagrams of $CSQCD_S$ map onto supercritical RFT, with the essential triple pomeron coupling given [16] by wee gluon anomaly interactions. It remains, however, a major challenge to carry out the mapping in detail. Assuming it can be done, the transition from CSQCD_S to QCD_S will necessarily give the Critical Pomeron. The reggeon condensate will disappear and the zero momentum chirality transitions (which can, equivalently, be regarded as Dirac sea shifts) in states and amplitudes will no longer be due to a semi-classical gauge field with fixed $SU(3)$ color. The transitions will remain and will be many in any scattering process, but they will correspond to random, dynamical, gauge field fluctuations within the color group. The transition from a fixed "magnetization" for the Dirac sea shifting gauge field, to a random, fluctuating, field characterizes the nature of the "critical phenomenon" that is associated with the high-energy behavior of QCD_S. In effect, the longitudinal massive vector meson interactions do not decouple entirely as the color symmetry breaking is removed. They remain as dynamical fluctuations, but only at zero light-cone momentum. Including such interactions (specifically) amounts to fixing the (Gribov) ambiguity in the light-cone quantization of QCD_S.

The quarks clearly have to be massless for the physics of QCD_S to be as we have described it. To add quark masses, and preserve the physics involved, appears non-trivial. The Dirac sea would have to undergo major shifts, of the kind envisaged by Gribov in his original confinement proposal [6], but in a random dynamical manner. In fact, the solution may be that effective quark masses are introduced via the bound-state masses originating from the embedding of QCD_S in GUT_S, that we discuss below.

QCD_S baryons will be formed as bound states of CSQCD_S nucleons and $SU(2)$ singlet quarks. Crucially, because there are no chiral symmetries mixing the two sectors, there will be no "hybrid" sextet/anti-triplet/anti-triplet combinations. Therefore, the only new baryons will be the sextet proton and the (stable) sextet neutron.

8. The Sextet QCD Scale and Electroweak Masses

Sextet antiquarks have the same $SU(3)$ triality as triplet quarks and so we define "Standard Model" couplings for sextet antiquarks (quarks) to be the same as triplet quarks (antiquarks). In an infinite momentum hadron, wee gluons should reproduce vacuum properties and, indeed, an anomaly interaction generates a mass¶ for an exchanged vector boson, as illustrated in Figure 6.

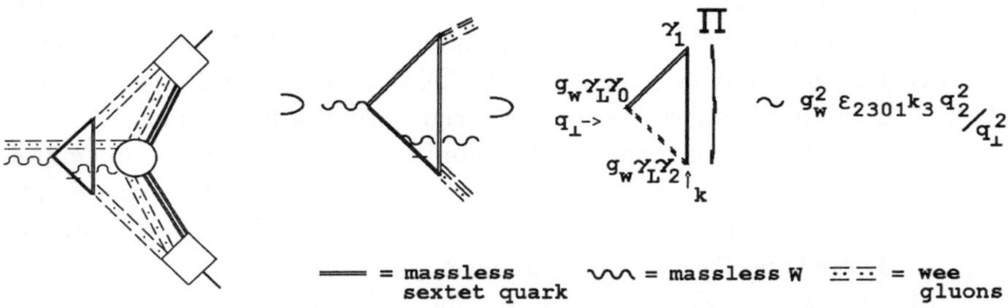

Figure 6. An Interaction Contributing to Vector Boson Mass Generation.

The sextet quark loop contribution is equivalent to the mixing‖ of a sextet pion (Π) with a W and gives $M_W^2 \sim g_W^2 \int dk\, k \equiv g_W^2 F_\Pi^2 \sim$ where k is a wee gluon momentum. The wee gluon origin implies this mass appears only in the S-Matrix. It also occurs only for vectors with a left-handed coupling. No photon (or gluon) mass is generated !

Both sextet and triplet pions contribute to the vector boson mass generation but, sextet pions dominate because of larger color factors. Assuming the Casimir Scaling rule holds ($C_6\ \alpha_s(F_\Pi^2) \sim C_3\ \alpha_s(F_\pi^2)$ with $C_6/C_3 \approx 3$), if α_s evolves sufficiently slowly (e.g. $\alpha_s(F_\pi^2) \sim 0.4$) then F_Π can consistently be the electroweak scale! The large wee gluon coupling to sextet quarks also implies that the wee gluon component of the pomeron couples very strongly ($\sim F_\Pi$) to sextet quarks. This strong coupling is a central element of our understanding of high-energy sextet cross-sections.

¶Similar interactions should also generate a pion mass in the even signature amplitude.
‖The anomaly chirality transition is necessary for this mixing.

9. New Strong Interaction Physics

In general, the larger color factors involved imply that the sextet sector will be a higher mass, stronger coupling, part of the strong interaction. As a result sextet states and cross-sections will dominate at high enough energy. At first sight, the proposition that the sought-after new physics "Beyond the Standard Model" is strong interaction physics seems very radical. However, it has a number of theoretical attractions and potentially answers a number of outstanding experimental problems, as the following list [1] illustrates.

(1) Electroweak symmetry breaking involves no new interaction besides the $SU(3)\otimes SU(2)\otimes U(1)$ gauge interactions of the Standard Model.
(2) The electroweak scale is a new QCD (sextet chiral) scale, which Casimir scaling shows to be the right order of magnitude.
(3) The spectrum is more limited than just color confinement and chiral symmetry breaking would imply, in better accord with experiment. Glueballs and quark resonances are excluded as asymptotic states. There is no BFKL pomeron and no odderon.
(4) New large cross-section physics above the electroweak scale gives a natural explanation for many cosmic ray phenomena. Big increases in average transverse momenta, and in the production of sextet neutrals, could give the apparent "knee" in the spectrum.
(5) The high-energy production of stable, neutral, sextet neutrons, in the early universe, provides a natural explanation for the dominance of dark matter - formed as nuclei, clumps, etc. from sextet neutrons.
(6) Being neutral and massive, sextet neutrons will avoid the GZK cut-off and so could be the mysterious, ultra-high energy, cosmic rays.
(7) Top quark production could be due to the η_6 and sextet quark effects could explain an excess, large E_T, jet cross-section at the Tevatron.

10. LHC Physics

If the cosmic ray spectrum knee is associated with an effective threshold for sextet quark physics (the energy is right, given that pomeron production is needed), large cross-section effects must appear very rapidly with energy to make the knee visible. Such effects should be apparent at the LHC, with dramatic new physics involved [1]. Considering anomaly pole amplitudes, we can show that the strong coupling of the pomeron to sextet states implies that hard double pomeron production of electroweak vector bosons will give jet cross-sections comparable with normal QCD jet (non-diffractive) cross-sections. In particular, the boson pair cross-section is estimated to be, roughly, twelve orders of magnitude larger than in the Standard model. Combining this estimate with pomeron regge theory gives a small transverse momentum cross-section that is correspondingly large. During the initial "soft physics" running period, it should be straightforward [17] to look for the vector boson pairs in the CMS central detector that should be produced in combination with scattered protons in the TOTEM Roman Pots.

If pomeron exchange amplitudes are large, then cut-pomeron amplitudes should also be large. As a result, we expect very large inclusive cross-sections for sextet states (multiple vector bosons, in particular) across most of the rapidity axis. This implies that jet cross-sections, at very large transverse momentum, will be orders of magnitude larger than expected. The production cross-section for sextet nucleon pairs should also be hadronic in size, although stable sextet neutrons (dark matter!) may be difficult to detect. If the sextet nucleon double pomeron cross-section is extraordinarily large, it might be detectable in the low luminosity run. If not, it might be seen by the high luminosity detectors [17] that will look for double pomeron production of the Standard Model Higgs particle.

11. GUT$_S$

Well above the electroweak scale, the QCD_S infra-red fixed point requires $\alpha_s \lesssim \frac{1}{34} \sim \alpha_{ew}$, implying that the sextet sector can, naturally, produce the decrease in α_s needed for unification. (Supersymmetry is not required !) A priori, unification could also determine how the $SU(2) \otimes U(1)$ sextet sector anomaly is canceled, as well as providing an origin for masses.

Many years ago (with Kyungsik Kang) we found [2] a remarkable, but puzzling, result. We looked at asymptotically free, anomaly-free. left-handed unified theories that contain the sextet sector, We discovered that a unique theory is selected, i.e. $SU(5)$ gauge theory with the fermion representation $5 + 15 + 40 + 45^*$ ($\equiv$ GUT$_S$). (In any higher unitary group the sextet sector requires a representation that is too large for asymptotic freedom.) Under $SU(3) \otimes SU(2) \otimes U(1)$

$$\mathbf{5} = (\mathbf{1},\mathbf{3},-\tfrac{1}{3}) + (\mathbf{2},\mathbf{1},\tfrac{1}{2}) \,, \qquad \mathbf{15} = (\mathbf{1},\mathbf{3},\mathbf{1}) + (\mathbf{3},\mathbf{2},\tfrac{1}{6}) + \{(\mathbf{6},\mathbf{1},-\tfrac{2}{3})\} \,,$$

$$\mathbf{40} = (\mathbf{1},\mathbf{2},-\tfrac{3}{2}) + (\mathbf{3},\mathbf{2},\tfrac{1}{6}) + (\mathbf{3}^*,\mathbf{1},-\tfrac{2}{3}) + (\mathbf{3}^*,\mathbf{3},-\tfrac{2}{3}) + \{(\mathbf{6}^*,\mathbf{2},\tfrac{1}{6})\} + (\mathbf{8},\mathbf{1},\mathbf{1}) \,,$$

$$\mathbf{45}^* = (\mathbf{1},\mathbf{2},-\tfrac{1}{2}) + (\mathbf{3}^*,\mathbf{1},\tfrac{1}{3}) + (\mathbf{3}^*,\mathbf{3},\tfrac{1}{3}) + (\mathbf{3},\mathbf{1},-\tfrac{4}{3}) + (\mathbf{3},\mathbf{2},\tfrac{7}{6}) + \{(\mathbf{6},\mathbf{1},\tfrac{1}{3})\} + (\mathbf{8},\mathbf{2},-\tfrac{1}{2}).$$

The triplet quark and lepton sectors, although not asked for, are remarkably close to the Standard Model. There are three "generations" of quarks/antiquarks, with quark charges $\frac{2}{3}$ and $-\frac{1}{3}$, and three "generations" of $SU(2)$ doublet ($SU(3)$ singlet) leptons. The puzzle is that the $SU(2) \otimes U(1)$ quantum numbers are almost, but not quite, right and there are also (apparently unwanted) color octet quarks with lepton-like electroweak quantum numbers. At the time [2], we considered various "anomalous fermion phenomena", but found no convincing dynamical route to the Standard Model.

12. A Massless Theory of Matter ?

In fact, GUT_S has similar properties to massless QCD_S. Asymptotic freedom is saturated and an infra-red fixed-point keeps the $SU(5)$ coupling very small. Also, the $SU(5)$ symmetry can be broken to $SU(4)$ with an asymptotically free scalar field and so the high-energy S-Matrix can be constructed via reggeon diagram anomaly interactions introduced by the symmetry breaking. Because infra-red divergences will confine $SU(5)$ color in the S-Matrix, *all elementary fermions will be massless and confined.* The massless Dirac sea will control the dynamics, but with a crucial difference from QCD_S. In GUT_S, left-handed fermion interactions will exponentiate the anomalous color parity divergences that produce the states and amplitudes of QCD_S. Therefore, these divergences *will only be produced by the* $SU(3)\otimes U(1)$ *vector part of the theory.* The left-handed vector bosons, with no SU(3) color, will aquire a mass, as in Section 8. Thus, just the interactions of the Standard Model will be selected within the GUT_S S-Matrix.

As yet, very little is certain, but our current ideas [18] about the S-Matrix construction, and other properties of GUT_S, can be listed as follows. (We use $SU(5) \rightarrow SU(3)_C \otimes SU(2)_L \otimes U(1) \rightarrow SU(2)_C \otimes SU(2)_L \otimes U(1)$.)

- *We begin with a $k_\perp$ cut-off and SU(5) broken to SU(2)$_C$.*
- *The states are SU(2)$_C$ singlet Goldstones (π's) which are $q\bar{q}$ pairs in an SU(2)$_C$ condensate, as in CSQCD$_S$.*
- *The π's contain q's that are* ***3's, 6's,*** *and* ***8's*** *under SU(3)$_C$. (**8's** are real wrt SU(3)$_C$, but give complex doublets wrt SU(2)$_C$.)*
- *Anomaly interactions, involving the SU(2)$_C$ condensate, generate $W^\pm$ and Z^0 masses via mixing (predominantly) with sextet π's.*
- *Similar interactions should generate π masses. The pattern, and how many parameters are involved, remains to be studied.*
- *Restoring SU(2)$_L$ symmetry gives SU(2)$_C$* x *SU(2)$_L$ invariant states.*
- *"Octet pions" form bound state leptons with elementary "leptons".*
- *After SU(2)$_L$ confinement, the right-handed sextet flavor dependence gives a (broken) SU(2) S-Matrix symmetry.*
- *With SU(3)$_C$ color, Pomeron is Critical and photon is massless.*

- *SU(3) octet quarks have no chiral anomaly and hence no pomeron coupling - hence no strong interaction! They appear in bound-state leptons as pairs of positive and negative energy massive states.*
- *Because the coupling is so small, the underlying SU(5) theory will be close to conformal and the smallest lepton masses will be very small.*
- *There are no spectrum symmetries that would prevent the S-matrix from having the full structure of the Standard Model.*
- *The (experimentally attractive) SU(5) value of the Weinberg angle should be valid for GUT_S, even though there is no proton decay !*
- *The **144** representation of SO(10) is anomaly free, asymptotically free, and contains GUT_S. There is, however, no "saturation" and so we can not construct a unitary, anomaly based, high-energy S-Matrix.*

13. Comments

The success of the Standard Model is a major scientific achievement. Nevertheless, it remains an enigma in the sense that there are a large number of parameters and there is no understanding of why it has been chosen by nature. The candidate theories/ideas that have so far been proposed for extending the model (including supersymmetry and superstrings) have failed to provide any understanding of the model's uniqueness and, as a result, have provided a wide range of possible extensions.

In the search to understand the origin of the Standard Model, the importance of asking that physical states give a unitary S-Matrix may been underestimated. When S-Matrix theory reigned supreme [5], before the dramatic rebirth of field theory that has led to the Standard Model, awareness of the difficulty involved led to the proposal that there is a unique S-Matrix that could, in principle, be bootstrapped from unitarity.

In an asymptotically free field theory, the wild divergence (non-summability) of perturbative interactions occurs in the infra-red region. Presumably, it is very difficult to keep this divergence out of the S-Matrix. Perhaps, there has to be a drastic reduction of the field theory degrees of freedom, in the states and in the interactions, that is only achievable (in very special circumstances) via infra-red anomalies of the massless Dirac sea. Could it be that the unitary S-Matrix is indeed unique but that to construct it, knowledge of the (also unique) underlying *massless field theory* is essential?

References

1. A.R. White, *Phys. Rev.* **D72**, 036007 (2005).
2. K. Kang and A.R. White, *Int. J. Mod. Phys.* **A2**, 409 (1987).
3. A.A. Migdal, A.M. Polyakov and K.A. Ter-Martirosyan, *Zh. Eksp. Teor. Fiz.* **67**, 848 (1975); H.D.I. Abarbanel and J.B. Bronzan, *Phys. Rev.* **D9**, 2397 (1974).
4. V.N. Gribov, I.Ya. Pomeranchuk and K.A. Ter-Martirosyan, *Phys. Rev.* **139**, B184 (1965).
5. A.R. White, "*The Past and Future of S-Matrix Theory*" in "*Scattering*" - eds. E.R. Pike and P. Sabatier, Academic Press, 2002, and references therein.
6. For details of Gribov's confinement picture and for references to the original papers see Y. L. Dokshitzer, presentation at this conference (2005).
7. V.S. Fadin, E.A. Kuraev and L.N. Lipatov, *Sov. Phys. JETP* **45**, 199 (1977); V.S. Fadin and L.N. Lipatov, *Nucl. Phys.* **B406**, *Nucl. Phys.* **B477**, 767 (1996) and further references therein; J. Bartels, V.S. Fadin and R. Fiore, *Nucl. Phys.* **B672**, 329 (2003) and references therein.
8. V.N. Gribov, *Soviet Phys. JETP* **26**, 414 (1968).
9. M. Moshe, *Phys. Rep.* **37**, 255 (1978).
10. A.R. White, *Int. J. Mod. Phys.* **A6**, 1859 (1991).
11. A.R. White, "*The Pomeron in QCD*", CERN-TH-2976 (1980).
12. D.J. Gross and F. Wilczek, *Phys. Rev.* **D8**, 3633 (1973).
13. W.J. Marciano, *Phys. Rev.* **D21**, 2425 (1980); see also E. Braaten, A.R. White and C.R. Willcox, *Int. J. Mod. Phys.*, **A1**, 693 (1986).
14. A.R. White, *Phys. Rev.* **D58**, 074008 (1998) describes the relevant multi-regge limit and the reggeon unitarity construction of multi-reggeon diagrams.
15. A.R. White, *Phys. Rev.* **D69**, 096002 (2004).
16. A.R. White, *Phys. Rev.* **D66**, 045009 (2002).
17. M.G. Albrow, hep-ex/0507095, hep-ph/0409308.
18. A.R. White, "*A Unique Massless Theory of Matter?*", to appear.

6D SUPERCONFORMAL THEORY AS THE THEORY OF EVERYTHING

A.V. SMILGA

SUBATECH, Université de Nantes
4 rue Alfred Kastler, BP 20722, Nantes 44307, France *
E-mail: smilga@subatech.in2p3.fr

We argue that the fundamental Theory of Everything is a conventional field theory defined in the *flat* multidimensional bulk. Our Universe should be obtained as a 3-brane classical solution in this theory. The renormalizability of the fundamental theory implies that it involves higher derivatives (HD). It should be supersymmetric (otherwise one cannot get rid of the huge induced cosmological term) and probably conformal (otherwise one can hardly cope with the problem of ghosts). We present arguments that in conformal HD theories the ghosts (which are inherent for HD theories) might be not so malignant. In particular, we present a nontrivial QM HD model where ghosts are *absent* and the spectrum has a well defined ground state. The requirement of superconformal invariance restricts the dimension of the bulk to be $D \leq 6$. We suggest that the TOE lives in six dimensions and enjoys the maximum $\mathcal{N} = (2, 0)$ superconformal symmetry. Unfortunately, no renormalizable field theory with this symmetry is presently known. We construct and discuss an $\mathcal{N} = (1, 0)$ 6D supersymmetric gauge theory with four derivatives in the action. This theory involves a dimensionless coupling constant and is renormalizable. At the tree level, the theory enjoys conformal symmetry, but the latter is broken by quantum anomaly. The sign of the β function corresponds to the Landau zero situation.

1. Motivation

Arguably, the most burning unresolved problem of modern theoretical physics is the absense of a satisfactory quantum theory of gravity. The *main* obstacle here is the geometric nature of gravity. Time is intertwined there with spatial coordinates and the notion of universal flat time is absent. As a result, in constrast to conventional field theory, one cannot write the (functional) Schrödinger equation, define the Hilbert space with unitary evolution operator, etc.

*On leave of absence from ITEP, Moscow, Russia.

As a matter of fact, Einstein gravity (and any other theory where the metric is considered as a fundamental dynamic variable) has problems also at the classical level. The equations of motion cannot be *always* formulated as a Cauchy problem. This leads to breaking of causality for some exotic configurations like Gödel universes or wormholes [1].

Even though these configurations are not realized in our world at the macroscopic level, their existence presents conceptual difficulties.

The modern paradigm is that a fundamental Theory of Everything is a form of string theory. If this is true, gravity has the status of an effective theory and one is not allowed to blame it for inconsistencies. But string theory also does not provide a satisfactory answer to all these troubling questions. Actually, they cannot even be posed there: we understand more or less well what is string theory only at the perturbative level (and even there we are not sure yet whether technical difficulties preventing one now to perform calculation of string amplitudes at the 2-loop level and higher can be efficiently resolved) while its non-perturbative formulation is simply absent.

This led us to suggest [2] that the TOE is a *field theory* living in *flat* higher-dimensional space. This higher-dimensional theory should involve 3-brane classical solutions, which might be associated with our Universe in the spirit of [3]. The gravity is induced there as an effective theory living on the brane. One can imagine a thin soap bubble. Its effective hamiltonian is

$$H^{\rm eff} \;=\; \sigma \int \sqrt{g}\, d^2x \;, \tag{1}$$

where σ is the surface tension. The hamiltonian (1) is geometric, but the fundamental theory of soap is not: it is formulated in flat $3D$ space and does not know anything about the metric etc. Of course, the analogy is not exact because the effective hamiltonian (1) does not have an Einstein form but looks rather as a cosmological term. The Einstein term and also the terms involving higher powers of curvature appear as corrections, however. In the observable world, the cosmological term is either zero or very small and one should think of a mechanism to get rid of it. One could succeed in that (if any) only if the fundamental theory is supersymmetric. Indeed, only supersymmetry can provide for the exact calcellation of quantum corrections to the energy density of the brane solution.

If we want the fundamental higher-dimensional theory to be renormalizable, the canonical dimension of the lagrangian should be larger than 4, i.e. it should involve higher derivatives. HD theories are known to have a problem of ghosts, which in many cases break unitarity and/or causality of the theory [4] [a]. However, a model study performed in Refs. [2,5] indicates that in some cases, namely, when the theory enjoys *exact* conformal invariance, the ghosts are not so malignant, a well defined ground state (the vacuum) might exist and the theory might enjoy a unitary S-matrix.

We conclude that the TOE should be superconformal theory. This restricts the number of dimensions D in the flat space-time where the theory is formulated by $D \leq 6$. Indeed, all superconformal algebras involving the super-Poincare algebra as a subalgebra are classified [6]. Their highest possible dimension is six, which allows for the minimal conformal superagebra (1,0) and the extended chiral conformal superalgebra (2,0).

Our hypothesis is that the TOE lives in six dimensions and enjoys the highest possible supersymmetry (2,0).

Unfortunately, no field theory with this symmetry group is actually known now. The corresponding lagrangian is not constructed, and only indirect results concerning scaling behavior of certain operators have been obtained so far [7]. In [8], we derived (using the formalism of harmonic superspace (HSS) [9]) the lagrangian for the $6D$ gauge theory with unextended $(1,0)$ superconformal symmetry. This theory is conformal at the classical level and renormalizable. However, it is not finite: the β function does not vanish there and conformal symmetry is broken at the quantum level by anomaly. In other words, this theory cannot be regarded as a viable candidate for the TOE. Its study represents, however, a necessary step before the problem of constructing and studying the $(2,0)$ theory could be tackled.

In the next section, we explain in more details *what* are the ghosts, why (if not dealt with) they make the theory sick, and also present a special QM HD model where the ghosts *are* tamed. In sect. 3 we derive the lagrangian of our superconformal $6D$ theory and calculate its beta function. The last section is devoted, as usual, to conclusions and speculations.

2. Ghost-free QM higher derivative model

To understand the nature of ghosts, one does not need to study field theories. It is clearly seen in toy models with finite number of degrees of freedom. Consider e.g. the lagrangian

$$L = \frac{1}{2}\ddot{q}^2 - \frac{\Omega^4}{2}q^2 . \tag{2}$$

[a] Physically, a ghost–ridden theory is simply a theory where the spectrum has no bottom and one cannot define what vacuum is.

It is straightforward to see that four independent solutions to the corresponding classical equations of motion are $q_{1,2}(t) = e^{\pm i\Omega t}$, $q_3(t) = e^{-\Omega t}$ and $q_4(t) = e^{\Omega t}$. The exponentially rising solution $q_4(t)$ displays instability of the classical vacuum $q = 0$. The quantum hamiltonian of such a system is not hermitian and the evolution operator $e^{-i\hat{H}t}$ is not unitary.

This vacuum instability is characteristic for all *massive* HD field theories — the dispersive equation has complex solutions in this case for small enough momenta. But for intrinsically massless (conformal) field theories the situation is different. Consider the lagrangian

$$L = \frac{1}{2}(\ddot{q} + \Omega^2 q)^2 - \frac{\alpha}{4}q^4 - \frac{\beta}{2}q^2\dot{q}^2 . \quad (3)$$

Its quadratic part can be obtained from the HD field theory lagrangian $\mathcal{L} = (1/2)\phi\Box^2\phi$ involving massless scalar field, when restricting it on the modes with a definite momentum $\vec{k}$ $(\Omega^2 = \vec{k}^2)$. If neglecting the nonlinear terms in (3), the solutions of the classical equations of motion $q(t) \sim e^{\pm i\Omega t}$ and $q(t) \sim te^{\pm i\Omega t}$ do not involve exponential instability, but include only comparatively "benign" oscillatory solutions with linearly rising amplitude.

We showed in [2] that, when nonlinear terms in Eq.(3) are included, an island of stability in the neighbourhood of the classical vacuum[b]

$$q = \dot{q} = \ddot{q} = q^{(3)} = 0 \quad (4)$$

exists in a certain range of the parameters α, β. In other words, when initial conditions are chosen at the vicinity of this point, the classical trajectories $q(t)$ do not grow, but display a decent oscillatory behaviour. This island is surrounded by the sea of instability, however. For generic initial conditions, the trajectories become singular: $q(t)$ and its derivatives reach infinity in a finite time.

Such a singular behaviour of classical trajectories often means trouble also in the quantum case. A well-known example when it does is the problem of $3D$ motion in the potential

$$V(r) = -\frac{\gamma}{r^2} . \quad (5)$$

The classical trajectories where the particle falls to the centre (reaches the singularity $r = 0$ in a finite time) are abundant. This occurs when

[b]Usually, the term classical vacuum is reserved for the point in the configuration (or phase) space with minimal energy. For HD theories and in particular for the theory (3) the classical energy functional is not bounded from below and by "classical vacuum" we simply mean a stationary solution to the classical equations of motion.

$l > \sqrt{2m\gamma}$, where l is the classical angular momentum. And it is also well known that, if $m\gamma > 1/4$, the quantum problem is not very well defined: the eigenstates with arbitrary negative energies exist and the hamiltonian does not have a ground state.

The bottomlessness of the quantum hamiltonian is not, however, a necessary corollary of the fact that the classical problem involves singular trajectories. In the problem (5), the latter are present for all positive γ, but the quantum ground state disappears only when γ exceeds the boundary value $1/(4m)$.

Our main observation here is that the system (3) exhibits a similar behaviour. If both α and β are nonnegative (and at least one of them is nonzero), the quantum hamiltonian has a bottom and the quantum problem is perfectly well defined even though some classical trajectories are singular.

2.1. *Free theory*

Before analyzing the full nonlinear system (3), let us study the dynamics of the truncated system with the lagrangian $L = (\ddot{q} + \Omega^2 q)^2/2$. As was observed in [10], this system displays a singular behavior. It is instructive to consider first the lagrangian

$$L = \frac{1}{2}\left[\ddot{q}^2 - (\Omega_1^2 + \Omega_2^2)\dot{q}^2 + \Omega_1^2\Omega_2^2 q^2\right] \tag{6}$$

and look what happens in the limit $\Omega_1 \to \Omega_2$. When $\Omega_1 > \Omega_2$, the spectrum of the theory (6) is

$$E_{nm} = \left(n + \frac{1}{2}\right)\Omega_1 - \left(m + \frac{1}{2}\right)\Omega_2 \tag{7}$$

with nonnegative integer n, m. On the other hand, when $\Omega_1 = \Omega_2 = \Omega$, the spectrum is

$$E_n = n\Omega \tag{8}$$

with generic integer n. In both cases, the quantum hamiltonian has no ground state, but in the limit of equal frequencies the number of degrees of freedom is apparently reduced in a remarkable way: instead of two quantum numbers n, m (the presence of two quantum numbers is natural — the phase space of the system (6) is 4–dimensional giving two pairs $(p_{1,2}, q_{1,2})$ of canonic variables), we are left with only one quantum number n.

This deficiency of the number of eigenstates compared to natural expectations would not surprise a mathematician. A generic 2×2 matrix has two different eigenvectors. But the Jordan cell $\begin{pmatrix} 1 & 1 \\ 0 & 1 \end{pmatrix}$ has only *one* eigenvector $\propto \begin{pmatrix} 1 \\ 0 \end{pmatrix}$. The statement is therefore that in the limit $\Omega_1 = \Omega_2$ our hamiltonian represents a kind of generalized Jordan cell.

Actually, the "lost" degrees of freedom reinstall themselves when taking into account nontrivial *time dynamics* of the degenerate system (6) with $\Omega_1 = \Omega_2$. The situation is rather similar to what has been unravelled back in the sixties when studying degenerate systems displaying "nonexponential decay" behavior (see e.g. [11]). We will not discuss here the nonstationary problem and concentrate on the *spectrum* of this system.

To begin with, let us construct the canonical hamiltonian corresponding to the lagrangian (6). This can be done using the general Ostrogradsky formalism [12c]. For a lagrangian like (6) involving $q, \dot{q}$, and $\ddot{q}$, it consists of introducing the new variable $x = \dot{q}$ and writing the hamiltonian $H(q, x; p_q, p_x)$ in such a way that the classical Hamilton equations of motion would coincide after excluding the variables x, p_x, p_q with the equations of motion

$$q^{(4)} + (\Omega_1^2 + \Omega_2^2)\ddot{q} + \Omega_1^2\Omega_2^2 q = 0 \tag{9}$$

derived from the lagrangian (6). This hamiltonian has the following form

$$H = p_q x + \frac{p_x^2}{2} + \frac{(\Omega_1^2 + \Omega_2^2)x^2}{2} - \frac{\Omega_1^2\Omega_2^2 q^2}{2} \,. \tag{10}$$

For example, the equation $\partial H/\partial p_q = \dot{q}$ gives the constraint $x = \dot{q}$, etc.

When $\Omega_1 \neq \Omega_2$, the quadratic hamiltonian (10) can be diagonalized by a certain canonical transformation $x, q, p_x, p_q \to a_{1,2}, a_{1,2}^*$ [5,10]. We obtain

$$H = \Omega_1 a_1^* a_1 - \Omega_2 a_2^* a_2 \,. \tag{11}$$

The classical dynamics of this hamiltonian is simply $a_1 \propto e^{-i\Omega_1 t}$, $a_2 \propto e^{i\Omega_2 t}$. Its quantization gives the spectrum (7). The negative sign of the second term in (11) implies the negative sign of the corresponding kinetic term, which is usually interpreted as the presence of the ghost states (the states with negative norm) in the spectrum. We prefer to keep the norm positive definite, with the creation and annihilation operators $a_{1,2}, a_{1,2}^\dagger$ (that correspond to the classical variables $a_{1,2}, a_{1,2}^*$) satisfying the usual commutation relations $[a_1, a_1^\dagger] = [a_2, a_2^\dagger] = 1$. However, irrespectively of whether the metric is kept positive definite or not and the world "ghost" is used or not,

[c]See e.g. Ref. [13] for its detailed pedagogical description.

the spectrum (7) does not have a ground state and, though the spectral problem for the free hamiltonian (11) is perfectly well defined, the absence of the ground state leads to a trouble, the falling to the centre phenomenon when switching on the interactions[d].

We are interested, however, not in the system (6) as such, but rather in this system in the limit $\Omega_1 = \Omega_2$. As was mentioned, this limit is singular. The best way to see what happens is to write down the explicit expressions for the wave functions of the states (7) and explore their behaviour in the equal frequency limit. This can be done by substituting the operators $-i\partial/\partial x$, $-i\partial/\partial q$ for p_x and p_q in Eq. (10) and searching for the solutions of the Schrödinger equation in the form

$$\Psi(q,x) = e^{-i\Omega_1\Omega_2 qx} \exp\left\{-\frac{\Delta}{2}\left(x^2 + \Omega_1\Omega_2 q^2\right)\right\}\phi(q,x) , \qquad (12)$$

where $\Delta = \Omega_1 - \Omega_2$. Then the operator acting on $\phi(q,x)$ is

$$\tilde{H} = -\frac{1}{2}\frac{\partial^2}{\partial x^2} + (\Delta x + i\Omega_1\Omega_2 q)\frac{\partial}{\partial x} - ix\frac{\partial}{\partial q} + \frac{\Delta}{2} . \qquad (13)$$

It is convenient to introduce

$$z = \Omega_1 q + ix , \qquad u = \Omega_2 q - ix , \qquad (14)$$

after which the operator (13) acquires the form

$$\tilde{H}(z,u) = \frac{1}{2}\left(\frac{\partial}{\partial z} - \frac{\partial}{\partial u}\right)^2 + \Omega_1 u\frac{\partial}{\partial u} - \Omega_2 z\frac{\partial}{\partial z} + \frac{\Delta}{2} . \qquad (15)$$

The holomorphicity of $\tilde{H}(z,u)$ means that its eigenstates are holomorphic functions $\phi(z,u)$. An obvious eigenfunction with the eigenvalue $\Delta/2$ is $\phi(z,u) = \text{const}$. Further, if assuming ϕ to be the function of only one holomorphic variable u or z, the equation $\tilde{H}\phi = E\phi$ acquires the same form as for the equation for the preexponential factor in the standard oscillator problem. Its solutions are Hermit polynomials,

$$\phi_n(u) = H_n(i\sqrt{\Omega_1}u) \equiv H_n^+, \qquad E_n = \frac{\Delta}{2} + n\Omega_1 ,$$
$$\phi_m(z) = H_m(\sqrt{\Omega_2}z) \equiv H_m^-, \qquad E_m = \frac{\Delta}{2} - m\Omega_2 . \qquad (16)$$

[d]A characteristic feature of this phenomenon is that some classical trajectories reach singularity in a finite time while the quantum spectrum involves a *continuum* of states with arbitrary low energies [14]. In our case, the "centre" is not a particular point in the configuration (phase) space but rather its boundary at infinity, but the physics is basically the same.

The solutions (16) correspond to excitations of only one of the oscillators while another one is in its ground state. For sure, there are also the states where both oscillators are excited. One can be directly convinced that the functions

$$\phi_{nm}(u,z) = \sum_{k=0}^{m} \left(\frac{i\Delta}{4\sqrt{\Omega_1\Omega_2}}\right)^k \frac{(n-m+k+1)!}{(m-k)!k!} H^+_{n-m+k} H^-_k , \qquad m \le n ,$$

$$\phi_{nm}(u,z) = \sum_{k=0}^{n} \left(\frac{i\Delta}{4\sqrt{\Omega_1\Omega_2}}\right)^k \frac{(m-n+k+1)!}{(n-k)!k!} H^+_k H^-_{m-n+k}, \qquad m > n \tag{17}$$

are the eigenfunctions of the operator (15) with the eigenvalues (7). Multiplying the polynomials (17) by the exponential factors as distated by Eq.(12), we arrive at the normalizable wave functions of the hamiltonian (10).

We are ready now to see what happens in the limit $\Omega_1 \to \Omega_2$ ($\Delta \to 0$). Two important observations are in order.

- The second exponential factor in (12) disappears and the wave functions cease to be normalizable.
- We see that in the limit $\Delta \to 0$, only the first terms survive in the sums (17) and we obtain

$$\lim_{\Delta\to 0} \phi_{nm} \sim H^+_{n-m} \quad , \quad m \le n$$
$$\lim_{\Delta\to 0} \phi_{nm} \sim H^-_{m-n} \quad , \quad m > n . \tag{18}$$

  In other words, the wave functions depend only on the difference $n - m$, which is the only relevant quantum number in the limit $\Omega_1 = \Omega_2$.

As this phenomenon is rather unusual and very important for us, let us spend few more words to clarify it. Suppose Ω_1 is very close to Ω_2, but still not equal. Then the spectrum includes the sets of nearly degenerate states. For example, the states $\Psi_{00}, \Psi_{11}, \Psi_{22}$, etc. have the energies $\Delta/2$, $3\Delta/2$, $5\Delta/2$, etc., which are very close. In the limit $\Delta \to 0$, the energy of all these states coincides, but rather than having an infinite number of degenerate states, we have only one state: the wave functions $\Psi_{00}, \Psi_{11}, \Psi_{22}$, etc. simply *coincide* in this limit by the same token as the eigenvectors of the matrix $\begin{pmatrix} 1 & 1 \\ \Delta & 1 \end{pmatrix}$ coincide in the limit $\Delta \to 0$.

2.2. *Interacting theory*

When $\Omega_1 = \Omega_2$, $u = \bar{z}$ and the operator (15) acquires the form

$$\tilde{H}(z,\bar{z}) = \frac{1}{2}\left(\frac{\partial}{\partial \bar{z}} - \frac{\partial}{\partial z}\right)^2 + \Omega\left(\bar{z}\frac{\partial}{\partial \bar{z}} - z\frac{\partial}{\partial z}\right) . \tag{19}$$

Its spectrum is bottomless. Let us deform (19) by adding there the quartic term $\alpha z^2 \bar{z}^2$ with positive α. Note first of all that it cannot be treated as a perturbation, however small α is: the wave functions are not normalizable and the matrix elements of $\alpha z^2 \bar{z}^2$ diverge. But one can use the variational approach. Let us take the Ansatz

$$|\text{var}\rangle = z^n e^{-Az\bar{z}} , \tag{20}$$

where A, n are the variational parameters. The matrix element of the unperturbed quadratic hamiltonian (19) over the state (20) is

$$\langle \text{var}|\tilde{H}|\text{var}\rangle = \frac{A(n+1)}{2} - \Omega n . \tag{21}$$

Obviously, by choosing n large enough and A small enough, one can make it as close to $-\infty$ as one wishes. The bottom is absent and one cannot reach it. For the deformed hamiltonian, the situation is different, however. We have

$$E^{\text{var}}(n,A) = \langle \text{var}|\tilde{H} + \alpha z^2\bar{z}^2|\text{var}\rangle = \frac{A(n+1)}{2} - \Omega n + \frac{\alpha(n+1)(n+2)}{4A^2} . \tag{22}$$

This function has a global minimum. It is reached when

$$A - \Omega - \frac{\alpha}{4A^2} = 0 \tag{23}$$

and $n = A^3/\alpha - 2$. For small $\alpha \ll \Omega^3$,

$$A \approx \Omega, \quad n \approx \frac{\Omega^3}{\alpha}, \quad \text{and} \quad E^{\text{var}} \approx -\frac{\Omega^4}{4\alpha} . \tag{24}$$

The smaller is α, the lower is the variational estimate for the ground state energy and the ground state energy itself. In the limit $\alpha \to 0$, the spectrum becomes bottomless. But for a finite α, the bottom exists. Note that in the interacting system, the spectrum is completely rearranged compared to the HD oscillator studied above and there is no reason to expect the peculiar Jordan-like degeneracy anymore. The eigenstates are conventional normalized functions and the solution of the time-dependent Schrödinger equation has the standard form.

Bearing in mind that $z = \Omega q + ix = \Omega q + i\dot{q}$, the deformation $\alpha z^2 \bar{z}^2$ amounts to a particular combination of the terms $\sim q^4$, $\sim q^2\dot{q}^2$, and $\sim \dot{q}^4$ in the hamiltonian. For the theory (3) with generic α, β, the algebra is somewhat more complicated, but the conclusion is the same: in the case when the form $\alpha q^4/4 + \beta q^2 x^2/2$ is positive definite, the system has a ground state.

The requirement of positive definiteness of the deformation is necessary. In the opposite case, choosing the Ansatz

$$|\text{var}\rangle \sim (\Omega q + ix)^n \exp\{-Aq^2 - Bx^2\}$$

and playing with A, B, one can always make the matrix element $\langle$ *var* $|$ *deformation* $|$ *var* $\rangle$ negative, which would add to the negative contribution $-\Omega n$ in the variational energy, rather than compensate it. The bottom is absent in this case.

3. Superconformal 6D theory

We start with reminding some basic facts of life for spinors in $SO(5,1)$ (or rather $Spin(5,1)$) . There are two different complex 4-component spinor representations, the $(1,0)$ spinors ψ^a and the $(0,1)$ spinors ξ_a. In the familiar $Spin(3,1)$ case, there are also two different spinor representations, which are transformed to each other under complex conjugation (on the other hand, complex conjugation leaves an Euclidean 4D spinor in the same representation). An essential distinguishing feature of $Spin(5,1)$ is that complex conjugation does *not* change the type of spinor represenation there (while it does for Euclidean 6D spinors, $Spin(6) \equiv SU(4)$).

Indeed, one can show that the spinor

$$\bar{\psi}^a = -C^a_{\dot{a}} \psi^{\dot{a}}, \tag{25}$$

is transformed in the same way as ψ^a. We defined $\psi^{\dot{a}} = (\psi^a)^*$ and introduced a symplectic charge-conjugation matrix C satisfying

$$C^a_{\dot{a}} C^{\dot{a}}_b = -\delta^a_b \ . \tag{26}$$

The operation $^-$ is the covariant conjugation. A somewhat unusual property $\bar{\bar{\psi}}^a = -\psi^a$ holds.

Bearing in mind, however, that ψ^a and $\bar{\psi}^a$ belong to the same representation, it is very convenient [15] to treat them on equal footing and introduce $\psi^a_{i=1,2} = (\psi^a, \bar{\psi}^a)$. The following relation holds:

$$\bar{\psi}^a_i = \psi^{ai} = \epsilon^{ij} \psi_{aj} \ . \tag{27}$$

We choose the antisymmetric representation of the 6D Weyl matrices

$$(\gamma^M)_{ab} = -(\gamma^M)_{ba} \quad \text{and} \quad \tilde{\gamma}^{ab}_M = \tfrac{1}{2}\varepsilon^{abcd}(\gamma_M)_{cd} \,, \tag{28}$$

where $M = 0, 1, \ldots, 5$ and ε^{abcd} is the totally antisymmetric symbol. The basic relations for these Weyl matrices are

$$(\gamma_M)_{ac}(\tilde{\gamma}_N)^{cb} + (\gamma_N)_{ac}(\tilde{\gamma}_M)^{cb} = -2\delta^b_a \eta_{MN} \,, \tag{29}$$

$$\varepsilon_{abcd} = \tfrac{1}{2}(\gamma^M)_{ab}(\gamma_M)_{cd} \,, \tag{30}$$

where η_{MN} is the metric of the 6D Minkowski space ($\eta_{00} = -\eta_{11} = \ldots = -\eta_{55} = 1$) and $\gamma_M = \eta_{MN}\gamma^N$.

The generators of the (1,0) spinor representation are $S^{MN} = -\frac{1}{2}\sigma^{MN}$, where

$$(\sigma^{MN})_a^b = \frac{1}{2}(\tilde{\gamma}^M \gamma^N - \tilde{\gamma}^N \gamma^M)_a^b \quad \text{and} \quad \overline{\sigma^{MN}} = \sigma^{MN} \,. \tag{31}$$

Supersymmetric field theories are most naturally formulated in the framework of superspace approach. The $6D$ superspace is more complicated than the 4-dimensional one. A simple-minded $6D$ superspace involves besides 6 bosonic coordinates 8 fermionic coordinates θ^a_i. However, one can effectively reduce the number of fermionic coordinates using the *harmonic superspace* approach and working with *Grassmann analytic* superfields [9]. We are not able to dwell on this in details and refer the reader to our paper [8]. Here we only present the results.

Let us remind first the form of the conventional quadratic in derivatives SYM action in 6 dimensons. It involves the $6D$ gauge field A_M, the gluino field ψ^a_i satisfying (27) and the triplet of auxiliary fields D_{ik}. The action reads

$$S = \frac{1}{f^2}\int d^6x \,\mathrm{Tr}\left\{-\frac{1}{2}F^2_{MN} - \frac{1}{2}\mathcal{D}^{ik}\mathcal{D}_{ik} + i\psi^k \gamma_M \nabla_M \psi_k\right\} \,, \tag{32}$$

where f is the coupling constant of canonical dimension -1 and ∇_M is the covariant derivative.

If going down to four dimensions, one reproduces the action for $\mathcal{N} = 2$ $4D$ SYM theory. A_M gives the $4D$ gauge field A_μ and the adjoint scalar, ψ^a_i gives two $4D$ gluino fields while the triplet of auxiliary fields can be decomposed into the real auxiliary field D of the 4-dimensional $\mathcal{N} = 1$ vector multiplet and the complex auxiliary field F of the adjoint chiral multiplet.

The action of the HD $6D$ gauge theory was derived in [8]. The result is

$$S = -\frac{1}{g^2}\int d^6x\,\mathrm{Tr}\Big\{\left(\nabla^M F_{ML}\right)^2 + i\psi^j\gamma^M\nabla_M(\nabla)^2\psi_j + \frac{1}{2}\left(\nabla_M\mathcal{D}_{jk}\right)^2$$
$$+\mathcal{D}_{lk}\mathcal{D}^{kj}\mathcal{D}_j{}^l - 2i\mathcal{D}_{jk}\left(\psi^j\gamma^M\nabla_M\psi^k - \nabla_M\psi^j\gamma^M\psi^k\right) + (\psi^j\gamma_M\psi_j)^2$$
$$+\frac{1}{2}\nabla_M\psi^i\gamma^M\sigma^{NS}[F_{NS},\psi_j] - 2\nabla^M F_{MN}\,\psi^j\gamma^N\psi_j\Big\}. \quad (33)$$

The lagrangian has the canonical dimension 6 and the coupling constant g is dimensionless.

Let us discuss this result. Note first of all that the quadratic terms in the lagrangian are obtained from (32) by adding the extra box operator (it enters with negative sign, this makes the kinetic terms positive definite in Minkowski space). It is immediately seen for the terms $\propto \mathcal{D}^2$ and for the fermions. This is true also for the gauge part due to the identity

$$\mathrm{Tr}\left\{(\nabla^M F_{MN})^2\right\} = -\frac{1}{2}\mathrm{Tr}\left\{F^{MN}\nabla^2 F_{MN}\right\} - 2i\mathrm{Tr}\left\{F_M{}^N F_{NS}F^{SM}\right\}. \quad (34)$$

The former auxiliary fields $\mathcal{D}^{ik}$ become dynamical. They carry canonical dimension 2 and their kinetic term involves two derivatives. There is a cubic term $\propto \mathcal{D}^3$. This sector of the theory reminds the renormalizable theory $(\phi^3)_6$. Gauge and fermion fields have the habitual canonical dimensions $[A_M] = 1$, $[\psi] = 3/2$. Their kinetic terms involve, correspondingly, 4 and 3 derivatives. The lagrangian involves also other interaction terms, all of them having the canonical dimension 6.

It is instructive to evaluate the number of on–shell degrees of freedom for this lagrangian. Consider first the gauge field. With the standard lagrangian $\propto \mathrm{Tr}\{F^2_{MN}\}$, a six–dimensional gauge field A_M has 4 on–shell d.o.f. for each color index. The simplest way to see this is to note that A_0 is not dynamical and we have to impose the Gauss law constraint on the remaining 5 spatial variables. For the higher-derivative theory, however, the presence of two extra derivatives doubles the number of d.o.f. and the correct counting is $2\times 5 = 10$ before imposing the Gauss law constraint and $10 - 1 = 9$ after that. In addition, there are 3 d.o.f. of the fields D_{ij} and we have all together 12 bosonic d.o.f. for each color index. The standard $6D$ Weyl fermion (with the lagrangian involving only one derivative) has 4 on–shell degrees of freedom. In our case, we have $4 \times 3 = 12$ fermionic d.o.f. due to the presence of three derivatives in the kinetic term. Not unexpectedly, the numbers of bosonic and fermionic degrees of freedom on mass shell coincide.

3.1. *Renormalization*

The lagrangian (33) does not involve dimensional parameters and is scale–invariant. A less trivial and rather remarkable fact is that the action is also invariant with respect to special conformal transformations and the full superconformal group. This is true at the classical level, but, unfortunately, conformal invariance of this theory is broken by quantum effects. To see this, let us calculate (at the one–loop level) the β function of our theory.

The simplest way to do this calculation is to evaluate 1–loop corrections to the structures $\sim (\partial_M \mathcal{D})^2$ and $\sim \mathcal{D}^3$. The relevant Feynman graphs are depicted in Figs. 1, 2.

For perturbative calculations, we absorb the factor $1/g$ in the definition of the fields. The relevant propagators are

$$\begin{aligned} \langle A_M^A A_N^B \rangle &= -\frac{i\eta_{MN}\delta^{AB}}{p^4}, \\ \langle \psi^{jA} \psi^{kB} \rangle &= -\frac{i\epsilon^{jk}\delta^{AB} p_N \tilde{\gamma}^N}{p^4}, \\ \langle \mathcal{D}_{ik}^A \mathcal{D}_{jl}^B \rangle &= -\frac{i\delta^{AB}}{p^2} \left(\epsilon_{ij}\epsilon_{kl} + \epsilon_{il}\epsilon_{kj} \right), \end{aligned} \tag{35}$$

where A, B are color indices, $A_M = A_M^A t^A$, etc. The vertices can be read out directly from the lagrangian.

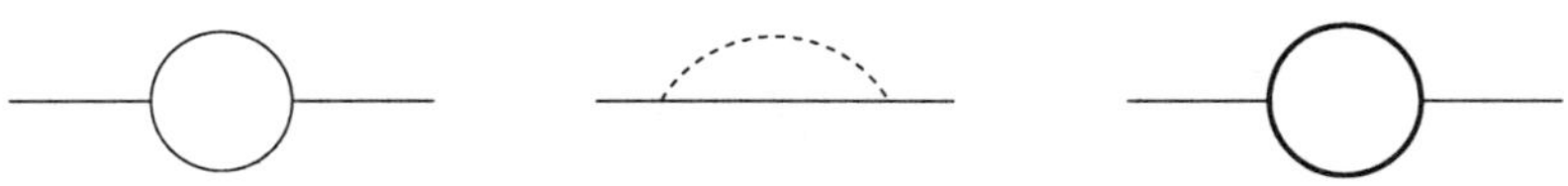

a) b) c)

Figure 1. Graphs contributing to the renormalization of the kinetic term. Thin solid lines stand for the particle $\mathcal{D}$, thick solid lines for fermions, and dashed lines for gauge bosons.

Consider first the graphs in Fig. 1. They involve logarithmic and quadratic divergences. The individual quadratically divergent contributions in the Wilsonean effective lagrangian are

$$\begin{aligned} \Delta\mathcal{L}_{1a}^{\text{eff}} &= -\frac{9c_V}{2} \text{Tr} \{\mathcal{D}_{jk}^2\} I, \\ \Delta\mathcal{L}_{1b}^{\text{eff}} &= \frac{c_V}{2} \text{Tr} \{\mathcal{D}_{jk}^2\} I, \\ \Delta\mathcal{L}_{1c}^{\text{eff}} &= 4c_V \text{Tr} \{\mathcal{D}_{jk}^2\} I, \end{aligned} \tag{36}$$

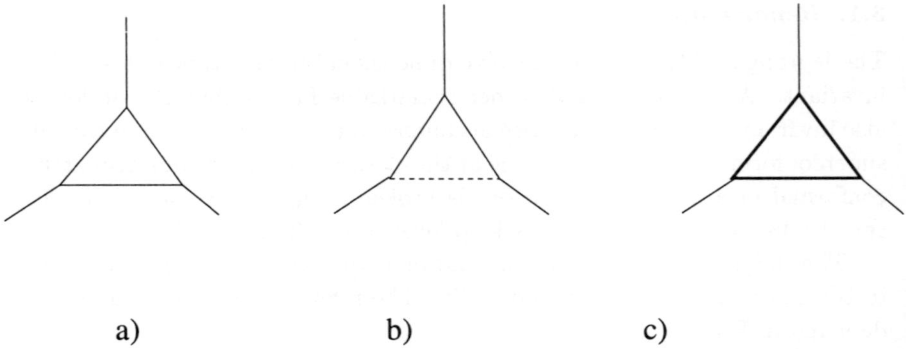

Figure 2. The same for the $\mathcal{D}^3$ vertex.

where c_V is the adjoint Casimir eigenvalue and

$$I = \int^{\Lambda} \frac{d^6 p_E}{(2\pi)^6 p_E^4} \,. \tag{37}$$

We see that the quadratic divergences cancel out in the sum of the three graphs. The logarithmic divergences in the sum of 2-point graphs are

$$\Delta \mathcal{L}^{\text{eff}}_{(2)} = -\frac{2g^2 c_V}{3} \text{Tr}\left\{(\partial_M \mathcal{D}_{jk})^2\right\} L \,, \tag{38}$$

where

$$L = \int_{\mu}^{\Lambda} \frac{d^6 p_E}{(2\pi)^6 p_E^6} = \frac{1}{64\pi^3} \ln \frac{\Lambda}{\mu} \tag{39}$$

and three terms in the parentheses correspond to the contributions of the graphs in Fig. 1a, b, c.

The 3-point graphs in Fig. 2 involve only logarithmic divergence. We obtain

$$\Delta \mathcal{L}^{\text{eff}}_{(3)} = \frac{14 g^3 c_V}{3} \text{Tr}\left\{\mathcal{D}_{lk} \mathcal{D}^{kj} \mathcal{D}^l_j\right\} L \,. \tag{40}$$

The full 1-loop effective lagrangian in the $\mathcal{D}$ sector is

$$\begin{aligned} \mathcal{L}^{\text{eff}}_{\mathcal{D}} &= -\frac{1}{2} \text{Tr}\left\{(\partial_M \mathcal{D}_{jk})^2\right\} \left(1 + \frac{4g^2 c_V}{3} L\right) \\ &\quad - g \text{Tr}\left\{\mathcal{D}_{lk} \mathcal{D}^{kj} \mathcal{D}^l_j\right\} \left(1 - \frac{14 g^2 c_V}{3} L\right) . \end{aligned} \tag{41}$$

Absorbing the renormalization factor of the kinetic term in the field redefinition, we finally obtain

$$g(\mu) \;=\; g_0\left(1-\frac{20g_0^2c_V}{3}L\right)=g_0\left(1-\frac{5g_0^2c_V}{48\pi^3}\ln\frac{\Lambda}{\mu}\right) \qquad (42)$$

for the effective charge renormalization.

The sign corresponds to the Landau zero situation, as in the conventional QED.

4. Discussion

Our study was motivated by the dream or rather by a sequence of dreams spelled out in the Introduction. By the reasons outlined there

(1) We *believe* that the TOE is a conventional field theory in multidimensional bulk
(2) We *believe* that our Universe represents a thin soap bubble — a classical 3-brane solution in this theory.
(3) If the theory claims to be truly fundamental, it should be renormalizable. For $D > 4$, this means the presence of higher derivatives in the action.
(4) We *believe* that for superconformal theories, a way to tackle the HD ghost trouble exists.
(5) We *believe* (but not so firmly, this is just the most attractive possibility) that the TOE enjoys the maximum $\mathcal{N} = 2$ superconformal symmetry in six dimensions.

Besides dreams, there are also some positive results. First, we constructed a QM HD model where the problem of ghosts *is* resolved. Second, we constructed the first nontrivial example of the renormalizable higher-dimensional supersymmetric gauge theory. It is $6D, \mathcal{N}{=}(1,0)$ gauge theory with four derivatives in the action and dimensionless coupling constant.

Our theory enjoys superconformal invariance at the classical level, but, unfortunately, the superconformal symmetry is anomalous in this case. As the result of this breaking, in accord with the arguments of [2], the quantum theory suffers from ghosts which can hardly be harmless.

Four-dimensional experience teaches us that nonsupersymmetric, $\mathcal{N} = 1$, and $\mathcal{N} = 2$ supersymmetric theories might be anomalous, but the maximum $\mathcal{N} = 4$ supersymmetric Yang-Mills theory is truly conformal — β function vanishes there. It is very natural therefore to *believe* that the unconstructed yet Holy Grail $\mathcal{N} = (2,0)$ maximum superconformal $6D$ theory is free from anomaly.

How can it look like ? The first idea coming to mind is to ape the $4D$ construction and to couple the $6D$ gauge supermultiplet to $6D$ hypermultiplets. Adding this term to (33) one might hope to obtain a theory which would enjoy extended superconformal symmetry. Unfortunately, this program meets serious technical difficulties and it is not clear at the moment whether it can be carried out.

The second possibility is that the $\mathcal{N} = (2,0)$ theory does not involve at all the gauge supermultiplet with the action (33), but depends on tensor rather than vector multiplets [7,16]. Unfortunately, to describe the tensor multiplet in the framework of HSS is not a trivial task which is not solved yet. As a result, no microscopic lagrangian for interacting (2,0) tensor multiplet is known today...

Finally, one cannot exclude a disapponting possibility that the (2,0) theory does not have a lagrangian formulation whatsoever.

But the hope dies last !

Acknowledgements

It is a pleasure to thank Julia Nyiri, Peter Levai, and Yura Dokshitzer for their honorable for me invitation to give a presentation at this meeting.

I wonder what Vladimir Naumovich would say if he were in the audience. I am afraid, his reaction would be negative. Gravity, strings, branes, higher dimensions, and higher derivatives were definitely outside the scope of his scientific interests. He had a very sharp "sense of physics" and was not interested at all by what he considered as mathematical excercises having little relation with the Nature. Theorists of younger generations might call him conservative.

A reasonal question might be - what is the point of presenting this talk at the Gribov memorial workshop ? I have two excuses:

- Gravity *is* part of the physical reality and the wish to get hold at some point of a consistent quantum theory of gravity was legitimate also from VN's viewpoint.
- I was trying to be as conservative as possible not introducing new revolutionary ideas like fundamental strings etc. and staying in the framework of conventional field theory. Only time will show whether this approach proves to be sound and associated dreams are realized.

References

1. See e.g. S. Hawking, *Comm. Math. Phys.* **87**, 395 (1982);
   C.J. Isham, *Canonical Quantum Gravity and the Problem of Time*, gr-qc/9210011, published in the *Proceedings of GIFT Int. Seminar on Theor. Physics*, Salamanca, 15-27 June, 1992;
   A. Carlini and I.D. Novikov, *Int. J. Phys.* **D5**, 445 (1996) [gr-qc/9607063].
2. A.V. Smilga, *Nucl. Phys.* **B706**, 598 (2005) [hep-th/0407231].
3. K. Akama, *Lect. Notes Phys.* **176**, 267 (1982);
   V.A. Rubakov and M.E. Shaposhnikov, *Phys. Lett.* **B125**, 136 (1983).
4. A. Pais and G.E. Uhlenbeck, *Phys. Rev.* **79**, 145 (1950).
5. A.V. Smilga, *Phys. Lett.* **B632**, 433 (2006).
6. S. Minwalla, *Adv. Theor. Math. Phys.* **2**, 781 (1998) [hep-th/9712074];
   A. Van Proeyen, "*Tools for supersymmetry*", hep-th/9910030.
7. P. Claus, R. Kallosh and A. Van Proeyen, *Nucl. Phys.* **B518**, 117 (1998) [hep-th/9711161];
   O. Aharony, M. Berkooz and N. Seiberg, *Adv. Theor. Math. Phys.* **2**, 119 (1998) [hep-th/9712117];
   F. Bastianelli, S. Frolov and A.A. Tseytlin, *JHEP* **0002**, 013 (2000) [hep-th/0001041];
   G. Arutyunov and E. Sokatchev, *Nucl. Phys.* **B635**, 3 (2002) [hep-th/0201145].
8. E.A. Ivanov, A.V. Smilga, and B.M. Zupnik, *Nucl. Phys.* **B726**, 131 (2005).
9. A. Galperin, E. Ivanov, V. Ogievetsky and E. Sokatchev, *Pis'ma ZhETF* **40**, 155 (1984) [*JETP Lett.* **40**, 912 (1984)];
   A.S. Galperin, E.A. Ivanov, S. Kalitzin, V.I. Ogievetsky and E.S. Sokatchev, *Class. Quantum Grav.* **1**, 469 (1984);
   A.S. Galperin, E.A. Ivanov, V.I. Ogievetsky and E.S. Sokatchev, *Harmonic superspace*, Cambridge University Press, 2001, 306 p.
10. P.D. Mannheim and A. Davidson, *Phys. Rev.* **A71**, 042110 (2005) [hep-th/0408104].
11. A. Peres, *Ann. Phys.* **129**, 33 (1980).
12. M. Ostrogradsky, *Mem. Ac. St. Petersburg* **VI 4**, 385 (1850).
13. T. Nakamura and S. Hamamoto, *Progr. Theor. Phys.* **95**, 469 (1996) [hep-th/9511219].
14. V. de Alfaro and T. Regge, *Potential Scattering*, Amsterdam (1965).
    For a recent discussion, see A.E. Shabad, hep-th/0403177.
15. P.S. Howe, G. Sierra and P.K. Townsend, *Nucl. Phys.* **B221**, 331 (1983).
16. E. Witten, "*Some comments on string dynamics*", In **Los Angeles 1995, Future perspectives in String Theory** 501 [hep-th/9507121];
    A. Strominger, *Phys. Lett.* **B383**, 44 (1996) [hep-th/9512059];
    N. Seiberg, *Nucl. Phys. Proc. Suppl.* **67B**, 158 (1998) [hep-th/9705117].

PHOTON-NEUTRINO INTERACTION
OR
OPTICAL ACTIVITY OF INTERGALACTIC SPACE

VICTOR NOVIKOV

ITEP, Moscow, Russia

The birefringence of a neutrino sea is discussed in the Standard Model. We demonstrate that the optical activity of a neutrino sea in the SM is dominated by the contribution of the two-loop diagrams that are five order of magnitudes larger than one-loop diagrams.

Vladimir Naumovich was fond of the Feynman diagram approach. For him diagrams were not something like boring perturbative terms. Behind the diagrams he saw a true space-time picture of the process. Diagrams represented the bird's-eye-view of the landscape of physics.

I am going to talk about a short paper where the clear understanding of the space-time picture helps to find an enhancement factor of the order of million. Though physical effects remain small even with this huge factor, the theoretical observation seems to be rather nice. I believe that B.H. would appreciate it.

My talk is based on a paper written in collaboration with G. Karl [1].

1. Introduction. Preliminaries

The photon-neutrino interaction is very feeble, involving only electrically neutral external particles. At a classical level this interaction vanishes (we assume that the neutrino magnetic moment μ_ν is zero). But in QFT the corresponding amplitudes are non-zero even for $\mu_\nu = 0$ due to virtual states with charged particles. Thus in Fermi Theory the amplitude is described by a triangle graph, with an electron running in the triangular loop (see Fig. 1).

The cross sections are exceedingly tiny. Therefore this interaction can only be of astrophysical interest. One of the possible applications involves

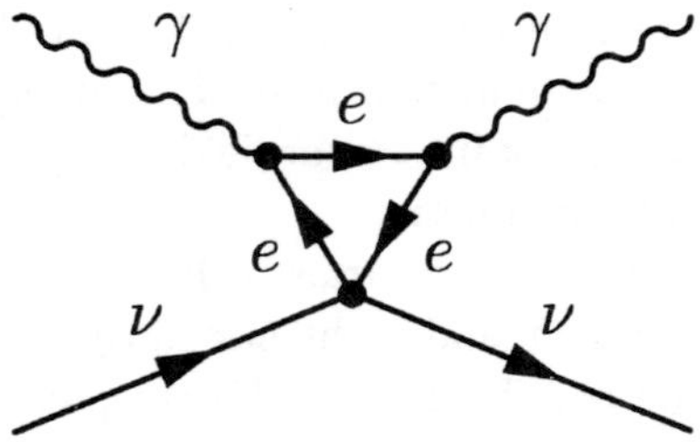

Figure 1. $\gamma\nu$ scattering in the four-fermion Model.

the propagation of light waves through a handed neutrino sea [2] which results in the optical activity (birefringence) of the intergalactic space.

1.1. *Prehistory*

Optical activity was known from the time of Napoleon's wars when Arago (1811) and Biot (1812) discovered that polarization plane of linearly polarized light rotates when light propagates through active media. Later (1817) Fresnel understood that this effect is connected with the fact that eigenstates of propagation in media are the states with definite circular polarization and that the speed of light is slightly different for the left-handed and right-handed polarizations of light, i.e. media are birefringent.

1.2. *Modern History*

In the last decade a number of authors raised a question on the possibility of the violation of Lorentz and CPT invariance. To get a taste of this activity, one can look at the paper written by Glashow and Coleman [3]. The most interesting scenario is that the speed of light in vacuum (i.e. in intergalactic space, in practice) is slightly different for different polarization states of light, i.e. now the vacuum is birefringent. It is clear that such an effect would indicate a small violation of Lorentz and CPT invariance at the fundamental level.

There are not many choices to get Lorenz violation at the level of a Lagrangian. If we wish to preserve gauge invariance and renormalizability of QED there is more or less one way to modify the QED Lagrangian. That is

$$L = -\frac{1}{4}F_{\mu\nu}F_{\mu\nu} + \Delta L, \quad \Delta L = \frac{1}{2}\epsilon_{\mu\nu\alpha\beta}b_\mu A_\nu F_{\alpha\beta} \ , \tag{1}$$

where A_μ, $F_{\alpha\beta}$ are the potential and the field-strength tensor for the electromagnetic field, and b_μ is some new field. The condensation of the vector field b_μ in vacuum, i.e. a constant vector field in eq. (1), generates a violation of Lorentz and CPT symmetry.

One can check that in the "rest" frame where $b_\mu = (b_0, 0, 0, 0)$ the index of vacuum refraction for left and right circularly polarized light $n_{+,-}$ is different from unity and is different for different polarization states:

$$n_+ = 1 + b_0/\omega, \;\; n_- = 1 - b_0/\omega \; . \tag{2}$$

Here ω is the light frequency.

For a linear polarized light this means that the plane of polarization rotates at angle $\Delta\phi$ when light propagates the distance l

$$\Delta\phi = \omega(n_+ - n_-)l \; . \tag{3}$$

Thus the rotation power $\Phi = \Delta\phi(l)/l$ is

$$\Phi = \omega(n_+ - n_-) = 2b_0 \; . \tag{4}$$

Non-zero b_0 (vacuum condensation of vector field b_μ) leads to a birefringence of vacua. That is how people work with vacuum birefringence at a fundamental level.

A less radical (and more solid) physical reason for the light to have different speeds of propagation in the intergalactic space for different polarization states is a P-odd interaction of light with neutrino sea. In this case the neutrino sea in the intergalactic space plays the role of optical active media, i.e. of a left-handed sugar.

The idea that the intergalactic space is a birefringent medium for the light due to the presence of a neutrino sea has been contemplated for a long time. Some thirty years ago, Royer [2] estimated in a $V - A$ theory an effect of the order of G_F

$$\Phi \sim \alpha G_F k_F^3, \tag{5}$$

where k_F is the Fermi momentum of the neutrino sea. Later, Stodolsky [a] noted that due to a theorem of Gell-Mann [5] there can be no such effect with massless neutrinos and a point-like coupling, i.e.

$$\Phi \sim 0. \tag{6}$$

[a]This observation was recorded in a review [4].

In the early 1980's data on propagation of radio waves through the intergalactic space put an upper bound on the possible optical activity of the neutrino sea [6,7].

More recently an evaluation was made for massless neutrino and for on-shell photons [8] in one-loop approximation within the Standard Model (SM). The effect is non-zero but extremely small

$$\Phi \sim \alpha G_F k_F^3 \frac{(k_F \omega)^2}{m_W^4}. \tag{7}$$

Here m_W is the mass of the W boson. Due to the Gell-Mann theorem there is an additional suppression factor of the order $(k_F\omega)^2/m_W^4$ to the naive estimate of Royer [b].

In my talk I demonstrate that these one-loop estimates for on-shell photon-neutrino scattering are irrelevant. The main contribution to the optical activity comes from the two-loop amplitude. The latter is larger than the one-loop estimates by a factor of 10^5 or more. There is simple physics behind this amazing result.

2. Photon-neutrino interaction

2.1. *Gell-Mann theorem*

According to Gell-Mann's theorem [5], point-like weak interactions and massless neutrino leads to vanishing amplitude for photon-neutrino effective interactions. The theorem is easy to understand in the cross channel, i.e. for the annihilation of a $\nu\bar{\nu}$ pair into $\gamma\gamma$ in the center of mass frame.

For point-like interaction the orbital moment of a pair of neutrinos is exactly zero L=0. Thus for massless left-handed neutrino and right-handed antineutrino the total angular momentum J=1. On the other hand, for two on-shell photons the states with J=1 are forbidden by Landau's theorem [10]. As a result, the transition between $\nu\bar{\nu}$ and $\gamma\gamma$ states is forbidden in a point-like limit.

To escape Gell-Mann's restriction, one needs non-local interactions in order to include higher orbital momenta of the $\nu\bar{\nu}$ pair into the annihilation process. In the Standard Model, where the W and Z bosons mediate the interaction, the probability amplitude for a neutrino pair to have a nonzero orbital moment is proportional to some power of a small factor $\sim p/m_W$,

[b]See also the calculations for off-mass shell photons in Ref. [9].

where p is the neutrino momentum. The factor $1/m_W$ measures the shortest separation of two neutrinos during their interaction (non-locality) in one-loop approximation. The loop changes from a triangle to a square, with the weak boson providing one of the sides (see Fig. 2).

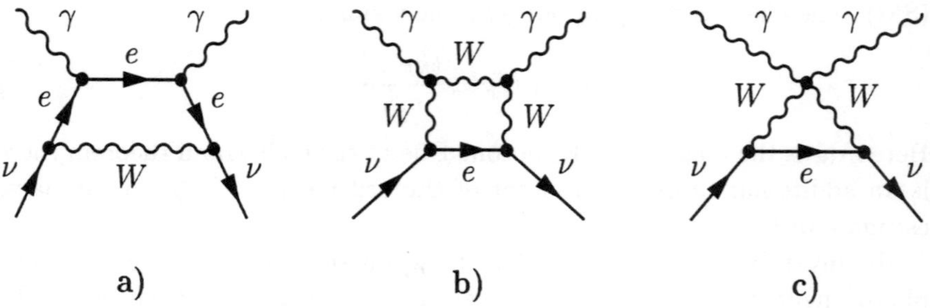

Figure 2. $\gamma\nu$ scattering in the Standard Model.

2.2. *Effective Lagrangian Approach*

The standard way to deal with low-energy scattering is to use the Effective Lagrangian approach.

The simplest example of the Effective Lagrangian is the four-fermion interactions of neutrinos ν with electrons e. For small fermion momenta one can forget about degrees of freedom associated with W and Z boson and can write the Effective Lagrangian only for fermionic fields:

$$L_{eff} = \frac{G_F}{\sqrt{2}}(\bar{\nu}\gamma_\alpha\nu)(\bar{e}\Gamma_\alpha e) \ , \tag{8}$$

where $\Gamma_\alpha = g_V\gamma_\alpha + g_A\gamma_\alpha\gamma_5$. In the SM $g_V = \frac{3}{2} - 2\sin^2\theta_W, g_A = \frac{3}{2}$. For momenta smaller than m_W (or m_Z) the effective Lagrangian is as good as the fundamental Lagrangian of the SM.

Consider now the process $\nu(p) + \gamma(k) \to \nu(p) + \gamma(k)$. The neutrino has no electric charge and the interaction between neutrino and photons takes place only at the level of loop diagrams. They are numerous and a bit complicated. For small momenta $(pk/m_W^2) \ll 1$ one can expand $\nu(p) + \gamma(k) \to \nu(p) + \gamma(k)$ amplitudes in the power series in this small parameter. The lowest terms of this expansion can be represented as a matrix element of the appropriate operators, i.e. of the particular terms of the Effective Lagrangian.

Each term of the Effective Lagrangian has to be a Lorenz-invariant combination of the gauge-invariant electromagnetic field tensor $F_{\mu\nu}$ and the left-handed neutrino field $\nu_L = \frac{1}{2}(1+\gamma_5)\nu$ and their derivatives. The operator may have high dimensions D. To preserve the correct dimension of the Lagrangian $[L] \sim [m]^4$ the coefficients in front of this operator should be proportional to the appropriate power of $(1/m)$, where m is the scale of mass that moves inside the loops. The actual calculation of the diagrams gives the numerical coefficients.

2.3. *P-even scattering amplitude*

Consider how all this works in the case of P-even $\nu\gamma$-scattering. In this case the amplitude should be the same for right-handed photons and left-handed photons. One has to construct appropriate Lorenz invariant operators from the fields $F_{\mu\nu}$ and ν_L. The simplest combination of the fields that satisfies all these conditions looks as follows:

$$L_{eff} \sim \frac{e^4}{m^4}[F_{\mu\alpha}F_{\mu\beta}]\bar{\nu}\gamma_\alpha\partial_\beta(1+\gamma_5)\nu + h.c. \ . \tag{9}$$

It has the dimension $D = 8$. The matrix element of L_{eff} for forward scattering gives the amplitude

$$T \sim \frac{e^4}{m^4}(pk)^2\epsilon(k)^*\epsilon(k) \ . \tag{10}$$

If we identify the parameter m in eq. (9) with the largest mass in the diagrams (i.e. with m_W), we reproduce the well-known result for the neutrino-photon amplitude [11] up to the numerical constant

$$T \sim G_F\alpha\frac{(pk)^2}{m_W^2}\epsilon(k)^*\epsilon(k) \ . \tag{11}$$

Of course it is important to calculate numerical coefficients. But even before any calculations from this simple exercise we have learned two lessons.

First, the non-zero term for $\gamma\nu$ scattering appears only in the second order in $(pk)^2$. It indicates immediately that any calculations that give an effect of the order $(G_F\alpha) \sim \alpha^2/m_W^2$ (such as in e.g. of Refs. [5,12]) are erroneous. It is impossible to violate the Gell-Mann theorem in the framework of the Effective Lagrangian approach. Appropriate operators do not exist.

Second, the actual calculations for the Effective Lagrangian are simplified enormously. Indeed now one expands every individual Feynman diagram up to the order of $(pk)^2$. To calculate the numerical factor one can put external momenta in remaining integrals to be equal zero, i.e. $p = 0$ and $k = 0$. For zero external momenta Feynman integration becomes trivial and the whole problem reduces to a pure algebraic (though rather boring) calculations of traces and products.

2.4. *Optical activity. P-odd scattering amplitude*

Now let us come back to P-odd effects in $\nu\gamma$ scattering and find appropriate operators in L_{eff} responsible for optical activity. The simplest Lagrangian that depend on P-odd combinations of photons amplitudes looks as follows

$$L_{eff} \sim \frac{1}{m^4}[F_{\mu\alpha}\tilde{F}_{\mu\beta}][\bar{\nu}\gamma_\alpha\partial_\beta(1+\gamma_5)\nu] + h.c. \ , \tag{12}$$

where $\tilde{F}_{\mu\nu} = \frac{1}{2}\epsilon_{\mu\nu\alpha\beta}F_{\alpha\beta}$.

The surprise is that this operator does not work in our case. Indeed the refraction index is proportional to the photon forward scattering amplitude. One can check that the matrix element of $F_{\mu\alpha}\tilde{F}_{\mu\beta}$ between photons with the same momenta and polarization (forward scattering) is identically zero. Thus the operators of $D = 8$ do not contribute into P-odd forward scattering and the effect is zero in $(pk)^2/m^4$ order!

Now we have to look for operators of higher dimension $D = 10$. One of these operators looks like

$$L_{eff} \sim \frac{1}{m^6}[F_{\mu\alpha}(\partial_\gamma\tilde{F}_{\mu\beta})][\bar{\nu}\gamma_\alpha\partial_\beta\partial_\gamma(1+\gamma_5)\nu] + h.c. \ . \tag{13}$$

With this L_{eff} the forward scattering amplitude of a photon of momentum k from a neutrino of momentum p is equal to

$$T = C(e^4/8\pi^2)(pk/m^2)^2\epsilon_{\mu\nu\alpha\beta}\epsilon_\mu(k)\epsilon^*_\nu(k)(p_\alpha k_\beta/m^2) \ . \tag{14}$$

This amplitude has different contributions to left-handed and to right-handed photon scatterings: $T_{LL} = -T_{RR}$.

2.5. *One-loop calculations of the P-odd effect*

The actual calculation of the coefficient C has been done in one loop-approximation in [8] with the results

$$T = C(e^4/8\pi^2 s^2)(pk/m_W^2)^2\epsilon_{\mu\nu\alpha\beta}\epsilon_\mu(k)\epsilon^*_\nu(k)(p_\alpha k_\beta/m_W^2) \ , \tag{15}$$

where

$$C = 4/3(\ln(m_W^2/m^2) - 11/3) \tag{16}$$

in the first reference in [8] and

$$C = 4/3(\ln(m_W^2/m^2) - 8/3) \tag{17}$$

in the second one [8]. The reason for that discrepancy is unknown and it would be interesting to understand whether there is a correct one-loop result.

The message of this talk is that for a P-odd effect there is an enormous enhancement factor in two-loop approximation. Thus any one-loop results (correct or erroneous) are irrelevant.

3. Two-loop calculations

3.1. *P-odd effect*

The physical reason for the dominance of the two-loop diagrams over the one-loop is simple.

To escape Gell-Mann's restriction one needs non-local interactions in order to include higher orbital momenta of the $\nu\bar{\nu}$ pair into an annihilation process. In the one loop approximation the only source of non-locality is the W boson exchange. Thus the one-loop probability amplitude $T^{(1)}$ for a neutrino pair to have nonzero orbital moment is $\sim p/m_W$. The factor $1/m_W$ measures the shortest separation of two neutrinos in one-loop approximation.

In the two-loop approximation there is a set of diagrams in which two neutrinos are emitted at a separation of the order of the electron Compton wavelength. This is due to $e^-e^+\nu$ in the intermediate state (see Fig. 3).

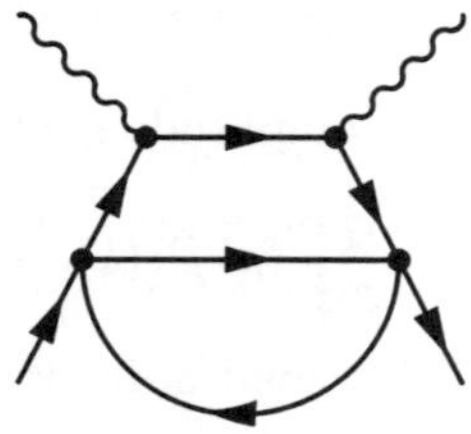

Figure 3. The sample of two-loop contribution into $\gamma\nu$ scattering

Thus in two-loop approximation the amplitude $T^{(2)}$ for a neutrino pair to have nonzero orbital moment is $\sim p/m_e$. Moving to the next order in electro-weak interaction we loose a small factor $\alpha/2\pi$, but win a large factor (m_W^2/m_e^2)! The net effect is

$$(T^{(2)}/T^{(1)}) \sim (\alpha/2\pi)(m_W^2/m_e^2) \sim 10^7. \quad (18)$$

The actual calculation is rather lengthy. The result is

$$T^{(2)} = (13/27)(G^2e^2/64\pi^4)(g_V^2 + g_A^2)(pk)^2\epsilon_{\mu\nu\alpha\beta}\epsilon_\mu(k)\epsilon_\nu^*(k)(p_\alpha k_\beta/m_e^2)\ . \quad (19)$$

The enhancement factor is

$$T^{(2)}/T^{(1)} = (\alpha/64\pi sin^2\theta_W)(m_W^2/m_e^2)(13/27)(g_V^2 + g_A^2)/C \sim 10^5\ , \quad (20)$$

where C is the one-loop coefficient from eq. (15). We have lost two orders of magnitude compared with the naive estimate. This happened mainly due to the large logarithmic factor in the one-loop coefficient C. Still, the effect is very large $\sim 10^5$!

3.2. *Two-loop estimates of P-even effect*

It is interesting to understand whether a similar enhancement factor appears for the main P-even amplitude, e.g. for the cross section of $\nu\bar{\nu}$ annihilation into two photons. The answer is negative.

The arguments are the following. Consider two-loop amplitudes with $(e^+e^-\nu)$ in the intermediate state between two external neutrino vertices. In local four-fermion approximation we expect that these diagrams are proportional to $G_F^2\alpha/m_e^2$. Thus to preserve the correct dimension in an effective Lagrangian, we need an operator of dimension $D = 10$. The appropriate effective Lagrangian is

$$L_{eff} \sim \frac{G_F^2\alpha}{m_e^2}[F_{\mu\alpha}(\partial_\gamma F_{\mu\beta})][\bar{\nu}\gamma_\alpha\partial_\beta\partial_\gamma(1+\gamma_5)\nu] + h.c. \quad (21)$$

For this L_{eff} the scattering amplitude is of the third order in (pk)

$$T \sim C\frac{G_F^2\alpha}{m_e^2}(pk)^3\epsilon(k)\epsilon^*(k). \quad (22)$$

Thus for P-even scattering second order loops give a correction of the order $(G_F M_W^2)(pk/m_e^2)$, i.e. a small correction to the one-loop result.

4. Numerical estimates

In spite of a huge missing factor 10^5 in the one-loop estimates of optical activity of the neutrino sea [13] we find that the physical effect is still tiny. It is rather unlikely that direct measurements of such a small rotation of the plane of polarization of light when it travels in the intergalactic space are ever possible. But may be correct estimates of the P-odd effect give hints for more subtle experiments. In any case it is nice to have a correct estimate.

5. Summary

To conclude:

If there is a neutrino sea, the Universe is slightly birefringent.

Acknowledgments

I would like to thank the organizers of "Gribov-75" Conference and especially Julia Nyíri, Yuri Dokshitzer, Peter Levai and Yulik Khriplovich for invitation, for their hospitality and for the excellent conference.

This research was partly supported by RFBR grant 05-02-17203.

Bibliography

1. G. Karl and V. Novikov, *Pis'ma v ZhETF* **81**, 313 (2005).
2. J. Royer, *Phys. Rev.* **174**, 1719 (1968).
3. S. Coleman and S. Glashow, *Phys. Rev.* **D59**, 116008 (1999);
   R. Jackiw and V.A. Kostelecky, *Phys. Rev. Lett.* **82**, 3572 (1999) and references therein.
4. G. Karl, *Canadian Journal of Physics* **54**, 568 (1976).
5. M. Gell-Mann, *Phys. Rev. Lett.* **6**, 70 (1961).
6. J.N. Clarke, P.P. Kronberg and M. Simard-Normandin, *Mon. Not. Roy.* **190**, 205 (1980).
7. J.N. Clarke, G. Karl and P.J.S. Watson, *Canadian Journal of Physics*, **60**, 1561 (1982).
8. G. Karl and V. Novikov, hep-ph/0009012;
   A. Abbasabadi and W. Repko, *Phys. Rev.* **D64**, 113007 (2001); *ibid.* **D67**, 073018 (2003).
9. S. Mohanty, J.F. Nieves and P. Pal, *Phys. Rev.* **D58**, 093007 (1998) 093007.
10. L.D. Landau, *Dokl. Akad. Nauk USSR* **60**, 207 (1948);
    C.N. Yang, *Phys. Rev.* **77**, 242 (1950).
11. M.J. Levine, *Nuovo Cimento* **A48**, 67 (1967).
12. S.G. Matinyan and N.N. Tsilosuni, *Zurn. Eksp. Teor. Fiz.* **41**, 1691 (1961); *Sov. Phys. JETP* **14**, 1195 (1961).
13. A. Abbasabadi and W. Repko, *Phys. Rev.* **D67**, 073018 (2003).

THE VIRUS OF RELATIVISTIC MASS IN THE YEAR OF PHYSICS*

L.B. OKUN

ITEP, Moscow, 117218, Russia

The "famous formula" $E = mc^2$ and the concept of "relativistic mass" increasing with velocity, which follows from it, are historical artifacts, contradicting the basic symmetry of Einstein's Special Relativity, the symmetry of 4-dimensional space-time. The relation discovered by Einstein is not $E = mc^2$, but $E_0 = mc^2$, where E_0 is the energy of a free body at rest introduced by Einstein in 1905. The source of the longevity of the "famous formula" is the irresponsible attitude of relativity theory experts to the task of explaining it to the non-experts.
The notion of "relativistic mass" presents a kind of pedagogical virus which very effectively infects new generations of students and professors and shows no signs of decline. Moreover in the Year of Physics it threatens to produce a real pandemia.

Before writing my first article against the "Einstein famous equation $E = mc^2$" I mentioned in 1987 this intention to Volodya Gribov in one of our daily conversations. We had a complete unanimity on the issue of physics. But Volodya was a better friend of mine than I myself. And his understanding of life was better than mine. Therefore he tried to dissuade me from wasting my time on a fight against an obviously wrong cliché, which I would inevitably loose. I discarded his wise advice and wrote my first two papers on the concept of mass [1,2].

The subject seemed important to me because it concerned the proper teaching of special relativity at high schools, colleges and universities and explaining its genuine meaning to a wide audience of non-physicists, the so-called "pedestrians" in popular science magazines and books.

The task looked also not absolutely formidable because a consistent presentation of relativity existed for a long time in the world-wide accepted text-book by Landau and Lifshitz [3], which was the basis of my own understanding, and in some other text-books.

*The paper was written for the Gribov Memorial Volume dedicated to the 75th birthday of V.N. Gribov.

The existence of countless texts, in which the essence of relativity was mutilated (or semi-mutilated) had two sides. On one hand, it looked discouraging, especially because among the authors of these texts there were many famous physicists, the fathers and greatest authorities of modern physics. On the other hand, it was a challenge. So I tried to explain clearly to the readers the beauty of four-dimensional space-time approach and the ugliness and inconsistency of "relativistic mass", an illegitimate child of relativistic and non-relativistic equations.

My optimism had increased when in 1992 Taylor and Wheeler in the second edition of the influential and popular "*Spacetime Physics*" [4] included a "Dialog: Use and Abuse of the Concept of Mass", in which they supported my articles [1,2]. A copy of this book is in my bookcase with a postcard sent to me in October 1991 by John Archibald Wheeler. The postcard has a photo of the famous Albert Einstein Memorial in front of the building of the National Academy of Sciences, Washington, DC. The bronze sculpture of Einstein includes a copy book with $E = mc^2$ on an open page.

Since that time I received hundreds of letters from physicists (both professors and students) stating their adherence to the four-dimensional formulation of relativity and to the Lorentz invariant concept of mass. In a few cases I helped the authors to correct erroneous explanations of the concept of mass in preparing new editions of their textbooks. However the number of proponents of relativistic mass seemed not to decrease.

A leading role in promoting the relativistic mass have played the books by Max Jammer [5,6]. Especially aggressive the proponents of relativistic mass became in connection with the World Year of Physics, which marks the 90th anniversary of fundamental articles published by Einstein in 1905.

The campaign started by the September 2004 issue of "*Scientific American*", full with "relativistic mass" equal to $m_0/\sqrt{1 - v^2/c^2}$, where m_0 is rest mass, and "the most famous equation $E = mc^2$". A letter to the editors, defending the four-dimentional approach and invariant mass had been rejected by the editor G. Collins who in April 2005 wrote: "Most important, we believe that tackling the issue head-on in the manner you and your coauthors want in the letters column of *Sci. Am.* would be very confusing to our general audience and it would make the subject seen all the more mysterious and impenetrable to them". Thus to avoid "head-on" collision of correct and false arguments the editors of *Sci. Am.* preferred to hide from the readers the correct viewpoint.

P. Rodgers – the Editor of European "*Physics World*" wrote in January 2005 in editorial [7]: "... $E = mc^2$ led to the remarkable conclusion that mass and energy are one and the same". Unlike G. Collins, P. Rodgers published a letter criticizing this statement and partly agreed with the criticism [8].

In September 2005 the bandwagon of relativistic mass was joined by "*The New York Times*", which published an article by B. Green [9].

The journalists were supported by renowned scientists, such as R. Penrose, who in a new thousand pages thick book had written [10]:

"In a clear sense mass and energy become completely equivalent to one another according to Einstein's most famous equation $E = mc^2$."

How many students, teachers and journalists will be infected by this sentence? How many readers had been infected by the famous book by S. Hawking [11], the second edition of which appeared in 2005? On the very first page of it Hawking wrote:

"Someone told me that each equation I included in the book would halve the sales. I therefore resolved not to have any equations at all. In the end, however, I did put in one equation, Einstein's famous equation $E = mc^2$. I hope that this will not scare off half of my potential readers."

I am sure that the usage of $E = mc^2$ had doubled the sales of his book, the buyers being attracted by the famous brand. But is it possible to estimate the damage done to their understanding of relativity theory and to the general level of the literature on relativity incurred by this case of spreading the virus.

Two recent preprints by Gary Oas [12,13] written in the framework of Educational Program for Gifted Youth at Stanford University were devoted to the use of relativistic mass. The author "urged, once again, that the use of the concept at all levels to be abandoned" [12]. The manuscript has been submitted for publication to the "*Americal Journal of Physics*", but was rejected as being "too lengthy" (it contains 12 pages!). A lengthy bibliography (on 30 pages) of books referring to special and/or general relativity is provided in Ref. [13] to give a background for discussions of the historical use of the concept of relativistic mass. It is easy to forecast the aggressive reaction of the virus infected community to this attempt to cure it.

I am grateful to the grant of Russian Ministry of Education and Science No. 2328.2003.2 for partial support of this work.

References

1. L.B. Okun, *Soviet Physics-Uspekhi* **32** (July 1989) p. 629-638.
2. L.B. Okun, *Physics Today*, June 1989, p. 31-36;
   http://www.physicstoday.org/vol-42/iss-6/vol42no6p31_36.pdf;
   L.B. Okun, *Physics Today*, May 1990, p. 147.
3. L.D. Landau and E.M. Lifshitz, *"The Classical Theory of Fields"*, translated from the second Russian ed., Pergamon, New York, 1951.
4. E.F. Taylor and J.A. Wheeler, *"Spacetime Physics"*, 2nd ed. Freeman and Company. New York, 1992, p. 246-252.
5. M. Jammer, *"Concepts of Mass in Classical and Modern Physics"*, Harvard, 1961.
6. M. Jammer, *"Concepts of Mass in Contemporary Physics and Philosophy"*, Princeton, 2000.
7. P. Rodgers, *Physics World* **18**, No. **1**, 13 (2005).
8. P. Rodgers, *Physics World* **18**, No. **10**, 20 (2005).
9. B. Green, *"That Famous Equation and You"*, *The New York Times*, Sept. 30, 2005.
10. R. Penrose, *"The road to reality. A complete guide to the laws of the universe"*, A. Knopf, New York, 2004, p. 434.
11. S. Hawking, *"A Brief History of Time"*, New York, 1988.
12. G. Oas, *"On the abuse and use of relativistic mass"*, arXiv: physics/0504110.
13. G. Oas, *"On the Use of Relativistic Mass in Various Published Works"*, arXiv: physics/0504111.

QUANTIZED BLACK HOLES, THEIR SPECTRUM AND RADIATION

I.B. KHRIPLOVICH
Budker Institute of Nuclear Physics
630090 Novosibirsk, Russia
E-mail: khriplovich@inp.nsk.su

Under quite natural general assumptions, the following results are obtained. The maximum entropy of a quantized surface is demonstrated to be proportional to the surface area in the classical limit. The general structure of the horizon spectrum is found. The discrete spectrum of thermal radiation of a black hole fits the Wien profile. The natural widths of the lines are much smaller than the distances between them. The total intensity of the thermal radiation is estimated. In the special case of loop quantum gravity, the value of the Barbero – Immirzi parameter is found. Other values for this parameter, obtained under an additional assumption that the horizon is described by a U(1) Chern – Simons theory, are demonstrated to be in conflict with the firmly established holographic bound.

The subject of my talk is unrelated to the main field of interests by Gribov: quarks, confinement, and strong interactions. It is related however to the brilliant prediction, made by Gribov in 1971 or 1972: black holes radiate. He precisely formulated this conclusion, and substantiated it with simple and clear physical arguments, 2 – 3 years before the famous paper by Hawking. One can only regret that Gribov did not publish this result, apparently he considered it self-evident.

1. Introduction

The idea of quantizing the horizon area of black holes was put forward by Bekenstein in the pioneering article [1]. He pointed out that reversible transformations of the horizon area of a nonextremal black hole found by Christodoulou and Ruffini [2,3] have an adiabatic nature. Of course, the quantization of an adiabatic invariant is perfectly natural, in accordance with the correspondence principle.

Later, the quantization of black holes was discussed by Mukhanov [4] and Kogan [5]. In particular, Kogan was the first to investigate this problem within the string approach.

Once the idea of area quantization is accepted, the general structure of the quantization condition for large quantum numbers N gets obvious, up to an overall numerical constant (written usually as $8\pi\,\gamma$). It should be [6]

$$A = 8\pi\,\gamma\, l_p^2\, N. \tag{1}$$

Indeed, the presence of the Planck length squared $l_p^2 = k\hbar/c^3$ is only natural in this quantization rule. Then, for the horizon area A to be finite in the classical limit, the power of N should be the same as that of $\hbar$ in l_p^2. This argument can be checked by considering any expectation value in quantum mechanics, nonvanishing in the classical limit.

It is worth mentioning that, contrary to widely spread beliefs, there are no compelling reasons to believe that the black hole spectrum (1) is equidistant.

A quite popular argument in favor of its equidistance is as follows [7] (see also Refs. [8,9]). On the one hand, the entropy S of horizon is related to its area A through the Bekenstein – Hawking relation

$$A = 4 l_p^2 S\,. \tag{2}$$

On the other hand, the entropy is nothing but $\ln g(n)$, where the statistical weight $g(n)$ of any quantum state n is an integer. In [7] this circumstance is used too naively, which results after simple reasoning not only in the equidistant spectrum (1), but also in the following allowed values for the numerical factor therein:

$$8\pi\gamma = 4\ln k, \quad k = 2,\, 3,\, \ldots\,.$$

It is well known, however, that the statistical weight being an integer, has no consequences for the entropy of macroscopic objects. A concrete error in arguments of [7] is pointed out in [10] (see also [11]).

There is also an observation, usually considered as an argument in favor of the equidistant area spectrum. It is due to Bekenstein, who demonstrated [12] that quantum effects result in the following lower bound on the change of the horizon area ΔA under an adiabatic process:

$$(\Delta A)_{\min} = \xi l_p^2\,, \tag{3}$$

where ξ is a numerical factor reflecting "the inherent fuzziness of the uncertainty relation"[a]. We will discuss this bound later and demonstrate that in fact it is a strong argument in favor of the discrete spectrum of the black hole radiation.

[a] As it should be expected, the right-hand-side of formula (3), corresponding to violation of adiabatic invariance, is proportional to $\hbar$, together with the Planck length squared l_p^2.

2. Structure of Quantized Area of Black Hole

Quantization condition (1) can be interpreted naturally as follows. The whole horizon area A consists of patches of typical size $\sim l_p^2$. Each of them can be characterized by a quantum number j, and the contribution a of a patch to the area depends on j, $a = a(j)$. Besides, a patch can possess a quantum number m, such that a is independent of it. (Of course, both j and m may refer in principle not to a single quantum number each, but to sets of them.) Then, formula (1) can be rewritten as

$$A = 8\pi\gamma\, l_p^2 \sum_{jm} a(j)\, \nu_{jm}\,, \tag{4}$$

where ν_{jm} is the number of patches with given j and m.

To derive general relations for the "occupation numbers" ν_{jm}, we will use the Bekenstein – Hawking relation (2) and the so-called holographic bound formulated in [13,14,15]. According to this bound, the entropy S of any spherical nonrotating system confined inside a sphere of area A is bounded by the relation

$$S \le \frac{A}{4l_p^2}\,, \tag{5}$$

with the equality attained only for a system which is a black hole.

A simple intuitive argument confirming this bound is as follows [15]. Let us allow the discussed system to collapse into a black hole. During the collapse the entropy increases from S to S_{bh}, and the resulting horizon area A_{bh} is certainly smaller than the initial confining one A. Now, with the account for the Bekenstein – Hawking relation (2) for a black hole, we arrive, through the obvious chain of (in)equalities

$$S \le S_{bh} = \frac{A_{bh}}{4l_p^2} \le \frac{A}{4l_p^2}\,,$$

at the discussed bound (5).

It should be pointed out that at least for spherically symmetric black holes, the holographic bound has been checked by careful analysis of various physical situations, and therefore its validity is firmly established.

The result (5) can be formulated otherwise. Among the spherical surfaces of a given area, it is the surface of a black hole horizon that has the largest entropy[b].

[b]Even irrespective of the holographic bound, the idea that the entropy of a black hole is maximum, looks quite natural and was used, e.g. in a model of quantum black hole originating from dust collapse [16].

We will consider now the "microcanonical" entropy S of a quantized surface defined as the logarithm of the number of states of this surface for a fixed area A (instead of fixed energy in common problems). Obviously, this number of states K depends on the assumption concerning the distinguishability of the patches. So, let us discuss first of all which of *a priori* possible assumptions is reasonable here from the physical point of view [10] (see also [11]).

One possibility, which at the first glance might look quite appealing, is that of complete indistinguishability of patches. It means that no permutation of any patches results in new states. Under this assumption, the total number of states created by $\nu_j = \sum_m \nu_{jm}$ patches of a given j is

$$K(j) = \frac{(\nu_j + g(j) - 1)!}{\nu_j!\,(g(j) - 1)!}\,; \tag{6}$$

here $g(j)$ is the total number of possible values of m for a given j [c]. Under the natural assumption $\nu_j \gg g(j)$, the partial contributions

$$s_j = \ln K_j = g(j)\ln \nu_j$$

to the black hole entropy $S = \sum_j s_j$ are parametrically smaller than the corresponding partial contributions

$$a(j)\nu_j$$

to $A/(4l_p^2)$, in obvious conflict with the Bekenstein – Hawking relation (2). Thus, with indistinguishable patches of the same j, one cannot make the entropy of a black hole proportional to its area (see also [18]).

Let us consider now the opposite assumption, that of completely distinguishable patches. In this case the total number of states is

$$K = \nu\,!\,, \quad \nu = \sum_j \nu_j = \sum_{jm} \nu_{jm}\,,$$

with the microcanonical entropy

$$S = \nu \ln \nu\,.$$

Obviously, here the maximum entropy for fixed $A \sim \sum_j a(j)\,\nu_j$ is attained with all $a(j)$ being as small as possible. Then, in the classical limit $\nu \gg 1$, the entropy of a black hole grows faster than its area: $A \sim \nu$, but

[c] Perhaps, the simplest derivation of expression (6) is to note that it is in fact the number of ways of distributing ν_j identical particles into $g(j)$ boxes. Then, the line of reasoning presented in [17], § 54, results in formula (6).

$S = \nu \ln \nu \sim A \ln A$. Thus, the assumption of complete distinguishability is in conflict with the holographic bound, and therefore should be discarded[d].

We go over to the third conceivable scheme, which is quite popular (see, for instance, Ref. [18]). According to it, the total number of states is

$$K = \prod_j g(j)^{\nu_j}, \tag{7}$$

with the entropy of the horizon surface

$$S = \sum_j \nu_j \ln g(j). \tag{8}$$

It can be easily demonstrated that this scheme corresponds to the following assumptions on the distinguishability of patches:

nonequal j,	any m	$\longrightarrow$	indistinguishable;
equal j,	nonequal m	$\longrightarrow$	distinguishable;
equal j,	equal m	$\longrightarrow$	indistinguishable.

The combination of the first two of them looks strange and unnatural (except the special case when only a single value of j is allowed for all patches).

Thus, we believe that the only reasonable set of assumptions on the distinguishability of patches, which may result in acceptable physical predictions (i.e. may comply both with the Bekenstein – Hawking relation and with the holographic bound) is as follows:

nonequal j,	any m	$\longrightarrow$	distinguishable;
equal j,	nonequal m	$\longrightarrow$	distinguishable;
equal j,	equal m	$\longrightarrow$	indistinguishable.

In this scheme, the number of states of the horizon surface, for a given number ν_{jm} of patches with quantum numbers j and m, is obviously

$$K = \nu\,! \prod_{jm} \frac{1}{\nu_{jm}\,!}\,, \quad \text{where} \quad \nu = \sum_j \nu_j\,, \quad \nu_j = \sum_m \nu_{jm}\,, \tag{9}$$

[d]There is no disagreement between this conclusion of ours and that of [18,19,20]: what is called complete distinguishability therein, corresponds to the last option considered by us below, which is the only reasonable one in our opinion.

and the corresponding entropy equals

$$S = \ln K = \ln(\nu\,!) - \sum_{jm} \ln(\nu_{jm}\,!)\,. \tag{10}$$

The structures of the last expression and of formula (4) are so different that in a general case the entropy certainly cannot be proportional to the area. However, this is the case for the maximum entropy in the classical limit.

In this limit, with all effective "occupation numbers" large, $\nu_{jm} \gg 1$, we use the Stirling approximation, so that the entropy is

$$S = \nu \ln \nu - \sum_{jm} \nu_{jm} \ln \nu_{jm}\,. \tag{11}$$

We calculate its maximum for a fixed area A, i.e. for a fixed sum[e]

$$N = \sum_{jm}^{\infty} a(j)\, \nu_{jm} = \text{const}\,. \tag{12}$$

The problem reduces to the solution of the system of equations

$$\ln \nu - \ln \nu_{jm} = \mu\, a(j)\,, \tag{13}$$

where μ is the Lagrange multiplier for the constraining relation (12). These equations can be rewritten as

$$\nu_{jm} = \nu\, e^{-\mu\, a(j)}\,, \tag{14}$$

or

$$\nu_j = \nu\, g(j)\, e^{-\mu\, a(j)}\,. \tag{15}$$

Now we sum expressions (15) over j, and with $\sum_j \nu_j = \nu$ arrive at the equation for μ:

$$\sum_j g(j)\, e^{-\mu\, a(j)} = 1. \tag{16}$$

Strictly speaking, the summation in formula (16) extends not to infinity, but to some j corresponding to the maximum contribution $a_{\max}$ to the area. The value of $a_{\max}$ follows from the obvious condition: none of ν_{jm} should be less than unity. Then, for $\nu \gg 1$ equation (14) gives

$$a_{\max} = \frac{\ln \nu}{\mu}\,. \tag{17}$$

[e]These calculations are a rather obvious generalization of those presented by us previously in [10,11,21,22] (and quoted below in section 4) for definite structures of $a(j)$ and $g(j)$ predicted in loop quantum gravity (see section 5).

It is well-known that the Stirling approximation for $n!$ has a reasonably good numerical accuracy even for $n = 1$. Due to it, formula (17) for $a_{\rm max}$ is not just an estimate, but has reasonably good numerical accuracy as well.

On the other hand, multiplying equation (13) by ν_{jm} and summing over jm, we arrive, with the constraint (12), at the following result for the maximum entropy for given N:

$$S_{\rm max} = \mu\, N = \frac{\mu}{8\pi\gamma l_p^2}\, A. \tag{18}$$

Thus, equation (4) for the quantized area can be rewritten as

$$A = 8\pi\gamma\, l_p^2\, \nu \sum_j g(j)\, a(j)\, e^{-\mu a(j)}\,, \tag{19}$$

where $\gamma = \mu/(2\pi)$, and the value of μ is found from equation (16).

3. Quantization of Mass of Rotating Black Hole

When discussing the radiation spectrum of quantized black holes, one should take into account the selection rules for angular momentum. Obviously, radiation of any particle with nonvanishing spin is impossible if both initial and final states of a black hole are spherically symmetric. Therefore, to find the radiation spectrum, the quantization rule for the mass of a Schwarzschild black hole should be generalized to that of a rotating Kerr black hole.

To derive the quantization rule for Kerr black hole, we come back to the thought experiment analyzed in [2,3]. Therein, under the adiabatic capture of a particle with an angular momentum j, the angular momentum J of a rotating black hole changes by a finite amount j, but the horizon area A does not change. Of course, under some other variation of parameters it is the angular momentum J that remains constant. In other words, we have here two independent adiabatic invariants, A and J, for a Kerr black hole with a mass M.

Such a situation is quite common in ordinary mechanics. For instance, the energy of a particle with mass m, bound in the Coulomb field $U(r) = -\alpha/r$, is

$$E = -\frac{m\alpha^2}{2\,(I_r + I_\phi)^2}\,, \tag{20}$$

where I_r and I_ϕ are adiabatic invariants for the radial and angular degree of freedom, respectively. Of course, the energy E is in a sense an adiabatic invariant also, but it is invariant only with respect to those variations of parameters under which both I_r and I_ϕ remain constant (or at least their sum). As to quantum mechanics, in it formula (20) goes over into

$$E = -\frac{m\alpha^2}{2\,\hbar^2\,(n_r + 1 + l)^2}\,, \tag{21}$$

where n_r and l are the radial and orbital quantum numbers, respectively.

This example prompts the solution of the quantization problem for a Kerr black hole. It is conveniently formulated in terms of the so-called irreducible mass M_{ir} of a black hole, related by definition to its horizon radius r_h and area A as follows:

$$r_h = 2kM_{ir} \quad \text{and} \quad A = 16\pi k^2 M_{ir}^2\,. \tag{22}$$

Together with the horizon area A, the irreducible mass is an adiabatic invariant. In accordance with (4) and (12), it is quantized as follows:

$$M_{ir}^2 = \frac{1}{2}\,m_p^2\gamma\,N\,, \tag{23}$$

where $m_p^2 = \hbar c/k$ is the Planck mass squared.

Of course, for a Schwarzschild black hole M_{ir} coincides with its ordinary mass M. However, for a Kerr black hole the situation is more interesting. Here

$$M^2 = M_{ir}^2 + \frac{J^2}{r_h^2} = M_{ir}^2 + \frac{J^2}{4k^2M_{ir}^2}\,, \tag{24}$$

where J is the internal angular momentum of a rotating black hole.

Now, with the account of equation (23), we arrive at the following quantization rule for the mass squared M^2 of a rotating black hole:

$$M^2 = \frac{1}{2}\,m_p^2\left[\gamma N + \frac{J(J+1)}{\gamma N}\right]. \tag{25}$$

Obviously, as long as a black hole is far away from an extremal one, i.e. while $\gamma N \gg J$, one can neglect the dependence of M^2 on J, and the angular momentum selection rules have practically no influence on the radiation spectrum of a black hole.

As to the mass and irreducible mass of a charged black hole, they are related as follows:

$$M = M_{ir} + \frac{q^2}{2r_h}\,; \tag{26}$$

here q is the charge of the black hole. This formula has a simple physical interpretation: the total mass (or total energy) M of a charged black hole consists of its irreducible mass M_{ir} and of the energy $q^2/2r_h$ of its electric field in the outer space $r > r_h$.

With $r_h = 2kM_{ir}$, relation (26) can be rewritten as

$$M^2 = M_{ir}^2 + \frac{q^4}{16k^2 M_{ir}^2} + \frac{q^2}{2k}. \tag{27}$$

Thus, for a charged black hole M^2 is quantized as follows:

$$M^2 = \frac{1}{2} m_p^2 \left[\gamma N + \frac{q^4}{4\gamma N} + q^2 \right]. \tag{28}$$

In fact, relations of this type (even in a more general form, for Kerr – Newman black holes, both charged and rotating) were presented already in the pioneering article [1], though with the equidistant quantization rule for M_{ir}^2, i.e. for the horizon area (see also [12]). More recently, the conclusion that the mass of a quantized black hole should be expressed via its quantized area and angular momentum, was made in the approach based on the notion of so-called isolated horizons [23,24].

I do not mention here those, quite numerous, attempts to quantize rotating and charged black holes which result in weird quantization rules for $\hat{J}^2$ and $e^2/\hbar c$.

4. Radiation Spectrum of Quantized Black Hole

It follows from expression (25) that for a rotating black hole the radiation frequency ω, which coincides with the loss ΔM of the black hole mass, is

$$\omega = \Delta M = T\mu\,\Delta N + \frac{1}{2kM}\,\frac{J+1/2}{\gamma N}\,\Delta J = T\Delta S \,+\, \Omega\,\Delta J\,, \tag{29}$$

where ΔN and ΔJ are the losses of the area quantum number N and of the angular momentum J, respectively; $\Delta S = \mu\,\Delta N$, and the effective angular velocity is $\Omega = (J+1/2)/(Mr_h^2)$.

For a charged black hole loosing charge Δq, the radiation frequency is

$$\omega = \Delta M = T\mu\,\Delta N + \frac{q}{r_h}\,\Delta q = T\Delta S + \phi\,\Delta q\,. \tag{30}$$

Obviously, equations (29), (30) are in fact common thermodynamic relations.

We will be interested mainly in the first, temperature terms in (29) and (30), dominating for black holes far from the extremal regime, i.e. for $J \ll \gamma N$, or $q^2 \ll 2\gamma N$. The natural assumption is that the temperature radiation occurs when a patch with a given value of j disappears, which means that

$$\Delta N_j = a(j) \quad \text{and} \quad \omega_j = T\mu\, a(j)\,. \tag{31}$$

Thus we arrive at the discrete spectrum with a finite number of lines. Their frequencies start at $\omega_{\text{min}} = T\mu\, a_{\text{min}}$, where a_{min} is the minimum value of $a(j)$, and terminate at $\omega_{\text{max}} = T \ln \nu$ (we recall here that $a_{\text{max}} = \ln \nu/\mu$). Thus, the number of lines is not so large, $\sim 10^2$, if the mass of black hole is comparable to that of the Sun. However, due to the exponential decrease of the radiation intensity with ω (see below), the existence of ω_{max} and finite number of lines are not of much importance.

To substantiate the made assumption, we come back to the lower bound (3) on the change of the horizon area under an adiabatic capture of a particle. The presence of the gap (3) in this process means that this threshold capture effectively consists of the increase by unity of the occupation number ν_{jm} corresponding to a_{min}. If the capture were accompanied by a reshuffle of few occupation numbers, the change of the area could be made in general as small as one wishes[f].

It is only natural to assume that in the radiation process as well, changing few occupation numbers, instead of one, is at least strongly suppressed. In this way we arrive at equations (31).

Our next assumption, at least as natural as this one, is that the probability of radiation of a quantum with frequency ω_j is proportional to the occupation number ν_j. Correspondingly, the radiation intensity I_j at this frequency ω_j is proportional to $\nu_j\, \omega_j$:

$$I_j \sim \nu_j\, \omega_j \sim \nu\, g(j)\, \omega_j\, e^{-\omega_j/T}. \tag{32}$$

We compare now this expression with the intensity of the black-body radiation in the Wien limit $\omega/T \gg 1$,

$$I(\omega)\, d\omega = A\, \frac{\omega^3}{4\pi^2}\, e^{-\omega/T} d\omega, \tag{33}$$

where A is the area of a spherical black body. First of all, our relation (32) for I_j reproduces naturally the exponential factor of the Wien spectrum. To reproduce the Wien profile completely, we have to supplement relation (32) with an obvious additional factor, a sort of "oscillator strength". Thus

[f]Except the case when $a(j)$ is a linear function of j, and, correspondingly, the area spectrum is equidistant. Generally speaking, this case cannot be excluded.

we arrive at the final expression for the discrete radiation spectrum of a black hole:

$$I_j\,\Delta\omega_j = I_j\,\frac{\partial a(j)}{\partial j}\,\Delta j = A\,T^4\,\frac{1}{4\pi^2}\,\mu^4 a^3(j)\,\frac{\partial a(j)}{\partial j}\,e^{-\mu\,a(j)}\,\Delta j\,. \tag{34}$$

Here Δj is the difference between two successive values of j. For instance, if only integer values of j are admissible, $\Delta j = 1$; if half-integer values are possible as well, $\Delta j = 1/2$.

One should mention that it was argued long ago [7], for the case of equidistant horizon quantization, that the discrete thermal radiation spectrum of a black hole should fit the Wien profile.

Of course, the discrete spectrum (34), together with the Wien spectrum (33), refers, strictly speaking, to high frequencies $\omega_j/T = \mu\,a(j) \gg 1$. However, one may hope that even for $\mu\,a_{\rm min} \sim 1$ the asymptotic value given by (34) does not differ too much from the true one.

On the other hand, when calculating here the radiation intensity, we have, in principle, to introduce the so-called grey factor, which is absent of course in the thermal radiation of a common black body. However, under the natural assumption $\omega_{\rm min}/T = \mu\,a_{\rm min} \sim 1$, the grey factor correction should not be as essential quantitatively even for the first line, and so much the more for higher ones.

Now, to estimate the total radiation intensity $I = \sum_j I_j\,\Delta\omega_j$, we approximate the sum over j by integral and thus obtain

$$I \simeq A\,T^4\,\frac{1}{4\pi^2}\int_{\mu a_{\rm min}}^{\infty} dx\,x^3 e^{-x}\,. \tag{35}$$

The integral obtained is almost independent of its lower limit $\mu a_{\rm min}$ if this limit remains on the order of unity: the value of the integral changes from 6 to 5.14 when $\mu a_{\rm min}$ changes from 0 to 2. Therefore, under the same natural assumption $\mu a_{\rm min} \sim 1$ the total radiation intensity is

$$I \approx 0.14\,A\,T^4\,. \tag{36}$$

The numerical coefficient in this expression is close to that in the total intensity of the common thermal radiation, i.e. to the Stefan-Boltzmann constant, which is $\pi^2/60 = 0.164$. This is quite natural since the Rayleigh-Jeans contribution to the total intensity, which is absent in the present spectrum, would be small anyway.

The emission probability for a quantum of frequency $\omega_j = T\mu\,a(j)$, i.e. the width of the corresponding line, is

$$\Gamma_j\,\Delta\omega_j = \frac{I_j\,\Delta\omega_j}{\omega_j} = A\,T^3\,\frac{1}{4\pi^2}\,\mu^3 a^2(j)\,\frac{\partial a(j)}{\partial j}\,e^{-\mu\,a(j)}\,\Delta j. \tag{37}$$

The ratio of this natural line width to the distance $\Delta\omega_j$ is

$$\Gamma_j = \frac{1}{16\pi^3}\,[\mu\,a(j)]^2\,e^{-\mu\,a(j)}. \tag{38}$$

With $[\mu\,a(j)]^2\,e^{-\mu\,a(j)} \leq 0.54$, this ratio is very small numerically: $\leq 10^{-3}$. Thus, the radiation spectrum of an isolated black hole is really discrete.

Equation (34) describes the thermal radiation of photons and gravitons, which have two polarizations. It applies as well to the thermal radiation of massless fermions. However, in the last case a proper account of the number of polarization states is necessary: e.g. for a two-component Dirac neutrino, with a single polarization, the numerical factor in (34) will be two times smaller.

As to the nonthermal radiation of extremal black holes, described by the terms with ΔJ and Δq in (29) and (30), these effects are due to tunneling (see relatively recent discussion of the subject, as well as detailed list of references, in [25,26]). The loss of a charge by a charged black hole is caused in fact by the Coulomb repulsion between the black hole and emitted particles with the same sign of charge. For a rotating black hole the cause is the interaction of angular momenta: particles (massless mainly), whose total angular momentum is parallel to that of a black hole, are repelled from it.

5. Black Holes in Loop Quantum Gravity

In this section we illustrate the general relations derived above with an example of a concrete model, that of loop quantum gravity (LQG) [27,28,29,30,31].

A quantized surface in LQG looks as follows. One ascribes to it a set of punctures (corresponding to patches of the previous sections). Each puncture is supplied with an integer or half-integer quantum number j:

$$j = 1/2, 1, 3/2, \ldots\,. \tag{39}$$

The projections m of these "angular momenta" run as usual from $-j$ to j. The area of a surface is

$$A = 8\pi\gamma\, l_p^2 \sum_{jm} \sqrt{j(j+1)}\,\nu_{jm}. \tag{40}$$

This is in fact a special case of the above general expressions with

$$a(j) = \sqrt{j(j+1)}, \quad g(j) = 2j+1. \tag{41}$$

It is worth mentioning here that though the area spectrum (40) is not equidistant, it is not far away from it. Indeed, even for the smallest quantum number $j = 1/2$, $\sqrt{j(j+1)}$ can be approximated by $j + 1/2$ with an accuracy 13%. And the approximation $\sqrt{j(j+1)} \approx j + 1/2$ gets better and better with growing j, i.e. spectrum (40) approaches an equidistant one more and more.

The numerical factor γ in (40) corresponds in LQG (it is called here Barbero – Immirzi parameter) to a family of inequivalent quantum theories, all of them being viable without such an input [32,33].

General relations derived above for arbitrary $a(j)$, $g(j)$ can be readily used in the present concrete case. In particular, "secular" equation (16) and its solution read now[g]

$$\sum_{j=1/2}^{\infty} (2j+1)\, e^{-\mu\sqrt{j(j+1)}} = 1\,, \quad \mu = 1.722. \tag{42}$$

Thus, the value of the Barbero-Immirzi parameter in LQG is

$$\gamma = \frac{\mu}{2\pi} = 0.274. \tag{43}$$

The minimum frequency is here

$$\omega_{\min} = \mu\, \frac{\sqrt{3}}{2}\, T = 1.491\, T\,, \tag{44}$$

One can find also in closed form corrections to the Bekenstein – Hawking relation (2). With the leading correction included, this relation looks as follows:

$$S = \frac{A}{4l_p^2} - \frac{1}{6\mu^2} \ln^3 \frac{A}{l_p^2}\,. \tag{45}$$

The existence of this $\ln^3$ correction was first pointed out in [34], though with a coefficient 1/3 instead of 1/6.

At last, the total radiation intensity of a black hole in LQG is

$$I = AT^4\, \frac{1}{8\pi^2}\, \mu^4 \sum_{j=1/2}^{\infty} [j(j+1)]^{3/2}\, e^{-\mu\sqrt{j(j+1)}} = 0.144\, AT^4. \tag{46}$$

We note that our conclusion of the discrete radiation spectrum of a black hole in LQG differs drastically from that of [35], according to which this spectrum is dense.

[g] These results were obtained by us few years ago [21] (see also [10,11,22]), and reproduced recently in [34], in somewhat different context.

Our last remark refers to the problem of Barbero – Immirzi parameter γ in LQG. The first attempts to fix its value, based on the analysis of the black hole entropy, were made in [36,37]. However, these attempts did not lead to concrete quantitative results.

Then it was argued in [38] that the horizon of a black hole should be described by a U(1) Chern – Simons theory, and characterized by punctures with quantum numbers $\pm 1/2$ only[h]. With these quantum numbers, one arrives easily at the equidistant area spectrum and at the value $\gamma_1 = \ln 2/(\pi\sqrt{3}) = 0.127$ for the Barbero – Immirzi (BI) parameter.

However, it was pointed out in [10] that this result violates the holographic bound. Indeed, common bulk states, where a surface area is given by expression (40), certainly exist in LQG. With smaller BI parameter, $\gamma_1 = 0.127 < \gamma = 0.274$, the entropy obtained in [38] is smaller than the maximum entropy corresponding to area (40). Therefore, in virtue of the firmly established (at least for spherically symmetric surfaces) holographic bound, the result of [38] cannot be correct.

Then, the result of [38] was revised in [40,41], still within the same idea of a U(1) Chern – Simons description of the horizon. Effectively, according to [40,41], one should ascribe to the punctures of the horizon arbitrary integer and half-integer j, but with only two maximum projections $\pm j$. No wonder that the equation for the BI parameter looks in [41] as

$$2 \sum_{j=1/2}^{\infty} e^{-\mu\sqrt{j(j+1)}} = 1 \tag{47}$$

with the solution $\gamma_2 = \mu_2/(2\pi) = 0.238$, instead of ours (42) with $\gamma = 0.274$ (see also the discussion of (47) in [34]). So, since $\gamma_2 = 0.238 < \gamma = 0.274$, here as well the holographic bound is violated.

The conclusion is obvious. Any restriction on the number of admissible states for the horizon in LQG, as compared to a generic quantized surface (with any j and $-j \le m \le j$), be it, for instance, the restriction to

$$j = 1/2\,, \quad m = \pm 1/2\,,$$

according to [38], or the restriction to

$$\text{any } j\,, \quad m = \pm j\,,$$

according to [40,41], results in a conflict with the holographic bound. Clearly, the schemes leading to such restrictions should be abandoned.

[h]This scheme coincides effectively with the so-called "it from bit" model formulated earlier by Wheeler [39].

Acknowledgements

I am grateful to A.A. Pomeransky and I.V. Volovich for useful discussions. Investigation was partially supported by Grant No. 05-02-16627 of the Russian Foundation for Basic Research.

References

1. J.D. Bekenstein, *Lett. Nuovo Cimento* **11**, 467 (1974).
2. D. Christodoulu, *Phys. Rev. Lett.* **25**, 1596 (1970).
3. D. Christodoulu and R. Ruffini, *Phys. Rev.* **D4**, 3552 (1971).
4. V.F. Mukhanov, *Pis'ma Zh. Eksp. Teor. Fiz.* **44**, 50 (1986) [*JETP Lett.* **44**, 63 (1986)].
5. Ya.I. Kogan, *Pis'ma Zh. Eksp. Teor. Fiz.* **44**, 209 (1986) [*JETP Lett.* **44**, 267 (1986)]; hep-th/9412232.
6. I.B. Khriplovich, *Phys. Lett.* **B431**, 19 (1998); gr-qc/9804004.
7. J.D. Bekenstein and V.F. Mukhanov, *Phys. Lett.* **B360**, 7 (1995); gr-qc/9505012.
8. J.D. Bekenstein and V.F. Mukhanov, in *Sixth Moscow Quantum Gravity Seminar*, Eds. V.A. Berezin, V.A. Rubakov, and D.V. Semicoz (World Scientific, Singapore, 1997).
9. J.D. Bekenstein, in *Proceedings of the Eighth Marcel Grossmann Meeting on General Relativity*, Eds. T. Piran and R. Ruffini (World Scientific, Singapore, 1999) 92; gr-qc/9710076.
10. I.B. Khriplovich, *Zh. Eksp. Teor. Fiz.* **126**, 527 (2004) [*Sov. Phys. JETP* **99**, 460 (2004)]; gr-qc/0404083.
11. I.B. Khriplovich, gr-qc/0409031 (2004).
12. J.D. Bekenstein, in *Cosmology and Gravitation*, Ed. M. Novello (Atlantisciences, France, 2000) 1; gr-qc/9808028.
13. J.D. Bekenstein, *Phys. Rev.* **D23**, 287 (1981).
14. G. 't Hooft, in *Salam Festschrift,* (Singapore, 1993); gr-qc/9310026 (1993).
15. L. Susskind, *J. Math. Phys.* **36**, 6377 (1995); gr-qc/9710007 (1997).
16. C. Vaz and L. Witten, *Phys. Rev.* **D64**, 084005 (2001); gr-qc/0104017.
17. L.D. Landau and E.M. Lifshitz, *Statistical Physics*, Pergamon Press, London, 1958.
18. A. Alekseev, A.P. Polychronakos and M. Smedback, *Phys. Lett.* **B574**, 296 (2003); hep-th/0004036 (2000).
19. A.P. Polychronakos, *Phys. Rev.* **D69**, 044010 (2004); hep-th/0304135 (2003).
20. G. Gour and V. Suneeta, *Class. Quant. Grav.* **21**, 3405 (2004); gr-qc/0401110 (2004).
21. R.V. Korkin and I.B. Khriplovich, *Zh. Eksp. Teor. Fiz.* **122**, 1 (2002) [*Sov. Phys. JETP* **95**, 1 (2002)]; gr-qc/0112074 (2001).
22. I.B. Khriplovich, *Zh. Eksp. Teor. Fiz.* **127**, 1223 (2005) [*Sov. Phys. JETP* **100**, 1075 (2005)]; gr-qc/0412121 (2004).
23. A. Ashtekar, S. Fairhurst and B. Krishnan, *Phys. Rev.* **D62**, 104025 (2000); gr-qc/0005083 (2000).

24. A. Ashtekar, C. Beetle and J. Lewandowski, *Class. Quantum Grav.* **19**, 1195 (2002); gr-qc/0111067 (2001).
25. I.B. Khriplovich, *Zh. Eksp. Teor. Fiz.* **115**, 1539 (1999) [*Sov. Phys. JETP* **88**, 845 (1999)]; gr-qc/9812060 (1998).
26. R.V. Korkin and I.B. Khriplovich, *Zh. Eksp. Teor. Fiz.* **121**, 531 (2002) [*Sov. Phys. JETP* **94**, 453 (2002)]; gr-qc/0107101 (2001).
27. C. Rovelli and L. Smolin, *Nucl. Phys.* **B442**, 593 (1995); erratum, *ibid.* **B456**, 753 (1995); gr-qc/9411005 (1994).
28. R. Loll, *Phys. Rev. Lett.* **75**, 3048 (1995); gr-qc/9506014 (1995); R. Loll, *Nucl. Phys.* **B460**, 143 (1996); gr-qc/9511030 (1995).
29. R. De Pietri and C. Rovelli, *Phys. Rev.* **D54**, 2664 (1996); gr-qc/9602023 (1996).
30. A. Ashtekar and J. Lewandowski, *Class. Quantum Grav.* **14**, 55 (1997); gr-qc/9602046 (1996).
31. S. Frittelli, L. Lehner and C. Rovelli, *Class. Quantum Grav.* **13**, 2921 (1996); gr-qc/9608043 (1996).
32. G. Immirzi, *Class. Quantum Grav.* **14**, L177 (1997); gr-qc/9701052 (1997).
33. C. Rovelli and T. Thiemann, *Phys. Rev.* **D57**, 1009 (1998); gr-qc/970559 (1997).
34. A. Ghosh and P. Mitra, *Phys. Lett.* **B616**, 114 (2005); gr-qc//0411035 (2004).
35. M. Barreira, M. Carfora and C. Rovelli, *Gen. Rel. Grav.* **28**, 1293 (1996); gr-qc/9603064 (1996).
36. C. Rovelli, *Phys. Rev. Lett.* **77**, 3288 (1996); gr-qc/9603063 (1996).
37. K.V. Krasnov, *Phys. Rev.* **D55**, 3505 (1997); gr-qc/9603025 (1996); K.V. Krasnov, *Gen. Rel. Grav.* **30**, 53 (1998); gr-qc/9605047 (1996); K.V. Krasnov, *Class. Quantum Grav.* **16**, 563 (1999); gr-qc/9710006 (1997).
38. A. Ashtekar, J. Baez, A. Corichi and K. Krasnov, *Phys. Rev. Lett.* **80**, 904 (1998); gr-qc/9710007 (1997).
39. J.A. Wheeler, in *Sakharov Memorial Lectures in Physics,* Moscow, vol. 2, p. 751 (1991).
40. J. Lewandowski and M. Domagala, *Class. Quantum Grav.* **21**, 5245 (2004); gr-qc/0407051 (2004).
41. K. Meissner, *Class. Quantum Grav.* **21**, 5245 (2004); gr-qc/0407052 (2004).

STRINGS AND BLACK HOLES

G. DOMOKOS AND S. KOVESI-DOMOKOS
Department of Physics and Astronomy
The Johns Hopkins University
Baltimore, MD 21218
E-mail:skd@jhu.edu

We give an elementary discussion of the asymptotic correspondence between highly excited strings and (mini)-black holes. A highly excited string can be assigned a temperature: *the Hagedorn temperature*, even though it is an isolated system. Some consequences of this result are discussed.

1. Introduction

Vladimir Naumovich Gribov – Volodya for his friends and colleagues – was a remarkable physicist of deep insights and rare skills in theoretical physics. He liked to contemplate puzzles in his field and solve them. His untimely death deprived the physics community of an esteemed colleague and of a valued friend. It is with deep sadness that we dedicate this article to his memory: perhaps, he would have liked to comment on the ideas described here.

It was quite a few years ago that Susskind [1] conjectured a correspondence between highly excited strings and black holes. His original suggestion was followed up by a large number of papers on the subject. In part, some authors considered the statistical mechanics of a gas of strings. To our knowledge, that idea was first conceived by Bowick and Wijewardhana [2] and it provided some insight into the physics of highly excited strings. A better understanding was provided by Amati and Russo [3] who considered a specific bosonic string model in the tree approximation and investigated its behavior in the limit of infinite excitations. They concluded that even an individual string radiated as a black (more precisely, grey) body at the Hagedorn temperature. The latter is expressed in terms of the string slope.

In this paper we argue that the result of Amati and Russo [3] is considerably more general than the context in which it was obtained. In fact, we show that any string model asymptotes to a thermal system radiating at the Hagedorn temperature, provided its microcanonical entropy at high excitations is a linear function of the excitation energy.

In the next section we demonstrate the equivalence, while in the section summarizing our conclusions we briefly indicate some phenomenological consequences.

2. Equivalence of the String and Black Hole Pictures

We illustrate our point on a simple example: the excitation of a high level string state. The inclusive single particle distribution considered by Amati and Russo [3] can be dealt with similarly. The levels of a string are labeled by a positive integer N such that the mass of the state is given by the formula

$$M^2 \sim NM_*^2 \qquad (N \gg 1)\,. \tag{1}$$

In eq. (1) M_* stands for the characteristic string scale. In general, for $N \gg 1$ the state is a highly degenerate one. Let now $\mathcal{O}$ be an operator which creates a string state of index N from an initial state $|\,i\rangle$, for instance in a neutrino-quark interaction. In that case, $|\,i\rangle$ can stand for a quark state and $\mathcal{O}$ for the fermionic current corresponding to the neutrino.)

Omitting trivial factors, the transition probability is given by:

$$P \propto \sum_\alpha \langle i\,|\mathcal{O}^\dagger|\,N,\alpha\rangle\langle N,\alpha\,|\mathcal{O}|\,i\rangle\delta\,(E - NM_*)\;. \tag{2}$$

In eq. (2) α stands for the collection of labels necessary for the full specification of states at level N: one has to sum over (most of) those, since the quantum numbers carried by $|\,i\rangle$ and $\mathcal{O}$ do not fully specify the substate at level N to which the transition takes place.

We recognize that the quantity in eq. (2),

$$\rho_M = \sum_\alpha |\,N,\alpha\rangle\langle N,\alpha\,|\,\delta\,(E - NM_*)\;, \tag{3}$$

is just the microcanonical density matrix of the final state. (In statistical mechanics, a smoothing of the δ function is necessary in order to define a level density; in the present context, such a smoothing is automatic if the finite width of resonances is taken into account.) The summation over α leads to the information loss discussed by Amati [4].

It is well known that a *large* microcanonical ensemble can be well approximated by a canonical one, see, *e.g.* Huang's textbook [5]. The argument given in textbooks typically runs as follows. One establishes that the fluctuations in a canonical ensemble asymptotically follow a normal distribution;

the width of the distribution gets narrower as the energy increases. In the limit of infinite energy the normalized distribution becomes a δ function.

A more direct approach is the following. Let E be the energy as before and H be the Hamiltonian of the system. One can write the microcanonical density matrix in operator form as follows

$$\rho_M = \delta(E - H) . \tag{4}$$

In the present context, the energy is the invariant center of mass energy of the interaction described by eq. (2), $E = \sqrt{s}$ and H is the mass operator of the string.

One then uses the following complex integral representation of the δ function:

$$\delta(E - H) = \frac{1}{2\pi i} \int_{L-i\infty}^{L+i\infty} dv \exp v (E - H) , \tag{5}$$

where an appropriate choice of L gives a convergence factor. Next, one takes the trace of both sides to obtain the partition functions. (This is sufficient: by adding external source terms to H, one obtains the generating functional of the correlation functions.) One has then (with $S(E)$ being the microcanonical entropy):

$$\exp S(E) = \frac{1}{2\pi i} \int_{L-i\infty}^{L+i\infty} dv \ \exp(vE) \ \exp(S_c(v)) . \tag{6}$$

We used the relation between the partition function and entropy on both sides of the previous equation. Obviously, $\exp S_c(v)$ is the canonical partition function at inverse temperature v. One evaluates the integral in the previous equation by means of the method of steepest descents. The location of the saddle point, $v = \beta$, is given by the equation:

$$E = -\left(\frac{\partial S_c}{\partial v}\right)_{v=\beta} , \tag{7}$$

which is the usual thermodynamic relation between temperature and energy. Using the preceding equation, the leading order contribution of the integral is given by

$$\exp(S_c(\beta) - E\beta) = \exp S(E) . \tag{8}$$

The quantity $S_c(\beta) - E\beta$ is recognized as the Legendre transform of the canonical entropy. Finally, by carrying out the Gaussian integration, one obtains a contribution to the entropy, which, up to an irrelevant additive constant equals to $-3/2 \ln E/M_*$. All these formulae are valid for $E/M_* \gg 1$.

An operational definition of a "large" microcanonical ensemble is that terms of $O(1/E)$ in the expression of the entropy are negligible[a].

Here we consider level densities of the form:

$$d(E) = \exp S(E) \sim C\,(E/M_*)^{-\gamma-3/2} \exp\left(\alpha(E/M_*)^{\delta}\right) . \qquad (9)$$

The quantities C, γ, α, δ depend on the specific model considered. The contribution of the Gaussian integral to the entropy has been taken into account. (In known string models, $\delta = 1$.) The inverse temperature is:

$$\beta = \frac{\partial S}{\partial E} \sim -\frac{\gamma + 3/2}{E} + \alpha\delta \left(\frac{E}{M_*}\right)^{\delta-1} M_*^{-1} . \qquad (10)$$

For $\delta = 1$, the temperature is asymptotically constant and, hence, it may be identified with a Hagedorn temperature, T_H. Notice, however that the expression of the inverse temperature, eq. (10) gives a temperature approaching T_H *from above.* From a physical point of view, this is unacceptable: the Hagedorn temperature is supposed to be a *maximal* temperature. This is due to the fact that the coefficient of $1/E$ in eq. (10) is negative. The problem is corrected by taking into account the contribution of Kaluza-Klein excitations to the entropy. Apart from an additive constant, the contribution is proportional to $(E/M_*)^n$, n being the number of compact dimensions. For instance, for an open superstring in the critical dimensionality, $n = 6$ and $\gamma = 9/2$, giving $-(\gamma + 3/2) + n = 0$. Thus, the approach to the Hagedorn temperature in this case is determined by higher order terms in the asymptotic expansion.

The expression of the Hagedorn temperature in terms of M_* is somewhat model dependent: in the calculations cited above, one has, $T_H = M_*/(a\pi)$, with a ranging, approximately, between 2 and 4 for various string models. For the purposes of numerical estimates in this work, we adopt $a = 3$: this is equivalent, of course, to fixing the parameter α in eq. (10). Equation (10) establishes the connection between the string and (microscopic) black hole pictures: in fact, it means that – in a sense – a large and highly degenerate isolated system serves as its own thermal reservoir.

The asymptotic estimate given here has far reaching consequences. In particular, the inclusive single particle distribution can be calculated using elementary statistical mechanics of a canonical ensemble as it was carried

[a] Amati and Russo [3] use an adaptation of the Darwin-Fowler method and discuss some terms of $O(1/N)$ in the framework of specific string models. For $N \gg 1$, the asymptotic approach used here and the Darwin-Fowler method yield identical results.

out in detail for specific string models by Amati and Russo, Ref. [3]. We have in the rest frame of a decaying resonance:

$$dn = \frac{d\sigma}{\sigma} = \frac{1}{\exp{(k_0/T)} \pm 1} \frac{d^3k}{k_0} L^2, \tag{11}$$

where σ is the total inelastic cross section and k_μ stands for the four momentum of the particle observed in the decay. At high excitations, we may put $T = T_H$. The sign $\pm$ is valid for Fermi/Bose statistics of the observed particle, respectively. The quantity L is a characteristic length of the system.

We now note that the expression of dn is Lorentz invariant. Consequently, one can replace the rest frame expression of k_0 by a Lorentz invariant quantity reducing to k_0 in that frame. This is elementary. One uses the invariant:

$$K = \frac{(P \cdot k)}{\sqrt{P^2}} \tag{12}$$

in place of k_0 in eq. (11). Here, P stands for the total four momentum of the decaying resonance. In a scattering process, such as

$$\nu + q \longrightarrow \text{resonance} \longrightarrow (k) + X,$$

one can approximate $\sqrt{P^2} \approx \sqrt{s}$. and the symbol "$(k)$" stands for the observed particle of four momentum k. The inclusive distribution, eq. (11) with the replacement $k_0 \longrightarrow K$ shows the expected properties if evaluated in the laboratory frame.

3. Conclusion

We found that the asymptotic equivalence of mini black holes and highly excited strings depends, in essence, only on an appropriate growth of the microcanonical entropy of the excited string. (If the asymptotic growth of the microcanonical entropy deviates from a linear one, the asymptotic temperature is either zero or infinity.) Some observable consequences were explored in Ref. [6].

It is quite remarkable that a rather general result emerges from simple considerations, like the asymptotic behavior of level densities in any kind of string theory, independently of the specific string model or of the strength of the string coupling.

References

1. L. Susskind, hep-th/9309145.
2. M.J. Bowick and L.C.R. Wijewardhana, *Phys. Rev. Lett.* **54**, 2485 (1985).
3. D. Amati and J.G. Russo, *Phys. Lett.* **B454**, 207 (1999).
4. D. Amati, *in Proceedings of the Abdus Salam Memorial Meeting*, edited by J. Ellis, F. Hussain, T. Kibble, G. Thompson and M. Virasoro, World Scientific Publishing Company, Singapore (1997).
5. K. Huang, "*Statistical Mechanics*", second edition, Wiley & Sons, (1987).
6. G. Domokos and S. Kovesi-Domokos, in *Nucl. Phys. B (Proceedings Supplement)*, special issue: Proceedings of the XIII International Symposium on Very High Energy Cosmic Ray Interactions. (2005).

AN UNEXPECTED USE OF THE AGK RULES IN BLACK-HOLE PHYSICS

G. VENEZIANO

Theory Division, CERN, CH-1211 Geneva 23, Switzerland
and
Collège de France, 11 place M. Berthelot, 75005 Paris, France

Previous results on trans-Planckian collisions in superstring theory are rewritten in terms of an explicitly unitary S-matrix whose domain of validity in energy–impact-parameter space borders with the black hole production threshold. Properties of the final state are analyzed by applying the famous AGK (Abramovskij–Gribov–Kancheli) rules and, as the region's boundary is approached, are found to resemble those expected from Hawking's evaporation process.

1. Introduction

The issue of a possible loss of information/quantum coherence in processes where a black hole (BH) is produced and then evaporates has been the subject of much debate since Hawking's observation [1] that black holes should emit an exactly thermal spectrum of light particles particles, since then known as Hawking radiation. Progress coming from string theory, in particular on the microscopic understanding of black-hole entropy [2] and on the AdS-CFT correspondence [3], have lent strong support [4] to the "no-information-loss" camp. In spite of these developments, several issues still remain unclear. One would like to understand, for instance, how exactly information is retrieved and what this implies on the properties of the pure final state that a given pure initial state generates through its unitary evolution. A few ideas on this issue have been put forward: for instance, it has been suggested [5] that black-hole formation and decay simply do not occur for a pure initial quantum state and would only emerge after tracing over initial states. Less radical solutions [6] suggest that information is recovered during the whole evaporation process, in subtle correlations among the final particles or that it is given back at the very end through some black-hole "remnants".

In a recent paper [7], using the fact that the threshold for BH formation, $E_{\rm th}$, can be made arbitrarily high by taking a sufficiently small string coupling, we have analysed the unitary S-matrix that describes the collision process in a large energy interval up to such a threshold. Using the (once?) famous AGK (Abramovskij–Gribov–Kancheli) rules [8], we were able to analyse the detailed structure of the final state and to see to what extent its properties start resembling more and more those predicted by Hawking's process as $E_{\rm th}$ is approached. The matching near $E_{\rm th}$ turned out to be very smooth. This is almost certainly related to the well-known correspondence [9,10] between strings and black holes at $M \sim E_{\rm th}$ (see below). Our findings in [7] gave further support to the belief that quantum coherence is fully recovered even when the region of "large" semi-classical black holes is reached.

There is a second, less fundamental, but phenomenologically interesting reason for studying high-energy collisions in the region $M_{\rm P} < E < E_{\rm th}$. Models have been proposed [11] in which gravity, being sensitive to some "large" extra dimensions of space, becomes strong at an effective energy scale M_D that is much much smaller than the "phenomenological" 4-dimensional Planck energy $M_{\rm P} \sim 10^{19}$ GeV. Assuming M_D to be not too far from the few TeV scale, a variety of interesting strong-gravity signals could be expected when the LHC will be turned on a few years from now.

One particularly dramatic signal [12] would be the production of TeV-scale black holes and their characteristic decay by the Hawking process. We expect black holes to be formed with unsuppressed rate provided the impact parameter b of the collision of the (essentially) point-like proton constituents (quarks and gluons) does not exceed the Schwarzschild radius $R_{\rm S}(E_{\rm cm})$ associated with the centre-of-mass energy. This intuition appears to be supported [13] by classical arguments on the formation of a closed trapped surface in the 2-particle collision process. It was pointed out, however, that the above cross-section estimates have to be amended by taking into account the finite tranverse size of the colliding objects [14]. Even in the most optimistic case, by embedding any quantum theory of gravity into string theory, the minimal transverse size of the colliding quanta is given by the string length parameter $l_s = M_s^{-1}$ and the formation of a black hole would require both $R_{\rm S}(E_{\rm cm}) > b$ and $R_{\rm S}(E_{\rm cm}) > l_s$. Given that, in (weakly coupled) string theory $l_s > l_D$[a] (where $l_D = M_D^{-1}$ is the true Planck length

[a]We denote by a subscript D quantities referring to a generic number D of "large" space-time dimensions, while reserving the label P for $D = 4$. We also use units in

of the theory), this effect will push the threshold of black hole production to some $E_{\text{th}} > M_D$. As a consequence, it is most likely that the energy reach of accelerators such as the LHC will be at best marginal for producing semi-classical black holes. This does not mean, however, that no new phenomena will be found below or just near E_{th}. Exploring the nature of such exotic phenomena was the other motivation behind the work in [7].

Here I will start by recalling the "phase-diagram" of super-Planckian collisions in string theory as it has emerged from work in the late eighties. I will then present an explicitly unitary ansatz for the S-matrix and argue for its validity in a large region of the phase diagram. Subsequently, by making use of the AGK rules, I will analyse the final state in the energy window $M_* \equiv \sqrt{M_s E_{\text{th}}} < E < E_{\text{th}}$ and show how the average energy of final particles smoothly *decreases* to the Hawking-temperature value (up to a constant of $O(1)$) as the total energy is *increased* towards E_{th}. This suggests the validity, within an energy window, of a sort of "antiscaling" behaviour for inclusive cross sections that would provide a smoking gun for the discovery of string/quantum–gravity effects at future colliders.

2. Three regimes in super-Planckian collisions

Collisions of light particles at super-Planckian energies ($E = \sqrt{s} \gg M_D$) have received considerable attention since the late eighties. While in [15] the focus was on $D = 4$ collisions in the field theory limit, two groups have carried out the analysis within superstring theory, possibly allowing for a number of "large" extra dimensions. In the approach due to Gross, Mende and Ooguri (GMO) [16] one starts from a genus-by-genus analysis of fixed angle scattering, and then attempts an all-genus resummation. In the work by Amati, Ciafaloni and myself (ACV) [17,18,19] one starts from an all-order eikonal description of small-angle scattering and then attempts to push the results towards larger and larger angles. In the approach by Fabbrichesi *et al.* [20] one relates, at arbitrary scattering angle, the semiclassical S-matrix to a boundary term and tries to estimate it in terms of classical solutions.

The picture that has emerged (see e.g. [21] for some reviews) is best explained by working in impact parameter ($b = 2J/E$) space, rather than in scattering angle (θ) or momentum transfer. We can thus represent the various regimes of super-Planckian collisions by appealing to an (E, b) plane or,

which $\hbar = c = 1$ and neglect numerical factors where they are not important (these are given explicitly in [7].

equivalently but more conveniently, to an (R_S, b) plane, where

$$R_S(E) \sim (G_D E)^{\frac{1}{D-3}} \tag{1}$$

is the Schwarzschild radius associated with the centre-of-mass energy E. Since both coordinates in this plane are lengths, we can also mark on its axes two (process-independent) lengths, the Planck length l_D and the string length l_s. We shall use the following definitions:

$$l_s = \sqrt{2\alpha'\hbar} = M_s^{-1}\ , \ G_D = l_D^{D-2} = M_D^{2-D}\ , \tag{2}$$

where α' is the open-string Regge-slope parameter (equal to twice that of the closed string). We will assume string theory to be (very) weakly coupled:

$$l_s = (g_s)^{-\frac{2}{D-2}} l_D \gg l_D\ ,\ \text{i.e.}\ M_D = M_s (g_s)^{-\frac{2}{D-2}} \gg M_s\ , \tag{3}$$

where g_s is the string coupling constant. We shall keep g_s (and hence $l_D/l_s = M_s/M_D$) fixed and very small.

By definition of super-Planckian energy, $R_S > l_D$. Since we also restrict ourselves to $b > l_D$, a small square near the origin is not considered. The rest of the diagram is divided essentially in three regions [21] (see Fig. 1):

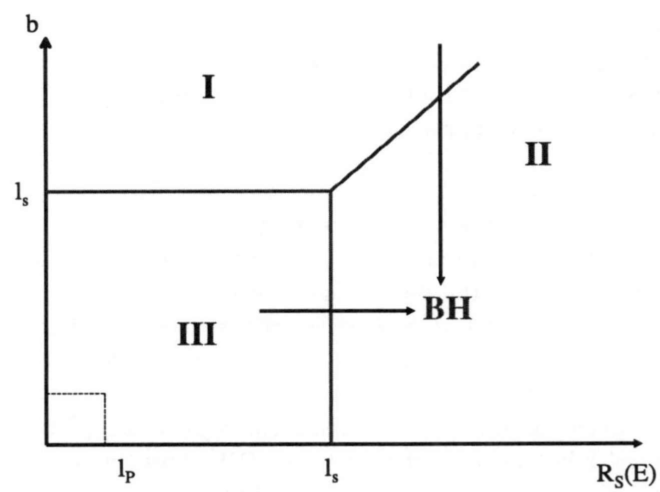

Figure 1. Phase diagram of trans-Planckian scattering showing three different kinematic regions and two different paths towards the regime of large BH production. Here $l_P = l_D$.

- The first region, characterized by $b > \max(l_s, R_S)$, is the easiest to analyse and corresponds to small-angle quasi-elastic scattering;

- The second region, $R_S > \max(b, l_s)$, is the most difficult: this is where we expect black-hole formation to show up. Unfortunately, in spite of much effort (see, e.g. [19,20]), not much progress has been achieved in the way of going through the $b = R_S > l_s$ boundary of Fig. 1;

- Finally, the third region ($b, R_S < l_s$), whose very existence depends on working in a string theory framework, has provided some very interesting insight [18] on how string effects may modify classical and quantum gravity expectations once the string scale is approached. The reason why progress could be made by ACV in this regime, unlike in the previous one, is that string-size effects intervene *before* large classical corrections make the problem intractable as $b \to R_S$. The physical reason for this is that string-size effects prevent the formation of a putative BH whose radius would be smaller than l_s. Instead, it is found [18] that a maximal deflection angle $\theta_{\max} \sim (R_S/l_s)^{D-3}$ is reached at an impact parameter of order l_s (the colliding strings grazing each other). If one considers scattering at angles larger than such $\theta_{\max}$, one finds (see e.g. [21]) an exponential suppression of the cross section, in very good agreement with the results obtained by GMO [16] through their very different approach.

Here we will concentrate our attention on this third regime. We will work at fixed impact parameter and gradually increase the energy to cross the boundary between the second and third regions ($b < l_s = R_S$). As the boundary is approached, one arrives at the threshold of BH formation. Actually, at $\sqrt{s} = E_{\text{th}}$, one expects to form "string-holes" i.e. states that can be seen alternatively either as strings or as black holes [9,10]. They lie, in an (M, g_s) plane, on the correspondence curve $M = M_s g_s^{-2}$ that separates string states (lying below such curve) from BH states (lying above it). Our aim here is to study details of the scattering process as this threshold is approached from below and to compare them with those expected from formation and evaporation of BHs as the same threshold is approached from above.

3. Explicitly unitary S-matrix in Regions I and III

The results of ACV, including various checks of inelastic unitarity, can be neatly summarized in terms of an explicitly unitary S-matrix whose validity can be justified well inside Regions I and III. It reads:

$$S = \exp(i\hat{I}) \ , \tag{4}$$

where the hermitian operator $\hat{I}$ is given by:

$$\hat{I} = (\hat{\delta} + \hat{\delta}^\dagger) + \sqrt{-2i(\hat{\delta} - \hat{\delta}^\dagger)}(C + C^\dagger) = \hat{I}^\dagger \ , \tag{5}$$

and the operators $\hat{\delta}$, $\hat{\delta}^\dagger$, C and $C^\dagger$ satisfy the commutation relations

$$[C, C^\dagger] = 1 \ , \ [\hat{\delta}, \hat{\delta}^\dagger] = [C, \hat{\delta}] = [C, \hat{\delta}^\dagger] = 0 \ . \tag{6}$$

Using well-known harmonic-oscillator formulae, Eqs. (4), (5) lead to the more convenient form:

$$S = \mathrm{e}^{2i\hat{\delta}} \mathrm{e}^{i\sqrt{-2i(\hat{\delta} - \hat{\delta}^\dagger)} \ C^\dagger} \mathrm{e}^{i\sqrt{-2i(\hat{\delta} - \hat{\delta}^\dagger)} \ C} \ . \tag{7}$$

The meaning of the C, $C^\dagger$ operators will be given below, while (with standard notation for normal ordering)

$$\hat{\delta} = \frac{1}{4\pi^2} \int_0^{2\pi} \int_0^{2\pi} d\sigma_u d\sigma_d : \delta(E, b + \hat{X}_u(\sigma_u) - \hat{X}_d(\sigma_d)) : \ , \tag{8}$$

where $\hat{X}_{u,d}$ are hermitian (and commuting) closed-string position operators taken at a fixed value of τ, and δ is the tree-level amplitude in impact parameter space:

$$\delta(E, b) = \frac{1}{(2\pi)^{D-2}} \int d^{D-2}q \frac{A(s,t)}{4s} \ \mathrm{e}^{-iqb} \ , \quad s = E^2, \quad t = -q^2 \ . \tag{9}$$

The explicit form of the tree-level amplitude is:

$$\frac{A(s,t)}{4s} = G_D \ \frac{\Gamma(-\alpha' t/4)}{\Gamma(1 + \alpha' t/4)} \left(\alpha' s\right)^{1 + \alpha' t/4} \mathrm{e}^{-i\pi\alpha' t/4} \ , \tag{10}$$

and its imaginary part is positive semi-definite,

$$\frac{\mathrm{Im} A(s,t)}{s} = G_D \ \Gamma^{-2}(1 + \alpha' t/4) \left(\alpha' s\right)^{1 + \alpha' t/4} \geq 0 \, , \tag{11}$$

an important feature in order to give meaning to (5). The imaginary part of δ and thus, from (8), the anti-hermitian part of $\hat{\delta}$ are given simply by:

$$\mathrm{Im}\delta = \frac{G_D \ s \ l_s^2}{(Y l_s)^{D-2}} \mathrm{e}^{-b^2/l_s^2 Y^2} \ , \tag{12}$$

where $Y = \sqrt{\log(\alpha' s)}$.

Equation (12) introduces a characteristic impact parameter $b_I = l_s Y$ above which δ is basically real and thus $\hat{\delta}$ is hermitian. We will be mainly concerned with the opposite region, $b \leq b_I$, but let us first recall some properties of the S-matrix in the former regime. At $b \gg b_I$, the behaviour of δ is dominated by the pole at $t = -q^2 = 0$ in $A(s,t)$ with the result:

$$\delta = \left(\frac{b_c}{b}\right)^{D-4} \equiv \frac{G_D s}{\Omega_{D-4}} b^{4-D} \,, \tag{13}$$

with $\Omega_d = \frac{2\pi^{d/2}}{\Gamma(d/2)}$ the solid angle in d dimensions. As is well known [17], the above formula reproduces, at very small angle, the classical relation between impact parameter and (centre-of-mass) scattering angle in the (Aichelbourg–Sexl) metric of a relativistic point-like source:

$$\theta = \frac{8\pi G_D \sqrt{s}}{\Omega_{D-2} b^{D-3}} \,. \tag{14}$$

As pointed out by ACV, however, the passage from δ to $\hat{\delta}$ introduces another impact parameter scale, b_{DE}, below which the elastic amplitude is suppressed in favour of channels where the initial particles are excited through graviton exchange. This is analogous to the phenomenon of diffractive excitation (DE) in hadronic physics, supposedly described in terms of (multi)-Pomeron exchange and Gribov's Reggeon calculus [22]. In our case this complication will be essentially ignored (see [7] for more discussion on this point).

Let us move instead to the region $b < b_I$. Here (modulo an irrelevant b-independent term) the real part of δ behaves as:

$$\mathrm{Re}\delta = -\frac{G_D \; s \; b^2}{(Y l_s)^{D-2}} \,. \tag{15}$$

If we were to neglect all non-elastic effects this phase shift would produce [17] the same relation between scattering angle and impact parameter that results from the geodesic motion of a particle in the metric of a homogeneous finite-size relativistic beam [23]:

$$\theta = \frac{16\pi G_D \; \rho \; b}{D-2} \;\; , \;\; \rho = \frac{E}{(Y l_s)^{D-2}} \,, \tag{16}$$

where $Y l_s$ plays the role of the transverse size of the beam.

Inelasticity now comes from two distinct sources:

- From replacing, as we already mentioned, $\mathrm{Re}\delta$ by $\frac{1}{2}(\hat{\delta} + \hat{\delta}^\dagger)$ in order to take account of diffractive excitation (as we mentioned for Region I).

- From the fact that δ has now acquired an imaginary part given by (12).

This latter effect is the one we shall concentrate our attention on. It is related to the possibility of "cutting" the exchanged gravi-reggeons, i.e. to the fact that, in string theory, because of (old Dolen–Horn–Schmit) duality, gravi-reggeon exchange in the t-channel is dual to some s-channel intermediate states. For open-string collisions these intermediate states correspond to a pair of excited open strings, while for closed-string scattering they represent a single excited closed string. String breaking will make these states yield showers of light particles. The operators $C, C^\dagger$ are just destruction and creation operators for a cut gravi-reggeon (CGR), i.e. for one such kind of final state. More generally, they should carry a label corresponding to the energy-momentum (and possible internal quantum numbers) of the CGR.

When a diagram with n exchanged gravi-reggeons is considered, the full imaginary part gets contributions from different final states, corresponding to cutting any number $m \leq n$ of gravi-reggeons. It is quite remarkable that, as we shall discuss in the next Section, precisely the AGK rules allow us to rewrite the full S-matrix in the simple form (7).

4. Application of the AGK rules

We want to identify the inelastic channels that saturate unitarity when $\hat{\delta}$ is not Hermitian, i.e. when $\mathrm{Im}\delta \neq 0$. Let us neglect DE and thus simply set $\hat{\delta} = \delta$ in our expression (7) for the S-matrix. Recalling that:

- $\mathrm{Im}\delta \neq 0$ comes from the imaginary part of the single (Reggeized) graviton exchange;
- The leading-eikonal formula $S = \exp(2i\delta)$ corresponds to summing over an infinite set of diagrams corresponding to the exchange of an increasing number $n = 0, 1, 2, \ldots$ of gravitons;
- The rules for reconstructing the imaginary part of a diagram with n (positive signature) exchanged Reggeons as a sum of contributions coming from "cutting" (i.e. taking the imaginary part of) m of them were given long ago by Abramowski, Gribov and Kancheli [8] (the so-called AGK rules) and were later justified by several authors (e.g. [24]).

These rules (when directly applied in impact parameter space) tells us that the full imaginary part of the n-graviton exchange graph is the result of a contribution

$$\sigma_m^n = (-1)^{n-m}\frac{(4\mathrm{Im}\delta)^n}{m!(n-m)!}\,,\ n=1,2,\ldots\ ,\ m=1,2,\cdots\leq n \tag{17}$$

due to cutting m out of n gravi-reggeons, and a contribution

$$\sigma_0^n = (-1)^n\frac{(4\mathrm{Im}\delta)^n}{n!} - 2\mathrm{Re}S^{(n)}\ ,\ n=1,2,\ldots\ , \tag{18}$$

when no gravi-reggeon is cut, where $S^{(n)}$ is the full n-GR exchange contribution to the S-matrix. As the symbol σ indicates, these are also to be interpreted as contributions to cross sections into inelastic channels corresponding to m cut gravi-reggeons. It is trivial to check that, for any given n, the sum of all contributions from $m=0$ to $m=n$ gives back twice the full imaginary part of $T^{(n)} \equiv i(1-S^{(n)})$, as it should.

Let us now construct the generating function of the n-CGR cross section:

$$\exp(W(z)) = \sum_{m=0}^{\infty} z^m\sigma_m = \sigma_0 + \sum_{m=1}^{\infty} z^m \sum_{n=m}^{\infty}\sigma_m^n\,. \tag{19}$$

This expression is easily computed to give:

$$W(z) = 4(z-1)\ \mathrm{Im}\delta\,. \tag{20}$$

We will now see that the same result follows from our operator formula for the S-matrix.

5. Properties of the final state in the energy window

In terms of our operators the generating function (19) can be written as

$$\exp(W(z)) \equiv \langle in|S^\dagger z^{N_{\mathrm{CGR}}}S|in\rangle\,,\ \ N_{\mathrm{CGR}} \equiv C^\dagger C\ , \tag{21}$$

where N_{CGR} counts the number of "cut gravi-reggeons". Evaluating $W(z)$ is now a simple exercise in harmonic oscillator operators, simplified by the well-known identity:

$$z^{N_{\mathrm{cgr}}} =: \exp\left((z-1)C^\dagger C\right) :\ . \tag{22}$$

Using properties of the coherent states one arrives at the simple result:

$$W(z) = 4(z-1)\ \mathrm{Im}\delta\ , \tag{23}$$

which is exactly Eq. (20) and implies that the distribution of N_{CGR} is exactly poissonian, with an average given by:

$$\langle N_{\mathrm{CGR}}\rangle = 4\mathrm{Im}\delta = \frac{G_D\ s\ l_s^2}{(Yl_s)^{D-2}} = O\left(\frac{s}{M_*^2}\right)\,, \tag{24}$$

where $M_* \sim M_s/g_s \sim \sqrt{M_s M_{th}}$ will turn out to play an important role in the following. Amusingly, M_* coincides, up to numerical factors, with the mass of $D0$-branes [25], possibly just a coincidence.

If we forbid particle production via CGRs, then we are penalized by an exponential factor (as is clear by taking the limit $z \to 0$ in Eqs. (21) and (23)). Thus, the elastic cross section is damped as[b]:

$$\sigma_{\rm el} \sim \exp(-8\ {\rm Im}\delta) = \exp\left[-\frac{G_D\ s\ l_s^2}{(Yl_s)^{D-2}}\right] . \tag{25}$$

Amusingly, such a suppression is quite similar to a factor $\exp(-S_{\rm bh})$ where $S_{\rm bh}$ is the Bekenstein–Hawking entropy of a black hole of mass $\sqrt{s}$. Such a suppression is known to appear in the elastic amplitude for the scattering of a particle off a classical BH when the impact parameter goes below that of classical capture (e.g. [26]). For $D = 4$ the agreement holds up to a numerical factor, while for $D > 4$ the functional dependence is different. For any value of D, however, agreement with a BH-type suppression holds (up to factors of $O(1)$) as one approaches $E_{\rm th}$.

Let us now look instead at the typical final state that roughly saturates the cross section. The total energy will be shared between $N_{\rm CGR}$ CGRs, giving an average energy per CGR:

$$\langle E\rangle_{\rm CGR} = \frac{\sqrt{s}}{\langle N_{\rm CGR}\rangle} \sim \frac{M_*^2}{\sqrt{s}} . \tag{26}$$

The average energy of the final states *decreases* as the energy is *increased* within our window. Once more, for $D = 4$ the functional dependence on $\sqrt{s}$ of $\langle E\rangle_{\rm CGR}$ follows that of a thermal stectrum with a Hawking temperature $T_H \sim \hbar/R_S$, while for $D > 4$ the functional dependence is different[c]. For all values of D, however, agreement (up to factors $O(1)$) with Hawking-temperature expectations occurs at $E_{\rm th}$. An "antiscaling" behaviour, corresponding to

$$\langle E\rangle_{\rm CGR}\sqrt{s} = M_*^2 = M_s^2 g_s^{-2} , \tag{27}$$

holds for any D in the energy window $M_* < E < E_{\rm th}$.

In Fig. 2 we show the centre-of-mass-energy dependence of $\langle E\rangle_{\rm CGR}$ in the above window and also, for completeness, that of $\langle M\rangle_{\rm DE}$. Note

[b]The exponent is actually twice its naive value as a result of not allowing DE final states either.

[c]Similar differences between $D = 4$ and $D > 4$ in the string–BH correspondence were noted in [10].

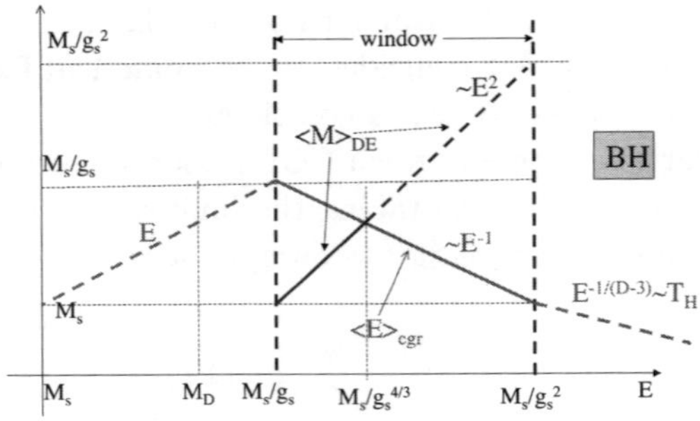

Figure 2. Energy dependence of $\langle M\rangle_{\rm DE}$ and $\langle E\rangle_{\rm CGR}$ in the energy window $M_s/g_s < E < M_s/g_s^2$ showing in particular the anti-scaling behaviour of the latter.

the opposite qualitative behaviour of $\langle E\rangle_{\rm CGR}$ and $\langle M\rangle_{\rm DE}$. The former decreases from M_* to M_s while the latter increases from M_s to a fraction of the total energy at $E_{\rm th}$. The two curves intersect when $E \sim M_s g_s^{-4/3}$, i.e. when $\langle M\rangle^2_{\rm DE} \sim M_s E$. When $\langle M\rangle^2_{\rm DE} \geq M_s E$, no rapidity gap will be left between the final particles of the two jets, making the distinction between DE and CGR states practically impossible (for this reason the solid line for $\langle M\rangle_{\rm DE}$ is not continued above $E \sim M_s g_s^{-4/3}$). All that matters, from there on, is the number of handles that are cut: each one of them corresponds to an excited closed string produced by the collision. Our S-matrix gives the final state, $|f\rangle = S|2\rangle$, as a coherent state of CGRs with a very large average occupation number.

These characteristic features of the final state would represent a clear signal of new string or quantum-gravity physics, already well below the threshold of BH production, provided the higher-dimensional model has $M_s \ll M_D$.

6. Implications for the information paradox

Which are the possible implications of our results when one crosses the threshold for BH production? At the moment, we can only put forward some educated guesses:

- The CGRs will start interacting with each other (see e.g. the so-called H-diagram of [19]), showers will develop, and the original Poisson distribution of the CGR should turn into an approximately thermal one (with BE or FD statistics) for the actual final particles.

- The eikonal phase δ will pick up an imaginary part over the one due to CGR production, particularly as we approach BH production from Region I. It is amusing to speculate on how the expected result at a point well inside Region II can be obtained by approaching it from Regions I and III. In Region I the general interpretation of the phase shift is in terms of some classical (retarded vs advance)-time delay [20]. When a critical $b \sim R_s$ is reached, the time delay diverges (for any D) like $R_s \log(b - R_S)$. As we go below this critical value of b, the time delay develops an imaginary part $\sim i\pi R_S$ (since there is no classical trajectory going off to future infinity); when inserted in the phase $iE\Delta t$, this gives:

$$|A_{\rm el}| \sim \exp(-\pi E R_s)\,, \tag{28}$$

namely the correct damping (up to numerical factors $O(1)$) for all D. How does this match the behaviour obtained by crossing the boundary horizontally rather than vertically? The answer is, again, that the l_s dependence in Eq. (25) should saturate when l_s approaches R_s, so that:

$$|A_{\rm el}| \sim \exp\left[-G_D \; s \; l_s^{4-D}\right] \to \exp\left[-G_D \; s \; R_s^{4-D}\right] \sim \exp(-ER_s)\,, \tag{29}$$

in agreement with (28). The same correspondence with BH physics would come out at the level of inclusive spectra if the l_s dependence saturates at $R_S = l_s$, giving:

$$\frac{s}{M_*^2} \to ER_S \sim S_{\rm BH}\;,\; T_{\rm eff} \to 1/R_S \sim T_H\;,\; x \to \omega/T_H\;. \tag{30}$$

I will conclude by summarizing the main points:

- we have been able to recast the main results of [17,18,19] in the form of an approximate, but exactly unitary, S-matrix whose range of validity covers a large region of the kinematic energy–angular-momentum plane;
- we have studied the nature of the dominant final states in a window of energy and impact parameter at whose boudary we expect black-hole formation to begin;
- we have found there a sort of precocious black-hole behaviour, in particular an "anti-scaling" dependence of the average energy of the final particles from the initial energy, quite reminiscent of the inverse relation between black-hole mass and temperature;

- this antiscaling behaviour introduces, through the variable $x = \omega E/M_*^2$, a new energy scale $M_* = M_s g_s^{-1}$, which, amusingly, is the scale of $D0$ branes;
- we believe that these results have a twofold application: a conceptual one in the search for an explicit resolution of the information paradox, and a more phenomenological one in the context of the expected string/quantum-gravity signals at collider physics, which are expected in models with large extra dimensions.

Our explicit results confirm, to a large extent, what had already been guessed at a more qualitative level, by many authors, including those of Refs. [17,20], and hopefully set a framework for starting a more quantitative study of super-Planckian collisions up to, and perhaps even above, the BH-production threshold.

Acknowledgements

I am very grateful to the organizers of this Symposium, and in particular to Julia Nyiri, for having kindly invited me to attend it and for having given me the opportunity to honour the memory of a great physicist and a dear friend and colleague through this modest contribution.

References

1. S. W. Hawking, *Nature* **248**, 30 (1974); *Commun. Math. Phys.* **43**, 199 (1975).
2. A. Strominger and C. Vafa, *Phys. Lett.* **B379**, 99 (1996); for a review see: J.R. David, G. Mandal and S.R. Wadia, "*Microsopic formulation of black holes in string theory*", *Phys. Rep.* **369**, 549 (2002) [hep-th/0203048].
3. J.M. Maldacena, *Adv. Theor. Math. Phys.* **2**, 231 (1998) 231;
   S.S. Gubser, I.R. Klebanov and A.M. Polyakov, *Phys. Lett.* **428**, 105 (1998);
   E. Witten, *Adv. Theor. Math. Phys.* **2**, 253 (1998).
4. J.M. Maldacena, *JHEP* **0304**, 021 (2003);
   J.L.F. Barbon and E. Rabinovici, *JHEP* **0311**, 047 (2003).
5. D. Amati, *Phys. Lett.* **B454**, 203 (1999);
   D. Amati and J.G. Russo, *Phys. Lett.* **B454**, 207 (1999).
6. See, for example, Y. Aharonov, A. Casher and S. Nussinov, *Phys. Lett.* **B191**, 51 (1987); S.B. Giddings, *Phys. Rev.* **D49**, 947 (1994);
   B.L. Hu, "*Correlation dynamics of quantum fields and black hole information paradox*", gr-qc/9511075.
7. G. Veneziano, JHEP **0411**, 001 (2004) [hep-th/0410166].
8. V.A. Abramovskij, V.N. Gribov and O.V. Kancheli, *Sov. J. Nucl. Phys.* **18**, 308 (1974).

9. G. Veneziano, *Europhys. Lett.* **2**, 133 (1986); *Proc. Meeting on "Hot hadronic matter: theory and experiments"*, Divonne, June 1994, eds. J. Letessier, H. Gutbrod and J. Rafelsky, NATO-ASI Series B: Physics (Plenum Press, New York, 1995), **346**, p. 63; L. Susskind, "*Some speculations about black hole entropy in string theory*", hep-th/9309145; E. Halyo, A. Rajaraman and L. Susskind, *Phys. Lett.* **B392**, 319 (1997); E. Halyo, B. Kol, A. Rajaraman and L. Susskind, *Phys. Lett.* **B401**, 15 (1997).
10. G.T. Horowitz and J. Polchinski, *Phys. Rev.* **D55**, 6189 (1997); *ibid.* **57**, 2557 (1998); T. Damour and G. Veneziano, *Nucl. Phys.* **B568**, 93 (2000).
11. I. Antoniadis, *Phys. Lett.* **B246**, 377 (1990); N. Arkani-Hamed, S. Dimopoulos and G. Dvali, *Phys. Lett.* **B429**, 263 (1998); L. Randall and R. Sundrum, *Phys. Rev. Lett.* **83**, 3370 (1999); *ibid.* **83**, 4690 (1999).
12. S. Dimopoulos and G. Landsberg, *Phys. Rev. Lett.* **87**, 161602 (2001); S.B. Giddings and S. Thomas, *Phys. Rev.* **D65**, 056010 (2002); G.F. Giudice, R. Rattazzi and J.D. Wells, *Nucl. Phys.* **B630**, 293 (2002).
13. R. Penrose, unpublished (1974); D.M. Eardley and S.B. Giddings, *Phys. Rev.* **D66**, 044011 (2002); S.B. Giddings and V.S. Rychkov, "*Black holes from colliding wavepackets*", hep-th/0409131.
14. E. Kohlprath and G. Veneziano, *JHEP* **0206**, 057 (2002).
15. G. 't Hooft, *Phys. Lett.* **B198**, 61 (1987);
I.J. Muzinich and M. Soldate, *Phys. Rev.* **D37**, 359 (1988).
16. D.J. Gross and P.F. Mende, *Phys. Lett.* **B197**, 129 (1987); *Nucl. Phys.* **B303**, 407 (1988); P.F. Mende and H. Ooguri, *Nucl. Phys.* **B339**, 641 (1990).
17. D. Amati, M. Ciafaloni and G. Veneziano, *Phys. Lett.* **B197**, 81 (1987); *Int. J. Mod. Phys.* **A3**, 1615 (1988).
18. D. Amati, M. Ciafaloni and G. Veneziano, *Phys. Lett.* **B216**, 41 (1989);
G. Veneziano, in *Proc. Superstrings 89*, Texas A&M University, 1989, eds. R. Arnowitt *et al.*, World Scientific, Singapore, 1990, p. 86.
19. D. Amati, M. Ciafaloni and G. Veneziano, *Nucl. Phys.* **B347**, 550 (1990); *ibid.* **B403**, 707 (1993).
20. M. Fabbrichesi, R. Pettorino, G. Veneziano and G.A. Vilkovisky, *Nucl. Phys.* **B419**, 147 (1994).
21. G. Veneziano "*Quantum string gravity near the Planck scale*", in *Proc. Pascos Conference*, Northeastern University, Boston, 1990, eds. P. Nath and S. Rencroft, World Scientific, Singapore, 1990, p. 486; "*A new approach to semiclassical gravitational scattering*", in *Proc. 2nd Paris Cosmology Colloquium*, Paris, 1994, eds. H.J. de Vega and N. Sanchez, World Scientific, Singapore, 1995, p. 322.
22. V. N. Gribov, *Sov. Phys.-JETP* **26**, 414 (1968).
23. G. Veneziano, *Mod. Phys. Lett.* **A2**, 899 (1987).
24. J. Koplik and A.H. Mueller, *Phys. Lett.* **B58**, 166 (1975).
25. See, e.g. J. Polchinski, *String Theory*, Vol. II (Cambridge University Press, 1998).
26. N. Sanchez, *Phys. Rev.* **D18**, 1030 (1978); *ibid.* **D18**, 1798 (1978).

MANY FACES OF DIMENSIONAL REDUCTION

A.T. FILIPPOV
Joint Institute for Nuclear Research,
Dubna, Moscow Region, RU-141980
E-mail: filippov@thsun1.jinr.ru

After a brief discussion of dimensional reductions leading to the 1+1 dimensional dilaton gravity theory we consider general properties of these theories and identify problems that arise in its further reductions to one dimensional theories – cosmological models, static states (in particular, black holes) and gravity-matter waves. To bypass shortcomings of the standard ('naive') reduction we propose to exploit more general ideas: 1. separating the space and time variables in generic models, 2. reductions of the moduli spaces in integrable models that may also be viewed as dimensional reductions. This allows us to clearly see a duality between static and cosmological solutions (that we call 'SC-duality') and to demonstrate a close relation of these objects to gravity-matter waves.

1. Introduction

Dimensional reduction in classical physics is a well known matter but, when one is working with gravity, some subtle points that may be ignored in standard physics become more important because geometric characteristics of the space-time become dynamical variables. This is quite clear in the Kaluza-Mandel-Klein-Fock dimensional reduction (KMKF reduction that usually but unjustly is called KK reduction), in which the metric coefficients become physical fields. This may look less clear in further reductions using cylindrical or spherical symmetries. Then the effective space-time becomes 1+1 dimensional, some higher dimensional metric coefficients become dynamical fields that mix with the original matter fields produced by reductions from higher dimensions.

There exist at least two sources of possible problems in dealing with this dimensional reduction. First, by analogy with the usual (naive) reduction, it may be tempting to simply make all the fields to depend only on one variable (space or time). Second, by analogy with the standard practice of classical gravity considerations, one may try to begin with reducing the number of the fields by gauge fixing[a]. Both seemingly innocent actions may in fact lead to incorrect results. In the first case, the solutions of the reduced theory are certainly the solutions of the higher dimensional one but some solutions may be lost. In the second case, the solutions of the reduced theory may not satisfy (and often do not satisfy) the higher dimensional equations of motion. Both these effects may be simply illustrated by considering the reduction of the 1+1 dimensional dilaton gravity to 1+0 ('cosmology') or 0+1 ('static state') dimensional one.

To avoid these problems we propose the following, more general approach to dimensional reduction. First, one should not fix the gauge before writing all the equations of motion. For example, the number of independent fields in the reduced theory should be not less than the number of the independent Einstein equations for the Ricci tensor plus the number of the equations for the matter fields. Second, instead of the naive reduction one should use the more general procedure of separating the space and time variables – 'separate and reduce!'. In the linear case this is a well defined procedure. When the equations of motion are essentially nonlinear, a separation is more difficult to find but there may exist more possibilities.

At the moment we cannot give a motivated classification of these possibilities and will only give simple examples using dimensional reductions from 1+1 dimensional dilaton gravity to the 1+0 and 0+1 dimensions. In particular, we may mention the earlier proposed reduction that may be called a 'dynamical dimensional reduction' or a 'moduli space reduction'. In this example, a class of reduced solutions of the two dimensional theory consists of those that essentially depend on two space-time variables (say, r and t) which nevertheless should be regarded as 'one dimensional solutions' in a well defined but somewhat unusual sense. Unfortunately, at the moment we can introduce this new dimensional reduction only for an explicitly integrable dilaton gravity theory[b]. Here we propose and briefly discuss an apparently new dimensional reduction based on separating space and time dependence of gravitational and matter fields in generic, not necessary integrable, dilaton gravity.

[a]The 1+1 dimensional theory we are speaking of is a gauge theory with two first class constraints and thus two gauge fixing conditions may be applied.

[b]Even more unfortunate circumstance is that this theory is not exactly a dimensional reduction of a higher dimensional gravity or supergravity theory although it may be considered as an approximation to some higher dimensional supergravity theory. In any case it certainly is a good model of higher dimensional gravity and supergravity having horizons and giving interesting cosmological models and gravity-matter waves.

To avoid misunderstanding, let us formulate the practical 'philosophy' behind the approach to gravity advocated in this paper. As distinct from the common tendency to concentrate on geometric and symmetry properties, we follow the Arnowitt-Deser-Misner approach in treatment gravity by using Lagrangian and Hamiltonian dynamics with constraints. Thus the 'geometric' variables are treated on the same footing with other dynamical variables and the aim is not only to derive the metric and other geometric properties of the space-time but to construct Lagrangians and Hamiltonians and to solve the dynamical equations which, eventually, should be quantized. To quantize such a complex nonlinear theory as gravity one should first find some simple explicitly integrable approximation, like the oscillator approximation in the standard QFT. Natural candidates for such 'gravitational oscillators' may be black holes, cosmological models and some simplest gravitational waves. One may argue that all these objects are somehow related to the Liouville equation rather than to the oscillator equation. Although one should not expect that such simplified models can give completely realistic description of gravity, cosmology or gravitational waves, they may serve as a tool to develop new intuition, which is so needed for understanding new data on the structure of our Universe. They also may give reasonable first approximations for constructing more realistic solutions and also give us some hints of how the three main gravitational objects are related physically (at the moment we find only mathematical relations). Using explicitly integrable models one may clearly see a duality between black holes and cosmologies as well as observe that they both are limiting cases of certain gravitational waves. The duality can also be seen in nonintegrable models while the 'triality' including some gravitational waves was at the moment observed only in integrable theories.

2. 1+1 Dimensional Dilaton Gravity

There exist a class of 1+1 dimensional dilaton gravity theories coupled to scalar matter fields that is a reliable model for some aspects of high dimensional black holes, cosmological models and branes. The connection between high and low dimensions has been demonstrated in different contexts of gravity and string theory and in some cases allowed to find general solution or some special classes of solutions in high dimensional theories. For a more detailed discussion of this connection and references see e.g. [1,2,3,4,5]. For example, spherically symmetric gravity coupled to Abelian gauge fields and massless scalar matter fields exactly reduces to a 1+1 dimensional dilaton gravity and can be explicitly solved if the scalar fields are constants independent of coordinates. Such solutions may describe some interesting physical objects – spherical static black holes, simplest cosmologies, etc. However, when the scalar matter fields, which presumably play a significant cosmological role, are not constant, few exact analytical solutions of high dimensional theories are known. Correspondingly, the generic two-dimensional models of dilaton gravity nontrivially coupled to scalar matter are usually not integrable.

To obtain integrable models of this sort one has to make some serious approximations, in other words, to deform the original two-dimensional model obtained by a direct dimensional reductions of realistic higher dimensional theories[c]. Nevertheless, these models may qualitatively describe some physically interesting solutions of higher dimensional gravity or supergravity theories related to the low energy limit of superstring theories.

The effective Lagrangian of the 1+1 dimensional dilaton gravity coupled to scalar fields ψ_n that may be obtained by dimensional reductions of a higher dimensional spherically symmetric (super)gravity can usually be (locally) transformed to the following form:

$$L = \sqrt{-g}\,[U(\varphi)R(g) + V(\varphi,\psi) + W(\varphi)(\nabla\varphi)^2 + \sum_n Z_{nm}\nabla\psi_n\nabla\psi_m]. \quad (1)$$

Here $g_{ij}(x^0, x^1)$ is a generic 1+1 metric with signature (-1,1), $g \equiv \det|g_{ij}|$ and $R \equiv R(g)$ is the Ricci curvature of the two-dimensional space-time,

$$ds^2 = g_{ij}\, dx^i\, dx^j\,, \quad (i,j = 0,1). \quad (2)$$

The effective potentials V and Z_{nm} depend on the dilaton $\varphi(x^0, x^1)$ and on $N-2$ scalar fields $\psi_n(x^0, x^1)$ (note that Z_{nm} define a negative definite quadratic form). They may depend on other parameters characterizing the parent higher dimensional theory, e.g. on charges introduced by solving the equations for the Abelian gauge fields, etc. Here we mainly consider the simplest kinetic terms, with $Z_{nm}(\varphi,\psi) = \delta_{nm}Z_n(\varphi)$, or even with

[c]Note that some important four-dimensional space-times having symmetries defined by two commuting Killing vectors may also be described by two-dimensional dilaton gravity. For example, the cylindrical gravitational waves are described by the 1+1 dimensional dilaton gravity coupled to one scalar field. The stationary axially symmetric pure gravity may be described by a 0+2 dimensional dilaton gravity coupled to one scalar field (this may be related to the previous cylindrical case by the analytic continuation of one space variable to imaginary values). Similar but more general dilaton gravity models were also obtained in string theory. Some of them may be solved by using modern mathematical methods developed in the soliton theory.

constant Z_n that are independent of the fields. The dilaton function $U(\varphi)$ is usually monotonic and one may choose (at least locally) $U(\varphi) = \varphi$ or $U(\varphi) = \exp(-2\varphi)$, etc. We also may use in Eq. (1) a Weyl transformation to exclude the gradient term for the dilaton, i.e. to make $W \equiv 0$.

Note that in two dimensional space-times all matter fields may eventually be reduced to different scalar fields although, for keeping traces of different symmetries, it may be convenient to retain gauge fields, spinor fields, etc. The Lagrangian Eq. (1) should be considered as an effective Lagrangian. In general, it is equivalent to the original one on the 'mass shell' but the solutions of the original equations may be completely recovered to construct the solutions of the higher dimensional 'parent' theory. For a detailed motivation and specific examples see [5], where references to other related papers may be found. Due to space limitations, only absolutely necessary references are given here.

To simplify derivations we will use the equations of motion in the light-cone metric, $ds^2 = -4f(u,v)\,du\,dv$ and with $U(\phi) \equiv \varphi$, $Z_{nm} = \delta_{nm} Z_n$, $W \equiv 0$. By first varying the Lagrangian in generic coordinates and then going to the light-cone ones we obtain the equations of motion

$$\partial_u \partial_v \varphi + f\, V(\varphi, \psi) = 0, \tag{3}$$

$$f \partial_i (\partial_i \varphi / f) = \sum Z_n (\partial_i \psi_n)^2 , \qquad (i = u, v) . \tag{4}$$

$$\partial_v (Z_n \partial_u \psi_n) + \partial_u (Z_n \partial_v \psi_n) + f V_{\psi_n}(\varphi, \psi) = \sum_m Z_{m,\psi_n}\, \partial_u \psi_m\, \partial_v \psi_m \ , \tag{5}$$

$$\partial_u \partial_v \ln |f| + f V_\varphi(\varphi, \psi) = \sum Z_{n,\varphi}\, \partial_u \psi_n\, \partial_v \psi_n \ , \tag{6}$$

where $V_\varphi = \partial_\varphi V$, $V_{\psi_n} = \partial_{\psi_n} V$, $Z_{n,\varphi} = \partial_\varphi Z_n$, and $Z_{m,\,\psi_n} = \partial_{\psi_n} Z_m$. These equations are not independent. Actually, (6) follows from (3) – (5). Alternatively, if (3), (4), (6) are satisfied, one of the equations (5) is also satisfied.

If the scalar fields are constant, $\psi = \psi_0$, these equations can be solved with a practically arbitrary potential V that should satisfy only one condition: $V_\psi(\varphi, \psi_0) = 0$, see Eq.(5). The constraints (4) can then be solved because their right-hand sides are identically zero. It is a simple exercise to prove that there exist chiral fields $a(u)$ and $b(v)$ such that $\varphi(u,v) \equiv \varphi(\tau)$ and $f(u,v) \equiv \varphi'(\tau)\, a'(u)\, b'(v)$, where $\tau \equiv a(u) + b(v)$ (the primes denote derivatives with respect to the corresponding argument). Using this result it is easy to prove that (3) has the integral $\varphi' + N(\varphi) = M$, where

$N(\varphi)$ is defined by the equation $N'(\varphi) = V(\varphi, \psi_0)$ and M is the constant (integral) of motion. The horizon, defined as a zero of the metric $h(\tau) \equiv M - N(\varphi)$, exists because the equation $M = N(\varphi)$ has at least one solution in some interval of values of M. These solutions are actually one-dimensional ('automatically' dimensionally reduced) and can be interpreted as black holes (Schwarzschild, Reissner Nordstrøm, etc.) or as cosmological models. These facts are known for a long time and were derived by many authors using different approaches (for references see e.g. the recent review [6]).

3. Dimensional Reduction

With this example in mind, it looks natural to introduce the following reduction to one dimensional theory: let φ and ψ depend only on $\tau \equiv a(u) + b(v)$ where τ may be interpreted either as the space or the time variable. Then we obtain both the 0+1 dimensional theory of static distributions of the scalar matter (including black holes) and 1+0 dimensional cosmological models. With nontrivial potentials $V(\varphi, \psi)$ these theories may be integrable if the equations of motion are equivalent to systems of the Liouville equations or to the Toda lattices. Analyzing their solutions (see simple examples in [3]) one can find that not all standard Friedmann cosmologies may be obtained in this way [5,7]. In view of the apparent symmetry ('duality') between the 1+0 and 0+1 reductions one may conclude that not all static solutions are obtained in this way. In other words, this simple (naive) procedure of dimensional reduction is incomplete!

For finding more general reductions we study possible separations of the space and time variables in the equations of motion. To illustrate connections of the equations (3)-(6) to higher dimensional theories let us consider a special example of the 4-dimensional spherically symmetric gravity coupled to one scalar field:

$$S_4 = \int d^4x \sqrt{-g_4}\,[R_4 + V_4(\psi) + Z_4(\psi)(\nabla\psi)^2]. \tag{7}$$

Here the most general spherically symmetric metric should be used,

$$ds_4^2 = e^{2\alpha}dr^2 + e^{2\beta}d\Omega^2(\theta,\phi) - e^{2\gamma}dt^2 + 2e^{2\delta}drdt\,, \tag{8}$$

where $\alpha, \beta, \gamma, \delta$ depend on t, r and $d\Omega^2(\theta,\phi)$ is the metric on the 2-dimensional sphere $S^{(2)}$. Integrating over the variables θ, ϕ one may find the reduced action[d] with the Lagrangian Eq. (1), where

$$U \Rightarrow e^{2\beta}, \;\; V \Rightarrow 2 + e^{2\beta} V_4, \;\; W(\nabla\phi)^2 \Rightarrow 2e^{2\beta}(\nabla\beta)^2, \;\; Z \Rightarrow Z_4(\psi) e^{2\beta}, \quad (9)$$

and the 2-dimensional metric is given by $e^{2\alpha}, e^{2\gamma}, e^{2\delta}$ (see e.g. [5]). Actually, the effective two dimensional Lagrangian also contains total derivatives that may be important in some problems but we will not discuss them here.

The equations of motion for this effective action can easily be derived and they coincide with Eqs. (3)-(6) if we pass to the light cone coordinates. It is not difficult to see (in fact, it is almost evident) that these EOM are identical to the Einstein equations (see e.g. [1]) To simplify the equations we write them in the limit of the diagonal metric (formally, one may take the limit $\delta \to -\infty$). Varying the action in α, β, γ and neglecting the δ-terms we obtain the Einstein equations for the diagonal components of the Einstein tensor. Varying the action in ψ we find the equation for ψ. Finally, by varying in δ we find one more equation corresponding to the non diagonal component of the Einstein tensor; it is not a consequence of other equations and is a combination of the two constraints (4).

The simplest way to write all necessary equations is to write the 2-dimensional effective action in the coordinates (8). Making first variations in δ we find (in the limit $\delta \to -\infty$) the constraint

$$4\dot{\beta}' - 2\dot{\beta}\beta' - 2\dot{\beta}\gamma' - 2\dot{\alpha}\beta' \;=\; Z_4 \dot{\psi}\psi', \quad (10)$$

where $\dot{\psi} \equiv \partial_t \psi$, $\psi' \equiv \partial_r \psi$, etc. The other equations can be derived (in the diagonal limit) from the effective Lagrangian

$$L_{\rm eff} \;=\; V_{\rm eff} + L_{\rm t} + L_{\rm r}, \quad (11)$$

where we omitted the δ-dependence and total derivative terms. The sum of the 'r-Lagrangian'

$$L_{\rm r} \;=\; e^{-\alpha+2\beta+\gamma}(2\beta'^2 + 2\beta'\gamma' + Z_4\psi'^2), \quad (12)$$

with the 't-Lagrangian'

$$L_{\rm t} \;=\; -e^{\alpha+2\beta-\gamma}(2\dot{\beta}^2 + 2\dot{\beta}\dot{\gamma} + Z_4\dot{\psi}^2), \quad (13)$$

[d] This derivation can easily be generalized to any dimension and any number of the scalar fields with more complex coupling potentials. One may similarly treat the pseudospherical, flat and cylindrical symmetries as well as any symmetry given by two Killing vectors.

as well as the constraint (10) are invariant under the substitution $\partial_r \Leftrightarrow i\partial_t$ and $\alpha \Leftrightarrow \gamma$. This means that the EOM are invariant under this transformation, as the effective potential [e],

$$V_{\rm eff} \;=\; V_4\, e^{\alpha+2\beta+\gamma} + 2k\, e^{\alpha+\gamma}, \quad k=0,\pm 1, \tag{14}$$

is naturally invariant. At the first sight this invariance may look trivial but one should recall that in higher dimensions there is no complete symmetry between space and time. Thus the simple relation between static and cosmological solutions suggested by this symmetry may give some new insight into both classes of objects. Even apart from any physical interpretation, this symmetry allows us to economize writing equations and it is extremely useful in considering separation of variables outlined below. In particular, these transformations allow one to derive cosmological solutions corresponding to static (black hole) solutions and vice versa. Although this is a special case of the formulated duality relation we call it 'static-cosmological' (SC) duality.

To illustrate how the separation of the variables looks like we write the three remaining equations (in addition to Eq. (9)):

$$[2e^{-2\alpha}(\beta'' + 2\beta'^2 - \beta'\alpha' + \beta'\gamma')] - [\alpha \Leftrightarrow \gamma,\ \partial_r \Rightarrow \partial_t] \;=\; V_{\rm eff}\,, \tag{15}$$

$$[2e^{-2\alpha}(\beta'' + \beta'^2 - \beta'\alpha' - \beta'\gamma')] + [\alpha \Leftrightarrow \gamma,\ \partial_r \Rightarrow \partial_t] \;=\; \frac{1}{2}Z_4(E_+ - E_-)\,, \tag{16}$$

where we denote

$$E_\pm \;\equiv\; e^{-2\alpha}\psi'^2 \;\pm\; e^{-2\gamma}\dot{\psi}^2. \tag{17}$$

The third equation has a similar structure

$$[e^{-2\alpha}(\gamma'' + \gamma'^2 - \gamma'\alpha' - \beta'^2] - [\alpha \Leftrightarrow \gamma,\ \partial_r \Rightarrow \partial_t] \;=\; -ke^{-2\beta} + \frac{1}{2}Z_4 E_-\,, \tag{18}$$

In practical procedures of the separation it may be convenient to also use the forth (dependent) equation and work with some other linear combinations of all equations.

Now it is clear that to separate the variables r and t one should first require that

$$\alpha = \alpha_0(t) + \alpha_1(r), \quad \beta = \beta_0(t) + \beta_1(r), \quad \gamma = \gamma_0(t) + \gamma_1(r), \tag{19}$$

[e] Here, in addition to the case of the spherical symmetry ($k=1$) we include the cases of pseudospherical ($k=-1$) and flat ($k=0$) symmetries.

and, depending on the analytic form of the potentials V_4 and Z_4,

$$\psi = \psi_0(t) + \psi_1(r), \quad \text{(a)} \quad \text{or} \quad \psi = \psi_0(t)\psi_1(r), \quad \text{(b)}\,. \tag{20}$$

Here we will not try to find and classify all possible cases of separation and mention only typical ones. If $\dot{\psi}\psi' = 0$, a separation is possible for generic potentials. If $\dot{\psi}\psi' \neq 0$, there are three obvious classes of the potentials that are favorable to the separation: 1. constant potentials V_4 and Z_4; 2. exponential V_4 and constant Z_4 (with the ansatz (20a); 3. power dependence of V_4 on ψ and constant Z_4 (with the ansatz (20b). Note that the case of the constant Z_4 and exponential V_4 is often met in standard reductions from higher dimensions.

Inserting ansatzes (19), (20) into the equations (15), (16), (18) one may find conditions for the separation. We regard possible separation as dimensional reductions in the sense that we try to clarify by the following example. Let us consider the case $\dot{\psi}\psi' = 0$. Then it is clear that the constraint (10) is rather restrictive. In fact one can see that it allows the separation (19) in the following three cases:

$$\begin{aligned} 1\,.\ \dot{\alpha_0} &= \dot{\beta_0}\,; \\ 2\,.\ \dot{\beta_0} &= {\beta_1}'\,; \\ 3\,.\ {\beta_1}' &= {\gamma_1}'\,. \end{aligned} \tag{21}$$

When no one of this conditions is satisfied, the separation is possible if

$$4.\ 1 + {\gamma_1}'/{\beta_1}' + \dot{\alpha_0}/\dot{\beta_0} = 0 \tag{22}$$

Other equations give further restrictions on the metric coefficient but one may see that they may be of somewhat different strength. Some of the restrictions may be so strong that they almost fix some parameters.

To make this point clearer suppose that we have chosen the last of the three conditions (21) that says that β_1 and γ_1 are some constants. Then the reduction idea is to consider α_0, β_0, γ_0, ψ_0 as dynamical functions of cosmological model while strongly restricted functions α_1, β_1, γ_1, (ψ_1 is a constant) then will define parameters of the background metric. Let us illustrate this idea by considering Eq. (15)

$$e^{-2\beta_1+2\alpha_0+2\beta_0} - e^{-2\gamma_1+2\alpha_0-2\gamma_0}(\ddot{\beta}_0 + ...) + e^{-2\alpha_1}(\beta_1'' + ...) = \frac{1}{2}V_4 e^{2\alpha_0}, \tag{23}$$

where we omitted some terms in brackets that may be recovered using Eq. (15). As β_1, γ_1 are constants, all terms in this equation depend on t only. It is not difficult to similarly treat two remaining equations and thus to find the complete system of equations for α_0, β_0, γ_0, ψ_0. Although α_1 remains an arbitrary function of r it can be transformed to any constant by an appropriate r-transformations. Of course one may see that we recovered the 'naive' cosmological reduction. The 'naive' static reduction may be reproduced simply by using our SC-duality. With less restrictive ansatzes we may construct other cosmological models and static configurations that are dual to them. It is interesting that there may exist some 'intermediate' cases that are more symmetric under the duality transformation. It would be premature to call them 'self-dual' before a detailed study of them will be completed.

4. Quasi Conclusion

In a separate publication I will present the list of all these reductions and their relation to black holes, cosmologies, and waves. Here I only mention the wave-like solution obtained in the integrable 1+1 dimensional gravity theory coupled to $N-2$ scalar matter fields [5,7]. The general solution of the model depends on the chiral moduli fields $\xi_n(u)$, $\eta_n(v)$ that move on the surfaces of the spheres $S^{(N-2)}$. The naive reduction to one-dimensional theories emerges when the moduli fields are constant and equal, $\xi_n = \eta_n$. When they are constant but otherwise arbitrary, we have a new class of reduced solutions that correspond to waves of scalar matter coupled to gravity. Under certain conditions these waves may be localized in space and time and thus may be regarded as a sort of solitary gravitational waves[f]. The very origin of these waves signals existence of a close relation between main gravitational objects – black holes, cosmologies and waves. This relation was studied in some detail for static states and cosmologies that may be considered as a sort of 'static-cosmological' duality (SC-duality). In the integrable models the transitions between static and cosmological states are possible and, moreover, the waves play a significant role in these transitions. This observation, which does not actually require integrability, may open a way to studies of real physical connections between these apparently diverse objects.

[f] A special solution of this kind has recently been found in [7] and is generalized and discussed in detail in the forthcoming paper [8]. Note that our solitary wave do not seem to have a relation to possible soliton-like states in the theories with the 'sigma-model' – like coupling of the scalar fields to gravity studied in Refs. [9,10,11,12,13] (see also a discussion in Ref. [7]).

In summary, one may identify at least three types of dimensional reduction: the 'standard' or 'naive' reduction supposes that functions of two variables depend on one variable only, the reduction by separating the variables, and the reduction in moduli spaces supposing that the moduli functions become constants. In all cases the important problem is to find the Lagrangians and Hamiltonians for the reduced systems. This is not difficult for the naive reduction and for the simple reductions based on separating the variables. It is not clear how to do this with the last, so to speak, 'moduli reduction'. In addition, it is not clear how to do such a reduction for not integrable systems.

I mentioned above the important problem of gauge fixing but have not really considered it here. It deserves a very careful and deep study using what we know about somewhat simpler nonabelian gauge systems not coupled to gravity. In this connection I often recall Gribov's classical work on gauge copies (Gribov's copies). This beautiful piece of ingenuity was also based on a dimensional reduction. Being apparently very simple and natural it gave a very strong result. Working with gauge theories, nonlinear waves and solitons I often recalled this lesson and I am sure it will be of great help in considering gravitational and cosmological theories. Naturally, in writing this short essay I tried to present my simple idea so as it would be more or less acceptable to Volodya. I am not sure that I succeeded but, regretfully, this is not possible to check in this world...

Acknowledgments

I very much appreciate the support of the Department of Theoretical Physics of the University of Turin and INFN (Section of Turin), of the MPI and the W. Heisenberg Institute (Munich), where some results were obtained. I am especially grateful to V. de Alfaro for his support for many years and very fruitful collaboration.

References

1. P. Thomi, B. Isaak and P. Hajicek, *Phys. Rev.* **D30**, 1168 (1984).
2. P. Breitenlohner, G. Gibbons and D. Maison, *Commun. Math. Phys.* **120**, 295 (1988).
3. A.T. Filippov, *Mod. Phys. Lett.* **A11**, 1691 (1996); *Int. J. Mod. Phys.* **A12**, 13 (1997).
4. A.T. Filippov, *Integrable models of black holes in supergravity and superstring theory*, *Phys. Part. Nucl.* **32S1**, 38 (2001); *Yadernaya Fizika* **65**, 997 (2002) [*Phys. Atom. Nucl.* **65**, 963 (2002)].

5. V. de Alfaro and A.T. Filippov, *Integrable Low Dimensional Models for Black Holes and Cosmologies from High Dimensional Theories*, arXiv:hep-th/0504101.
6. D. Grumiller, W. Kummer and D. Vassilevich, *Phys. Rep.* **369**, 327 (2002).
7. A.T. Filippov, *Integrable Models of 1+1 Dimensional Dilaton Gravity Coupled to Scalar Matter*, arXiv:hep-th/0505060.
8. V. de Alfaro and A.T. Filippov, in preparation.
9. V.A. Belinskii and V.E. Zakharov, *Sov. Phys. JETP* **48**, 985 (1978).
10. D. Maison, *Phys. Rev. Lett.* **41**, 521 (1978).
11. H. Nicolai, D. Korotkin and H. Samtleben, *Integrable classical and quantum gravity*, arXiv:hep-th/9612065.
12. A.T. Filippov and V.G. Ivanov *Phys. Atom. Nucl.* **61**, 1639 (1998); arXiv:hep-th/9803059.
13. G. Alekseev, *Monodromy data parametrization of the spaces of local solutions of integrable reductions of Einstein's field equations*, arXiv:gr-qc/0503043.

QUANTUM EINSTEIN GRAVITY AND ASYMPTOTIC SAFETY

PÉTER FORGÁCS

LMPT, CNRS-UMR 6083, Univ. de Tours
Parc de Grandmont, Tours, 37200, FRANCE
and
KFKI RMKI, Theoretical Physics Department
P.O. Box 49, Budapest, 1525, HUNGARY

I review some of the salient points of the construction of a perturbative quantum theory of the dimensionally reduced pure Einstein gravity from 4 to 2 dimensions assuming that the 4 dimensional (4D) metrics admit two commuting Killing vectors. The dimensionally reduced theory corresponds to an O(1,2) symmetric σ-model coupled to two scalar fields in flat spacetime. It inherits the lack of standard perturbative renormalizability from 4D gravity, however, it turns out that strict cutoff independence can be achieved to all loop orders in a space of Lagrangians differing only by a field dependent conformal factor. The renormalization group flow possesses a unique non-Gaussian fixed point at which the trace anomaly vanishes. The existence of this non-Gaussian fixed point is compatible with Weinberg's "asymptotic safety" scenario.

1. Introduction

It is a great honor for me to give a talk at this conference to commemorate the 75th birthday of the outstanding physicist, Volodya Gribov, in front of an exceptional audience gathered together in Budapest to pay tribute to his memory. I had the privilege to be a colleague of Volodya for quite some years and during that time I could profit from some of his unique physical insights in the course of numerous discussions with him. Since I got to know Volodya to a certain degree, I am now pretty sure that if he were here he would very likely express a high degree of scepticism to the subject of this talk, which is about a relatively straightforward quantum field theoretic approach to such a speculative and conceptually unclear subject as Quantum Gravity. And I do not even have to offer any results of clear physical significance! Volodya would be of course totally right, and the results I present here should considered only as a feasibility study of a not totally straightforward approach to construct a theory which could eventually be relevant to Quantum Gravity (even if it is not quite clear what it would exactly be).

It is well known that the theory that is often referred to as "Quantum Einstein Gravity" (QEG) is based on the classical action

$$S[g] = -\frac{1}{16\pi G}\int d^4x\,\sqrt{-g}\;[R(g) + 2\Lambda]\;, \tag{1}$$

where g stands for the 4D metric tensor, Λ resp. G denotes the cosmological resp. Newton's constant, is *not renormalizable* in the standard graviton loop expansion around a fixed background spacetime [1,2,3]. Therefore perturbation theory (PT) around the Gaussian fixed-point can only be used to define an "effective", procedure-dependent QEG. Nevertheless, it is still possible to extract some *unambiguous* quantum gravity corrections from this effective theory, for example one can compute the leading large distance correction to Newton's potential [4]:

$$V(r) = -G\frac{m_1 m_2}{r}\left[1 + 3\frac{G(m_1+m_2)}{c^2 r} + \frac{41}{10\pi}\frac{l_{\rm Pl}^2}{r^2} + \ldots\right], \tag{2}$$

where $l_{\rm Pl}$ denotes Planck's length. Quantitatively such corrections are way too small, however, to be physically relevant, (recall that $l_{\rm Pl} = \sqrt{G\hbar/c^3} \approx 10^{-33}$cm). Taken at face value (2) would indicate that the effective coupling, $G(k)$ increases for increasing momentum scales, e.g.

$$G(k) \approx \frac{G_0}{1 - \alpha G_0 k^2} \qquad k \sim \frac{\xi}{r}\,, \tag{3}$$

where α is a (positive) number, k is the momentum scale, and ξ is a constant. The non-renormalizability of the Einstein-Hilbert action in a perturbative expansion around a fixed background spacetime is often taken to indicate that a quantum theory based on it is meaningful only as an effective field theory with a domain of applicability that is limited both in the infrared and in the ultraviolet (UV) regime.

S. Weinberg has put forward the idea that one could maybe define QEQ in a Wilsonian framework if the averaged renormalization group flow in a suitable "theory space" had a hidden UV stable fixed point, with a finite dimensional critical manifold [1]. This would be equivalent to quantum Einstein gravity being non-perturbatively renormalizable – a scenario whose verification or refutation is clearly of outstanding theoretical significance. Some version of this "asymptotic safety" scenario arguably underlies also the dynamical triangulations approach to quantum gravity and, at least indirectly, the canonical quantization program of A. Ashtekar and its ramifications [5]. The scenario lay mostly dormant for a long time, presumably for the very considerable conceptual and technical difficulties needed to be overcome to produce evidence either in favour or against it. Since the last few years, however, in a series of works important evidence for its viability has been reported by M. Reuter and his collaborators [6,7]. Their setting is that of the 'exact renormalization group equations' (ERGE), i.e. the functional RG flow equations based on effective average actions. Due to the complexity of the problem only the (extended) Einstein-Hilbert truncation has been studied, corresponding to a projection of the averaging flow into a 2 (or 3) dimensional subspace in which a non-Gaussian fixed point is seen [7].

In this talk, based on joint work with Max Niedermaier [8], I report on a complementary body of evidence in favour of the asymptotic safety scenario based on quantization of a truncation of Einstein's gravity i.e. on restricted phase space quantization. In particular considering spacetimes with two Killing vectors simplifies sufficiently the problem of quantization, so that rather complete results can be obtained on the problem of a non-Gaussian fixed point. The reduced system turns out to inherit the non-renormalizability of full Einstein gravity (beyond one loop). The renormalization flow will therefore move in an *infinite* dimensional space of Lagrangians constrained only by the initial symmetry requirement. This space can be parameterized by functions $\mathbf{h}(\,\cdot\,)$ of one real variable which generalize the notion of a renormalized coupling. Functions connected by the renormalization flow $\mu \rightarrow \bar{\mathbf{h}}(\,\cdot\,,\mu)$ defined below are regarded as equivalent. In this infinite dimensional theory space one can obtain UV renormalizability in the sense of strict cutoff independence, and to search for fixed points. Remarkably there is unique fix point and moreover the renormalization flow appears to be driven locally toward the fixed point.

Of course one cannot expect that truncation of classical degrees of freedom and quantization are commuting operations. It is, however, hard to imagine that a symmetry reduction of classical degrees of freedom would make the UV renormalization properties of the original theory worse. In other words our analysis also provides a complementary test of the asymptotic safety scenario: If quantum Einstein gravity indeed has a hidden fixed point it should emerge already in the symmetry reduced quantum theory (but clearly not the other way round).

2. Dimensional Reduction Of Gravity

Let me start by sketching the dimensional reduction of gravity from 4D to 2D [9,10]. One is interested in the solutions of Einstein's equations with two commuting Killing vectors. Starting with the standard Einstein-Hilbert action in 4 dimensions, with $\Lambda = 0$ its Lagrangian is given by Eq. (1). Assuming that the 4D metric, g, admits two commuting Killing vectors, K_1, K_2, the 4D line element can be parameterized in an adapted coordinate system (e.g. when $K_1 = \partial/\partial x_2$, $K_2 = \partial/\partial x_3$) by functions depending on $x^\mu = (x^0, x^1)$ only. The ansatz for the line element then reads

$$ds^2 = \gamma_{\mu\nu}(x)dx^\mu dx^\nu - \rho(x)M_{ab}(x)dx^a dx^b\,, \quad x^a = (x^2\,, x^3)\,, \tag{4}$$

where $\gamma_{\mu\nu}(x)$ is a 2D metric with Minkowskian signature if both Killing vectors are spacelike, and Euclidian if one Killing vector is spacelike and one is timelike. The above 2 Killing vector reduction of gravity covers a variety of physically interesting situations: If one of the Killing vectors is timelike these are stationary axisymmetric spacetimes, containing among others in particular all the prominent (Schwarzschild and Kerr) black hole solutions. When both Killing vectors are spacelike the corresponding spacetimes comprise cylindrical gravitational waves, colliding plane gravitational waves, as well as generalized Gowdy cosmologies. To avoid keeping track of $\pm$ signs in the formulae I concentrate on the case of two spacelike Killing vectors. $M_{ab}(x)$ in Eq. (4) is a symmetric 2×2 matrix which in our case can be normalized to have determinant $\det M = 1$. The matrix M can be conveniently parameterized in terms of 'hyperbolic spins' n^j, $j = 0, 1, 2$, where $n \cdot n \equiv (n^0)^2 - (n^1)^2 - (n^2)^2 = 1$, in the following way:

$$M = \begin{pmatrix} n^0 + n^2 & -n^1\,, \\ -n^1 & n^0 - n^2 \end{pmatrix}\,. \tag{5}$$

The fields, $n = (n^0, n^1, n^2)$, are in fact mappings from the 2D spacetime manifold into the hyperboloid $H = \{n = (n^0, n^1, n^2) \in \mathbb{R}^3 \,|\, n \cdot n = (n^0)^2 - (n^1)^2 - (n^2)^2 = 1\}$. Assuming that the 2D metric, $\gamma_{\mu\nu}(x)$, is conformally flat (conformal gauge):

$$\gamma_{\mu\nu} = \eta_{\mu\nu}\, e^\sigma\,, \quad \eta_{\mu\nu} = \begin{pmatrix} 1 & 0 \\ 0 & -1 \end{pmatrix}\,. \tag{6}$$

In the conformal gauge $\gamma_{\mu\nu} = e^\sigma \eta_{\mu\nu}$ the Einstein-Hilbert action (1) can be recast into that of a flat space O(1,2) non-linear σ-model coupled to two scalar fields, σ, ρ:

$$L(n, \rho, \sigma) = -\frac{1}{2\lambda}[\rho\partial^\mu n \cdot \partial_\nu n + \partial^\mu \rho \partial_\mu (2\sigma + \ln \rho)]\,, \tag{7}$$

where $n \cdot n = 1$ and λ is Newton's constant per unit volume of the internal space. (I apologize for the improper use of "σ" standing for two completely different things: non-linear σ-models and at the same time "σ" also denotes the single (Liouville-type) scalar field of the the 2D metric tensor in the conformal gauge. I can only hope that my abuse of notation does not lead to confusion.) The reduced Einstein-Hilbert action (7) contains 2D gravity (described by the scalar field σ) which is non-minimally coupled via a "radion" field, ρ, to a 2D matter system. As we have seen the radion field is related to the determinant of the internal metric, the matter system being a noncompact $O(1,2)$ nonlinear σ-model. It is important that both the reduced Einstein equations and the induced symplectic structure follow from a two-dimensional action principle, which we take as the basis for developing the quantum theory. The phase space is then characterized by the equations of motion and the symplectic structure following from the 2D flat space action corresponding to the Lagrangian (7). The symplectic structure associated to the phase space (the set of all solutions) of the variational eqs. following from (7) is the one induced by the standard symplectic structure of general relativity. Furthermore, to account for the original 2D diffeomorphism invariance, the weak vanishing of the hamiltonian and the diffeomorphism constraints $\mathcal{H}_0 \approx 0$, $\mathcal{H}_1 \approx 0$ has to be imposed. The latter ones are given by $\mathcal{H}_0 = T_{00}$ and $\mathcal{H}_1 = T_{01}$, where $T_{\mu\nu}$ denotes the classical energy momentum tensor derived from (7). The trace T^{μ}_{μ} vanishes on-shell. Classical observables are quantities which Poisson commute with these constraints and are weakly nonvanishing.

A remarkable feature of the 2-Killing reduced gravity is that its field equations can be expressed as compatibility conditions of an associated linear problem (Lax-pair), i.e. this subsector of Einstein's equations is in some sense "integrable" [11]. This makes this reduced system a particularly interesting laboratory for studying quantum aspects of general relativity [12].

3. Quantization of the Dimensionally Reduced System

In order to quantize (à la Dirac) the system (7), the vector n as well as ρ, σ are considered as independent quantum fields whose dynamics is governed by the action (7). The constraints are to be imposed subsequently as composite operators in terms of which the projection onto the physical state space is formulated. Clearly the key issue to be addressed is that of the *renormalizability* of an action functional motivated by the Lagrangian (7) together with its symmetries (extended by suitable sources needed to define composite operators). In contrast to the standard nonlinear σ-models with O(n)-type symmetries without coupling to gravity, it turns out that the quantum field theory based on (7) is *not* UV renormalizable. Renormalizability in the strict quantum field theoretical sense typically presupposes that the bare and the renormalized (source extended) action have the same functional form, only the arguments of that functional (fields, sources, and coupling constants) get renormalized. Though the bare action is motivated by the classical one it can be very different from it. In any case the form of the (bare=renormalized) action functional is meant to be known beforehand. The 2-Killing vector reduction of general relativity has been known for some time to be renormalizable at the 1-loop level [13]. This may not be that surprising since Einstein's gravity in 4D is known to be 1-loop finite [2]. We have found that the reduced gravity system given by (7) is not renormalizable in the above "strict" sense beyond one loop in PT. This implies that the reduced system still exhibits a rather important conceptual and technical problem present when trying to quantize full Einstein gravity [2,3]. The non-renormalizability arises because the dependence of the bare Lagrangian (7) on the dimensionless two dimensional scalar radion field, ρ, is not constrained by any Noether symmetry. Thus it can – and does – enter the counter terms in a different way as in the bare action, no matter how the latter is chosen.

The starting point to define a quantum theory based on the bare Lagrangian (7) is the observation that one *can define* a renormalized quantum theory in an *enlarged space* of Lagrangians, $L_{\mathbf{h}}(\rho)$, which differ by an overall conformal factor, $\mathbf{h}(\rho)$, which depends exclusively on ρ. More precisely we consider the following family of Lagrangians $L_{\mathbf{h}}$ labelled by $\mathbf{h}(\,\cdot\,)$ which are real valued functions:

$$L_{\mathbf{h}}(n,\rho,\sigma) = -\frac{1}{2\lambda}\mathbf{h}(\rho)\left[\partial^\mu n\cdot\partial_\mu n + \rho^{-1}\partial^\mu\rho\partial_\mu(2\sigma+\ln\rho)\right]. \qquad (8)$$

We have shown that to *all orders* in the loop expansion nonlinear field renormalizations exist such that for any prescribed bare $\mathbf{h}_B(\,\cdot\,)$ there exists such nonlinear field renormalizations

$$(n_B,\rho_B,\sigma_B) \longrightarrow (n,\rho,\sigma)\,, \qquad (9)$$

that

$$L_{\mathbf{h}_B}(n_B,\rho_B\,.\sigma_B) = L_{\mathbf{h}}(n,\rho,\sigma)\,. \qquad (10)$$

In general, however, $\mathbf{h}_B(\,\cdot\,)\neq\mathbf{h}(\,\cdot\,)$. A subscript '$_B$' denotes the bare fields while the plain symbols refer to the renormalized ones. The fact that $\mathbf{h}_B(\,\cdot\,)$ and $\mathbf{h}(\,\cdot\,)$ differ marks the deviation from conventional renormalizability.

To perform explicit computations we convert the O(1,2) σ-model coupled to 2D gravity and to the radion field, (7), to a (pseudo)Riemannian σ-model by promoting the fields ρ and σ to extra coordinates on a 4 dimensional auxiliary target space. This way our reduced system (7) becomes an effective, purely metric σ-model of the standard form with a new 4D target space. In order to do this we note that the target space of the O(1,2) σ-model is a 2D hyperboloid (of constant negative curvature) which can be parameterized by two unconstrained fields, Γ, Δ, defined as being the canonical coordinates of its metric:

$$ds_H^2 = (d\Delta^2 + d\Gamma^2)/\Delta^2 \,. \tag{11}$$

The fields Γ, Δ are related to the hyperbolic spins as

$$n^0 = \frac{1+\Delta^2+\Gamma^2}{2\Delta}\,, \quad n^1 = -\frac{\Gamma}{\Delta}\,, \quad n^2 = \frac{-1+\Delta^2+\Gamma^2}{2\Delta}\,. \tag{12}$$

Then introducing the following 4 fields:

$$\phi^1 = \Delta\,, \quad \phi^2 = \Gamma\,, \quad \phi^3 = \rho\,, \quad \phi^4 = \sigma\,, \tag{13}$$

and defining an auxiliary σ-model metric, $g_{ij}(\phi)$ as:

$$g_{ij}(\phi) = \begin{pmatrix} \rho/\Delta^2 & 0 & 0 & 0 \\ 0 & \rho/\Delta^2 & 0 & 0 \\ 0 & 0 & a/\rho & b \\ 0 & 0 & b & 0 \end{pmatrix}\,, \tag{14}$$

where a, b are constants, the standard σ-model Lagrangian with the 4D auxiliary target-space metric (14) with coordinates ϕ^j

$$\mathcal{L} = \frac{1}{2\lambda} g_{ij}(\phi)\,\partial\phi^i\partial\phi^j \qquad i,j = 1,\dots 4 \tag{15}$$

reproduces the Lagrangian (7) (for $a = b = -1$). Now one can apply the well developed machinery of the nonlinear σ-model renormalization technology for the theory given by (15). To be specific we have used dimensional regularization, minimal subtraction and the covariant background field expansion, with adaptations from [14,15].

Assuming that the bare and the renormalized fields are related by

$$\phi^j_{\scriptscriptstyle B} = \phi^j + \frac{1}{2-d}\sum_{l\geq 1}\left(\frac{\lambda}{2\pi}\right)^l \phi^j_l(\phi) + \dots\,, \tag{16}$$

we require that the functions $\phi^j_l(\phi)$ do not contain derivatives of the fields ϕ^j. The counter terms to the metric can be suitably parameterized as

$$g^{\scriptscriptstyle B}_{ij} = \mu^{d-2}\left[g_{ij} + T_{ij}(g)\right]\,, \tag{17}$$

where μ is the renormalization scale and $T_{ij}(g)$ denotes the counter term tensor. In PT the bare metric of a generic Riemannian σ-model is expanded in terms of the renormalized one as

$$g^{B}_{ij}(\phi_{B}) = \mu^{d-2}\left[g_{ij}(\phi) + \frac{1}{2-d}\sum_{l\geq 1}\left(\frac{\lambda}{2\pi}\right)^{l} T^{(1,l)}_{ij}(g) + \dots\right] . \tag{18}$$

Explicit results for the loop expansion of the metric counter terms for general metrics are known, here I quote the two loop results only, see e.g. Ref. [15]:

$$\begin{aligned} T^{(1,1)}_{ij}(g) &= R_{ij} , \\ T^{(1,2)}_{ij}(g) &= \frac{1}{4} R_{iklm}\, R_{j}{}^{klm} . \end{aligned} \tag{19}$$

It turns out that to all loop orders the counter tensors $T^{(1,l)}_{ij}(g)$ are constrained to be of the form

$$T^{(1,l)}_{ij}(g) = \frac{1}{\mathbf{h}^{l-1}}\,\mathrm{diag}\left(\frac{\zeta_l}{\Delta^2}, \frac{\zeta_l}{\Delta^2}, \frac{S_l(\mathbf{h})}{\rho^2}, 0\right) , \qquad \forall\, l \geq 1 , \tag{20}$$

where $S_l(\mathbf{h})$ is a differential polynomial in $\mathbf{h}$, ζ_l are the coefficients of the counter terms of the flat space O(1,2) symmetric (hyperbolic) σ-model. The first two are explicitly given as:

$$\begin{aligned} T^{(1,1)}_{ij}(g) &: \quad \zeta_1 = -1 , \quad S_1(\mathbf{h}) = -(\rho\partial_\rho)^2 \ln \mathbf{h} + \tfrac{1}{2}(\rho\partial_\rho \ln \mathbf{h})^2 , \\ T^{(1,2)}_{ij}(g) &: \quad \zeta_2 = 1/2 , \quad S_2(\mathbf{h}) = 0 . \end{aligned} \tag{21}$$

Modulo signs, the coefficients $l\zeta_l$ coincide with the known β-function coefficients in the O(3) nonlinear sigma model computed in the minimal subtraction scheme.

Motivated by the form of the counter tensors we allow for a change of the target space metric by a singular ρ-dependent prefactor. Explicitly we assume that the loop expansion of the renormalized metric, $g_{ij}(\phi)$, in terms of the renormalized fields is of the form

$$g^{B}_{ij}(\phi) = \mu^{d-2}\, g_{ij}(\phi)\left[1 + \frac{1}{2-d}\sum_{l\geq 1}\left(\frac{\lambda}{2\pi}\right)^{l} H_l(\rho) + \dots\right] , \tag{22}$$

where the functions $H_l(\rho)$ are to be adjusted. (The dots indicate higher pole contributions.) The parameters a, b and the coupling λ do not get renormalized. (Here the coupling λ merely plays the role of a loop counting parameter.) The function $\mathbf{h}$ is part of the specification of the quantum theory, the renormalization is carried out for generic functions $\mathbf{h}$. Combining Eqs. (16), (18) and (22), one arrives at the finiteness conditions

$$\mathcal{L}_{\phi_l} g_{ij} + H_l(\rho) g_{ij} = T_{ij}^{(1,l)}(g) \,, \qquad l \geq 1 \,, \tag{23}$$

where $\mathcal{L}_v g_{ij}$ is the Lie derivative. The ρ-dependence in the $H_l(\rho)$ term marks the difference to a renormalization in the "strict" quantum field theoretical sense. See Refs. [16,17] for quite an analogous use of the finiteness conditions (23), to define suitable renormalized σ-model target space metrics, although in somewhat different contexts (string theory and target space duality).

In the original Ernst system where $\mathbf{h}(\rho) = \rho$ the "renormalization functions", $H_l(\rho)$, are given as:

$$H_1(\rho) = 0 \,, \quad H_2(\rho) = 1/(4\rho^2)$$

In particular this implies that the Ernst system is renormalizable in the strict sense only at the 1-loop level.

The function $\mathbf{h}(\rho, \lambda)$ plays the role of a generalized (renormalized) coupling, therefore its loop expansion can be written as:

$$\mathbf{h}(\rho, \lambda) = h_0(\rho) + \frac{\lambda}{2\pi} h_1(\rho) + \left(\frac{\lambda}{2\pi}\right)^2 h_2(\rho) + \ldots \,, \tag{24}$$

where the functions $h_l = h_l[h_0]$, $l = 1, \ldots$, are local functionals of h_0. We shall assume $h_0(\rho) \sim \rho^p$, $p > 0$, as this ensures 1-loop renormalizability ($H_1(\rho) \equiv 0$ for $h_0(\rho) \sim \rho^p$). Eq. (24) can be interpreted as a quantum deformation of the classical conformal factor $h_0(\rho)$. In Weinberg's terminology [1] one replaces the inessential "coupling" $h_0(\rho) \sim \rho^p$ by an "essential" coupling function $\mathbf{h}(\,\cdot\,, \lambda)$. Roughly speaking an inessential coupling is one whose flow is affected by field redefinitions and which may continue to run even at a fixed point. A simple test is to compute the variation of the Lagrangian with respect to the bare quantity. If it comes out a total divergence modulo the equations of motion the quantity is an "inessential coupling". Applied to the Einstein-Hilbert action, this criterion disqualifies Newton's constant as an inessential coupling [1]. It is only if one includes a cosmological constant term and/or higher order curvature scalars that the ratios of their prefactors become "essential couplings" – in the space of which one can search for a fixed point [7]. In the context of the 2-Killing reduction the deformation (24) achieves precisely the same: Since $\mathbf{h}\frac{\partial}{\partial \mathbf{h}} L = L = -\lambda \frac{\partial}{\partial \lambda} L$ both the reduced Newton constant λ and $\mathbf{h}(\,\cdot\,)$ will be inessential if L can be written as a total divergence on-shell. One can see that this is the case if and only if $\mathbf{h}(\rho) \sim \rho^p$ or $\mathbf{h}(\rho) \sim \ln \rho$. Since we insist on 1-loop renormalizability and exclude trivial deformations in (24) the set of functions $\mathbf{h}(\,\cdot\,,\lambda)$ qualifies as an *essential coupling*.

To be explicit in the $\overline{\text{MS}}$ scheme one finds:

$$\mathbf{h}_B(\,\cdot\,) = \mu^{d-2}\mathbf{h}(\,\cdot\,)\left[1 + \frac{1}{2-d}\sum_{l\geq 1}\left(\frac{\lambda}{2\pi}\right)^l H_l[h](\,\cdot\,) + \ldots\right] \tag{25}$$

The deformation (24) is necessary because $\mathbf{h}(\,\cdot\,)$ is subject to a nontrivial flow equation $\mu\frac{d}{d\mu}\bar{\mathbf{h}} = \beta_{\mathbf{h}}(\bar{\mathbf{h}}/\lambda)$, where μ is the renormalization scale and $\mu \to \bar{\mathbf{h}}(\,\cdot\,,\mu)$ is the 'running' coupling function. To avoid a possible misunderstanding let us stress that the way how $\bar{\mathbf{h}}(\,\cdot\,,\mu)$ depends on its first argument changes with μ, *not* just the value $\mathbf{h}(\rho(x))$, which is still a function of spacetime and hence not a good coupling. The specific form of the corresponding functional *beta*-function $\beta_{\mathbf{h}}(\mathbf{h})$ does not allow one to preserve $\mathbf{h}(\rho) \sim \rho^p$, even if this was the inital choice for some normalization scale μ_0. However one can impose $\mathbf{h}(\rho)/\rho^p \to 1$ for $\rho \to \infty$ as a boundary condition; it is preserved under the flow and guarantees that the flow is solely driven by the counter terms, as it should.

Remarkably $\beta_{\mathbf{h}}(\mathbf{h})$ can be obtained in closed form and is given by

$$\beta_{\mathbf{h}}(\mathbf{h}/\lambda) = -\rho\partial_\rho\left[\frac{\mathbf{h}(\rho)}{\lambda}\int_\rho^\infty \frac{du}{u}\frac{\mathbf{h}(u)}{\lambda}\beta_\lambda\Big(\frac{\lambda}{\mathbf{h}(u)}\Big)\right]\,. \tag{26}$$

Here $\beta_\lambda(\lambda)$ is the conventional beta function of the O(1, 2) nonlinear sigma-model without coupling to gravity, computed in the minimal subtraction scheme. $\beta_{\mathbf{h}}(\mathbf{h})$ can thus be regarded as a "gravitationally dressed" version of $\beta_\lambda(\lambda)$, akin to the phenomenon found in [18] in a different context. With $\beta_{\mathbf{h}}(\mathbf{h})$ known explicitly we can now search for a fixed point function $\mathbf{h}^*(\,\cdot\,)$, satisfying $\beta_{\mathbf{h}}(\mathbf{h}^*) = 0$. Most importantly it exists! Subject to the previously mentioned asymptotic boundary condition $\mathbf{h}^*(\,\cdot\,)$ is unique and up to three loops it can be written as

$$\mathbf{h}^*(\rho) = \rho^p - \frac{\lambda}{2\pi}\frac{2\zeta_2}{\zeta_1} - \left(\frac{\lambda}{2\pi}\right)^2\frac{3\zeta_3}{2\zeta_1}\rho^{-p} + \ldots\,, \tag{27}$$

where $l\zeta_l$, $l \geq 1$, denote as before the O(1, 2) beta function coefficients: $\zeta_1 = -1$, $\zeta_2 = 1/2$, $\zeta_3 = -5/12$ [19,20]. Note that both the functional form and the numerical coefficients in $\mathbf{h}^*(\rho)$ are fixed by the "conformal renormalizability" condition (10). Moreover, although we use a loopwise expansion to construct the fixed point (27) it can be regarded as "non-Gaussian" in the following sense: There exists a function $\mathbf{h}^{\text{ren}}(\,\cdot\,)$ for which

$\mathbf{h}_B^{\rm ren}(\,\cdot\,) = \mathbf{h}^{\rm ren}(\,\cdot\,)$, so that one almost recovers conventional renormalizability. However beyond one loop $\mathbf{h}^{\rm ren}(\rho)$ *differs* from $\mathbf{h}^*(\rho)$, in particular $\beta_{\mathbf{h}}(\mathbf{h}^{\rm ren}) \neq 0$. Thus if one was to explore the vicinity of this conventionally renormalizable Lagrangian the fixed point (27) could not be seen and, in view of Eq. (31) below, there was little chance to impose the constraints.

From a Wilsonian viewpoint the linearized flow around a fixed point encodes information about the critical manifold and the rate of approach to it. The functional flow induced by (26) can be written as:

$$\mu \longrightarrow \mathbf{h}^*(\cdot) + \delta\bar{\mathbf{h}}(\,\cdot\,,\mu) + O(\lambda^2)\,, \tag{28}$$

and from the flow equations one finds:

$$\delta\bar{\mathbf{h}}(\,\cdot\,,\mu) = \frac{\lambda}{2\pi}\delta_1(\varrho,t) + \left(\frac{\lambda}{2\pi}\right)^2 \delta_2(\varrho,t) + \dots \tag{29}$$

where $\varrho = \rho^p$, $t = \ln(\mu/\mu_0)$. In fact more recently Niedermaier [21] has shown the following:

UV stability: Given smooth initial $\delta_i(\varrho,0)$ with $\delta_i(\infty,t) = 0\,,\ i \geq 1$, the solution of the linearized flow equation is unique and satisfies:
$\delta_i(\varrho,t) \to 0 \quad$ for $t \to \infty$.

The existence of an ultraviolet fixed point in a non-renormalizable theory therefore is in the spirit of the asymptotic safety scenario. To quote from Ref. [1]: "A theory is said to be asymptotically safe if the 'essential' coupling parameters approach a fixed point as the momentum scale of their renormalization point goes to infinity." There are also some important differences however; due to the complicated form of $\beta_{\mathbf{h}}(\mathbf{h})$ no preferred parametrization of the running $\bar{\mathbf{h}}(\,\cdot\,,\mu)$ in terms of numerical running couplings exist; the solutions $\delta\bar{\mathbf{h}}$ can be grouped into classes with qualitatively different rate of decay (it is not always powerlike). In particular there is no obvious way to define the dimension of the critical manifold. Another somewhat annoying point is that according the results of Ref. [4] the expected anti-screening of the gravitational interactions cannot be seen in PT in terms of a graviton loop expansion (c.f. Equations (2),(3) for the graviton loop corrections to Newton's potential).

Ultimately the reason why (10) works is that $L_{\mathbf{h}}$ possesses two conformal symmetries in *field space* which together with the obvious strict symmetries characterize the Lagrangian up to a ρ-dependent prefactor. In fact one can show that there exist currents C_μ and D_μ obeying the on-shell identities

$$\partial^\mu C_\mu = \rho\partial_\rho \ln\mathbf{h}\cdot L_{\mathbf{h}}\,, \qquad \partial^\mu D_\mu = -\ln\rho\cdot\rho\partial_\rho\ln\mathbf{h}\cdot L_{\mathbf{h}}\,. \tag{30}$$

They can be promoted to Ward identities for the corresponding composite operators and could also serve to characterize the theory in a regularization independent way. Here I only mention the following points concerning composite operators:

1. Arbitrary functions of ρ are finite as composite operators; no additional renormalizations are needed.

2. The O(1, 2) Noether currents J^i_μ, $i = 0, 1, 2$, can be defined as finite composite operators obeying $[\![J^i_\mu(\mathbf{h}, n)]\!] = J^i_\mu(\mathbf{h}_B, n_B)$, where $[\![\,\cdot\,]\!]$ denotes the normal product in the above scheme.

3. The energy momentum operator can be defined as a composite operator $[\![T_{\mu\nu}]\!]$; it is unique up to an improvement term of the form $\Delta T_{\mu\nu} = (\partial_\mu \partial_\nu - \eta_{\mu\nu}\partial^2)\Phi$, where the bare and the renormalized improvement potential are of the form $\Phi_B = f_B(\rho_B) + f_0 \sigma_B$ and $\Phi = f(\rho) + f_0 \sigma$, respectively. Here f_0 is a finite constant and in general $f_B(\,\cdot\,) \neq f(\,\cdot\,)$.

In order for $\mathbf{h}^*$ to qualify as a good fixed point one should also have a chance to impose the constraints when $\mathbf{h} = \mathbf{h}^*$. From point 3. above follows that $T_{00} = \mathcal{H}_0$ and $T_{01} = \mathcal{H}_1$ can be defined as composite operators. Clearly a necessary condition for the quantum constraints to have a sufficiently large 'kernel' on the state space generated by $\rho, \partial_\mu \sigma, J^i_\mu$ is that the trace anomaly vanishes. This is to say there should exist an improvement potential $\Phi = f(\rho) + f_0 \sigma$ such that when $[\![T^\mu{}_\mu]\!]$ is computed as a composite operator in the above scheme it vanishes modulo contributions proportional to the equations of motion operator. The counter terms relating $f_B(\rho_B)$ to $f(\rho)$ and those relating ρ_B to ρ depend on $\mathbf{h}$. This triggers a non-autonomous inhomogeneous flow equation for a running $\bar{f}(\,\cdot\,, \mu)$. One might expect that when $\bar{\mathbf{h}}(\,\cdot\,, \mu)$ becomes stationary also $\bar{f}(\,\cdot\,, \mu)$ becomes stationary and defines the proper improvement potential at the fixed point. This is indeed the case, moreover

$$\mu \frac{d}{d\mu} \bar{\mathbf{h}}(\,\cdot\,, \mu) = 0 = \mu \frac{d}{d\mu} \bar{f}(\,\cdot\,, \mu) \quad \Longleftrightarrow \quad [\![T^\mu_\mu]\!] = 0\,, \tag{31}$$

to all loop orders. That is, the trace anomaly of the improved energy momentum tensor vanishes precisely at the fixed point of the functional flow. The proof links the stationarity of $\bar{f}(\,\cdot\,, \mu)$ to a variant of the Curci-Paffuti relation [22]. The combinations $[\![\mathcal{H}_0 \pm \mathcal{H}_1]\!]$ are then expected to generate a 2D conformal algebra with formal central charge $c = 4$. The value $c = 4$ in itself is not of direct significance because the (sub-) state space generated by $\rho, \partial_\mu \sigma$ has indefinite metric. The latter feature is not an artifact, it is directly related to the notorious "conformal factor problem" of 4D quantum Einstein gravity.

Relating the present Lagrangian-based perturbative quantum theory to other, non-perturbative approaches, such as the bootstrap construction of Ref. [23] for example, is clearly an important problem. A concrete possibility to make such a connection would be for example the construction of the first nonlocal quantum observable at the fixed point (following Lüscher's strategy in the O(3) σ-model [24]). Our perturbative Lagrangian construction on the other hand provides a direct link to 4D quantum Einstein gravity; in particular the fixed point in the truncated theory established here appears to be a necessary prerequisite for a fixed point in full quantum Einstein gravity.

Acknowledgments

I would like to thank the organizers, in particular Júlia and Péter for all their efforts to organize such a memorable conference in memoriam Volodya, and for giving me the possibility to present the above results in the presence of an exceptional audience. I should also thank Professor Khriplovich for enjoyable and enlightening discussions about quantum gravity corrections to Newton's law. This research has been partly supported by the OTKA grant TS44665.

References

1. S. Weinberg, in: *General relativity, an Einstein Centenary survey*, Eds. S. Hawking, W. Israel, Cambridge University Press, 1979.
2. G. 't Hooft and M. Veltman, *Ann. Inst. Poincaré* **20**, 69 (1974).
3. M. Goroff and A. Sagnotti, *Nucl. Phys.* **B266**, 709 (1986).
4. N.E.J. Bjerrum-Bohr, J.F. Donoghue and B.R. Holstein, *Phys. Rev.* **D67**, 084033 (2003);
   I.B. Khriplovich and G.G. Kirilin, *J. Exp. Theor. Phys.* **98**, 1063 (2004) [*Zh. Eksp. Teor. Fiz.* **125**, 1219 (2004)], arXiv:gr-qc/0402018.
5. C. Rovelli, *Quantum Gravity*, Cambridge University Press, (2004), Cambridge, UK.
6. M. Reuter, *Phys. Rev.* **D57**, 971 (1998).
7. O. Lauscher and M. Reuter, *Phys. Rev.* **D65**, 025013 (2002); *Class. Quant. Grav.* **19**, 483 (2002).
8. P. Forgács and M. Niedermaier, arXiv:hep-th/0207028;
   M. Niedermaier, *JHEP* **12**, 066 (2002).
9. R. Matzner and C. Misner, *Phys. Rev.* **154**, 1229 (1967).
10. F. Ernst, *Phys. Rev.* **167**, 1175 (1968).

11. D. Maison, *Phys. Rev. Lett.* **41**, 521 (1978);
V.A. Belinsky and V.E. Zakharov, *Sov. Phys. JETP* **48**, 985 (1978); *ibid.* **50**, 1 (1979)
C. Klein and O. Richter, *Ernst Equation and Riemann Surfaces*, Lecture Notes in Physics, Vol. 685, Springer, 2005.
12. H. Nicolai and D. Korotkin, *Nucl. Phys.* **B475**, 397 (1996).
13. B. de Wit, M. T. Grisaru, H. Nicolai and E. Rabinovici, *Phys. Lett.* **B286**, 78 (1992).
14. D. Friedan, *Ann. Phys.* **163**, 138 (1985).
15. H. Osborn, *Nucl. Phys.* **B294**, 595 (1987).
16. A. Tseytlin, *Phys. Rev.* **D47**, 3421 (1993) [arXiv:hep-th/9211061];
A. Tseytlin, *Nucl. Phys.* **B390**, 153 (1993) [arXiv:hep-th/9209023].
17. J. Balog, P. Forgács, Z. Horváth and L. Palla, *Phys. Lett.* **B388**, 121 (1996) [arXiv:hep-th/9606187].
18. I. Klebanov, I. Kogan and A. Polyakov, *Phys. Rev. Lett.* **71**, 3243 (1993).
19. S. Hikami, *Phys. Lett.* **B98**, 208 (1981).
20. F. Wegner, *Nucl. Phys.* **B316**, 663 (1989).
21. M. Niedermaier, *Nucl. Phys.* **B673**, 131 (2003).
22. G. Curci and G. Paffuti, *Nucl. Phys.* **B286**, 399 (1987).
23. M. Niedermaier and H. Samtleben, *Nucl. Phys.* **579**, 437 (2000).
24. M. Lüscher and K. Pohlmeyer, *Nucl. Phys.* **B137**, 46 (1978);
M. Lüscher, *Nucl. Phys.* **B135**, 1 (1978).

CONCLUDING REMARKS

CONCLUDING REMARKS

YURI L. DOKSHITZER*
LPTHE, University Paris-VI
4 place Jussieu,
75005 Paris, France
E-mail: yuri@lpthe.jussieu.fr

As a concluding remark to the Memorial Budapest Conference Gribov–75 I recall the peculiar atmosphere in which in 1973 certain physical issues have been discussed that did not lose their actuality 30+ years later.

Gribov–75

The Memorial Budapest Conference brought together theorists of all ages, origins and professional interests. What all those who came and/or contributed to this book had in common, I believe, was deep respect to the giant of a man in theoretical physics of the 20th century, Vladimir Naumovich Gribov. Some participants never met him; others consider BH[a] their spiritual godfather (in a pre–Puzo/Coppola meaning of the word).

The concluding talk that I had an honour to give at the conference on May 24, 2005 was *nothing personal*: I spoke about the Gribov programme of attacking the colour confinement problem. A similar talk I presented a year earlier at the Non-Perturbative QCD Paris workshop, in the proceedings of which the key points of Gribov's light quark confinement theory can be found [1]. In much greater detail, Gribov's approach to non-Abelian gauge field dynamics, and to QCD confinement in particular, was displayed in the review that we worked on together with Dima Kharzeev [2]. Moreover, the subject matters of the Gribov light quark confinement scenario are well covered in the present book by Carlo Ewerz [3].

So, seeing no need to repeat *strictly business* issues I decided to try to contribute to the personal memos collected in this volume. They make

*On leave of absence from PNPI, 188350 Gatchina, St. Petersburg, Russia.
[a][beh ush]

for amazing reading. It is telling how one and the same picture stuck in everyone's memory, namely that of BH at the blackboard!

- "...*I saw a slim, black-haired young man running down the stairs, beginning to argue with the speaker ...*" (A. Frenkel [4])

•• "...*I remember giving seminars during which he came up to the blackboard to comment a point and engaging a discussion in which it was no longer clear if the seminar was mine or his and which lasted hours ...* (D. Amati [5])

••• "*V.N. jumped to the blackboard, ... took the chalk from the speaker, and started to explain why and where he, the speaker, was not right.*" (A. Belavin [6])

The picture was space–time invariant too: the above snapshots were made in Dubna (early 60's), Leningrad (60's–70's) and Chernogolovka (80's), correspondingly. The environment did not matter either, be it a conference (•), a regular seminar (••) or a scientific council (•••). Whatever the circumstance, Gribov was always ready and would not hesitate to attack, for physics' sake.[b] It is this very readiness to discuss physics and freely exchange ideas, in the search for Truth, that determined the upbringing of those nicknamed Gribov (or Leningrad) school. And the Gribov theory seminar (first, in the Ioffe institute in Leningrad and then in Gatchina LNPI) was the training and testing ground for the pack. It was an institution of its own. I cannot resist the temptation to recall here this unique establishment,[c] for the benefit of those who did not have the fortune of passing

> ... through the beneficial ordeal of a notorious "Gribov seminar". This was a legendary seminar. It had no time limit and would go on as long as was necessary to establish the truth. Some visitors hated it and would never repeat that most dreadful experience of their life; others loved it: finding the truth of the matter was at stake and the speaker would be the first to benefit. [...]
>
> To understand the spirit of the seminar you have to accept the notion of "aggressive friendliness". No merits counted, no excuses were given: a newcomer and a renowned academician were treated

[b]Unfortunate as it is for them, some of the Gribov school offspring learned and unconditionally adopted these manners but too well.

[c]I must apologise for the extensive self-citation of the text known as "*not-an-obituary*" that was written for the January 1998 issue of *Physics World* [7].

equally, that is equally amicably and aggressively. After 5 minutes of a smooth introduction BH would jump to the blackboard and make his three points: what this guy is trying to tell us, why it is "all wrong", and how the problem had to be approached. This would trigger a hot discussion involving all the audience (including the speaker; though, markedly, there were historical exceptions when a speaker would leave the seminar room [d]).

BH as a speaker would be given the same friendly treatment. That is, the story goes, how Lev Lipatov, now a world-known theoretician and an academician, became a co-author of the famous Gribov & Lipatov work of 1970–71 which laid the basis of a field-theoretical description of deep inelastic scattering and e^+e^- annihilation. Gribov was presenting his work, and the young man made a couple of "killing" comments. Gribov got stuck trying to answer Lipatov's questions: "Lev, you are a co-author already, help me", was the solution.

Many a difficult problem was cracked in this fashion, at the blackboard, in the noisy (and, in early days, smoky) atmosphere of PTI/LNPI seminars.

But wait... That characteristic combination of noise and smoke. When did I encounter it for the first time?... Right, it sure was

February 1973

when the annual LNPI Winter School was held at a seashore resort in Zelenogorsk, some 40 miles north of Leningrad, in a fancy sanatorium for trade union *apparatchiks*. Apart from retired trade-unionists, halls and indoor gardens of the sanatorium were inhabited by caged parrots. The most resistant among them survived the invasion. Was it noise or smoke (or the combination) that was responsible for the fate of less lucky ones remained unclear. Whether the manager of the establishment kept his job after he let those wild physicists in, I cannot tell. (In any case, the institute had to look for another venue to run the next year school.)

Not only parrots were stressed then. The physicists felt pretty much excited too, and behaved correspondingly. Pre-planned morning and af-

[d] I don't feel like divulging the hero's identity without his consent. To give a hint, he was playing harmonica sitting, legs outside, on a window-seal while the discussion was continuing in the adjacent room. (Should this not be enough a hint, it was a future Nobel laureate.)

ternoon lectures and talks were followed by informal evening seminars and improvised brainstorming discussions that ran deep into after midnight.[e] It was then that unforgettable chess battles between Sasha Migdal and Lev Lipatov started that were taking place at 2–3 a.m. on the floor of a barely heated sport hall. The external temperature did not pose a problem though: both the players and the spectator crowd were comfortable, thanks to preceding hot physics discussions as well as to other means of internal heating. (This event became a tradition which died out years later when one of the two contestants abandoned chess for programmable pocket calculators.)

The reasons for excitement were many. The construction by V.N. Gribov of Reggeon Field Theory [8] engaged *cross-breeding* of condensed matter and high energy particle physics. Never before was (and never again was to be) the *cross-brooding* of respective theorists so mutually beneficial. Ingenious diagram techniques developed by V.N. Gribov and A.A. Migdal for analyzing multi-Pomeron effects in the strong coupling regime [9] contributed to understanding second order phase transitions. On the other hand, high energy physics was "paid back" right away: Sasha Polyakov gave an artistic presentation of the schizophrenic picture of the cascade production of hadrons in e^+e^- annihilation *jets* in a field theory, inspired by critical phenomena and turbulence in particular [10]. Violation of the Bjorken scaling in structure functions of deep inelastic lepton-hadron scattering (DIS) and in inclusive hadron distributions (fragmentation functions) of e^+e^- annihilation had been scrutinized in *logarithmic* field theories by Gribov and Lipatov [11,12]. The net that Polyakov cast was a wider one. Having formulated the problem "*less definitely that is more precisely*" [13] he came up with a prophetic prediction of the multiplicity fluctuation pattern (*KNO scaling*) in field theory [14]. One year later L.N. Lipatov explicitly reformulated their earlier results with V.N. Gribov in terms of parton dynamics, the formulation that potentially embodied all — inclusive and exclusive — information about evolution of parton spectra and fluctuations in perturbation theory. The only missing brick in the wall was QCD.

Meantime, with Asymptotic Freedom just around the corner, by 1973 DIS and e^+e^- annihilation into hadrons (now known as *hard* processes)

---

[e] I am grateful to Sasha Belavin for reminding me the name of "lipatniks" that these night sessions bore. I seem to remember a characteristic piece of a poem that immortalized the atmosphere of *lipatniks*: Belavin argues desperately/Chopin melts in smoke/ Lipatov is walking randomly/ lightly bouncing off the walls ...

and "classical" high energy hadron scattering phenomena (*soft* physics) were coming closer and closer together.

At that very School in Zelenogorsk V.N. Gribov unveiled the s-channel structure of t-channel reggeon exchange and described his unified approach to hard and soft hadron phenomena. His lecture triggered an immediate response. During breaks theorists were condensing here and there in the halls forming gesticulating clouds. The most adhesive drop was obviously formed by Oleg Kancheli, Sergei Matinyan and BH himself.[f] Both Sergei Gaikovich and Oleg are extremely gentle people (and none of them a smoker). Nevertheless, if you closed your eyes and let yourself be guided by your ears (or nose), following the gradient would have inevitably lead you to that particular three-body system. The poor birds[g] did not die in vain: the discussion resulted in extending the Gribov analysis and incorporating into the picture multiplicity fluctuations in soft scattering processes, due to multi-Pomeron exchanges. The famous AGK cutting rules that elegantly linked together quasi-elastic and inelastic processes with shadowing corrections [15] were conceived in this atmosphere. And we were fortunate to witness real physics being done before our very (undergraduate) eyes!

Treasure chest

Gribov's seminal 1973 lecture [16] became accessible to a broader (not yet fluent in Russian) world only recently when it was translated by Nyíri, V. Petrov and Strikman, put on the web and published in the Collection of Gribov works in 2002 [17]. However, today an unsuspecting reader is likely to bounce off the very first sentence, after having found that the author is going "*to describe electromagnetic interactions of hadrons* [...] *without the introduction of quarks or other exotic objects*." Hold on! If you consider it a mistake, this one belongs to those "*'mistakes'* [that] *become great new ideas*" [18].

Gribov interpreted the internal structure of long-living fluctuations inside an energetic projectile in terms of partonic wave function. Logarithmic hadron multiplicity in soft ("minimum bias") inelastic hadron interactions appeared naturally as a result of breaking up the coherence of the parton fluctuation after a slow parton interacted with the target.

---

[f] I see a witty and ironic Karen Avetovich Ter-Martirosyan in this group too, but I don't remember with certainty whether what I see now is what really happened.

[g] If statistics were kept, that day would have shown the peak in parrot mortality.

The asymptotically constant total cross section (Pomeranchuk pole hypothesis) corresponds to the situation when the density of slow partons in the momentum space — the result of an interplay of parton production and recombination processes — *reaches equilibrium*, the constant value $\phi(k_\perp)$ that does not depend on the rapidity of the hadron projectile.

The analysis of the space–time development of parton fluctuations led to the understanding of the phenomenon of *shrinkage of the diffractive cone* (which was predicted from t-channel Regge-inspired dynamics [19]) as originating from *parton diffusion* in the impact paremeter space, in the s-channel language. Diffusion leads to logarithmic increase with energy of the hadron radius. The Pomeron hypothesis in the impact parameter space $\rho_\perp$ translates then into the picture of a hadron whose *radius swells* with energy (while its *transparency increases*): $R \propto \sqrt{\ln s}$ (while $\phi(\rho_\perp) \propto R^{-1}$).

In general, the parton diffusion picture naturally explains the Froissart bound on the total cross section, $\sigma_{\rm tot} < \ln^2 s$, as well as on the maximal growth of the interaction radius, $R \propto \ln s$.

The 25-page writeup of the Gribov lecture is a treasure chest.[h] And one that remains barely half-open.

Feynman–Gribov puzzle

In particular, few are aware of the fact that the Gribov approach to partons violently collided with that of Feynman: to the east from the notorious "iron curtain" (once a solid line, now dot-dashed) the words *Feynman–Gribov puzzle* would hardly make an eyebrow rise. Which is a pity. Because behind the contradicting expectations about the structure of the fragmentation region of the target proton in DIS essential physics was hidden.

Feynman prophesied the *uniform rapidity plateau* stretching from $y = -\ln Q$ (tip of the current fragmentation) all the way to $y = +\ln(Q/x)$, while Gribov was betting on a *hole* at positive rapidities $0 < y < \ln Q$ (*wee* partons in the proton fragmentation region; with rapidities referring to the Breit frame). No wonder. Feynman, who from the very start had kept quarks in mind, simply did not want fractional electric charge to be left flying in the current hemisphere. And he was right. Gribov, on the contrary, could not care less about quarks at that stage but knew quantum mechanics too well. He saw, and showed in his lecture, that inside the partonic wave

[h] If it were legitimate to compare the scientific expertise of the "Leningrad school" with an *artichoke*, that lecture would take the place of the artichoke's *heart*.

function of the fast proton, (wee) partons with positive rapidities smaller than that of the hit quark, $y < \ln Q$, are undisturbed by the DIS interaction. These partons remain coherently faithful to the parent proton, therefore the corresponding part of the initial state parton fluctuation must collapse back leaving the *gap* ("Gribov hole") in the target fragmentation region. And Gribov was right too.

So who was *more right*? The puzzle lay for more than a decade, hibernating, till QCD finally reconciled the two positions [20]. There *is no gap* in the rapidity distribution of final hadrons in DIS, so Feynman was right. Ironically, the particle density turned out to be *maximal* in the region where Gribov expected the gap. One more to the collection of Gribov's mistakes? No. Not entirely, and not at all. In fact, hadrons that fill in the Gribov gap in QCD do not originate from *partons* seen (and defined by the orthodox parton picture) as constituents of a target hadron. This "anti-Gribov" filling is splinters of a coherent gluon field that is driven by the colour charge transfer due to the loss of a (colour triplet) quark by the target. Materialising into accompanying hadrons via gluon bremsstrahlung, the intensity of this field depends exclusively on the momentum transfer Q^2 and is insensitive to the internal structure of the target hadron and independent of its energy. Given that neither Feynman nor Gribov had in mind the presence and the rôle of a specific conserved current (colour) as a source of additional final state particles, it is up to the reader to decide who won the argument.

$\sigma_{pp} \longrightarrow \sigma_{pd} \longrightarrow \sigma_{p\,Merc}$

Another jewel from the 1973 treasure chest is even less known, while *even more relevant*, for theorists' preoccupation today. It concerns *the* problem of strong interactions, namely the high energy behaviour of total hadron scattering cross sections. This part of the discourse is based on yet another "mistake", the supposition that parton transverse momenta are limited from above (as in $\lambda\phi^3$ field theory). In such a world Bjorken scaling would be *exact* (it is the parton fluctuations with large transverse momenta, up to Q^2, that induce logarithmic scaling violations). Gribov showed that, closely related with the Bjorken scaling, is another amusing phenomenon [21], namely, asymptotic convergence of the total scattering cross sections of *all* strongly interacting objects. Be it a proton or a deuteron (or a Mercedes) that enters the collision. This asymptotic limit is truly academic: at imaginable energies we wouldn't have a slightest chance to see even *pd* and *pp* cross sections equalise, spare the thought of Mercedes. Nevertheless, to appreciate the physics underlying the Gribov argument, it is essential for understanding the problem of hadron interactions at high energies.

QCD presently cannot deal with the *true* Pomeron, that is predict the nature of the leading singularity of t-channel partial waves in the complex angular momentum plane. What we have at our disposal is the *perturbative* or *hard* "BFKL Pomeron". It is not the Pomeron at all since it is not about $\sigma_{\rm tot}$ though it may apply all right to hard interactions like DIS. As long as the corresponding hard cross sections are *small* they are allowed to increase as fast with energy as the "BFKL Pomeron" with an intercept exceeding the maximal Froissart value $j=1$ predicted (and they apparently do). It is the Gribov diffusion–73 that pushed above unity the intercept of the interacting two-gluon system in multi-Regge kinematics. However, that very diffusion (which in pQCD occurs in $\ln k_\perp$) makes the "hard Pomeron" infrared unstable [i].

The *true Pomeron* [j] problem remains essentially non-perturbative and cannot be solved without understanding confinement. Approaching the problem from the perturbative side, we increase collision energy and talk about overlapping partons, large gluon densities, screening and saturation effects, trying to grasp how the s-channel unitarity gets imposed, bringing the Froissart bound onto the stage.

The *Pomeranchuk pole* hypothesis implies *factorization.* In physical terms, this means that asymptotically the density (and the interaction properties) of *wee partons* in the wave function of the fast projectile should become independent of the collision energy. In other words, parton clouds attached to fast constituent of the incident hadron must fully overlap producing a universal "collectivized" distribution. Equality of all $\sigma_{\rm tot}$'s is one little step away from this blueprint. And this is a step worth watching.

Inelastic diffraction

Consider an inelastic diffraction process, that is, a transformation of the incident particle p_a into some final state system with total momentum P_b

---

[i] The position and the nature of the leading singularity, derived perturbatively, changes drastically depending on the assumption one makes about the behaviour of the QCD coupling at small momentum scales [22,23].

[j] It is a pity one has to use extra adjectives and/or italics to deliver the message. It would have been so much better if the beautiful BFKL construction bore some other name. Like HERon, for example, standing for High Energy Regime of DIS (HERA) cross sections. But fighting established usage/mis-usage of words comes close the top of the list of hopeless endeavours.

in the course of scattering with momentum transfer $q = P_b - p_a$. Intimately related with the equality of total cross sections is another consequence of the parton saturation. It reads:

Inelastic diffraction vanishes in the forward direction. $(*)$

Forward refers to vanishing *transversal components* of the total momentum transfer [k]. At large energy $E = p_{a0}$ it suffices to transfer to the projectile a miniscule *longitudinal* momentum $q_z \propto (P_b^2 - m_a^2)/E$ in order to produce a (finite invariant mass) system P_b^2 in the final state. Suffices *kinematically*, that is. According to $(*)$, however, *dynamics* prevents inelastic channels "open up" unless the projectile acquires in addition a *transverse* kick, $\mathbf{q}_\perp \neq 0$. This claim totally disregards details of the b state which might be a multi-particle system with a complex internal structure. The only thing that matters is the *total transferred momentum* and, more specifically, its transverse component. This makes $(*)$ sound strange in its generality. It is this generality, however, which makes this prediction really universal and necessary to keep in mind when we are discussing high energy scattering phenomena, and modelling parton screening and saturation.

Knowing the answer definitely helps to find the proof. The qualitative argument runs as follows. Inelastic diffraction is a process of releasing a virtual fluctuation inside the incident particle. The $\mathbf{q}_\perp \to 0$ limit corresponds to the picture of the particle traversing an almost constant field. Such a field cannot resolve an internal structure of the fast projectile and therefore should not break up the coherence of the initial state.

High energy QED processes do respect $(*)$. Take for example dissociation of a photon with four-momentum $k = (E; E, 0, 0)$ into a e^+e^- pair, $k+q = p_1+p_2$, in a static Coulomb field ($q_0 = 0$). The transition amplitude (for the sake of simplicity we take electrons to be spin zero particles)

$$A^{\mu\nu} = \frac{(2p_1 - k)^\mu (2p_2 - q)^\nu}{m^2 - (p_1 - k)^2} + \frac{(2p_2 - k)^\mu (2p_1 - q)^\nu}{m^2 - (p_2 - k)^2} - 2g^{\mu\nu}$$

is explicitly gauge invariant: $k^\mu A_{\mu\nu} = 0$, $A_{\mu\nu} q^\nu = 0$. Sandwiching the amplitude between the physical (*transversal*) photon polarisation vector $e_\mu^\lambda(k)$ and the Coulomb potential, $a_\nu = \delta_{\nu 0}/q^2$, the matrix element reads

$$M = e_\mu^\lambda \, A^{\mu\nu} \, a_\nu = \frac{2\mathbf{e}_\perp^\lambda}{q^2} \cdot \left(\mathbf{p}_{\perp 1} \frac{p_{02}}{(p_1 k)} + \mathbf{p}_{\perp 2} \frac{p_{01}}{(p_2 k)} \right).$$

[k] Talking about specific component of the four-vector q is crucial and should not scare those who respect Lorentz invariance: remember, at high energies $4 = 2 + 2$.

Observing that $E(p_i k)^{-1} = (p_{0i} - p_{zi})^{-1} = (p_{0i} + p_{zi})(m^2 + \mathbf{p}_{\perp i}^2)^{-1}$ we approximate, in the high energy limit, $p_{0i} + p_{zi} \simeq 2p_{0i}$ to arrive at

$$M \simeq \frac{4p_{01}p_{02}}{E} \frac{\mathbf{e}_\perp^\lambda}{q^2} \cdot \left(\frac{\mathbf{p}_{\perp 1}}{m^2 + \mathbf{p}_{\perp 1}^2} + \frac{\mathbf{p}_{\perp 2}}{m^2 + \mathbf{p}_{\perp 2}^2} \right).$$

Bearing in mind that $\mathbf{p}_{\perp 1} + \mathbf{p}_{\perp 2} = \mathbf{q}_\perp$ we conclude that M indeed vanishes at $\mathbf{q}_\perp \to \mathbf{0}$.

Let us remark that dropping the electron mass and substituting $p_2 = q - p_1$ we get the structure

$$\frac{\mathbf{p}_\perp}{\mathbf{p}_\perp^2} - \frac{(\mathbf{p}-\mathbf{q})_\perp}{(\mathbf{p}-\mathbf{q})_\perp^2}.$$

This is nothing but the famous pQCD "BFKL kernel" which describes small-x evolution of t-channel two-gluon exchange, with $\mathbf{p}$ the transverse momentum of a real s-channel gluon and $\mathbf{q}$ — the momentum transfer along the "ladder". The BFKL kernel also respects (∗): no momentum transfer — no gluon production.

In gauge theories (QED, QCD) the vanishing of inelastic diffraction can be looked upon as a consequence of current conservation (gauge invariance). Gribov, however, had to demonstrate this property from general principles, without appealing to concrete microscopic dynamics which was unknown back in 1973 and, by the way, remains basically unknown today when we address soft hadron interactions. He chose deuteron as a example of a composite object to verify his field theoretical intuition on a quantum mechanical system.

The Gribov analysis of the forward dissociation of a deuteron is a marvel. To the best of my knowledge, he never published it. He only used this as a physical analogy in the study of the so-called "weak coupling regime" of the reggeon field theory, the one that resulted in equality of total cross sections. He told the deuteron story at Leningrad University when he decided to share with students, in a two-semester course, *everything he learned himself* about strong hadron interactions before leaving the field. This particular lecture I remember as the one of the two that made the audience shiver [1] I very much hope that in this "Strong Interactions" lecture course (to be published by Cambridge University Press in 2007) an inimitable voice of Gribov the lecturer will be heard and his unique approach to physics will become accessible to young researchers of today.

[1]The other one, earlier in the course, was about the inevitability of Mandelstam cuts.

Quark apology

In 1973 Gribov mocked quarks "strange objects"; in the following twenty years, the logic of his search led him to courting light quarks as key players in the colour confinement drama. Had there been an (higher) authority to keep record of theorists' mistakes, Gribov's neglect of quarks in early 70's would have been definitely written off the books.

Gribov realised very early on that the Pomeron problem will not be solved without understanding the structure of the hadron vacuum. He needed a mechanism for preventing soft parton fluctuation from growing catastrophically large with energy, to stabilise wee parton densities. He found inspiration in non-Abelian gauge theory. The fact that it possessed an intrinsic instability due to catastrophic growth of the interaction strength, was a positive sign: a quantum field theory could offer, in principle, a mechanism for stabilising the Pomeron at a price of solving its own infrared problem — QCD hadron confinement. Gribov turned to QCD and, as apprentice, made a major discovery that changed forever our perception of non-Abelian QFTs (Gribov copies). But that was another story, another Leningrad Winter School, 1976 and far away from Zelenogorsk.

Acknowledgements

I am deeply thankful to Julia Nyiri and Peter Levai for all the effort they put into organising the Memorial Gribov Conference and making it a success. I would like to express my gratitude to Gavin Salam for constructive criticism of the manuscript.

References

1. "GRIBOV LIGHT QUARK CONFINEMENT SCENARIO", Yu.L. Dokshitzer, *plenary talk at the 8th Workshop on Nonperturbative QCD*, Paris, France, 7–11 June 2004; *Int. J. Mod. Phys.* **A20**, 4363 (2005)
2. "THE GRIBOV CONCEPTION OF QUANTUM CHROMODYNAMICS", Yu.L. Dokshitzer, D.E. Kharzeev, *Ann. Rev. Nucl. Part. Sci.* **54**, 487 (2004)
3. "GRIBOV'S PICTURE OF CONFINEMENT AND CHIRAL SYMMETRY BREAKING", C. Ewerz, this volume
4. "REMEMBERING VOLODYA", A. Frenkel, this volume
5. "MEMORIES OF VOLODYA", D. Amati, this volume
6. "MEMORY SKETCH TO V.N. GRIBOV'S PORTRAIT", A. Belavin, this volume
7. "V.N. GRIBOV 1930–1997", Yu.L. Dokshitzer, *Physics World*, Jan 1998; also in: *Continuous advances in QCD*, Minneapolis 1998; e-print Archive: physics/9801025

8. "A REGGEON DIAGRAM TECHNIQUE", V.N. Gribov, *Zh. Eksp. Teor. Fiz.* **53**, 654 (1967) [*Sov. Phys. JETP* **26**, 414 (1968)]
9. "STRONG COUPLING IN THE POMERANCHUK POLE PROBLEM", V.N. Gribov and A.A. Migdal, *Zh. Eksp. Teor. Fiz.* **55**, 1498 (1968) [*Sov. Phys. JETP* **28**, 784 (1969)]
10. "DESCRIPTION OF LEPTON HADRON REACTIONS IN QUANTUM FIELD THEORY", A.M. Polyakov, *Zh. Eksp. Teor. Fiz.* **61**, 2193 (1972) [*Sov. Phys. JETP* **34**, 1177 (1972)]
11. "DEEP INELASTIC *ep* SCATTERING IN PERTURBATION THEORY", V.N. Gribov and L.N. Lipatov, *Yad. Fiz.* **15**, 781 (1972) [*Sov. J. Nucl. Phys.* **15**, 438 (1972)]
12. "e^+e^- PAIR ANNIHILATION AND DEEP INELASTIC *ep* SCATTERING IN PERTURBATION THEORY", V.N. Gribov and L.N. Lipatov, *Yad. Fiz.* **15**, 1218 (1972) [*Sov. J. Nucl. Phys.* **15**, 675 (1972)]
13. M.A. Birman, *Lectures on functional analysis*, Leningrad University, 1972
14. "THE PARTON MODEL AND PERTURBATION THEORY", L.N. Lipatov, *Yad. Fiz.* **20**, 181 (1974) [*Sov. J. Nucl. Phys.* **20**, 94 (1975)]
15. "CHARACTER OF INCLUSIVE SPECTRA AND FLUCTUATIONS PRODUCED IN INELASTIC PROCESSES BY MULTI-POMERON EXCHANGE", V.A. Abramovsky, V.N. Gribov and O.V. Kancheli, *Yad. Fiz.* **18**, 595 (1973) [*Sov. J. Nucl. Phys.* **18**, 308 (1974)]
16. "SPACE–TIME DESCRIPTION OF THE HADRON INTERACTION AT HIGH ENERGIES", V.N. Gribov, in: *Proc. of the 8th Winter School on Nuclear and Elementary Particle Physics*, p. 5, Leningrad; *Proc. of the first ITEP Winter School*, vol. 1, p. 65, Moscow, 1973 (in Russian); hep-ph/0006158
17. "V.N. GRIBOV. GAUGE THEORIES AND QUARK CONFINEMENT", Phasis, Moscow, 2002
18. "THE HIGH ENERGY GRIBOV; SOME RECOLLECTIONS", L. McLerran, this volume
19. "POSSIBLE ASYMPTOTIC BEHAVIOR OF ELASTIC SCATTERING", V.N. Gribov, *Pisma Zh. Eksp. Teor. Fiz.* [*JETP Lett.*] **41**, 667 (1961)
20. "THE STRUCTURE OF A FINITE HADRONIC STATE IN DEEP INELASTIC SCATTERING", L.V. Gribov et al., *Pisma Zh. Eksp. Teor. Fiz.* **45**, 405 (1987) [*JETP Lett.* **45**, 515 (1987)]; *Phys. Lett.* **B202**, 276 (1988)
21. "PROPERTIES OF POMERANCHUK POLES, DIFFRACTION SCATTERING AND ASYMPTOTIC EQUALITY OF TOTAL CROSS-SECTIONS", V.N. Gribov, *Yad. Fiz.* **17**, 603 (1973) [*Sov. J. Nucl. Phys.* **17**, 313 (1973)]
22. "LIMITATIONS ON USING THE OPERATOR PRODUCT EXPANSION AT SMALL VALUES OF X", A.H. Mueller, *Phys. Lett.* **B396**, 251 (1997)
23. "MODEL (IN)DEPENDENT FEATURES OF THE HARD POMERON", G. Camici and M. Ciafaloni, *Phys. Lett.* **B395**, 118 (1997)

LIST OF PARTICIPANTS

Pierre van Baal
Instituut-Lorentz for Theoretical Physics,
P.O.Box 9506, NL-2300 RA Leiden, The Netherlands
e-mail: vanbaal@lorentz.LeidenUniv.nl

Jochen Bartels
II Institut für Theoretische Physik, Universität Hamburg,
Luruper Chaussee 149, 22761 Hamburg, Germany
e-mail: bartels@mail.desy.de

Alexander Belavin
Landau Institute for Theoretical Physics,
Chernogolovka, 142432 Russia
e-mail: belavin@itp.ac.ru

Tamás Csörgő
Department of Theoretical Physics,
KFKI Research Institute for Particle and Nuclear Physics,
29-33 Konkoly Thege Str., H-1121 Budapest, Hungary
e-mail: csorgo@sunserv.kfki.hu

Yuri Dokshitzer
Laboratoire de Physique Theorique et Haute Energies (LPTHE),
Universite Pierre et Marie Curie (Paris VI),
e-mail: yuri@lpthe.jussieu.fr

Carlo Ewerz
Dipartimento di Fisica,
Universita di Milano and INFN, Sezione di Milano,
Via Celoria 16, 20133 Milano, Italy
and
ECT, Strada delle Tabarelle 286
I-38050 Villazzano (Trento), Italy
e-mail: ewerz@ect.it

George Fai
Center for Nuclear Research, Department of Physics,
Kent State University, Kent OH-44242, USA
e-mail: fai@cnr8.physics.kent.edu

László Fehér
Department of Theoretical Physics,
KFKI Research Institute for Particle and Nuclear Physics,
29-33 Konkoly Thege Str., H-1121 Budapest, Hungary
e-mail: lfeher@rmki.kfki.hu

Péter Forgács
Department of Theoretical Physics,
KFKI Research Institute for Particle and Nuclear Physics,
29-33 Konkoly Thege Str., H-1121 Budapest, Hungary
e-mail: forgacs@rmki.kfki.hu

Leonid Frankfurt
Department of Physics and Astronomy,
Tel-Aviv University,
Ramat Aviv, Tel-Aviv, Israel
e-mail: frankfur@lev.tau.ac.il

Andor Frenkel
Department of Theoretical Physics,
KFKI Research Institute for Particle and Nuclear Physics,
29-33 Konkoly Thege Str., H-1121 Budapest, Hungary
e-mail: frenkel@rmki.kfki.hu

Vahtang Gogohia
Department of Theoretical Physics,
KFKI Research Institute for Particle and Nuclear Physics,
29-33 Konkoly Thege Str., H-1121 Budapest, Hungary
e-mail: gogohia@rmki.kfki.hu

Gösta Gustafson
Department of Theoretical Physics, Lund University,
Sölvegatan 14 A, Lund, S-22362, Sweden
e-mail: Gosta.Gustafson@thep.lu.se

Peter Hasenfratz
Institute for Theoretical Physics, University of Bern,
Sidlerstr. 5., Bern, CH-3012, Switzwerland
e-mail: hasenfra@itp.unibe.ch

Valery Khoze
Institute for Particle Physics Phenomenology,
University of Durham, Durham, DH1 3LE, UK
e-mail: V.A.Khoze@durham.ac.uk

Iosif B. Khriplovich
Budker Institute of Nuclear Physics,
630090 Novosibirsk, Russia
e-mail: khriplovich@inp.nsk.su

Mikhail Kobrinsky
Nuclear Safety Institute (IBRAE),
Russian Academy of Sciences, Moscow, Russia
e-mail: mnk@ibrae.ac.ru

Gregory Korchemsky
Laboratoire de Physique Théorique,
Université de Paris-Sud XI,
91405 Orsay Cedex, France
e-mail: Gregory.Korchemsky@th.u-psud.fr

Péter Lévai
Department of Theoretical Physics,
KFKI Research Institute for Particle and Nuclear Physics,
29-33 Konkoly Thege Str., H-1121 Budapest, Hungary
e-mail: plevai@rmki.kfki.hu

Bao-An Li
Department of Chemistry and Physics, Arkansas State University,
P.O.Box 419, State University, AR 72467-0419, USA
e-mail: Bali@astate.edu

Giuseppe Marchesini
Dipartimento di Fisica, Universitá di Milano-Bicocca and INFN,
Sezione di Milano, Italy
e-mail: Giuseppe.Marchesini@mib.infn.it

Ferenc Niedermayer
Institute for Theoretical Physics, University of Bern,
Sidlerstr. 5, Bern, CH-3012, Switzerland
e-mail: niederma@itp.unibe.ch

Nikolai N. Nikolaev
Institut für Kernphysik, Forschungszentrum Jülich,
Jülich, D-52425 Germany
and
L.D. Landau Institute for Theoretical Physics,
Chernogolovka, 142432 Russia
e-mail: N.Nikolaev@fz-juelich.de

Victor Novikov
ITEP, Moscow, 117218 Russia
e-mail: novikov@itep.ru

Júlia Nyíri
Department of Theoretical Physics,
KFKI Research Institute for Particle and Nuclear Physics,
29-33 Konkoly Thege Str., H-1121 Budapest, Hungary
e-mail: nyiri@rmki.kfki.hu

András Patkós
Department of Atomic Physics, Eötvös University,
Budapest, H-1117 Hungary
e-mail: patkos@ludens.elte.hu

Herbert Petry
Rheinische Friedrich-Wilhelms-Universität Bonn,
Helmholtz-Institut für Strahlen- und Kernphysik,
Nussallee 14-16, Bonn, D-53115 Germany
e-mail: petry@itkp.uni-bonn.de

Ján Pisút
Department of Theoretical Physics,
Faculty of Mathematics, Physics and Informatics, Comenius University,
Bratislava, SK-84248, Slovakia
e-mail: pisut@fmph.uniba.sk

János Polonyi
Laboratoire de Physique Theorique, Universite Louis Pasteur,
3 rue de l'Universite, 67084 Strasbourg Cedex, France
e-mail: polonyi@physique.u-strasbg.fr

Mikhail Shifman
William I. Fine Theoretical Physics Institute, University of Minnesota,
Minneapolis, MN 55455 USA
e-mail: shifman@umn.edu

Andrei Smilga
SUBATECH, Université de Nantes,
4 rue Alfred Kastler, BP 20722, Nantes 44307, France
e-mail: smilga@subatech.in2p3.fr

Mark Strikman
104 Davey Laboratory, Penn State University,
University Park, PA 16802, USA
e-mail: strikman@phys.psu.edu

Zsolt Szép
Department of Atomic Physics, Eötvös University,
Budapest, H-1117 Hungary
e-mail: szepzs@achilles.elte.hu

Zoltán Szőkefalvi-Nagy
KFKI Research Institute for Particle and Nuclear Physics,
29-33 Konkoly Thege Str., H-1121 Budapest, Hungary
e-mail: sznagy@rmki.kfki.hu

Kálmán Tóth
Department of Theoretical Physics,
KFKI Research Institute for Particle and Nuclear Physics,
29-33 Konkoly Thege Str., H-1121 Budapest, Hungary
e-mail: ktoth@rmki.kfki.hu

Arkady Vainshtein
William I. Fine Theoretical Physics Institute, University of Minnesota,
Minneapolis, MN 55455 USA
e-mail: vainshte@umn.edu

Péter Vecsernyés
Department of Theoretical Physics,
KFKI Research Institute for Particle and Nuclear Physics,
29-33 Konkoly Thege Str., H-1121 Budapest, Hungary
e-mail: vecser@rmki.kfki.hu

Bryan Webber
University of Cambridge, Cavendish Laboratory,
Madingley Road, Cambridge CB3 OHE, UK
e-mail: webber@hep.phy.cam.ac.uk

Alan White
Argonne National Laboratory,
9700 South Cass, Il 60439, USA
e-mail: arw@hep.anl.gov